Introduction to Probability

Second Edition

Introduction to Probability

Second Edition

George G. Roussas

Department of Statistics
University of California, Davis

ELSEVIER

AMSTERDAM • BOSTON • HEIDELBERG • LONDON • NEW YORK
OXFORD • PARIS • SAN DIEGO • SAN FRANCISCO • SINGAPORE
SYDNEY • TOKYO

Academic Press is an imprint of Elsevier

Introduction to Probability, 2nd Edition

Roussas, George

Original English language edition published by Elsevier, Academic Press

ISBN : 978-0-12-800041-0

Reprint ISBN : 979-11-5688-012-7

Elsevier Korea L.L.C

4FL, Chunwoo Bldg, 534 Itaewon-dong,

Yongsan-gu, Seoul 140-861, S. Korea

First Published December 2014

Print in Taiwan

To the memory of David Blackwell and Lucien Le Cam

Contents

Preface

Overview

This book is an introductory textbook in probability. No prior knowledge in probability is required; however, previous exposure to an elementary precalculus course in probability would prove beneficial in that the student would not see the basic concepts discussed here for the first time.

The mathematical prerequisite is a year of calculus. Familiarity with the basic concepts of linear algebra would also be helpful in certain instances. Often students are exposed to such basic concepts within the calculus framework. Elementary differential and integral calculus will suffice for the majority of the book. In some parts of Chapters 7–11 the concept of a multiple integral is used. In Chapter 11, the student is expected to be at least vaguely familiar with the basic techniques of changing variables in a single or a multiple integral.

Chapter Descriptions

The material discussed in this book is enough for a one-semester course in introductory probability. This would include a rather detailed discussion of Chapters 1–12 except, perhaps, for the derivations of the probability density functions following Definitions 1 and 2 in Chapter 11. It could also include a cursory discussion of Chapter 13.

Most of the material in Chapters 1–12—with a quick description of the basic concepts in Chapter 13—can also be covered in a one-quarter course in introductory probability. In such a case, the instructor would omit the derivations of the probability density functions mentioned above, as well as Sections 9.4, 10.3, 11.3, and 12.2.3.

A chapter-by-chapter description follows. Chapter 1 consists of 16 examples selected from a broad variety of applications. Their purpose is to impress upon the student the breadth of applications of probability, and draw attention to the wide range of situations in which probability questions are pertinent. At this stage, one could not possibly provide answers to the questions posed without the methodology developed in the subsequent chapters. Answers to most of these questions are given in the form of examples and exercises throughout the remaining chapters. In Chapter 2, the concept of a random experiment is introduced, along with related concepts and some fundamental results. The concept of a random variable is also introduced here, along with the basics in counting. Chapter 3 is devoted to the introduction of the concept of probability and the discussion of some basic properties and results, including the distribution of a random variable.

Conditional probability, related results, and independence are covered in Chapter 4. The quantities of expectation, variance, moment-generating function, median, and mode of a random variable are discussed in Chapter 5, along with some basic probability inequalities.

The next chapter, Chapter 6, is devoted to the discussion of some of the commonly used discrete and continuous distributions.

When two random variables are involved, one talks about their joint distribution, as well as marginal and conditional probability density functions and conditional expectation and variance. The relevant material is discussed in Chapter 7. The discussion is pursued in Chapter 8 with the introduction of the concepts of covariance and correlation coefficient of two random variables.

The generalization of concepts in Chapter 7, when more than two random variables are involved, is taken up in Chapter 9, which concludes with the discussion of two popular multivariate distributions and the citation of a third such distribution. Independence of events is suitably carried over to random variables. This is done in Chapter 10, in which some consequences of independence are also discussed. In addition, this chapter includes a result, Theorem 6 in Section 10.3, of significant importance in statistics.

The next chapter, Chapter 11, concerns itself with the problem of determining the distribution of a random variable into which a given random variable is transformed. The same problem is considered when two or more random variables are transformed into a set of new random variables. The relevant results are mostly simply stated, as their justification is based on the change of variables in a single or a multiple integral, which is a calculus problem. The last three sections of the chapter are concerned with three classes of special but important transformations.

The book is essentially concluded with Chapter 12, in which two of the most important results in probability are studied, namely, the weak law of large numbers and the central limit theorem. Some applications of these theorems are presented, and the chapter is concluded with further results that are basically a combination of the weak law of large numbers and the central limit theorem. Not only are these additional results of probabilistic interest, they are also of substantial statistical importance.

As previously mentioned, the last chapter of the book provides an overview of statistical inference.

Features

This book has a number of features that may be summarized as follows. It starts with a brief chapter consisting exclusively of examples that are meant to provide motivation for studying probability.

It lays out a substantial amount of material—organized in twelve chapters—in a logical and consistent manner.

Before entering into the discussion of the concept of probability, it gathers together all needed fundamental concepts and results, including the basics in counting.

The concept of a random variable and its distribution, along with the usual numerical characteristics attached to it, are all introduced early on so that fragmentation in definitions is avoided. Thus, when discussing some special discrete and continuous random variables in Chapter 6, we are also in a position to present their usual numerical characteristics, such as expectation, variance, moment-generating function, etc.

Generalizations of certain concepts from one to more than one random variable and various extensions are split into two parts in order to minimize confusion and enhance understanding. We do these things for two random variables first, then for more than two random variables. Independence of random variables is studied systematically within the framework dictated by the level of the book. In particular, the reproductive property of certain distributions is fully exploited.

All necessary results pertaining to transformation of random variables are gathered together in one chapter, Chapter 11, rather than discussing them in a fragmented manner. This also allows for the justification of the distribution of order statistics as an application of a previously stated theorem. The study of linear transformations provides the tool of establishing Theorem 7 in Section 11.3, a result of great importance in statistical inference.

In Chapter 12, some important limit theorems are discussed, preeminently the weak law of large numbers and the central limit theorem. The strong law of large numbers is not touched upon, as not even an outline of its proof is feasible at the level of an introductory probability textbook.

The book concludes with an overview of the basics in statistical inference. This feature was selected over others, such as elements of Markov chains, of Poisson processes, and so on, in order to provide a window into the popular subject matter of statistics. At any rate, no justice could be done to the discussion of Markov chains, of Poisson processes, and so on, in an introductory textbook.

The book contains more than 150 examples discussed in great detail and more than 450 exercises suitably placed at the end of sections. Also, it contains at least 60 figures and diagrams that facilitate discussion and understanding. In the appendix, one can find a table of selected discrete and continuous distributions, along with some of their numerical characteristics, a table of some formulas used often in the book, a list of some notation and abbreviations, and often extensive answers to the even-numbered exercises.

Concluding Comments

An *Answers Manual*, with extensive discussion of the solutions of all exercises in the book, is available at booksite.elsevier.com/9780128000410 for the instructor.

A table of selected discrete and continuous distributions, along with some of their numerical characteristics, can also be found on the inside covers of the book. Finally, the appendix contains tables for the binomial, Poisson, normal, and chi-square distributions.

The expression log x (logarithm of x), whenever it occurs, always stands for the natural logarithm of x (the logarithm of x with base e). The rule for the use of decimal numbers is that we retain three decimal digits, the last of which is rounded up to the next higher number (if the fourth decimal is greater or equal to 5). An exemption to this rule is made when the division is exact, and when the numbers are read out of tables.

On several occasions, the reader is referred to proofs for more comprehensive treatment of some topics in the book *A Course in Mathematical Statistics*, 2nd edition (1997), Academic Press, by G.G. Roussas.

Thanks are due to my project assistant, Carol Ramirez, for preparing a beautiful set of typed chapters out of a collection of messy manuscripts.

Preface to the Second Edition

This is a revised version of the first edition of the book, copyrighted by Elsevier Inc, 2007.

The revision consists in correcting typographical errors, some factual oversights, rearranging some of the material, and adding a handful of new section exercises.

Also, a substantial number of hints is added to exercises to facilitate their solutions, and extensive cross-references are made as to where several of the examples mentioned are discussed in detail. Furthermore, a brief discussion of the central limit theorem is included early on in Chapter 6, Section 6.2.3. It is done so, because of its great importance in probability theory, and on the suggestion of a reviewer of the book.

Finally, a substantial number of exercises have been added at the end of each chapter. However, because of inadequate time to have them included in the book itself, these exercises will be posted (booksite.elsevier.com/9780128000410), and their answers will be made available to those instructors using this book as a textbook.

In all other respects, the book is unchanged, and the overview, chapter description, features, and guiding comments in the preface of the first edition remain the same.

The revision was skillfully implemented by my associates Michael McAssey and Chu Shing (Randy) Lai, to whom I express herewith my sincere appreciation.

George Roussas
Davis, California
June 2013

CHAPTER

Some Motivating Examples

1

This chapter consists of a single section that is devoted to presenting a number of examples (16 to be precise), drawn from a broad spectrum of human activities. Their purpose is to demonstrate the wide applicability of probability (and statistics). In each case, several relevant questions are posed, which, however, cannot be answered here. Most of these questions are dealt with in subsequent chapters. In the formulation of these examples, certain terms, such as at random, average, data fit by a line, event, probability (estimated probability, probability model), rate of success, sample, and sampling (sample size), are used. These terms are presently to be understood in their everyday sense, and will be defined precisely later on.

Example 1. In a certain state of the Union, n landfills are classified according to their concentration of three hazardous chemicals: arsenic, barium, and mercury. Suppose that the concentration of each one of the three chemicals is characterized as either high or low. Then some of the questions that can be posed are as follows:

(i) If a landfill is chosen at random from among the n, what is the probability it is of a specific configuration? In particular, what is the probability that it has:

  (a) High concentration of barium?
  (b) High concentration of mercury and low concentration of both arsenic and barium?
  (c) High concentration of any two of the chemicals and low concentration of the third?
  (d) High concentration of any one of the chemicals and low concentration of the other two?

(ii) How can one check whether the proportions of the landfills falling into each one of the eight possible configurations (regarding the levels of concentration) agree with a priori stipulated numbers?

(See some brief comments in Chapter 2, and Example 1 in Chapter 3.)

Example 2. Suppose a disease is present in $100p_1\%$ $(0 < p_1 < 1)$ of a population. A diagnostic test is available but is yet to be perfected. The test shows $100p_2\%$ false positives $(0 < p_2 < 1)$ and $100p_3\%$ false negatives $(0 < p_3 < 1)$. That is, for a patient not having the disease, the test shows positive (+) with probability p_2 and negative (−) with probability $1 - p_2$. For a patient having the disease, the test shows

Introduction to Probability, Second Edition. http://dx.doi.org/10.1016/B978-0-12-800041-0.00001-8

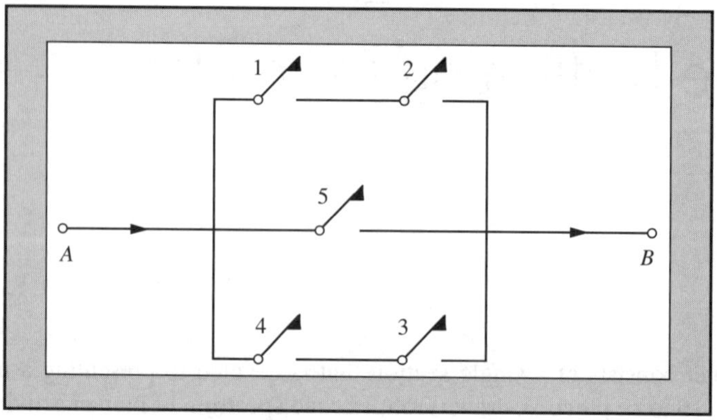

FIGURE 1.1
Transmitting electric current from point A to point B.

"−" with probability p_3 and "+" with probability $1-p_3$. A person is chosen at random from the target population, is subject to the test, and let D be the event that the person is diseased and N be the event that the person is not diseased. Then, it is clear that some important questions are as follows: In terms of p_1, p_2, and p_3:

(i) Determine the probabilities of the following configurations: D and $+$, D and $-$, N and $+$, N and $-$.
(ii) Also, determine the probability that a person will test + or the probability the person will test −.
(iii) If the person chosen tests +, what is the probability that he/she is diseased? What is the probability that he/she is diseased, if the person tests −? (See some brief comments in Chapter 2, and Example 10 in Chapter 4.)

Example 3. In the circuit drawn above (Figure 1.1), suppose that switch $i = 1, \ldots, 5$ turns on with probability p_i and independently of the remaining switches. What is the probability of having current transferred from point A to point B?
(This example is discussed in the framework of Proposition 2(ii) of Chapter 3.)

Example 4. A travel insurance policy pays $1,000 to a customer in case of a loss due to theft or damage on a 5-day trip. If the risk of such a loss is assessed to be 1 in 200, what is a fair premium for this policy?
(This example is discussed in Example 1 of Chapter 5.)

Example 5. Jones claims to have extrasensory perception (ESP). In order to test the claim, a psychologist shows Jones five cards that carry different pictures. Then Jones is blindfolded and the psychologist selects one card and asks Jones to identify the picture. This process is repeated n times. Suppose, in reality, that Jones has no ESP but responds by sheer guesses.

(i) Decide on a suitable probability model describing the number of correct responses.

(ii) What is the probability that at most $n/5$ responses are correct?
(iii) What is the probability that at least $n/2$ responses are correct?

(See a brief discussion in Chapter 2, and Theorem 1 in Chapter 6.)

Example 6. A government agency wishes to assess the prevailing rate of unemployment in a particular county. It is felt that this assessment can be done quickly and effectively by sampling a small fraction n, say, of the labor force in the county. The obvious questions to be considered here are:

(i) What is a suitable probability model describing the number of unemployed?
(ii) What is an estimate of the rate of unemployment?

Example 7. Suppose that, for a particular cancer, chemotherapy provides a 5-year survival rate of 75% if the disease could be detected at an early stage. Suppose further that n patients, diagnosed to have this form of cancer at an early stage, are just starting the chemotherapy. Finally, let X be the number of patients among the n who survive 5 years.
Then the following are some of the relevant questions that can be asked:

(i) What are the possible values of X, and what are the probabilities that each one of these values is taken on?
(ii) What is the probability that X takes values between two specified numbers a and b, say?
(iii) What is the average number of patients to survive 5 years, and what is the variation around this average?

Example 8. An advertisement manager for a radio station claims that over $100p\%$ $(0 < p < 1)$ of all young adults in the city listen to a weekend music program. To establish this conjecture, a random sample of size n is taken from among the target population and those who listen to the weekend music program are counted.

(i) Decide on a suitable probability model describing the number of young adults who listen to the weekend music program.
(ii) On the basis of the collected data, check whether the claim made is supported or not.
(iii) How large a sample size n should be taken to ensure that the estimated average and the true proportion do not differ in absolute value by more than a specified number with prescribed (high) probability?

Example 9. When the output of a production process is stable at an acceptable standard, it is said to be "in control." Suppose that a production process has been in control for some time and that the proportion of defectives has been p. As a means of monitoring the process, the production staff will sample n items. Occurrence of k or more defectives will be considered strong evidence for "out of control."

(i) Decide on a suitable probability model describing the number X of defectives; what are the possible values of X, and what is the probability that each of these values is taken on?
(ii) On the basis of the data collected, check whether or not the process is out of control.
(iii) How large a sample size n should be taken to ensure that the estimated proportion of defectives will not differ in absolute value from the true proportion of defectives by more than a specified quantity with prescribed (high) probability?

Example 10. At a given road intersection, suppose that X is the number of cars passing by until an observer spots a particular make of a car (e.g., a Mercedes).

Then some of the questions one may ask are as follows:

(i) What are the possible values of X?
(ii) What is the probability that each one of these values is taken on?
(iii) How many cars would the observer expect to observe until the first Mercedes appears?

(See a brief discussion in Chapter 2, and Section 6.1.2 in Chapter 6.)

Example 11. A city health department wishes to determine whether the mean bacteria count per unit volume of water at a lake beach is within the safety level of 200. A researcher collected n water samples of unit volume and recorded the bacteria counts.

Relevant questions here are:

(i) What is the appropriate probability model describing the number X of bacteria in a unit volume of water; what are the possible values of X, and what is the probability that each one of these values is taken on?
(ii) Do the data collected indicate that there is no cause for concern?

(See a brief discussion in Chapter 2, and Section 6.1.3 in Chapter 6.)

Example 12. Measurements of the acidity (pH) of rain samples were recorded at n sites in an industrial region.

(i) Decide on a suitable probability model describing the number X of the acidity of rain measured.
(ii) On the basis of the measurements taken, provide an estimate of the average acidity of rain in that region.

(See a brief discussion in Chapter 2, and Section 6.2.4 in Chapter 6.)

Example 13. To study the growth of pine trees at an early state, a nursery worker records n measurements of the heights of 1-year-old red pine seedlings.

(i) Decide on a suitable probability model describing the heights X of the pine seedlings.

(ii) On the basis of the n measurements taken, determine the average height of the pine seedlings.
(iii) Also, check whether these measurements support the stipulation that the average height is a specified number.

(As in Example 12.)

Example 14. It is claimed that a new treatment is more effective than the standard treatment for prolonging the lives of terminal cancer patients. The standard treatment has been in use for a long time, and from records in medical journals the mean survival period is known to have a certain numerical value (in years). The new treatment is administered to n patients, and their duration of survival is recorded.

(i) Decide on suitable probability models describing the survival times X and Y under the old and the new treatments, respectively.
(ii) On the basis of the existing journal information and the data gathered, check whether or not the claim made is supported.

(As in Example 12.)

Example 15. The lifetime of a new equipment just put in service is an unknown quantity X. Some important relevant questions are the following:

(i) What is a suitable model describing the lifetime of the equipment?
(ii) What is the probability that the lifetime will be at least t_0, a prescribed amount of time units?
(iii) What is the expected lifetime of the equipment?
(iv) What is the expected cost for operating said equipment?

(See a brief discussion in Chapter 2, and Sections 6.2.1 and 6.2.2 in Chapter 6.)

Example 16. It is known that human blood is classified in four types denoted by A, B, AB, and O. Suppose that the blood of n persons who have volunteered to donate blood at a plasma center has been classified in these four categories. Then a number of questions can be posed, some of which are:

(i) What is the appropriate probability model to describe the distribution of the blood types of the n persons into the four types?
(ii) What is the estimated probability that a person, chosen at random from among the n, has a specified blood type (e.g., O)?
(iii) What are the proportions of the n persons falling into each one of the four categories?
(iv) How can one check whether the observed proportions are in agreement with a priori stipulated numbers?

(See Section 9.2 in Chapter 9.)

As it has already been mentioned, some further comments on Examples 1–16 are presented in the next chapter, Chapter 2.

CHAPTER

Some Fundamental Concepts

2

This chapter consists of four sections. In the first section, the fundamental concepts of a random experiment, sample point, sample space, and event are introduced and illustrated by several examples. In the second section, the usual set-theoretic type of operations on events are defined, and some basic properties and results are recorded. In the third section, the very important concept of a random variable is introduced and illustrated by means of examples. The closing section is devoted to the discussion of some basic concepts and results in counting, including permutations and combinations.

2.1 Some Fundamental Concepts

One of the most basic concepts in probability (and statistics) is that of a *random experiment*. Although a more precise definition is possible, we will restrict ourselves here to understanding a random experiment as a procedure that is carried out under a certain set of conditions; it can be repeated any number of times under the same set of conditions; and upon the completion of the procedure, certain results are observed. The results obtained are denoted by s and are called *sample points*. The set of all possible sample points is denoted by $\mathcal{S}$ and is called a *sample space*. Subsets of $\mathcal{S}$ are called *events* and are denoted by capital letters A, B, C, etc. An event consisting of one sample point only, $\{s\}$, is called a *simple* event, and *composite* otherwise. An event A *occurs* (or *happens*) if the outcome of the random experiment (that is, the sample point s) belongs in A, $s \in A$; A *does not occur* (or *does not happen*) if $s \notin A$. The event $\mathcal{S}$ always occurs and is called the *sure* or *certain* event. On the other hand, the event $\varnothing$ never happens and is called the *impossible* event. Of course, the relation $A \subseteq B$ between two events A and B means that the event B *occurs whenever A does*, but not necessarily the opposite. (See Figure 2.1 for the *Venn diagram* depicting the relation $A \subseteq B$.) The events A and B are *equal* if both $A \subseteq B$ and $B \subseteq A$.

Some random experiments are given in the following, along with corresponding sample spaces and some events.

Example 1. Tossing three distinct coins once.

Then, with H and T standing for "heads" and "tails," respectively, a sample space is:

$$\mathcal{S} = \{HHH, HHT, HTH, THH, HTT, THT, TTH, TTT\}.$$

Introduction to Probability, Second Edition. http://dx.doi.org/10.1016/B978-0-12-800041-0.00002-X

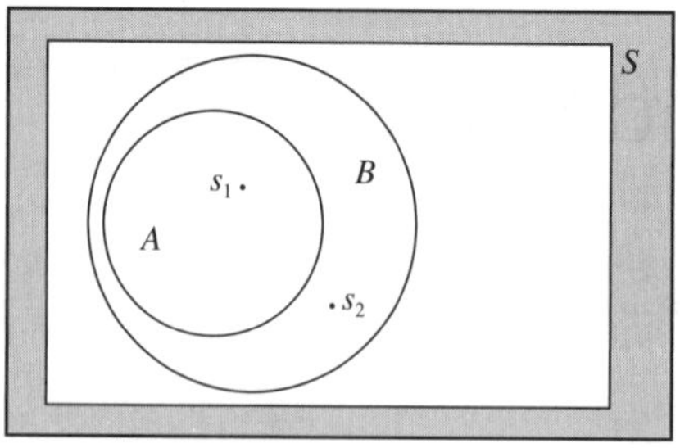

FIGURE 2.1
$A \subseteq B$; in fact, $A \subset B$, because $s_2 \in B$, but $s_2 \notin A$.

The event A = "no more than 1 H occurs" is given by:

$$A = \{TTT, HTT, THT, TTH\}.$$

Example 2. Rolling once two distinct dice.

Then a sample space is:

$$\mathcal{S} = \{(1, 1), (1, 2), \ldots, (1, 6), \ldots, (6, 1), (6, 2), \ldots, (6, 6)\},$$

and the event B = "the sum of numbers on the upper faces is ≤ 5" is:

$$B = \{(1, 1), (1, 2), (1, 3), (1, 4), (2, 1), (2, 2), (2, 3), (3, 1), (3, 2), (4, 1)\}.$$

Example 3. Drawing a card from a well-shuffled standard deck of 52 cards.

Denoting by C, D, H, and S clubs, diamonds, hearts, and spades, respectively, by J, Q, K Jack, Queen, and King, and using 1 for aces, the sample space is given by:

$$\mathcal{S} = \{1_C, \ldots, 1_S, \ldots, 10_C, \ldots, 10_S, \ldots, K_C, \ldots, K_S\}.$$

An event A may be described by: A = "red and face card," so that

$$A = \{J_D, J_H, Q_D, Q_H, K_D, K_H\}.$$

Example 4. Recording the gender of children of two-children families.

With b and g standing for boy and girl, and with the first letter on the left denoting the older child, a sample space is: $\mathcal{S} = \{bb, bg, gb, gg\}$. An event B may be: B = "children of both genders." Then $B = \{bg, gb\}$.

Example 5. Ranking five horses in a horse race.

Then the suitable sample space $\mathcal{S}$ consists of 120 sample points, corresponding to the 120 permutations of the numbers 1, 2, 3, 4, 5. (We exclude ties.) The event A = "horse #3 comes second" consists of the 24 sample points, where 3 always occurs in the second place. (See also the Corollary to Theorem 1 in Section 2.4.)

Example 6. Tossing a coin repeatedly until H appears for the first time. The suitable sample space here is:

$$\mathcal{S} = \{H, TH, TTH, \ldots, TT\ldots TH, \ldots\}.$$

Then the event $A =$ "the 1st H does not occur before the 10th tossing" is given by:

$$A = \{\underbrace{T\ldots T}_{9}H,\ \underbrace{T\ldots T}_{10}H, \ldots\}.$$

Example 7. Recording the number of traffic accidents that occur in a specified location within a certain period of time.

The appropriate sample space here is $\mathcal{S} = \{0, 1, \ldots, M\}$ for a suitable number M. If M is sufficiently large, then $\mathcal{S}$ is taken to be: $\mathcal{S} = \{0, 1, \ldots\}$.

Example 8. Recording the number of particles emitted by a certain radioactive source within a specified period of time.

As in the previous example, $\mathcal{S}$ is taken to be: $\mathcal{S} = \{0, 1, \ldots, M\}$, where M is often a large number, and then as before $\mathcal{S}$ is modified to be: $\mathcal{S} = \{0, 1, \ldots\}$.

Example 9. Recording the lifetime of an electronic device, or of an electrical appliance, etc.

Here $\mathcal{S}$ is the interval $(0, T]$ for some reasonable value of T; that is, $\mathcal{S}=(0, T]$. Sometimes, for justifiable reasons, we take $\mathcal{S} = (0, \infty)$.

Example 10. Recording the distance from the bull's eye of the point where a dart, aiming at the bull's eye, actually hits the plane. Here it is clear that $\mathcal{S} = [0, \infty)$.

Example 11. Measuring the dosage of a certain medication, administered to a patient, until a positive reaction is observed.

Here $\mathcal{S} = (0, D]$ for some suitable D (not rendering the medication lethal!).

Example 12. Recording the yearly income of a target population.

If the incomes are measured in dollars and cents, the outcomes are fractional numbers in an interval $[0, M]$ for some reasonable M. Again, for reasons similar to those cited in Example 9, $\mathcal{S}$ is often taken to be $\mathcal{S} = [0, \infty)$.

Example 13. Waiting until the time the Dow Jones Industrial Average Index reaches or surpasses a specified level.

Here, with reasonable qualifications, we may choose to take $\mathcal{S} = (0, \infty)$.

Example 14. Waiting until the lifetime of a new piece of equipment just put in service expires.

Here $\mathcal{S}=(0, T]$ for some suitable (presumably large) T, which for mathematical convenience may be taken to be ∞; i.e., $\mathcal{S} = (0, \infty)$.

Examples 1–16 in Chapter 1, suitably interpreted, may also serve as further illustrations of random experiments. Most examples described previously will be revisited on various occasions.

For instance, in Example 1 in Chapter 1 and in self-explanatory notation, a suitable sample space is:

$$\mathcal{S} = \{A_{\mathrm{h}}B_{\mathrm{h}}M_{\mathrm{h}},\ A_{\mathrm{h}}B_{\mathrm{h}}M_{\ell},\ A_{\mathrm{h}}B_{\ell}M_{\mathrm{h}},\ A_{\ell}B_{\mathrm{h}}M_{\mathrm{h}},\ A_{\mathrm{h}}B_{\ell}M_{\ell},\\ A_{\ell}B_{\mathrm{h}}M_{\ell},\ A_{\ell}B_{\ell}M_{\mathrm{h}},\ A_{\ell}B_{\ell}M_{\ell}\}.$$

Then the events A = "no chemical occurs at high level" and B = "at least two chemicals occur at high levels" are given by:

$$A = \{A_{\ell}B_{\ell}M_{\ell}\}, \quad B = \{A_{\mathrm{h}}B_{\mathrm{h}}M_{\ell},\ A_{\mathrm{h}}B_{\ell}M_{\mathrm{h}},\ A_{\ell}B_{\mathrm{h}}M_{\mathrm{h}},\ A_{\mathrm{h}}B_{\mathrm{h}}M_{\mathrm{h}}\}.$$

(See also Example 1 in Chapter 3.)

In Example 2 in Chapter 1, a patient is classified according to the result of the test, giving rise to the following sample space:

$$\mathcal{S} = \{D+,\ D-,\ N+,\ N-\},$$

where D and N stand for the events "patient has the disease" and "patient does not have the disease," respectively. Then the event A = "false diagnosis of test" is given by: $A = \{D-, N+\}$. (See also Example 10 in Chapter 4.)

Example 3 of Chapter 1 is discussed in the framework of Proposition 2(ii) of Chapter 3. Example 4 of Chapter 1 is discussed in Example 1 of Chapter 5. In Example 5 in Chapter 1, the suitable probability model is the so-called binomial model. The sample space $\mathcal{S}$ is the set of 2^n points, each point consisting of a sequence of n S's and F's, S standing for success (on behalf of Jones) and F standing for failure. Then the questions posed can be answered easily. (See also Theorem 1 in Chapter 6.)

Examples 6–9 in Chapter 1 can be discussed in the same framework as that of Example 5 with obvious modifications in notation.

In Example 10 in Chapter 1, a suitable sample space is:

$$\mathcal{S} = \{M,\ M^cM,\ M^cM^cM, \ldots,\ M^c \cdots M^cM, \ldots\},$$

where M stands for the passing by of a Mercedes car. Then the events A and B, where A = "Mercedes was the 5th car which passed by" and B = "Mercedes was spotted after the first 3 cars passed by" are given by:

$$A = \{M^cM^cM^cM^cM\} \text{ and } B = \{M^cM^cM^cM,\ M^cM^cM^cM^cM, \ldots\}.$$

(See also Section 6.1.2 in Chapter 6.)

In Example 11 in Chapter 1, a suitable sample space is: $\mathcal{S} = \{0, 1, \ldots, M\}$ for an appropriately large (integer) M; for mathematical convenience, $\mathcal{S}$ is often taken to be: $\mathcal{S} = \{0, 1, 2, \ldots\}$. (See also Section 6.1.3 in Chapter 6.)

In Examples 12–14 in Chapter 1, a suitable sample space is $\mathcal{S} = (0, T]$ for some reasonable value of T. (See also Section 6.2.4 in Chapter 6.)

In Example 15 in Chapter 1, a suitable sample space is either $\mathcal{S} = (0, T]$ for a suitable T, or $\mathcal{S} = (0, \infty)$ for mathematical convenience. (See also 6.2.1 and 6.2.2 in Chapter 6.)

In Example 16 in Chapter 1, a suitable sample space $\mathcal{S}$ is the set of 4^n points, each point consisting of a sequence of n symbols A, B, AB, and O. The underlying probability model is the so-called multinomial model, and the questions posed can be discussed by available methodology. Actually, there is no need even to refer to the sample space $\mathcal{S}$. All one has to do is to consider the outcomes in the n trials and then classify the n outcomes into four categories A, B, AB, and O. (See Section 9.2 in Chapter 9.)

In many cases, questions posed can be discussed without reference to any explicit sample space. This is the case, for instance, in Example 14 in Chapter 1.

In the examples discussed previously, we have seen sample spaces consisting of finitely many sample points (Examples 1–5 here and 1, 2, 5, and 6–9 in Chapter 1), sample spaces consisting of countably infinite many points (for example, as many as the positive integers) (Examples 6–8 here and 10 and 11 in Chapter 1 if we replace M by ∞ for mathematical convenience in 7, 8, and 11), and sample spaces consisting of as many sample points as there are in a nondegenerate finite or infinite interval in the real line, which interval may also be the entire real line (Examples 9–14 here and 12, 13, and 15 in Chapter 1). Sample spaces with countably many points (i.e., either finitely many or countably infinite many) are referred to as *discrete* sample spaces. Sample spaces with sample points as many as the numbers in a nondegenerate finite or infinite interval in the real line $\Re = (-\infty, \infty)$ are referred to as *continuous* sample spaces.

2.2 Some Fundamental Results

Returning now to events, when one is dealing with them, one may perform the same operations as those with sets. Thus, the *complement* of the event A, denoted by A^c, is the event defined by: $A^c = \{s \in \mathcal{S};\, s \notin A\}$. The event A^c is presented by the Venn diagram in Figure 2.2. So A^c occurs whenever A does not, and vice versa.

The *union* of the events $A_1, \ldots, A_n$, denoted by $A_1 \cup \cdots \cup A_n$ or $\bigcup_{j=1}^{n} A_j$, is the event defined by $\bigcup_{j=1}^{n} A_j = \{s \in \mathcal{S};\, s \in A_j, \textit{for at least one } j = 1, \ldots, n\}$. So the event $\bigcup_{j=1}^{n} A_j$ occurs whenever at least one of A_j, $j = 1, \ldots, n$ occurs. For $n = 2$, $A_1 \cup A_2$ is presented in Figure 2.3. The definition extends to an infinite number of events. Thus, for countably infinite many events A_j, $j = 1, 2, \ldots$, one has $\bigcup_{j=1}^{\infty} A_j = \{s \in \mathcal{S};\, s \in A_j, \textit{for at least one } j = 1, 2, \ldots\}$.

The *intersection* of the events A_j, $j = 1, \ldots, n$ is the event denoted by $A_1 \cap \cdots \cap A_n$ or $\bigcap_{j=1}^{n} A_j$ and is defined by $\bigcap_{j=1}^{n} A_j = \{s \in \mathcal{S};\, s \in A_j, \textit{for all } j = 1, \ldots, n\}$. Thus, $\bigcap_{j=1}^{n} A_j$ occurs whenever all A_j, $j = 1, \ldots, n$ occur

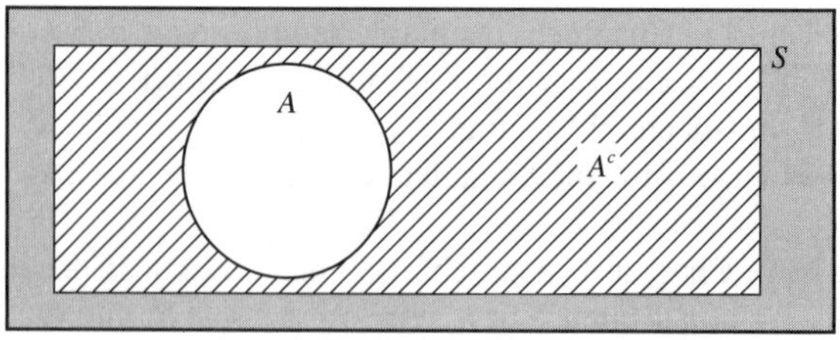

FIGURE 2.2
A^c is the shaded region.

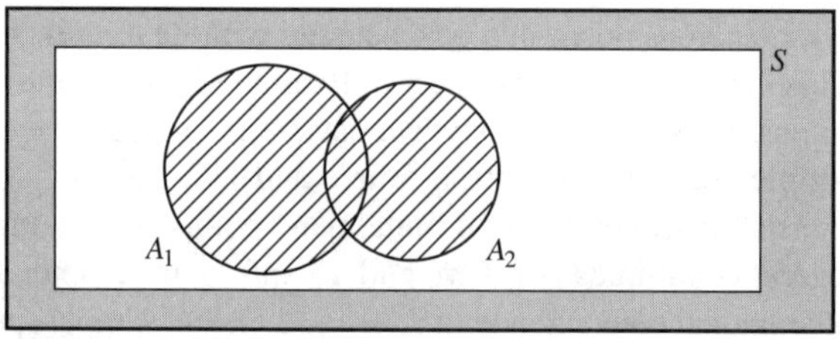

FIGURE 2.3
$A_1 \cup A_2$ is the shaded region.

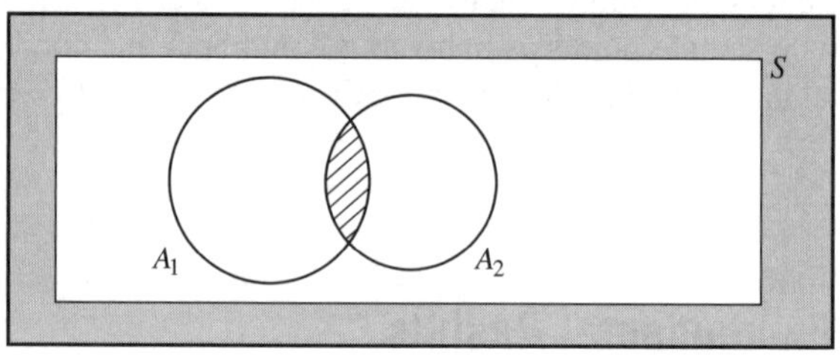

FIGURE 2.4
$A_1 \cap A_2$ is the shaded region.

simultaneously. For $n = 2$, $A_1 \cap A_2$ is presented in Figure 2.4. This definition extends to an infinite number of events. Thus, for countably infinite many events A_j, $j = 1, 2, \ldots$, one has $\bigcap_{j=1}^{\infty} A_j = \{s \in \mathcal{S};\, s \in A_j, \textit{for all } j = 1, 2, \ldots\}$.

If $A_1 \cap A_2 = \varnothing$, the events A_1 and A_2 are called *disjoint* (see Figure 2.5). The events A_j, $j = 1, 2, \ldots$, are said to be *mutually* or *pairwise disjoint*, if $A_i \cap A_j = \varnothing$ whenever $i \neq j$.

The *differences* $A_1 - A_2$ and $A_2 - A_1$ are the events defined by $A_1 - A_2 = \{s \in \mathcal{S};\, s \in A_1, s \notin A_2\}$, $A_2 - A_1 = \{s \in \mathcal{S};\, s \in A_2, s \notin A_1\}$ (see Figure 2.6).

From the definition of the preceding operations, the following properties follow immediately, and they are listed here as a proposition for reference.

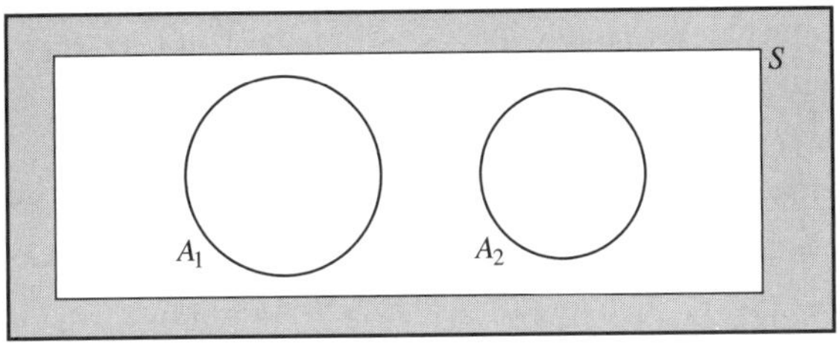

FIGURE 2.5
A_1 and A_2 are disjoint; that is, $A_i \cap A_j = \varnothing$.

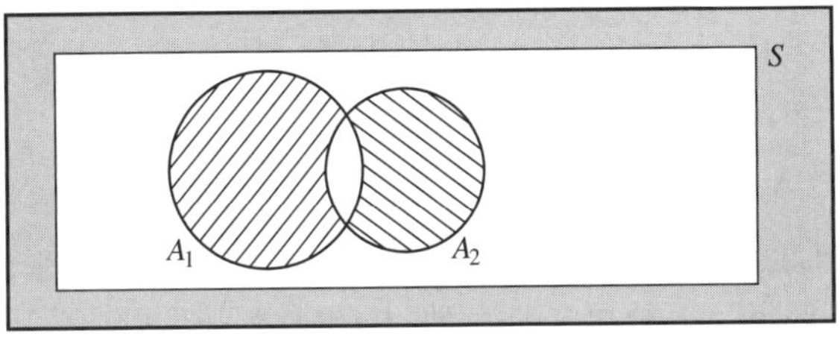

FIGURE 2.6
$A_1 - A_2$ is ////, $A_2 - A_1$ is \\\\ .

Proposition 1.

(i) $\mathcal{S}^c = \varnothing$, $\varnothing^c = \mathcal{S}$, $(A^c)^c = A$.
(ii) $\mathcal{S} \cup A = \mathcal{S}$, $\varnothing \cup A = A$, $A \cup A^c = \mathcal{S}$, $A \cup A = A$.
(iii) $\mathcal{S} \cap A = A$, $\varnothing \cap A = \varnothing$, $A \cap A^c = \varnothing$, $A \cap A = A$.

The previous statements are all obvious, as is the following: $\varnothing \subseteq A$ for every event A in $\mathcal{S}$. Also, on the operations of forming unions and/or intersections, we have the following properties (see also Exercise 2.13).

Proposition 2.

(i) $$\left.\begin{array}{l} A_1 \cup (A_2 \cup A_3) = (A_1 \cup A_2) \cup A_3 \\ A_1 \cap (A_2 \cap A_3) = (A_1 \cap A_2) \cap A_3 \end{array}\right\} \quad \text{(associative laws)}$$

(ii) $$\left.\begin{array}{l} A_1 \cup A_2 = A_2 \cup A_1 \\ A_1 \cap A_2 = A_2 \cap A_1 \end{array}\right\} \quad \text{(commutative laws)}$$

(iii) $$\left.\begin{array}{l} A \cap (\cup_j A_j) = \cup_j (A \cap A_j) \\ A \cup (\cap_j A_j) = \cap_j (A \cup A_j) \end{array}\right\} \quad \text{(distributive laws)}$$

Remark 1. In the last relations, as well as elsewhere, when the range of the index j is not indicated explicitly, it is assumed to be either a finite set, such as $\{1, \ldots, n\}$, or a countably infinite set, such as $\{1, 2, \ldots\}$.

For the purpose of demonstrating some of the set-theoretic operations just defined, let us consider some concrete examples.

Example 15. Consider the sample space $\mathcal{S} = \{s_1, s_2, s_3, s_4, s_5, s_6, s_7, s_8\}$ and define the events A_1, A_2, and A_3 as follows: $A_1 = \{s_1, s_2, s_3\}$, $A_2 = \{s_2, s_3, s_4, s_5\}$, $A_3 = \{s_3, s_4, s_5, s_8\}$. Then observe that:

$$\begin{aligned}
&A_1^c = \{s_4, s_5, s_6, s_7, s_8\}, \quad A_2^c = \{s_1, s_6, s_7, s_8\}, \quad A_3^c = \{s_1, s_2, s_6, s_7\};\\
&A_1 \cup A_2 = \{s_1, s_2, s_3, s_4, s_5\}, \quad A_1 \cup A_3 = \{s_1, s_2, s_3, s_4, s_5, s_8\},\\
&A_2 \cup A_3 = \{s_2, s_3, s_4, s_5, s_8\}, \quad A_1 \cup A_2 \cup A_3 = \{s_1, s_2, s_3, s_4, s_5, s_8\};\\
&A_1 \cap A_2 = \{s_2, s_3\}, \quad A_1 \cap A_3 = \{s_3\}, \quad A_2 \cap A_3 = \{s_3, s_4, s_5\},\\
&\quad A_1 \cap A_2 \cap A_3 = \{s_3\};\\
&A_1 - A_2 = \{s_1\}, \quad A_2 - A_1 = \{s_4, s_5\}, \quad A_1 - A_3 = \{s_1, s_2\},\\
&A_3 - A_1 = \{s_4, s_5, s_8\}, \quad A_2 - A_3 = \{s_2\}, \quad A_3 - A_2 = \{s_8\};\\
&(A_1^c)^c = \{s_1, s_2, s_3\}(=A_1), \; (A_2^c)^c = \{s_2, s_3, s_4, s_5\}(=A_2),\\
&(A_3^c)^c = \{s_3, s_4, s_5, s_8\}(=A_3).
\end{aligned}$$

An identity and DeMorgan's laws stated subsequently are of significant importance. Their justifications are left as exercises (see Exercises 2.14 and 2.15).

Proposition 3 (An Identity). $\cup_j A_j = A_1 \cup (A_1^c \cap A_2) \cup (A_1^c \cap A_2^c \cap A_3) \cup \cdots \cup (A_1^c \cap A_2^c \cap \cdots \cap A_{n-1}^c \cap A_n) \cup \cdots$

The significance of this identity is that the events on the right-hand side are pairwise disjoint, whereas the original events A_j, $j \geq 1$, need not be so.

Proof (partial). As an indication of how one argues in establishing such an identity, suppose we wish to show that $A_1 \cup A_2 = A_1 \cup (A_1^c \cap A_2)$. Let s belong to the left-hand side. Then, if $s \in A_1$, it clearly belongs to the right-hand side, whereas if $s \notin A_1$, then $s \in A_2$ and hence $s \in (A_1^c \cap A_2)$, so that s belongs to the right-hand side again. Now, let s belong to the right-hand side, and suppose that $s \in A_1$. Then, clearly, s belongs to the left-hand side. If $s \notin A_1$, then $s \in (A_1^c \cap A_2)$, so that $s \in A_2$, and hence belongs to the left-hand side again (see also Exercise 2.14). ▲

Example 16. From Example 15, we have:

$$A_1 = \{s_1, s_2, s_3\}, \quad A_1^c \cap A_2 = \{s_4, s_5\}, \quad A_1^c \cap A_2^c \cap A_3 = \{s_8\}.$$

Note that A_1, $A_1^c \cap A_2$, $A_1^c \cap A_2^c \cap A_3$ are pairwise disjoint. Now $A_1 \cup (A_1^c \cap A_2) \cup (A_1^c \cap A_2^c \cap A_3) = \{s_1, s_2, s_3, s_4, s_5, s_8\}$, which is equal to $A_1 \cup A_2 \cup A_3$; that is,

$$A_1 \cup A_2 \cup A_3 = A_1 \cup (A_1^c \cap A_2) \cup (A_1^c \cap A_2^c \cap A_3),$$

as the preceding identity states.

Proposition 4 (DeMorgan's Laws). $(\cup_j A_j)^c = \cap_j A_j^c$, $(\cap_j A_j)^c = \cup_j A_j^c$.

Proof (partial). As an illustration, consider the case $(A_1 \cap A_2)^c = A_1^c \cup A_2^c$. Let $s \in (A_1 \cap A_2)^c$. Then $s \notin (A_1 \cap A_2)$ so that $s \notin A_1$ or $s \notin A_2$. If $s \notin A_1$, then $s \in A_1^c$ and hence s belongs to the right-hand side. Similarly, if $s \notin A_2$. Next, let s

belong to the right-hand side, so that $s \in A_1^c$ or $s \in A_2^c$. If $s \in A_1^c$, then $s \notin A_1$, so that $s \notin (A_1 \cap A_2)$ and hence $s \in (A_1 \cap A_2)^c$. So, s belongs to the left-hand side, and similarly if $s \in A_2^c$ (see also Exercise 2.15). ▲

Example 17. Again from Example 15, one has:

$$\begin{aligned}
&(A_1 \cup A_2)^c = \{s_6, s_7, s_8\}, \quad A_1^c \cap A_2^c = \{s_6, s_7, s_8\};\\
&(A_1 \cup A_2 \cup A_3)^c = \{s_6, s_7\}, \quad A_1^c \cap A_2^c \cap A_3^c = \{s_6, s_7\};\\
&(A_1 \cap A_2)^c = \{s_1, s_4, s_5, s_6, s_7, s_8\}, \quad A_1^c \cup A_2^c = \{s_1, s_4, s_5, s_6, s_7, s_8\};\\
&(A_1 \cap A_2 \cap A_3)^c = \{s_1, s_2, s_4, s_5, s_6, s_7, s_8\},\\
&A_1^c \cup A_2^c \cup A_3^c = \{s_1, s_2, s_4, s_5, s_6, s_7, s_8\},
\end{aligned}$$

so that

$$\begin{aligned}
(A_1 \cup A_2)^c = A_1^c \cap A_2^c, \quad (A_1 \cup A_2 \cup A_3)^c = A_1^c \cap A_2^c \cap A_3^c,\\
(A_1 \cap A_2)^c = A_1^c \cup A_2^c, \quad (A_1 \cap A_2 \cap A_3)^c = A_1^c \cup A_2^c \cup A_3^c,
\end{aligned}$$

as DeMorgan's laws state.

As a further demonstration of how complements, unions, and intersections of events are used for the expression of new events, consider the following example.

Example 18. In terms of the events A_1, A_2, and A_3 (in some sample space $\mathcal{S}$) and, perhaps, their complements, unions, and intersections, express the following events:

$$\begin{aligned}
&D_i = \text{“}A_i \text{ does not occur,” } i = 1, 2, 3, \text{ so that } D_1 = A_1^c, D_2 = A_2^c, D_3 = A_3^c;\\
&E = \text{“all } A_1, A_2, A_3 \text{ occur,” so that } E = A_1 \cap A_2 \cap A_3;\\
&F = \text{“none of } A_1, A_2, A_3 \text{ occurs,” so that } F = A_1^c \cap A_2^c \cap A_3^c;\\
&G = \text{“at least one of } A_1, A_2, A_3 \text{ occurs,” so that } G = A_1 \cup A_2 \cup A_3;\\
&H = \text{“exactly two of } A_1, A_2, A_3 \text{ occur,” so that } H = \left(A_1 \cap A_2 \cap A_3^c\right)\\
&\qquad \cup \left(A_1 \cap A_2^c \cap A_3\right) \cup \left(A_1^c \cap A_2 \cap A_3\right);\\
&I = \text{“exactly one of } A_1, A_2, A_3 \text{ occurs,” so that } I = \left(A_1 \cap A_2^c \cap A_3^c\right) \cup\\
&\qquad \left(A_1^c \cap A_2 \cap A_3^c\right) \cup \left(A_1^c \cap A_2^c \cap A_3\right).
\end{aligned}$$

It also follows that:

$$\begin{aligned}
G &= \text{“exactly one of } A_1, A_2, A_3 \text{ occurs”} \cup \text{“exactly two of } A_1, A_2, A_3 \text{ occur”}\\
&\qquad \cup \text{“all } A_1, A_2, A_3 \text{ occur”}\\
&= I \cup H \cup E.
\end{aligned}$$

This section is concluded with the concept of a monotone sequence of events. Namely, the sequence of events $\{A_n\}$, $n \geq 1$, is said to be *monotone* if either $A_1 \subseteq A_2 \subseteq \cdots$ (*increasing*) or $A_1 \supseteq A_2 \supseteq \cdots$ (*decreasing*). In case of an increasing sequence, the union $\bigcup_{j=1}^{\infty} A_j$ is called the *limit* of the sequence, and in case of a decreasing sequence, the intersection $\bigcap_{j=1}^{\infty} A_j$ is called its *limit*.

The concept of the limit is also defined, under certain conditions, for nonmonotone sequences of events, but we are not going to enter into it here. The interested reader is referred to Definition 1 in Chapter 1, of the book *A Course in Mathematical Statistics*, 2nd edition (1997), Academic Press, by G. G. Roussas.

Exercises.

2.1 An airport limousine departs from a certain airport with three passengers to be delivered to any one of three hotels denoted by H_1, H_2, H_3. Let (x_1, x_2, x_3) denote the number of passengers (not which ones!) left at hotels H_1, H_2, and H_3, respectively.

(i) Write out the sample space $\mathcal{S}$ of all possible deliveries.

(ii) Consider the events A, B, C, D, and E, defined as follows, and express them in terms of sample points.

A = "one passenger in each hotel"
B = "all passengers in H_1"
C = "all passengers in one hotel"
D = "at least two passengers in H_1"
E = "fewer passengers in H_1 than in each one of H_2 and H_3."

2.2 A machine dispenses balls that are either red or black or green. Suppose we operate the machine three successive times and record the color of the balls dispensed, to be denoted by r, b, and g for the respective colors.

(i) Write out an appropriate sample space $\mathcal{S}$ for this experiment.

(ii) Consider the events A, B, and C, defined as follows, and express them by means of sample points.

A = "all three colors appear"
B = "only two colors appear"
C = "at least two colors appear."

2.3 A university library has five copies of a textbook to be used in a certain class. Of these copies, numbers 1 through 3 are of the 1st edition, and numbers 4 and 5 are of the 2nd edition. Two of these copies are chosen at random to be placed on a 2-hour reserve.

(i) Write out an appropriate sample space $\mathcal{S}$.

(ii) Consider the events A, B, C, and D, defined as follows, and express them in terms of sample points.

A = "both books are of the 1st edition"
B = "both books are of the 2nd edition"
C = "one book of each edition"
D = "no book is of the 2nd edition."

2.4 A large automobile dealership sells three brands of American cars, denoted by a_1, a_2, a_3; two brands of Asian cars, denoted by b_1, b_2; and one brand of a

European car, denoted by c. We observe the cars sold in two consecutive sales. Then:

(i) Write out an appropriate sample space for this experiment.
(ii) Express the events defined as follows in terms of sample points:

A = "American brands in both sales"
B = "American brand in the first sale and Asian brand in the second sale"
C = "American brand in one sale and Asian brand in the other sale"
D = "European brand in one sale and Asian brand in the other sale."

Hint. For part (i), denote by (x_1, x_2) the typical sample point, where x_1 and x_2 stand for one of $a_1, a_2, a_3; b_1, b_2; c$.

2.5 Of two gas stations I and II located at a certain intersection, I has five gas pumps and II has six gas pumps. On a given time of a day, observe the numbers x and y of pumps (not which ones!) in use in stations I and II, respectively.

(i) Write out the sample space $\mathcal{S}$ for this experiment.
(ii) Consider the events A, B, C, and D, defined as follows, and express them in terms of sample points.

A = "only three pumps are in use in station I"
B = "the number of pumps in use in both stations is the same"
C = "the number of pumps in use in station II is larger than that in station I"
D = "the total number of pumps in use in both stations is not greater than 4."

2.6 At a certain busy airport, denote by A, B, C, and D the events defined as follows:

A = "at least 5 planes are waiting to land"
B = "at most 3 planes are waiting to land"
C = "at most 2 planes are waiting to land"
D = "exactly 2 planes are waiting to land."

In terms of the events A, B, C, and D and, perhaps, their complements, express the following events:

E = "at most 4 planes are waiting to land"
F = "at most 1 plane is waiting to land"
G = "exactly 3 planes are waiting to land"
H = "exactly 4 planes are waiting to land"
I = "at least 4 planes are waiting to land."

2.7 Let $\mathcal{S} = \{(x, y) \in \Re^2 = \Re \times \Re; -3 \leq x \leq 3,\ 0 \leq y \leq 4,\ x \text{ and } y \textit{ integers}\}$, and define the events A, B, C, and D as follows:

$$A = \{(x, y) \in \mathcal{S}; x = y\}, \quad B = \{(x, y) \in \mathcal{S}; x = -y\},$$
$$C = \{(x, y) \in \mathcal{S}; x^2 = y^2\}, \quad D = \{(x, y) \in \mathcal{S}; x^2 + y^2 \leq 5\}.$$

(i) List explicitly the members of S.
(ii) List the members of the events just defined.

2.8 In terms of the events A_1, A_2, A_3 in a sample space $\mathcal{S}$ and, perhaps, their complements, express the following events:

(i) $B_0 = \{s \in \mathcal{S}; s$ belongs to none of $A_1, A_2, A_3\}$.
(ii) $B_1 = \{s \in \mathcal{S}; s$ belongs to exactly one of $A_1, A_2, A_3\}$.
(iii) $B_2 = \{s \in \mathcal{S}; s$ belongs to exactly two of $A_1, A_2, A_3\}$.
(iv) $B_3 = \{s \in \mathcal{S}; s$ belongs to all of $A_1, A_2, A_3\}$.
(v) $C = \{s \in \mathcal{S}; s$ belongs to at most two of $A_1, A_2, A_3\}$.
(vi) $D = \{s \in \mathcal{S}; s$ belongs to at least one of $A_1, A_2, A_3\}$.

2.9 If for three events A, B, and C it happens that either $A \cup B \cup C = A$ or $A \cap B \cap C = A$, what conclusions can you draw? That is, how are the events A, B, and C related? What about the case that both $A \cup B \cup C = A$ and $A \cap B \cap C = A$ hold simultaneously?

2.10 Show that A is the impossible event (that is, $A = \varnothing$), if and only if $(A \cap B^c) \cup (A^c \cap B) = B$ for every event B.

2.11 Let A, B, and C be arbitrary events in $\mathcal{S}$. Determine whether each of the following statements is correct or incorrect.

(i) $(A - B) \cup B = (A \cap B^c) \cup B = B$.
(ii) $(A \cup B) - A = (A \cup B) \cap A^c = B$.
(iii) $(A \cap B) \cap (A - B) = (A \cap B) \cap (A \cap B^c) = \varnothing$.
(iv) $(A \cup B) \cap (B \cup C) \cap (C \cup A) = (A \cap B) \cup (B \cap C) \cup (C \cap A)$.

Hint. For part (iv), you may wish to use Proposition 2 (iii).

2.12 For any three events A, B, and C in a sample space $\mathcal{S}$, show the transitive property relative to $\subseteq$ holds (i.e., $A \subseteq B$ and $B \subseteq C$ imply that $A \subseteq C$).

2.13 Establish the distributive laws; namely $A \cap (\cup_j A_j) = \cup_j (A \cap A_j)$ and $A \cup (\cap_j A_j) = \cap_j (A \cup A_j)$.

Hint. Show that the event of either side is contained in the event of the other side.

2.14 Establish the identity:

$$\cup_j A_j = A_1 \cup \left(A_1^c \cap A_2\right) \cup \left(A_1^c \cap A_2^c \cap A_3\right) \cup \cdots$$

Hint. As in Exercise 2.13.

2.15 Establish DeMorgan's laws, namely:

$$(\cup_j A_j)^c = \cap_j A_j^c \quad \text{and} \quad (\cap_j A_j)^c = \cup_j A_j^c.$$

Hint. As in Exercise 2.13.

2.16 Let $\mathcal{S} = \Re$ and, for $n = 1, 2, \ldots$, define the events A_n and B_n by:

$$A_n = \left\{x \in \Re; -5 + \frac{1}{n} < x < 20 - \frac{1}{n}\right\}, \qquad B_n = \left\{x \in \Re; 0 < x < 7 + \frac{3}{n}\right\}.$$

(i) Show that the sequence $\{A_n\}$ is increasing and the sequence $\{B_n\}$ is decreasing.

(ii) Identify the limits, $\lim_{n\to\infty} A_n = \bigcup_{n=1}^{\infty} A_n$ and $\lim_{n\to\infty} B_n = \bigcap_{n=1}^{\infty} B_n$.
Remark: See discussion following Example 18.

2.3 Random Variables

For every random experiment, there is at least one sample space appropriate for the random experiment under consideration. In many cases, however, much of the work can be done without reference to an explicit sample space. Instead, what are used extensively are random variables and their distributions. These quantities will be studied extensively in subsequent chapters. What is presented in this section is the introduction of the concept of a random variable.

Formally, a *random variable* (r.v.) is simply a function defined on a sample space $\mathcal{S}$ and taking values in the real line $\Re = (-\infty, \infty)$. Random variables are denoted by capital letters, such as X, Y, Z, with or without subscripts. Thus, the value of the r.v. X at the sample point s is $X(s)$, and the set of all values of X, that is, the *range* of X, is usually denoted by $X(\mathcal{S})$. The only difference between an r.v. and a function in the usual calculus sense is that the domain of an r.v. is a sample space $\mathcal{S}$, which may be an abstract set, unlike the usual concept of a function, whose domain is a subset of $\Re$ or of a Euclidean space of higher dimension. The usage of the term "random variable" employed here rather than that of a function may be explained by the fact that an r.v. is associated with the outcomes of a random experiment. Thus, one may argue that $X(s)$ is not known until the random experiment is actually carried out and s becomes available. Of course, on the same sample space, one may define many distinct r.v.'s (see Figure 2.7).

In reference to Example 1, instead of the sample space $\mathcal{S}$ exhibited there, one may be interested in the number of heads appearing each time the experiment is

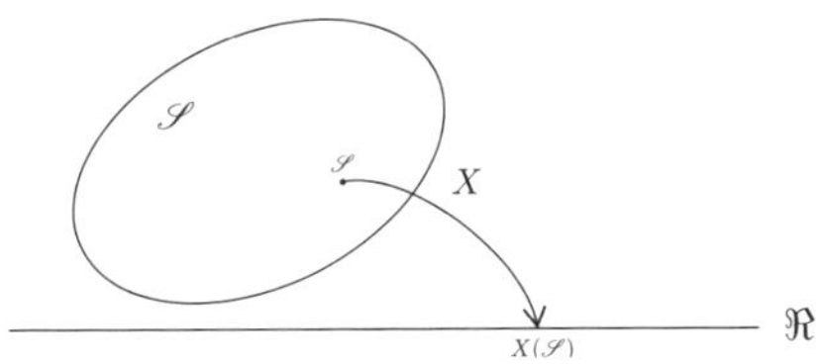

FIGURE 2.7
The r.v. X maps (transfers) the sample space $\mathcal{S}$ into the real line $\Re$.

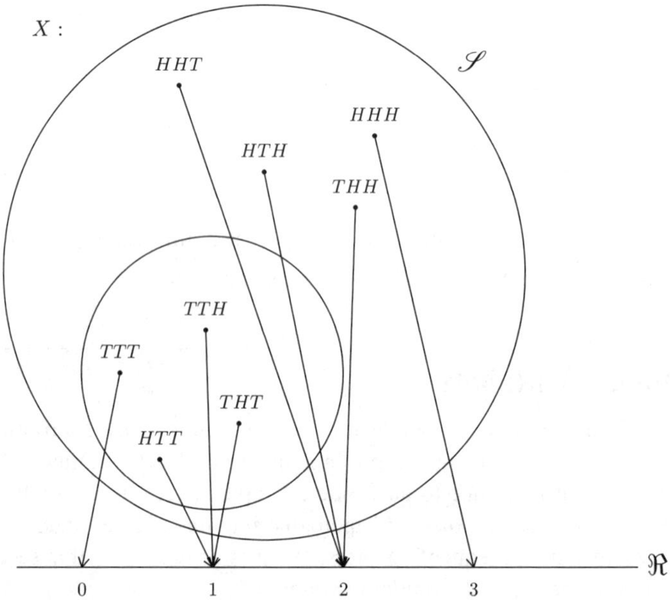

FIGURE 2.8

For $B = \{0, 1\}$, we have: $(X \in B) = (X \leq 1) = X^{-1}(B) = \{TTT, HTT, THT, TTH\}$.

carried out. This leads to the definition of the r.v. X by: $X(s) = \#$ of H's in s. Thus, $X(HHH) = 3, X(HHT) = X(HTH) = X(THH) = 2, X(HTT) = X(THT) = X(TTH) = 1$, and $X(TTT) = 0$, so that $X(\mathcal{S}) = \{0, 1, 2, 3\}$. The notation $(X \leq 1)$ stands for the event $\{s \in \mathcal{S};\ X(s) \leq 1\} = \{TTT, HTT, THT, TTH\}$. In the general case and for $B \subseteq \Re$, the notation $(X \in B)$ stands for the event A in the sample space $\mathcal{S}$ defined by: $A = \{s \in \mathcal{S};\ X(s) \in B\}$. It is also denoted by $X^{-1}(B)$ (see Figure 2.8).

In reference to Example 2, an r.v. X of interest may be defined by $X(s) =$ sum of the numbers in the pair s. Thus, $X((1, 1)) = 2, X((1, 2)) = X((2, 1)) = 3, \ldots, X((6, 6)) = 12$, and $X(\mathcal{S}) = \{2, 3, \ldots, 12\}$. Also,

$$X^{-1}(\{7\}) = \{s \in \mathcal{S};\ X(s) = 7\} = \{(1, 6), (2, 5), (3, 4), (4, 3), (5, 2), (6, 1)\}.$$

Similarly for Examples 3–5.

In reference to Example 6, a natural r.v. X is defined to denote the number of tosses needed until the first head occurs. Thus, $X(H) = 1, X(TH) = 2, \ldots, X(\underbrace{T \ldots T}_{n-1} H) = n, \ldots$, so that $X(\mathcal{S}) = \{1, 2, \ldots\}$. Also, $(X > 4) = (X \geq 5) = \{TTTTH, TTTTTH, \ldots\}$ (see Figure 2.9).

In reference to Examples 7 and 8, an obvious r.v. X is: $X(s) = s, s = 0, 1, \ldots$

In reference to Example 9, an r.v. X of interest is $X(s) = s, s \in \mathcal{S}$, and similarly for Examples 10, 12, and 13. In Example 11, $X(s) = s, s \in (0, D]$.

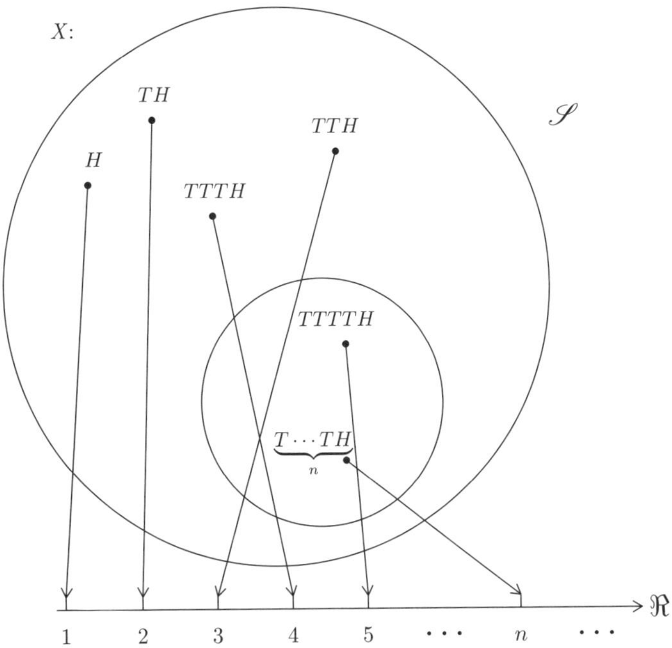

FIGURE 2.9

For $B = \{5, 6, \ldots\}$, we have: $(X \in B) = (X \geq 5) = X^{-1}(B) = \{TTTTH, TTTTTH, \ldots, \underbrace{T\ldots T}_{n-1}H, \ldots\}$.

In reference to Example 14 (with $\mathcal{S} = (0, \infty)$), let X be the r.v. denoting the cost of operating said equipment up to time s, and, to be specific, suppose that $X(s) = 2(1 - 0.5e^{-0.2s})$, $s > 0$. Then the range of X is $(1, 2)$, and for $B = [1.25, 1.75]$, we have: $(X \in B) = (1.25 \leq X \leq 1.75) = X^{-1}([1.25, 1.75]) = [5\log(4/3), 5\log 4]$. This is so because by routine manipulation, $1.25 \leq 2(1 - 0.5e^{-0.2s}) \leq 1.75$ is equivalent to $5\log(4/3) \leq s \leq 5\log 4$, where as always, log stands for the natural logarithm (see Figure 2.10).

In reference to Example 10 in Chapter 1, an r.v. X may be defined thus: $X(s) =$ the position of M in s. Then, clearly, $X(\mathcal{S}) = \{1, 2, \ldots\}$.

In reference to Example 16 in Chapter 1, the r.v.'s of obvious interest are: $X_A = \#$ of those persons, out of n, having blood type A, and similarly for X_B, X_{AB}, X_O.

From the preceding examples, two kinds of r.v.'s emerge: random variables, which take on countably many values, such as those defined in conjunction with Examples 1–5, 6–8 here, and 16 in Chapter 1, and r.v.'s, which take on all values in a nondegenerate (finite or not) interval in $\Re$. Such are r.v.'s defined in conjunction with Examples 9–14. Random variables of the former kind are called discrete r.v.'s (or r.v.'s of the discrete type), and r.v.'s of the latter type are called continuous r.v.'s (or r.v.'s of the continuous type).

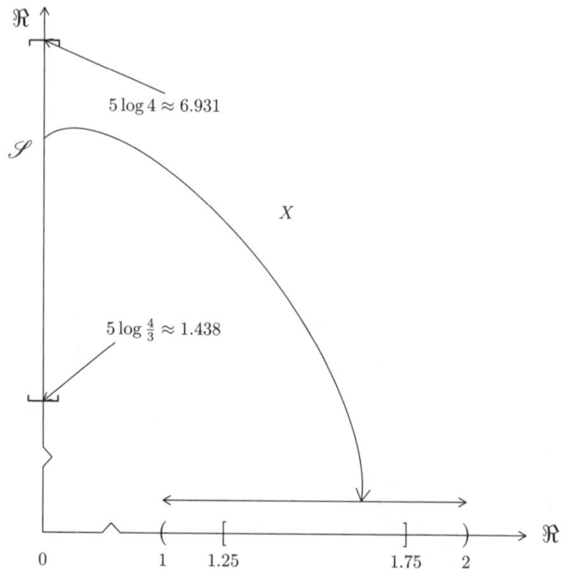

FIGURE 2.10

The interval $[5\log(4/3), 5\log 4] \simeq [1.438, 6.931]$ is the set of all sample points s mapped onto $[1.25, 1.75]$ under the r.v. X; it is $X^{-1}([1.25, 1.75])$.

More generally, an r.v. X is called *discrete* (or *of the discrete type*), if X takes on countably many values; that is, either finitely many values such as $x_1, \ldots, x_n$ or countably infinite many values such as $x_0, x_1, \ldots$ or $x_1, x_2, \ldots$. On the other hand, X is called *continuous* (or *of the continuous type*) if X takes all values in a proper interval $I \subseteq \Re$. Although there are other kinds of r.v.'s, in this book we will restrict ourselves to discrete and continuous r.v.'s as just defined.

The study of r.v.'s is one of the main objectives of this book.

Exercises.

3.1 In reference to Exercise 2.1, define the r.v.'s $X_i, i = 1, 2, 3$ as follows: $X_i =$ # of passengers delivered to hotel H_i.
Determine the values of each $X_i, i = 1, 2, 3$, and specify the values of the sum $X_1 + X_2 + X_3$.

3.2 In reference to Exercise 2.2, define the r.v.'s X and Y as follows: $X =$ # of red balls dispensed, $Y =$ # of balls other than red dispensed.
Determine the values of X and Y, and specify the values of the sum $X + Y$.

3.3 In reference to Exercise 2.5, define the r.v.'s X and Y as follows: $X =$ # of pumps in use in station I, $Y =$ # of pumps in use in station II.
Determine the values of X and Y, and of the sum $X + Y$.

3.4 In reference to Exercise 2.7, define the r.v. X by: $X((x, y)) = x + y$.
Determine the values of X, as well as the following events: $(X \leq 2)$, $(3 < X \leq 5)$, $(X > 6)$.

3.5 Consider a year with 365 days, which are numbered serially from 1 to 365. Ten of those numbers are chosen at random and without repetition, and let X be the r.v. denoting the largest number drawn.
Determine the values of X.

3.6 A four-sided die has the numbers 1 through 4 written on its sides, one on each side. If the die is rolled twice:

(i) Write out a suitable sample space $\mathcal{S}$.
(ii) If X is the r.v. denoting the sum of numbers appearing, determine the values of X.
(iii) Determine the events: $(X \leq 3)$, $(2 \leq X < 5)$, $(X > 8)$.

Hint. For part (i), the typical sample point is a pair (x, y), where x and y run through the values 1, 2, 3, 4.

3.7 From a certain target population, n individuals are chosen at random and their blood types are determined. Let X_1, X_2, X_3, and X_4 be the r.v.'s denoting the number of individuals having blood types A, B, AB, and O, respectively.
Determine the values of each one of these r.v.'s, as well as the values of the sum $X_1 + X_2 + X_3 + X_4$.

3.8 A bus is expected to arrive at a specified bus stop any time between 8:00 and 8:15 a.m., and let X be the r.v. denoting the actual time of arrival of the bus.

(i) Determine the suitable sample space $\mathcal{S}$ for the experiment of observing the arrival of the bus.
(ii) What are the values of the r.v. X?
(iii) Determine the event: "The bus arrives within 5 min before the expiration of the expected time of arrival."

2.4 Basic Concepts and Results in Counting

In this brief section, some basic concepts and results are discussed regarding the way of counting the total number of outcomes of an experiment, or the total number of different ways we can carry out a task. Although many readers will undoubtedly be familiar with parts of or the entire material in this section, it would be advisable, nevertheless, to invest some time here in introducing and adopting some notation, establishing some basic results, and then using them in computing probabilities in the classical probability framework in Chapter 3.

Problems of counting arise in a great number of different situations. Here are some of them. In each one of these situations, we are asked to compute the number of different ways that something or other can be done. Here are a few illustrative cases.

Example 19.

(i) Attire yourself by selecting a T-shirt, a pair of trousers, a pair of shoes, and a cap out of n_1 T-shirts, n_2 pairs of trousers, n_3 pairs of shoes, and n_4 caps (e.g., $n_1 = 4, n_2 = 3, n_3 = n_4 = 2$).

(ii) Form all k-digit numbers by selecting the k digits out of n available numbers (e.g., $k = 2, n = 4$ such as $\{1, 3, 5, 7\}$).
(iii) Form all California automobile license plates by using one number, three letters and then three numbers in the prescribed order.
(iv) Form all possible codes by using a given set of symbols (e.g., form all "words" of length 10 by using the digits 0 and 1).
(v) Place k books on the shelf of a bookcase in all possible ways.
(vi) Place the birthdays of k individuals in the 365 days of a year in all possible ways.
(vii) Place k letters into k addressed envelopes (one letter to each envelope).
(viii) Count all possible outcomes when tossing k distinct dice.
(ix) Select k cards out of a standard deck of playing cards (e.g., for $k = 5$, each selection is a poker hand).
(x) Form all possible k-member committees out of n available individuals.

The calculation of the numbers asked for in situations (i) through (x) just outlined is in actuality a simple application of the so-called *fundamental principle of counting*, stated next in the form of a theorem.

Theorem 1 (Fundamental Principle of Counting).
Suppose a task is completed in k stages by carrying out a number of subtasks in each one of the k stages. If the numbers of these subtasks are $n_1, \ldots, n_k$ for the k stages, respectively, then the total number of different ways the overall task is completed is: $n_1 \times \cdots \times n_k$.

Thus, in (i) above the number of different attires is: $4 \times 3 \times 2 \times 2 = 48$.

In (ii), the number of all 2-digit numbers formed by using 1, 3, 5, 7 is: $4 \times 4 = 16$ (11, 13, 15, 17; 31, 33, 35, 37; 51, 53, 55, 57; 71, 73, 75, 77).

In (iii), the number of all possible license plates (by using indiscriminately all 10 digits from 0 through 9 and all 26 letters of the English alphabet, although this is not the case in practice) is: $10 \times (26 \times 26 \times 26) \times (10 \times 10 \times 10) = 175{,}760{,}000$.

In (iv), the number of all possible "words" is found by taking $k = 10$ and $n_1 = \cdots = n_{10} = 2$ to obtain: $2^{10} = 1{,}024$.

In (v), all possible arrangements are obtained by taking $n_1 = k, n_2 = k-1, \ldots, n_k = k-(k-1) = 1$ to get: $k(k-1)\ldots 1 = 1 \ldots (k-1)k$. For example, for $k = 10$, the number of arrangements is: 3,628,800.

In (vi), the required number is obtained by taking $n_1 = \cdots = n_k = 365$ to get: 365^k. For example, for $k = 3$, we have $365^3 = 48{,}627{,}125$.

In (vii), the required number is: $k(k-1)\ldots 1 = 1 \ldots (k-1)k$ obtained by taking $n_1 = k, n_2 = k-1, \ldots, n_k = k-(k-1) = 1$.

In (viii), the required number is: 6^k obtained by taking $n_1 = \cdots = n_k = 6$. For example, for $k = 3$, we have $6^3 = 216$, and for $k = 10$, we have $6^{10} = 60{,}466{,}176$.

In (ix), the number of poker hands is: $\frac{52 \times 51 \times 50 \times 49 \times 48}{120} = 2{,}598{,}960$. The numerator is obtained by taking $n_1 = 52, n_2 = 51, n_3 = 50, n_4 = 49, n_5 = 48$. The division by $120(= 1 \times 2 \times 3 \times 4 \times 5)$ accounts for elimination of hands consisting of the same cards but drawn in a different order.

Finally, in (x), the required number is: $\frac{n(n-1)\cdots(n-k+1)}{1\times2\times\cdots\times k}$, by arguing as in (ix). For example, for $n = 10$ and $k = 3$, we have: $\frac{10\times9\times8}{1\times2\times3} = 120$.

In all of the situations (i) through (x), the required numbers were calculated by the appropriate application of Theorem 1. Furthermore, in many cases, as clearly exemplified by cases (ii), (iii), (v), (vii), (ix), and (x), the task performed consisted of selecting and arranging a number of objects out of a set of available objects. In so doing, the order in which the objects appear in the arrangement may be of significance, as is, indeed, the case in situations (ii), (iii), (iv), (v), (vi), and (vii), or it may be just irrelevant, as happens, for example, in cases (ix) and (x). This observation leads us to the concepts of permutations and combinations. More precisely, we have

Definition 1. An *ordered* arrangement of k objects taken from a set of n (distinct) objects ($1 \leq k \leq n$) is a *permutation* of the n objects taken k at a time. An *unordered* arrangement of k objects taken from a set of n (distinct) objects is a *combination* of the n objects taken k at a time.

The question then arises as to how many permutations and how many combinations there are. The answer to this question is given next.

Corollary 2.1 (to Theorem 1).

(i) *The number of ordered arrangements of a set of n objects taken k at a time ($1 \leq k \leq n$) is n^k when repetitions are allowed. When no repetitions are allowed, this number becomes the permutations of n objects taken k at a time, is denoted by $P_{n,k}$, and is given by:*

$$P_{n,k} = n(n-1)\cdots(n-k+1). \tag{2.1}$$

In particular, for $k = n$,

$$P_{n,n} = n(n-1)\cdots 1 = 1\cdots(n-1)n = n!,$$

where the notation $n!$ is read "n factorial."

(ii) *The number of combinations (i.e., the number of unordered and without repetition arrangements) of n objects taken k at a time ($1 \leq k \leq n$) is denoted by $\binom{n}{k}$ and is given by:*

$$\binom{n}{k} = \frac{P_{n,k}}{k!} = \frac{n!}{k!(n-k)!}. \tag{2.2}$$

Remark 2. Whether permutations or combinations are appropriate in a given problem follows from the nature of the problem. For instance, in (ii), permutations rather than combinations are appropriate as, for example, 13 and 31 are distinct entities. The same is true of cases (iii)–(viii), whereas combinations are appropriate for cases (ix) and (x).

As an example, in part (ii), $P_{4,2} = 4 \times 3 = 12$ (leave out the numbers with identical digits 11, 22, 33, and 44), and in part (ix), $\binom{52}{5} = \frac{52!}{5!47!} = 2{,}598{,}960$, after cancellations and by carrying out the arithmetic.

Remark 3. In (2.2), set $k = n$. Then the left-hand side is clearly 1, and the right-hand side is $\frac{n!}{n! \times 0!} = \frac{1}{0!}$. In order for this to be 1, we *define* $0! = 1$. From formula (2.2), it also follows that $\binom{n}{0} = 1$.

In computing the permutations (factorial) $P_{n,n} = n!$, the assumption was made that the n objects were distinct. If this fails to be true, the number $n!$ will have to be adjusted suitably. More precisely, we have the following result.

Proposition 5. *Consider n objects that are divided into k groups $(1 \leq k \leq n)$ with the property that the m_i members of the ith group are identical and distinct from the members of the remaining groups, $m_1 + \cdots + m_k = n$. Then the number of distinct arrangements of the n objects is $n!/(m_1! \times \cdots \times m_k!)$*

Proof. One way of generating all distinct arrangements of the n objects is to select m_i positions out of n available in $\binom{n}{m_i}$ possible ways and place there the m_i identical objects, $i = 1, \ldots, k$. Then, by Theorem 1, the total number of arrangements is:

$$\binom{n}{m_1} \times \binom{n - m_1}{m_2} \times \cdots \times \binom{n - m_1 - \cdots - m_{k-1}}{m_k} =$$

$$\frac{n!}{m_1!(n - m_1)!} \times \frac{(n - m_1)!}{m_2!(n - m_1 - m_2)!} \times \cdots \times \frac{(n - m_1 - \cdots - m_{k-1})!}{m_k!(n - m_1 - \cdots - m_{k-1} - m_k)!} =$$

$$\frac{n!}{m_1! \times m_2! \times \cdots \times m_k!}, \quad \text{since} (n - m_1 - \cdots - m_{k-1} - m_k)! = 0! = 1.$$

An alternative way to look at this problem would be to consider the $n!$ arrangements of the n objects, and then make the $m_i!$ arrangements within the ith group, $i = 1, \ldots, k$, which leave the overall arrangement unchanged. Thus, the number of distinct arrangements of the n objects is $n!/(m_1! \times m_2! \times \cdots \times m_k!)$. ▲

This section is concluded with the justification of Theorem 1 and its corollary and some applications of these results.

Proof of Theorem 1. It is done by induction. For $k = 2$, all one has to do is to pair out each one of the n_1 ways of carrying out the subtask at stage 1 with each one of the n_2 ways of carrying out the subtask at stage 2 in order to obtain $n_1 \times n_2$ for the number of ways of completing the task. Next, make the induction hypothesis that the conclusion is true for $k = m$ and establish it for $k = m + 1$. So, in the first m stages, the total number of ways of doing the job is: $n_1 \times \cdots \times n_m$, and there is still the final $(m + 1)$ st stage for completing the task. Clearly, all we have to do here is to combine each one of the $n_1 \times \cdots \times n_m$ ways of doing the job in the first m stages with each one of the n_{m+1} ways of carrying out the subtask in the $(m+1)$-st stage to obtain the number $n_1 \times \cdots \times n_m \times n_{m+1}$ of completing the task. ▲

Proof of the Corollary.

(i) Here, we are forming an ordered arrangement of objects in k stages by selecting one object at each stage from among the n available objects (because repetitions are allowed). Thus, the theorem applies with $n_1 = \cdots = n_k = n$ and gives the result n^k. When repetitions are not allowed, the only thing that changes from the case just considered is that: $n_1 = n, n_2 = n-1, \ldots, n_k = n-(k-1) = n-k+1$, and formula (2.1) follows.

(ii) Let $\binom{n}{k}$ be the number of combinations (unordered and without repetition arrangements) of the n objects taken k at a time. From each one of these unordered arrangements, we obtain $k!$ ordered arrangements by permutation of the k objects. Then $k! \times \binom{n}{k}$ is the total number of ordered arrangements of the n objects taken k at a time, which is $P_{n,k}$, by part (i); i.e., $k! \times \binom{n}{k} = P_{n,k}$. Solving for $\binom{n}{k}$, we obtain the first expression in (2.2). The second expression follows immediately by multiplying by $(n-k)\ldots 1$ and dividing by $1\ldots(n-k) = (n-k)!$. ▲

There are many interesting variations and deeper results based on Theorem 1 and its corollary. Some of them may be found in Sections 2.4 and 2.6 of Chapter 2 of the book *A Course in Mathematical Statistics*, 2nd edition (1997), Academic Press, by G. G. Roussas.

Example 20. The faculty in an academic department at UC-Davis consists of 4 assistant professors, 6 associate professors, and 5 full professors. Also, it has 30 graduate students. An ad hoc committee of 5 is to be formed to study a certain curricular matter.

(i) What is the number of all possible committees consisting of faculty alone?
(ii) How many committees can be formed if 2 graduate students are to be included and all academic ranks are to be represented?

Discussion. It is clear that combinations are the appropriate tool here. Then we have:

(i) This number is: $\binom{15}{5} = \frac{15!}{5!10!} = \frac{11\times12\times13\times14\times15}{1\times2\times3\times4\times5} = 3{,}003.$

(ii) Here the number is: $\binom{30}{2}\binom{4}{1}\binom{6}{1}\binom{5}{1} = \frac{30!}{2!28!}\times4\times6\times5 = \frac{29\times30}{2}\times120 =$ 52,200.

Example 21. In how many ways can one distribute 5 gifts to 15 persons if no person receives more than one gift?

Discussion. The answer is $\binom{15}{5} = \frac{15!}{5!10!} = 3003$.

Example 22. How many 5-letter words can be formed by using the 26 letters of the English alphabet if: (i) No restrictions are imposed; (ii) All 5 letters are to be distinct.

Discussion.

(i) The answer here is $26^5 = 11{,}881{,}376$.
(ii) In this case, the answer is $P_{26,5} = 26 \times 25 \times 24 \times 23 \times 22 = 7{,}893{,}600$.

Example 23. Each one of 10 workers is to be assigned to one of 10 different jobs. How many assignments are possible?

Discussion. Clearly, the answer is $10! = 1 \times 2 \times \cdots \times 10 = 3{,}628{,}800$.

Example 24. By using 3 A's, 2 E's, 1 H, 2 L's, 2 S's , and 1 T, one can form the word TALLAHASSEE, the name of the capital city of the state of Florida. How many other distinct words can be formed?

Discussion. There are 11 letters altogether. Then, by Proposition 5, the total number of words is:

$$\frac{11!}{3! \times 2! \times 1! \times 2! \times 2! \times 1!} = 4 \times 5 \times 6 \times 7 \times 9 \times 10 \times 11 = 831{,}600.$$

Therefore there are another 831,599 distinct words in addition to TALLAHASSEE.

Example 25. Use Proposition 5 in order to show that the multinomial expansion of $(x_1 + \cdots + x_k)^n$ is given by:

$$\sum \frac{n!}{m_1! \times \cdots \times m_k!} x_1^{m_1} \times \cdots \times x_k^{m_k},$$

where the summation extends over all nonnegative integers $m_1, \ldots, m_k$ with $m_1 + \cdots + m_k = n$.

Discussion. The multinomial expansion of $(x_1 + \cdots + x_k)^n$ is the summation of all possible terms of the form $x_1^{m_1} \times \cdots \times x_k^{m_k}$, where the m_i's are as described. However, this amounts to computing the number of distinct arrangements of n objects divided into k groups with m_i identical objects (x_i) in the ith group, $i = 1, \ldots, k$. This number was seen to be $n!/m_1! \times \cdots \times m_k!$

Exercises.

4.1 Telephone numbers at UC-Davis consist of 7 digits, the first 3 of which are 752. It is estimated that about 15,000 different telephone numbers are needed to serve the university's needs.
Are there enough telephone numbers available for this purpose? Justify your answer.

4.2 An experimenter is studying the effects of temperature, pressure, and a catalyst on the yield of a certain chemical reaction. Three different temperatures, four different pressures, and five different catalysts are under consideration.

- **(i)** If any particular experimental run involves the use of a single temperature, pressure, and catalyst, how many experimental runs are possible?
- **(ii)** How many experimental runs are there that involve use of the lowest temperature and the two lowest pressures?
- **(iii)** How many experimental runs are possible if a specified catalyst is to be used?

4.3

- **(i)** Given that a zip code consists of a 5-digit number, where the digits are selected from among the numbers $0, 1, \ldots, 9$, calculate the number of all different zip codes.
- **(ii)** If X is the r.v. defined by: $X(\text{zip code}) = \#$ of nonzero digits in the zip code, which are the possible values of X?
- **(iii)** Give 3 zip codes and the respective values of X.

4.4 State how many 5-digit numbers can be formed by using the numbers 1, 2, 3, 4, and 5, so that odd positions are occupied by odd numbers and even positions are occupied by even numbers, if:

- **(i)** Repetitions are allowed.
- **(ii)** Repetitions are not allowed.

4.5 How many 3-digit numbers can be formed by using the numbers: 0, 1, 2, 3, 4, 5, 6, 7, 8, and 9, and satisfying one of the following requirements:

- **(i)** No restrictions are imposed.
- **(ii)** All 3 digits are distinct.
- **(iii)** All 3-digit numbers start with 1 and end with 0.

4.6 On a straight line, there are n spots to be filled in by either a dot or a dash. What is the number of the distinct groups of resulting symbols? What is this number if $n = 5, 10, 15, 20$, and 25?

4.7 For any integers m and n with $0 \leq m \leq n$, show that $\binom{n}{m} = \binom{n}{n-m}$ either by writing out each side in terms of factorials or by using a suitable argument without writing out anything.

4.8 Show that $\binom{n+1}{m+1} / \binom{n}{m} = \frac{n+1}{m+1}$.

Hint. Write out each expression in terms of factorials.

4.9 If M and m are positive integers with $m \leq M$, show that:

$$\binom{M}{m} = \binom{M-1}{m} + \binom{M-1}{m-1},$$

by recalling that $\binom{k}{x} = 0$ for $x > k$.

Hint. As in Exercise 4.8, starting with the right-hand side.

4.10 Without any calculations and by recalling that $\binom{k}{x} = 0$ for $x > k$, show that:

$$\sum_{x=0}^{r} \binom{m}{x}\binom{n}{r-x} = \binom{m+n}{r}.$$

4.11 The binomial expansion formula states that for any x and y real and n, a positive integer:

$$(x+y)^n = \sum_{k=0}^{n} \binom{n}{k} x^k y^{n-k}.$$

(i) Justify this formula by using relation (2.2).

(ii) Use this formula in order to show that:

$$\sum_{k=0}^{n} \binom{n}{k} = 2^n \quad \text{and} \quad \sum_{k=0}^{n} (-1)^k \binom{n}{k} = 0.$$

Hint. For part (i), see also Example 25.

4.12 In the plane, there are n points such that no three of them lie on a straight line. How many triangles can be formed? What is this number for $n = 10$?

4.13 Beethoven wrote 9 symphonies, Mozart wrote 27 piano concertos, and Schubert wrote 15 string quartets.

(i) If a university radio station announcer wishes to play first a Beethoven symphony, then a Mozart concerto, and then a Schubert string quartet, in how many ways can this be done?

(ii) What is the number in part (i) if the three pieces are played in all possible orderings?

4.14 If n countries exchange ambassadors, how many ambassadors are involved? What is this number for $n = 10, 50, 100$?

4.15 Suppose automobile license plate numbers are formed by first selecting first one digit out of the numbers $0, 1, \ldots, 9$, then 3 letters out of the 26 letters of the English alphabet, and then 3 digits out of the numbers $0, 1, \ldots, 9$. How many license plate numbers can be formed:

(i) Without any restrictions imposed.

(ii) The 3 letters are distinct.

(iii) The 4 digits are distinct.

(iv) Both the 3 letters and the 4 digits are distinct.

CHAPTER

The Concept of Probability and Basic Results

3

This chapter consists of three sections. The first section is devoted to the definition of the concept of probability. We start with the simplest case, where complete symmetry occurs, proceed with the definition by means of relative frequency, and conclude with the axiomatic definition of probability. The defining properties of probability are illustrated by way of examples. In the second section, a number of basic properties, resulting from the definition, are stated and justified. Some of them are illustrated by means of examples. The section is concluded with two theorems, which are stated but not proved. In the third section, the probability distribution function of a random variable (r.v.) is introduced. Also, the distribution function and the probability density function of an r.v. are defined, and we explain how they determine the distribution of the r.v.

3.1 Definition of Probability

When a random experiment is entertained, one of the first questions that arises is: What is the probability that a certain event occurs? For instance, in reference to Example 1 in Chapter 2, one may ask: What is the probability that exactly one head occurs? In other words, what is the probability of the event $B = \{HTT, THT, TTH\}$? The answer to this question is almost automatic and is $3/8$. The relevant reasoning goes like this: Assuming that the three coins are balanced, the probability of each one of the 8 outcomes, considered as simple events, must be $1/8$. Since the event B consists of 3 sample points, it can occur in 3 different ways, and hence its probability must be $3/8$.

This is exactly the intuitive reasoning employed in defining the concept of probability when two requirements are met: First, the sample space $\mathcal{S}$ has finitely many outcomes, $\mathcal{S} = \{s_1, \ldots, s_n\}$, say, and second, each one of these outcomes is "equally likely" to occur or has the same chance of appearing whenever the relevant random experiment is carried out. This reasoning is based on the underlying symmetry. Thus, one is led to stipulating that each one of the (simple) events $\{s_i\}$, $i = 1, \ldots, n$ has probability $1/n$. Then the next step, that of defining the probability of a composite event A, is simple; if A consists of m sample points, $A = \{s_{i_1}, \ldots, s_{i_m}\}$, say $(1 \leq m \leq n)$ (or none at all, in which case $m = 0$), then the probability of A must be m/n. The notation used is: $P(\{s_1\}) = \cdots = P(\{s_n\}) = \frac{1}{n}$ and $P(A) = \frac{m}{n}$. Actually, this is the so-called *classical* definition of probability. That is,

Introduction to Probability, Second Edition. http://dx.doi.org/10.1016/B978-0-12-800041-0.00003-1

Classical Definition of Probability

Let $\mathcal{S}$ be a sample space, associated with a certain random experiment and consisting of finitely many sample points n, say, each of which is equally likely to occur whenever the random experiment is carried out. Then the probability of any event A, consisting of m sample points $(0 \leq m \leq n)$, is given by $P(A) = \frac{m}{n}$.

In reference to Example 1 in Chapter 2, $P(A) = \frac{4}{8} = \frac{1}{2} = 0.5$. In Example 2, Chapter 2 (when the two dice are unbiased), $P(X = 7) = \frac{6}{36} = \frac{1}{6} \simeq 0.167$, where the r.v. X and the event $(X = 7)$ are defined in Section 2.3.

From the preceding (classical) definition of probability, the following simple properties are immediate: For any event A, $P(A) \geq 0$; $P(\mathcal{S}) = 1$; if two events A_1 and A_2 are disjoint $(A_1 \cap A_2 = \varnothing)$, then $P(A_1 \cup A_2) = P(A_1) + P(A_2)$. This is so because if $A_1 = \{s_{i_1}, \ldots, s_{i_k}\}$, $A_2 = \{s_{j_1}, \ldots, s_{j_\ell}\}$, where all $s_{i_1}, \ldots, s_{i_k}$ are distinct from all $s_{j_1}, \ldots, s_{j_\ell}$, then $A_1 \cup A_2 = \{s_{i_1}, \ldots, s_{i_k} s_{j_1}, \ldots, s_{j_\ell}\}$ and $P(A_1 \cup A_2) = \frac{k+\ell}{n} = \frac{k}{n} + \frac{\ell}{n} = P(A_1) + P(A_2)$.

In many cases, the stipulations made in defining the probability as above are not met, either because $\mathcal{S}$ has not finitely many points (as is the case in Examples 6, 7, and 8 (by replacing C and M by ∞), and 9–14, all in Chapter 2), or because the (finitely many outcomes) are not equally likely. This happens, for instance, in Example 1, Chapter 2, when the coins are not balanced, and in Example 2, Chapter 2, when the dice are biased. Strictly speaking, it also happens in Example 4 in the same chapter. In situations like this, the way out is provided by the so-called *relative frequency* definition of probability. Specifically, suppose a random experiment is carried out a large number of times N, and let $N(A)$ be the *frequency* of an event A, the number of times A occurs (out of N). Then the *relative frequency* of A is $\frac{N(A)}{N}$. Next, suppose that, as $N \to \infty$, the relative frequencies $\frac{N(A)}{N}$ oscillate around some number (necessarily between 0 and 1). More precisely, suppose that $\frac{N(A)}{N}$ converges, as $N \to \infty$, to some number. Then this number is called the *probability* of A and is denoted by $P(A)$. That is, $P(A) = \lim_{N\to\infty} \frac{N(A)}{N}$.

Relative Frequency Definition of Probability

Let $N(A)$ be the number of times an event A occurs in N repetitions of a random experiment, and assume that the relative frequency of A, $\frac{N(A)}{N}$, converges to a limit as $N \to \infty$. This limit is denoted by $P(A)$ and is called the probability of A.

At this point, it is to be observed that empirical data show that the relative frequency definition of probability and the classical definition of probability agree in the framework in which the classical definition applies.

From the relative frequency definition of probability and the usual properties of limits, it is immediate that: $P(A) \geq 0$ for every event A; $P(\mathcal{S}) = 1$; and for A_1, A_2 with $A_1 \cap A_2 = \varnothing$,

$$
\begin{aligned}
P(A_1 \cup A_2) &= \lim_{N\to\infty} \frac{N(A_1 \cup A_2)}{N} = \lim_{N\to\infty} \left(\frac{N(A_1)}{N} + \frac{N(A_2)}{N} \right) \\
&= \lim_{N\to\infty} \frac{N(A_1)}{N} + \lim_{N\to\infty} \frac{N(A_2)}{N} = P(A_1) + P(A_2);
\end{aligned}
$$

that is, $P(A_1 \cup A_2) = P(A_1) + P(A_2)$, provided $A_1 \cap A_2 = \varnothing$. These three properties were also seen to be true in the classical definition of probability. Furthermore, it is immediate that under either definition of probability, $P(A_1 \cup \cdots \cup A_k) = P(A_1) + \cdots + P(A_k)$, provided the events are pairwise disjoint; $A_i \cap A_j = \varnothing, i \neq j$.

The above two definitions of probability certainly give substance to the concept of probability in a way consonant with our intuition about what probability should be. However, for the purpose of cultivating the concept and deriving deep probabilistic results, one must define the concept of probability in terms of some basic properties, which would not contradict what we have seen so far. This line of thought leads to the so-called axiomatic definition of probability due to Kolmogorov.

Axiomatic Definition of Probability

Probability is a function, denoted by P, defined for each event of a sample space $\mathcal{S}$, taking on values in the real line $\Re$, and satisfying the following three properties:

(P1) $P(A) \geq 0$ for every event A (nonnegativity of P).
(P2) $P(\mathcal{S}) = 1$ (P is normed).
(P3) For countably infinite many pairwise disjoint events A_i, $i = 1, 2, \ldots, A_i \cap A_j = \varnothing, i \neq j$, it holds:

$$P(A_1 \cup A_2 \cup \cdots) = P(A_1) + P(A_2) + \cdots; \quad \text{or} \quad P\left(\bigcup_{i=1}^{\infty} A_i\right) = \sum_{i=1}^{\infty} P(A_i)$$

(sigma-additivity (σ-additivity) of P).

Comments on the Axiomatic Definition

1. Properties (P1) and (P2) are the same as the ones we have seen earlier, whereas property (P3) is new. What we have seen above was its so-called *finitely-additive* version; that is, $P(\bigcup_{i=1}^{n} A_i) = \sum_{i=1}^{n} P(A_i)$, provided $A_i \cap A_j = \varnothing, i \neq j$. It will be seen below that finite-additivity is implied by σ-additivity (see Proposition 1(ii)) but not the other way around. Thus, if we are to talk about the probability of the union of countably infinite many pairwise disjoint events, property (P3) must be stipulated. Furthermore, the need for such a union of events is illustrated as follows: In reference to Example 6 in Chapter 2, calculate the probability that the first head does not occur before the nth tossing. By setting $A_i = \{\underbrace{T \ldots T}_{i-1} H\}$, $i = n, n+1, \ldots$, what we are actually after here is $P(A_n \cup A_{n+1} \cup \cdots)$ with $A_i \cap A_j = \varnothing, i \neq j, i$ and $j \geq n$.
2. Property (P3) is superfluous (reduced to finite-additivity) when the sample space $\mathcal{S}$ is finite, which implies that the total number of events is finite.
3. Finite-additivity is implied by additivity for two events, $P(A_1 \cup A_2) = P(A_1) + P(A_2)$, $A_1 \cap A_2 = \varnothing$, by way of induction.

Here are two examples in calculating probabilities.

Example 1. In reference to Example 1 in Chapter 1, take $n = 58$, and suppose we have the following configuration:

	Barium			
	High Mercury		Low Mercury	
Arsenic	High	Low	High	Low
High	1	3	5	9
Low	4	8	10	18

Calculate the probabilities mentioned in (i) (a)–(d).

Discussion. For simplicity, denote by B_h the event that the site selected has a high barium concentration, and likewise for other events below. Then:

(i)(a) $B_h = (A_h \cap B_h \cap M_h) \cup (A_h \cap B_h \cap M_\ell) \cup (A_\ell \cap B_h \cap M_h) \cup (A_\ell \cap B_h \cap M_\ell)$ and the events on the right-hand side are pairwise disjoint. Therefore (by Proposition 1(ii) below):

$$P(B_h) = P(A_h \cap B_h \cap M_h) + P(A_h \cap B_h \cap M_\ell) + P(A_\ell \cap B_h \cap M_h) + P(A_\ell \cap B_h \cap M_\ell) = \frac{1}{58} + \frac{3}{58} + \frac{4}{58} + \frac{8}{58} = \frac{16}{58} = \frac{8}{29} \simeq 0.276.$$

(i)(b) Here $P(M_h \cap A_\ell \cap B_\ell) = P(A_\ell \cap B_\ell \cap M_h) = \frac{10}{58} = \frac{5}{29} \simeq 0.172.$

(i)(c) Here the required probability is as in (a):

$$P(A_h \cap B_h \cap M_\ell) + P(A_h \cap B_\ell \cap M_h) + P(A_\ell \cap B_h \cap M_h) = \frac{12}{58} = \frac{6}{29} \simeq 0.207.$$

(i)(d) As above,

$$P(A_h \cap B_\ell \cap M_\ell) + P(A_\ell \cap B_h \cap M_\ell) + P(A_\ell \cap B_\ell \cap M_h) = \frac{27}{58} \simeq 0.466.$$

Example 2. In ranking five horses in a horse race (Example 5 in Chapter 2), calculate the probability that horse #3 finishes at least second.

Discussion. Let A_i be the event that horse #3 finishes in the ith position, $i = 1, \ldots, 5$. Then the required event is $A_1 \cup A_2$, where A_1, A_2 are disjoint. Thus,

$$P(A_1 \cup A_2) = P(A_1) + P(A_2) = \frac{24}{120} + \frac{24}{120} = \frac{2}{5} = 0.4.$$

Example 3. In tossing a coin repeatedly until H appears for the first time (Example 6 in Chapter 2), suppose that $P\{\underbrace{T \ldots T}_{i-1} H\} = P(A_i) = q^{i-1}p$ for some $0 < p < 1$

and $q = 1 - p$ (in anticipation of Definition 4 in Chapter 4). Then (see also #4 in Table 6 in the Appendix):

$$P\left(\bigcup_{i=n}^{\infty} A_i\right) = \sum_{i=n}^{\infty} P(A_i) = \sum_{i=n}^{\infty} q^{i-1}p = p\sum_{i=n}^{\infty} q^{i-1} = p\frac{q^{n-1}}{1-q} = p\frac{q^{n-1}}{p} = q^{n-1}.$$

For instance, for $p = 1/2$ and $n = 3$, this probability is $\frac{1}{4} = 0.25$. That is, when tossing a fair coin, the probability that the first head does not appear either the first or the second time (and therefore it appears either the third time or the fourth time, etc.) is 0.25. For $n = 10$, this probability is approximately $0.00195 \simeq 0.002$.

3.2 Some Basic Properties and Results

The defining properties (P1)–(P3) of a probability function cited in the previous section imply a number of almost anticipated basic results, which are listed here in the form of two propositions. These propositions are proved, and illustrated by means of examples. It is to be emphasized that the justification of these propositions requires only properties (P1)–(P3) and nothing else beyond them.

Proposition 1. The defining properties (P1)–(P3) of a probability function imply the following results:

- **(i)** $P(\varnothing) = 0$.
- **(ii)** For any pairwise disjoint events $A_1, \ldots, A_n$, $P(\bigcup_{i=1}^{n} A_i) = \sum_{i=1}^{n} P(A_i)$.
- **(iii)** For any event A, $P(A^c) = 1 - P(A)$.
- **(iv)** $A_1 \subseteq A_2$ implies $P(A_1) \leq P(A_2)$ and $P(A_2 - A_1) = P(A_2) - P(A_1)$.
- **(v)** $0 \leq P(A) \leq 1$ for every event A.

Proof.

- **(i)** From the obvious fact that $\mathcal{S} = \mathcal{S} \cup \varnothing \cup \varnothing \cup \cdots$ and property (P3),

$$P(\mathcal{S}) = P(\mathcal{S} \cup \varnothing \cup \varnothing \cup \cdots) = P(\mathcal{S}) + P(\varnothing) + P(\varnothing) + \cdots$$

  or $P(\varnothing) + P(\varnothing) + \cdots = 0$. By (P1), this can happen only when $P(\varnothing) = 0$. (Of course, that the impossible event has probability 0 does not come as a surprise. Any reasonable definition of probability should imply it.)
- **(ii)** Take $A_i = \varnothing$ for $i \geq n + 1$, consider the following obvious relation, and use (P3) and part (i) to obtain:

$$P\left(\bigcup_{i=1}^{n} A_i\right) = P\left(\bigcup_{i=1}^{\infty} A_i\right) = \sum_{i=1}^{\infty} P(A_i) = \sum_{i=1}^{n} P(A_i).$$

- **(iii)** From (P2) and part (ii), $P(A \cup A^c) = P(\mathcal{S}) = 1$ or $P(A) + P(A^c) = 1$, so that $P(A^c) = 1 - P(A)$.

(iv) The relation $A_1 \subseteq A_2$ clearly implies $A_2 = A_1 \cup (A_2 - A_1)$, so that, by part (ii), $P(A_2) = P(A_1) + P(A_2 - A_1)$. Solving for $P(A_2 - A_1)$, we obtain $P(A_2 - A_1) = P(A_2) - P(A_1)$, so that, by (P1), $P(A_1) \leq P(A_2)$. (At this point it must be pointed out that $P(A_2 - A_1)$ *need* not be $P(A_2) - P(A_1)$, if A_1 is *not* contained in A_2.)

(v) Clearly, $\varnothing \subseteq A \subseteq S$ for any event A. Then (P1), part (i) and part (iv) give: $0 = P(\varnothing) \leq P(A) \leq P(S) = 1$. ▲

Proposition 2. The defining properties (P1)–(P3) of a probability function also imply the following results:

(i) For any two events A_1 and A_2:

$$P(A_1 \cup A_2) = P(A_1) + P(A_2) - P(A_1 \cap A_2).$$

(ii) For any three events A_1, A_2, and A_3:

$$\begin{aligned} P(A_1 \cup A_2 \cup A_3) = {} & P(A_1) + P(A_2) + P(A_3) - [P(A_1 \cap A_2) \\ & + P(A_1 \cap A_3) + P(A_2 \cap A_3)] + P(A_1 \cap A_2 \cap A_3). \end{aligned}$$

(For more than three events, see Theorem 1 below.)

(iii) For any events $A_1, A_2, \ldots$,

$$P\left(\bigcup_{i=1}^{\infty} A_i\right) \leq \sum_{i=1}^{\infty} P(A_i) \qquad (\sigma\text{-subadditivity}),$$

and

$$P\left(\bigcup_{i=1}^{n} A_i\right) \leq \sum_{i=1}^{n} P(A_i) \qquad (\text{finite-subadditivity}).$$

Proof.

(i) It is clear (by means of a Venn diagram, for example) that

$$A_1 \cup A_2 = A_1 \cup \left(A_2 \cap A_1^c\right) = A_1 \cup (A_2 - A_1 \cap A_2).$$

Then, by means of Proposition 1(ii),(iv):

$$P(A_1 \cup A_2) = P(A_1) + P(A_2 - A_1 \cap A_2) = P(A_1) + P(A_2) - P(A_1 \cap A_2).$$

(ii) Apply part (i) to obtain:

$$\begin{aligned} P(A_1 \cup A_2 \cup A_3) &= P[(A_1 \cup A_2) \cup A_3] = P(A_1 \cup A_2) + P(A_3) \\ &\quad - P[(A_1 \cup A_2) \cap A_3] \\ &= P(A_1) + P(A_2) - P(A_1 \cap A_2) + P(A_3) \\ &\quad - P[(A_1 \cap A_3) \cup (A_2 \cap A_3)] \\ &= P(A_1) + P(A_2) + P(A_3) - P(A_1 \cap A_2) \\ &\quad - [P(A_1 \cap A_3) + P(A_2 \cap A_3) - P(A_1 \cap A_2 \cap A_3)] \end{aligned}$$

$$= P(A_1) + P(A_2) + P(A_3) - P(A_1 \cap A_2) - P(A_1 \cap A_3)$$
$$-P(A_2 \cap A_3) + P(A_1 \cap A_2 \cap A_3).$$

(iii) By Proposition 3 in Chapter 2 and (P3):

$$P\left(\bigcup_{i=1}^{\infty} A_i\right) = P\big[A_1 \cup (A_1^c \cap A_2) \cup \ldots \cup (A_1^c \cap \ldots \cap A_{n-1}^c \cap A_n) \cup \ldots\big]$$
$$= P(A_1) + P(A_1^c \cap A_2) + \cdots + P(A_1^c \cap \ldots \cap A_{n-1}^c \cap A_n) + \cdots$$
$$\leq P(A_1) + P(A_2) + \cdots + P(A_n) + \cdots$$

(by Proposition 1(iv) in Chapter 3).

For the finite case:

$$P\left(\bigcup_{i=1}^{n} A_i\right) = P\big[A_1 \cup (A_1^c \cap A_2) \cup \cdots \cup (A_1^c \cap \cdots \cap A_{n-1}^c \cap A_n)\big]$$
$$= P(A_1) + P(A_1^c \cap A_2) + \cdots + P(A_1^c \cap \cdots \cap A_{n-1}^c \cap A_n)$$
$$\leq P(A_1) + P(A_2) + \cdots + P(A_n). \quad \blacktriangle$$

Next, some examples are presented to illustrate some of the properties listed in Propositions 1 and 2.

Example 4.

(i) For two events A and B, suppose that $P(A) = 0.3$, $P(B) = 0.5$, and $P(A \cup B) = 0.6$. Calculate $P(A \cap B)$.
(ii) If $P(A) = 0.6$, $P(B) = 0.3$, $P(A \cap B^c) = 0.4$, and $B \subset C$, calculate $P(A \cup B^c \cup C^c)$.

Discussion.

(i) From $P(A \cup B) = P(A) + P(B) - P(A \cap B)$, we get $P(A \cap B) = P(A) + P(B) - P(A \cup B) = 0.3 + 0.5 - 0.6 = 0.2$.
(ii) The relation $B \subset C$ implies $C^c \subset B^c$ and hence $A \cup B^c \cup C^c = A \cup B^c$. Then $P(A \cup B^c \cup C^c) = P(A \cup B^c) = P(A) + P(B^c) - P(A \cap B^c) = 0.6 + (1 - 0.3) - 0.4 = 0.9$.

Example 5. Let A and B be the respective events that two contracts I and II, say, are completed by certain deadlines, and suppose that: P(at least one contract is completed by its deadline) $= 0.9$ and P(both contracts are completed by their deadlines) $= 0.5$. Calculate the probability: P(exactly one contract is completed by its deadline).

Discussion. The assumptions made are translated as follows: $P(A \cup B) = 0.9$ and $P(A \cap B) = 0.5$. What we wish to calculate is: $P((A \cap B^c) \cup (A^c \cap B))$. However, it is easily seen (for example, by means of a Venn diagram) that

$$(A \cap B^c) \cup (A^c \cap B) = (A \cup B) - (A \cap B).$$

Therefore, by Proposition 1(iv),

$$\begin{aligned} P((A \cap B^c) \cup (A^c \cap B)) &= P((A \cup B) - (A \cap B)) \\ &= P(A \cup B) - P(A \cap B) \\ &= 0.9 - 0.5 = 0.4. \end{aligned}$$

Example 6.

(i) For three events A, B, and C, suppose that $P(A \cap B) = P(A \cap C)$ and $P(B \cap C) = 0$. Then show that $P(A \cup B \cup C) = P(A) + P(B) + P(C) - 2P(A \cap B)$.
(ii) For any two events A and B, show that $P(A^c \cap B^c) = 1 - P(A) - P(B) + P(A \cap B)$.

Discussion.

(i) We have $P(A \cup B \cup C) = P(A) + P(B) + P(C) - P(A \cap B) - P(A \cap C) - P(B \cap C) + P(A \cap B \cap C)$. But $A \cap B \cap C \subset B \cap C$, so that $P(A \cap B \cap C) \leq P(B \cap C) = 0$, and therefore $P(A \cup B \cup C) = P(A) + P(B) + P(C) - 2P(A \cap B)$.
(ii) Indeed, $P(A^c \cap B^c) = P((A \cup B)^c) = 1 - P(A \cup B) = 1 - P(A) - P(B) + P(A \cap B)$.

Example 7. In ranking five horses in a horse race (Example 5 in Chapter 2), what is the probability that horse #3 will finish either first or second or third?

Discussion. Denote by B the required event and let A_i = "horse #3 finishes in the ith place," $i = 1, 2, 3$. Then the events A_1, A_2, A_3 are pairwise disjoint, and therefore:

$$P(B) = P(A_1 \cup A_2 \cup A_3) = P(A_1) + P(A_2) + P(A_3).$$

But $P(A_1) = P(A_2) = P(A_3) = \frac{24}{120} = 0.2$, so that $P(B) = 0.6$.

Example 8. Consider a well-shuffled deck of 52 cards (Example 3 in Chapter 2), and suppose we draw at random three cards. What is the probability that at least one is an ace?

Discussion. Let A be the required event, and let A_i be defined by: A_i = "exactly i cards are aces," $i = 0, 1, 2, 3$. Then, clearly, $P(A) = P(A_1 \cup A_2 \cup A_3)$. Instead, we may choose to calculate $P(A)$ through $P(A) = 1 - P(A_0)$, where

$$P(A_0) = \frac{\binom{48}{3}}{\binom{52}{3}} = \frac{48 \times 47 \times 46}{52 \times 51 \times 50} = \frac{4,324}{5,525}, \quad \text{so that } P(A) = \frac{1,201}{5,525} \simeq 0.217.$$

Example 9. It happens that 4 hotels in a certain large city have the same name, for example, Grand Hotel. Four persons make an appointment to meet at the Grand Hotel. If each one of the 4 persons chooses the hotel at random, calculate the following probabilities:

(i) All 4 choose the same hotel.
(ii) All 4 choose different hotels.

Discussion.

(i) If A = "all 4 choose the same hotel," then $P(A) = \frac{n(A)}{n(\mathcal{S})}$, where $n(A)$ is the number of sample points in A. Here, $n(\mathcal{S}) = 4 \times 4 \times 4 \times 4 = 4^4$, by Theorem 1 in Chapter 2 applied with $k = 4$ and $n_1 = n_2 = n_3 = n_4 = 4$, and $n(A) = 4 \times 1 \times 1 \times 1 = 4$, by Theorem 1 again applied with $n_1 = 4, n_2 = n_3 = n_4 = 1$. Thus, $P(A) = \frac{4}{4^4} = \frac{1}{4^3} = \frac{1}{64} = 0.015625 \simeq 0.016$.

(ii) If B = "all 4 choose different hotels," then, by the first part of the corollary to Theorem 1 in Chapter 2, $n(B) = P_{4,4} = 4!$, so that $P(B) = \frac{4!}{4^4} = \frac{1\times2\times3}{4^3} = \frac{3}{32} = 0.09375 \simeq 0.094$. Alternatively, $n(\mathcal{S}) = \binom{4}{1}\times4\times\binom{3}{1}\times4\times\binom{2}{1}\times4\times\binom{1}{1}\times4 = 4! \times 4^4$; $n(A) = \binom{4}{1} \times 4 \times \binom{3}{1} \times 1 \times \binom{2}{1} \times 1 \times \binom{1}{1} \times 1 = 4! \times 4$; $n(B) = \binom{4}{1} \times 4 \times \binom{3}{1} \times 3 \times \binom{2}{1} \times 2 \times \binom{1}{1} \times 1 = 4! \times 4!$, so that:

$$P(A) = \frac{4! \times 4}{4! \times 4^4} = \frac{1}{4^3}, \quad \text{as before, and}$$

$$P(B) = \frac{(4!)^2}{4! \times 4^4} = \frac{4!}{4^4}, \quad \text{again as before.}$$

Example 10 (See also Example 20 in Chapter 2). The faculty in an academic department at UC-Davis consists of 4 assistant professors, 6 associate professors, and 5 full professors. Also, it has 30 graduate students. An ad hoc committee of 5 is to be formed to study a certain curricular matter.

(i) What is the number of all possible committees consisting of faculty alone?
(ii) How many committees can be formed if 2 graduate students are to be included and all academic ranks are to be represented?
(iii) If the committee is to be formed at random, what is the probability that the faculty will not be represented?

Discussion. By Example 20 in Chapter 2, we have for parts (i) and (ii):

(i) This number is: $\binom{15}{5} = \frac{15!}{5!10!} = \frac{11\times12\times13\times14\times15}{1\times2\times3\times4\times5} = 3,003$.

(ii) Here the number is: $\binom{30}{2}\binom{4}{1}\binom{6}{1}\binom{5}{1} = \frac{30!}{2!28!} \times 4 \times 6 \times 5 = \frac{29\times30}{2} \times 120 = 52,200$.

(iii) The required probability is:

$$\frac{\binom{30}{5}\binom{15}{0}}{\binom{45}{5}} = \frac{\binom{30}{5}}{\binom{45}{5}} = \frac{30!/5!25!}{45!/5!40!} = \frac{26 \times 27 \times 28 \times 29 \times 30}{41 \times 42 \times 43 \times 44 \times 45} = \frac{2,262}{19,393} \simeq 0.117.$$

Example 11. What is the probability that a poker hand contains 4 pictures, including at least 2 Jacks? It is recalled here that there are 12 pictures consisting of 4 Jacks, 4 Queens, and 4 Kings.

Discussion. A poker hand can be selected in $\binom{52}{5}$ ways. The event described, call it A, consists of the following number of sample points: $n(A) = n(J_2)+n(J_3)+n(J_4)$, where J_i = "the poker hand contains exactly i Jacks," $i = 2, 3, 4$. But

$$n(J_2) = \binom{4}{2}\binom{8}{2}\binom{40}{1}, \quad n(J_3) = \binom{4}{3}\binom{8}{1}\binom{40}{1},$$
$$n(J_3) = \binom{4}{4}\binom{8}{0}\binom{40}{1},$$

so that

$$P(A) = \frac{[\binom{4}{2}\binom{8}{2} + \binom{4}{3}\binom{8}{1} + \binom{4}{4}\binom{8}{0}]\binom{52}{5}}{\binom{52}{5}} = \frac{8,040}{2,598,960} \simeq 0.003.$$

(For the calculation of $\binom{52}{5}$ see Example 19(ix) and the discussion following Remark 2 in Chapter 2.)

This section is concluded with two very useful results stated as theorems. The first is a generalization of Proposition 2(ii) to more than three events, and the second is akin to the concept of continuity of a function as it applies to a probability function.

Theorem 1. The probability of the union of any n events, $A_1, \ldots, A_n$, is given by:

$$P\left(\bigcup_{j=1}^{n} A_j\right) = \sum_{j=1}^{n} P(A_j) - \sum_{1\le j_1<j_2\le n} P(A_{j_1} \cap A_{j_2})$$
$$+ \sum_{1\le j_1<j_2<j_3\le n} P(A_{j_1} \cap A_{j_2} \cap A_{j_3}) - \cdots$$
$$+(-1)^{n+1} P(A_1 \cap \ldots \cap A_n).$$

Although its proof (which is by induction) will not be presented, the pattern of the right-hand side above follows that of part (ii) in Proposition 2 and it is clear. First, sum up the probabilities of the individual events, then subtract the probabilities of the intersections of the events, taken two at a time (in the ascending order of indices), then add the probabilities of the intersections of the events, taken three at a time as before, and continue like this until you add or subtract (depending on n) the probability of the intersection of all n events.

Recall that if $A_1 \subseteq A_2 \subseteq \ldots$, then $\lim_{n\to\infty} A_n = \bigcup_{n=1}^{\infty} A_n$, and if $A_1 \supseteq A_2 \supseteq \ldots$, then $\lim_{n\to\infty} A_n = \bigcap_{n=1}^{\infty} A_n$.

Theorem 2. For any monotone sequence of events $\{A_n\}$, $n \geq 1$, it holds

$$P(\lim_{n\to\infty} A_n) = \lim_{n\to\infty} P(A_n).$$

This theorem will be employed in many instances, and its use will be then pointed out.

The interested reader may find proofs of Theorem 1 and 2 in Chapter 2 of the book *A Course in Mathematical Statistics*, 2nd edition (1997), Academic Press, by G. G. Roussas.

Exercises.

2.1

(i) If $P(A) = 0.4$, $P(B) = 0.6$, and $P(A \cup B) = 0.7$, calculate $P(A \cap B)$.

(ii) By a simple example show that $P(A - B)$ need not be equal to $P(A) - P(B)$ if B does not imply A.

2.2 If for two events A and B, it so happens that $P(A) = \frac{3}{4}$ and $P(B) = \frac{3}{8}$, show that:

$$P(A \cup B) \geq \frac{3}{4}, \quad \text{and} \quad \frac{1}{8} \leq P(A \cap B) \leq \frac{3}{8}.$$

2.3 If for the events A, B, and C, it so happens that $P(A) = P(B) = P(C) = 1$, then show that:

$$P(A \cap B) = P(A \cap C) = P(B \cap C) = P(A \cap B \cap C) = 1.$$

Hint. Use Proposition 1(iv) and Proposition 2(i), (ii).

2.4 If the events A, B, and C are related as follows: $A \subset B \subset C$ and $P(A) = \frac{1}{4}$, $P(B) = \frac{5}{12}$, and $P(C) = \frac{7}{12}$, compute the probabilities of the following events:

$$A^c \cap B, \quad A^c \cap C, \quad B^c \cap C, \quad A \cap B^c \cap C^c, \quad A^c \cap B^c \cap C^c.$$

Hint. Use Proposition 1(iii), (iv) here, and Proposition 4 in Chapter 2.

2.5 Let $\mathcal{S}$ be the set of all outcomes when flipping a fair coin four times, so that all 16 outcomes are equally likely. Define the events A and B by:

$A = \{s \in \mathcal{S};\ s$ contains more T's than H's$\}$,

$B = \{s \in \mathcal{S};$ there are both H's and T's in s, and every T precedes every $H\}$.

Compute the probabilities $P(A)$, $P(B)$.

2.6 Let $\mathcal{S} = \{x\ integer;\ 1 \leq x \leq 200\}$, and define the events A, B, and C as follows:

$$A = \{x \in \mathcal{S};\ x \text{ is divisible by } 7\}$$
$$B = \{x \in \mathcal{S};\ x = 3n + 10, \text{ for some positive integer } n\}$$
$$C = \{x \in \mathcal{S};\ x^2 + 1 \leq 375\}.$$

Calculate the probabilities $P(A)$, $P(B)$, and $P(C)$.

2.7 If two fair (and distinct) dice are rolled once, what is the probability that the total number of spots shown is:

(i) Equal to 5?
(ii) Divisible by 3?

2.8 Students at a certain college subscribe to three newsmagazines A, B, and C according to the following proportions:

$$A : 20\%, \qquad B : 15\%, \qquad C : 10\%,$$

both A and B: 5%, both A and C: 4%, both B and C: 3%, all three A, B, and C: 2%. If a student is chosen at random, what is the probability he/she subscribes to none of the newsmagazines?

Hint. Use Proposition 4 in Chapter 2, and Proposition 2(ii) here.

2.9 A high school senior applies for admission to two colleges A and B, and suppose that: $P(\text{admitted at } A) = p_1$, $P(\text{rejected by } B) = p_2$, and $P(\text{rejected by at least one, } A \text{ or } B) = p_3$.

(i) Calculate the probability that the student is admitted by at least one college; express it in terms of p_1, p_2, p_3.
(ii) Find the numerical value of the probability in part (i), if $p_1 = 0.6$, $p_2 = 0.2$, and $p_3 = 0.3$.

2.10 An airport limousine service has two vans, the smaller of which can carry 6 passengers and the larger 9 passengers. Let x and y be the respective *numbers* of passengers carried by the smaller and the larger van in a given trip, so that a suitable sample space $\mathcal{S}$ is given by:

$$\mathcal{S} = \{(x, y);\ x = 0, \ldots, 6 \ \text{ and } \ y = 0, 1, \ldots, 9\}.$$

Also, suppose that for all values of x and y, the probabilities $P(\{(x, y)\})$ are equal. Finally, define the events A, B, and C as follows:

A = "the two vans together carry either 4 or 6 or 10 passengers"

B = "the larger van carries twice as many passengers as the smaller van"

C = "the two vans carry different numbers of passengers."

Calculate the probabilities: $P(A)$, $P(B)$, and $P(C)$.

2.11 A child's set of blocks consists of 2 red, 4 blue, and 5 yellow blocks. The blocks can be distinguished only by color. If the child lines up the blocks in a row at random, calculate the following probabilities:

(i) Red blocks appear at both ends.
(ii) All yellow blocks are adjacent.
(iii) Blue blocks appear at both ends.

Hint. Use Proposition 5 in Chapter 2.

2.12 Suppose that the letters C, E, F, F, I, and O are written on six chips and placed into a box. Then the six chips are mixed and drawn one by one without replacement. What is the probability that the word "OFFICE" is formed?

Hint. Use Proposition 5 in Chapter 2.

2.13 A course in English composition is taken by 10 freshmen, 15 sophomores, 30 juniors, and 5 seniors. If 10 students are chosen at random, calculate the probability that this group will consist of 2 freshmen, 3 sophomores, 4 juniors, and 1 senior.

2.14 From among n eligible draftees, m are to be drafted in such a way that all possible combinations are equally likely to occur. What is the probability that a specified man is not drafted (expressed in terms of m and n)?

2.15 From 10 positive and 6 negative numbers, 3 numbers are chosen at random and without repetitions. What is the probability that their product is a negative number?

2.16 A shipment of 2,000 light bulbs contains 200 defective items and 1,800 good items. Five hundred bulbs are chosen at random and are tested, and the entire shipment is rejected if more than 25 bulbs from among those tested are found to be defective. What is the probability that the shipment will be accepted? (Just write down the right formula.)

2.17 Three cards are drawn at random and without replacement from a standard deck of 52 playing cards. Compute the probabilities $P(A_i), i = 1, \ldots, 4$, where the events $A_i, i = 1, \ldots, 4$ are defined as follows: A_1 = "all 3 cards are black," A_2 = "exactly 1 card is an ace," A_3 = "1 card is a diamond, 1 card is a heart, and 1 card is a club," A_4 = "at least 2 cards are red."

2.18 A student committee of 12 people is to be formed from among 100 freshmen (40 male + 60 female), 80 sophomores (30 male and 50 female), 70 juniors (24 male and 46 female), and 40 seniors (12 male and 28 female). Calculate the following probabilities:

(i) Seven students are female and 5 are male.
(ii) The committee consists of the same number of students from each class.
(iii) The committee consists of 2 female students and 1 male student from each class.
(iv) The committee includes at least 1 senior (one of whom, regardless, will serve as the chairperson of the committee).

The following tabular form of the data facilitates the calculations:

Class\\ Gender	Male	Female	Totals
Freshman	40	60	100
Sophomore	30	50	80
Junior	24	46	70
Senior	12	28	40
Totals	106	184	290

2.19 From a class of 50 students, of whom 30 are computer science majors and 20 other majors, 5 students are chosen at random to form an advisory committee.

(i) How many such committees can be formed?
(ii) How many such committees include 3 computer science majors?
(iii) What is the probability that such a committee includes 3 computer science majors?

2.20 Let S and L be the events that a patient's visit to a primary care physician's office results in a referral to a specialist and for laboratory work, respectively. Suppose that $P(S) = 0.25$, $P(L) = 0.35$, and that the probability that there is no referral to either a specialist or for laboratory work is 0.45. Calculate the probability that there is a referral:

(i) To both a specialist and for laboratory work.
(ii) To either a specialist or for laboratory work.

Hint. For part (i), use Proposition 1(iii), and for part (ii), use Proposition 2(ii).

2.21 In reference to Exercise 4.15 in Chapter 2, suppose one license plate number is chosen at random. What is the probability it is $5\,A\,\cdot\,\cdot\,1\,2\,\cdot$ for each one of the cases (i)–(iv) described there?

3.3 Distribution of a Random Variable

The paramount feature of an r.v. X that we are interested in is its probability distribution function or just distribution. That is, the probability by which X takes values in any set B, subset of the real line $\Re$. Recalling that $(X \in B)$ stands for the event $\{s \in \mathcal{S};\ X(s) \in B\}$, the focus of our interest is:

$$P(X \in B) = P(\{s \in \mathcal{S};\ X(s) \in B\}),\ \ B \subseteq \Re. \tag{3.1}$$

Assessing probabilities as the ones in relation (3.1) is the best one can do in absence of certainty. In this section, the concept of the probability distribution function of an r.v. is defined, as well as those of the distribution function and probability density function of an r.v. Also, some comments are made on their relationships.

Definition 1.

(i) The *probability distribution function* (or just *distribution*) of an r.v. X is a set function P_X which assigns values to subsets B of $\Re$ according to relation (3.1); the value assigned to B is denoted by $P_X(B)$.

(ii) By taking B to be an interval of the form $(-\infty, x]$; i.e., $B = (-\infty, x]$, relation (3.1) becomes

$$P(X \in (-\infty, x]) = P(\{s \in \mathcal{S};\ X(s) \le x\}) = P(X \le x),$$

and it defines a point function denoted by F_X and called the *distribution function* (d.f.) of X.

Remark 1.

(i) The distribution of the r.v. X is a set function defined on subsets of $\Re$. As such, it is seen that it is, actually, a probability function (on subsets of $\Re$). The details of the justification are left as an exercise (see Exercise 3.23).

(ii) From Definition 1 it follows that if we know $P_X(B)$ for all B in $\Re$, then we certainly know $F_X(x)$ for all $x \in \Re$. Somewhat surprisingly, the converse is also true; its justification is well beyond the level of this book. It does provide, however, a justification of why we occupy ourselves at all with $F_X(x)$, $x \in \Re$.

The d.f. F_X of any r.v. has four basic properties summarized in the following proposition.

Proposition 3. The d.f. of an r.v. X, F_X, has the following properties:

(i) $0 \le F_X(x) \le 1$, $x \in \Re$.

(ii) F_X is nondecreasing; i.e., for $x_1 < x_2$, $F_X(x_1) \le F_X(x_2)$.

(iii) F_X is continuous from the right; that is, as $n \to \infty$, $x_n \downarrow x$ implies $F_X(x_n) \to F_X(x)$.

(iv) $F_X(\infty) = 1$ and $F_X(-\infty) = 0$, where $F_X(\infty) = \lim_{n\to\infty} F_X(x_n)$, $x_n \uparrow \infty$, and $F_X(-\infty) = \lim_{n\to\infty} F_X(x_n)$, $x_n \downarrow -\infty$.

The justification of this proposition is left as an exercise (see Exercise 3.24). Figures 3.1 and 3.2 below present typical cases of d.f.'s.

In particular, Figure 3.3 presents the d.f. of the (discrete) r.v. X distributed as follows:

x	−14	−6	5	9	24
P(X = x)	0.17	0.28	0.22	0.22	0.11

The entity, however, which facilitates truly (at least in principle) the actual calculation of probabilities associated with an r.v. X is the so-called probability density function of X. At this point, the discrete and the continuous case are treated separately.

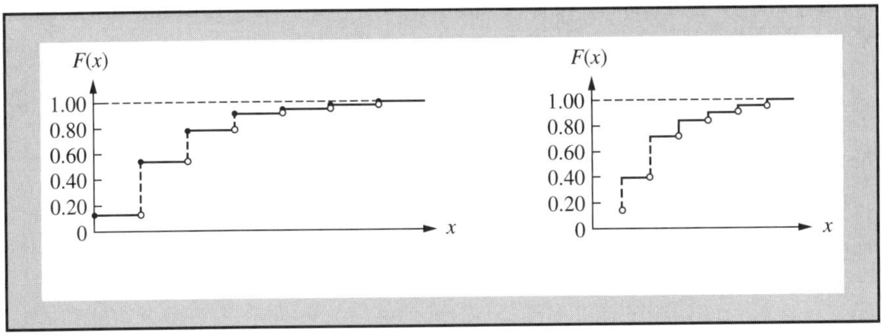

FIGURE 3.1
Examples of graphs of d.f.'s.

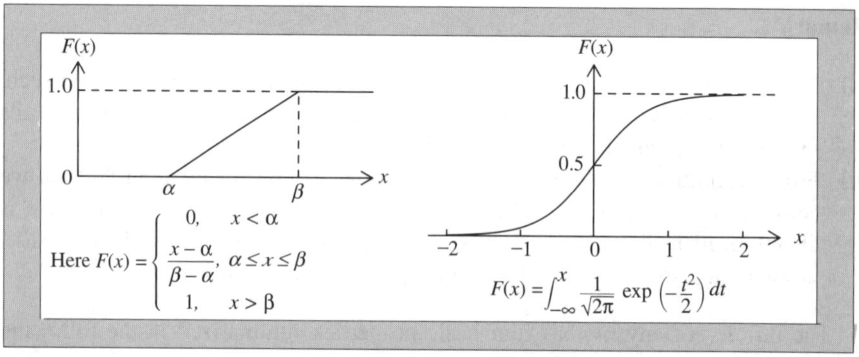

FIGURE 3.2

Examples of graphs of d.f.'s.

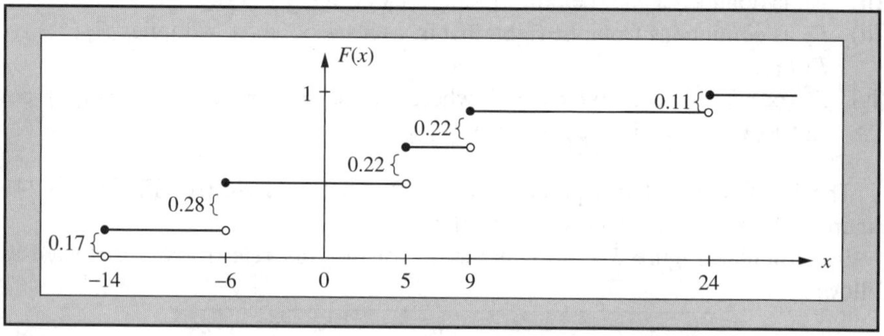

FIGURE 3.3

Example of d.f.

Definition 2. Let X be a (discrete) r.v. taking on the values $x_i, i \geq 1$ (finitely or infinitely many) with respective probabilities $P(X = x_i),\ i \geq 1$. Define the function f_X as follows:

$$f_X(x) = \begin{cases} P(X = x_i) & \text{if } x = x_i,\ i \geq 1 \\ 0 & \text{otherwise.} \end{cases} \tag{3.2}$$

The function f_X is called the *probability density function* (p.d.f.) of the r.v. X.

The following properties are immediate from the definition.

Proposition 4. Let f_X be as in relation (3.2). Then:

(i) $f_X(x) \geq 0$ for all $x \in \Re$.

(ii) For any $B \subseteq \Re$, $P(X \in B) = \sum_{x_i \in B} f_X(x_i)$.

(iii) In particular,

$$F_X(x) = \sum_{x_i \le x} f_X(x_i),\ x \in \Re,\ and \sum_{x_i \in \Re} f_X(x_i) = 1.$$

(iv) Assuming that $x_i < x_{i+1}, i \ge 1$, it follows that

$$f_X(x_{i+1}) = F_X(x_{i+1}) - F_X(x_i),\ i \ge 1,\ f_X(x_1) = F_X(x_1).$$

Its simple justification is left as an exercise (see Exercise 3.25).

Definition 3. Let X be an r.v. of the continuous type, and suppose there exists a function f_X such that:

$$f_X(x) \ge 0 \text{ for all } x \in \Re,\ P(X \in B) = \int_B f_X(x)dx,\ B \subseteq \Re, \text{ and}$$
$$\int_{\Re} f_X(x)dx = 1. \tag{3.3}$$

Then the function f_X is called the *probability density function* (p.d.f.) of the r.v. X.

From Definition 3 and a result from calculus, the following properties are immediate.

Proposition 5. Let f_X be as in Definition 3. Then:

(i) $F_X(x) = \int_{-\infty}^{x} f_X(t)dt$ for all $x \in \Re$ (by taking $B = (-\infty, x]$ in relation (3.3)).
(ii) $\frac{d}{dx}F_X(x) = f_X(x)$ (for all $x \in \Re$ for which $f_X(x)$ is continuous, as is well known from calculus).

Thus, by Proposition 4(ii) and relation (3.3), the calculation of the probability $P(X \in B), B \subseteq \Re$, is reduced to a summation over B (for the case the r.v. X is discrete), or an integration over B (when the r.v. X is of the continuous type). Although integration over B can be given meaning for $B \subseteq \Re$ other than intervals, at this level, B will consist either of a finite or infinite interval, or at most of finitely many such intervals. Thus, $\int_B f_X(x)dx$ will be a familiar operation.

Remark 2.

(i) It can be seen that if we are given a function F which satisfies properties (i)–(iv) in Proposition 3, then we can always construct an r.v. X such that $F_X(x) = F(x),\ x \in \Re$. A somewhat sophisticated construction of such an r.v. is proved in Theorem 9 in Chapter 11.
(ii) Part (i), along with Proposition 4(iii) and Proposition 5(ii) justify the following question: *When is a given function f a candidate for a p.d.f. of an r.v.?* The answer is this: First, there must be $f(x) \ge 0$ for all $x \in \Re$; and second, either $\sum_{x_i} f(x_i) = 1$ where $f(x_i),\ i \ge 1$, are the (countably many) values which are not 0, or $\int_{-\infty}^{\infty} f_X(x)dx = 1$ otherwise.

(iii) If the r.v. X is of the continuous type, relation (3.3) gives for $B = \{x\}$: $P(X \in \{x\}) = P(X = x) = \int_{\{x\}} f_X(t)dt$. However, $\int_{\{x\}} f_X(t)dt = 0$. Thus, for a continuous r.v. X, $P(X = x) = 0$ for all $x \in \Re$. Yet, $\int_{\Re} f_X(x)dx = 1$ by Proposition 5(ii). Why are these two statements not contradictory? (See Exercise 3.26.)

Let us conclude this section with the following concrete examples.

Example 12. The number of light switch turn-ons at which the first failure occurs is an r.v. X whose p.d.f. is given by: $f(x) = c\left(\frac{9}{10}\right)^{x-1}$, $x = 1, 2, \ldots$ (and 0 otherwise).

(i) Determine the constant c.
(ii) Calculate the probability that the first failure will not occur until after the 10th turn-on.
(iii) Determine the corresponding d.f. F.

Hint. Refer to #4 in Table 6 in the Appendix.

Discussion.

(i) The constant c is determined through the relationship:

$$\sum_{x=1}^{\infty} c\left(\frac{9}{10}\right)^{x-1} = 1.$$

However,

$$\sum_{x=1}^{\infty} c\left(\frac{9}{10}\right)^{x-1} = c\sum_{x=1}^{\infty}\left(\frac{9}{10}\right)^{x-1} = c\left[1 + \left(\frac{9}{10}\right) + \left(\frac{9}{10}\right)^2 + \cdots\right]$$
$$= c\frac{1}{1-\frac{9}{10}} = 10c,$$

so that $c = \frac{1}{10}$.

(ii) Here

$$P(X > 10) = P(X \geq 11) = c\sum_{x=11}^{\infty}\left(\frac{9}{10}\right)^{x-1} = c\left[\left(\frac{9}{10}\right)^{10} + \left(\frac{9}{10}\right)^{11} + \cdots\right]$$
$$= c\frac{\left(\frac{9}{10}\right)^{10}}{1-\frac{9}{10}} = c\times 10\left(\frac{9}{10}\right)^{10} = \frac{1}{10}\times 10\left(\frac{9}{10}\right)^{10} = (0.9)^{10} \simeq 0.349.$$

(iii) First, for $x < 1$, $F(x) = 0$. Next, for $x \geq 1$,

$$F(x) = \sum_{t=1}^{x} c\left(\frac{9}{10}\right)^{t-1} = 1 - \sum_{t=x+1}^{\infty} c\times\left(\frac{9}{10}\right)^{t-1}$$
$$= 1 - c\sum_{t=x+1}^{\infty}\left(\frac{9}{10}\right)^{t-1} = 1 - \frac{1}{10}\times\frac{\left(\frac{9}{10}\right)^{x}}{1-\frac{9}{10}} = 1 - \left(\frac{9}{10}\right)^{x}.$$

Thus, $F(x) = 0$ for $x < 1$, and $F(x) = 1 - \left(\frac{9}{10}\right)^x$ for $x \geq 1$.

Example 13. The recorded temperature in an engine is an r.v. X whose p.d.f. is given by: $f(x) = n(1-x)^{n-1}$, $0 \leq x \leq 1$ (and 0 otherwise), where $n \geq 1$ is a known integer.

(i) Show that f is, indeed, a p.d.f.
(ii) Determine the corresponding d.f. F.

Discussion.

(i) Because $f(x) \geq 0$ for all x, we simply have to check that $\int_{-\infty}^{\infty} f(x)\,dx = 1$. To this end, $\int_{-\infty}^{\infty} f(x)dx = \int_0^1 n(1-x)^{n-1}\,dx = -n\frac{(1-x)^n}{n}\Big|_0^1 = -(1-x)^n\Big|_0^1 = 1$.
(ii) First, $F(x) = 0$ for $x < 0$, whereas for $0 \leq x \leq 1$, $F(x) = \int_0^x n(1-t)^{n-1}dt = -(1-t)^n\Big|_0^x$ (from part (i)), and this is equal to: $-(1-x)^n + 1 = 1 - (1-x)^n$. Thus,

$$F(x) = \begin{cases} 0, & x < 0 \\ 1 - (1-x)^n, & 0 \leq x \leq 1 \\ 1, & x > 1. \end{cases}$$

Exercises.

3.1 A sample space describing a three-children family is as follows:

$$\mathcal{S} = \{bbb, bbg, bgb, gbb, bgg, gbg, ggb, ggg\},$$

and assume that all eight outcomes are equally likely to occur. Next, let X be the r.v. denoting the number of girls in such a family. Then:

(i) Determine the set of all possible values of X.
(ii) Determine the p.d.f. of X.
(iii) Calculate the probabilities: $P(X \geq 2)$, $P(X \leq 2)$.

3.2 An r.v. X has d.f. F given by:

$$F(x) = \begin{cases} 0, & x \leq 0 \\ 2c\left(x^2 - \frac{1}{3}x^3\right), & 0 < x \leq 2 \\ 1, & x > 2. \end{cases}$$

(i) Determine the corresponding p.d.f. f.
(ii) Determine the constant c.

Hint. For part (i), use Proposition 5(iii), and for part (ii), use Remark 2(ii).

3.3 The r.v. X has d.f. F given by:

$$F(x) = \begin{cases} 0, & x \leq 0 \\ x^3 - x^2 + x, & 0 < x \leq 1 \\ 1, & x > 1. \end{cases}$$

(i) Determine the corresponding p.d.f. f.
(ii) Calculate the probability $P(X > \frac{1}{2})$.

Hint. As in Exercise 3.2.

3.4 The r.v. X has d.f. F given by:

$$F(x) = \begin{cases} 0, & x < 4 \\ 0.1, & 4 \leq x < 5 \\ 0.4, & 5 \leq x < 6 \\ 0.7, & 6 \leq x < 8 \\ 0.9, & 8 \leq x < 9 \\ 1, & x \geq 9. \end{cases}$$

(i) Draw the graph of F.
(ii) Calculate the probabilities:

$$P(X \leq 6.5), \quad P(X > 8.1), \quad P(5 < x < 8).$$

3.5 Let X be an r.v. with p.d.f. $f(x) = cx^{-(c+1)}$, for $x \geq 1$, where c is a positive constant.

(i) Determine the constant c, so that f is, indeed, a p.d.f.
(ii) Determine the corresponding d.f. F.

Hint. For part (i), use Remark 2(ii), and for part (ii), use Proposition 5(i).

3.6 Let X be an r.v. with p.d.f. $f(x) = cx + d$, for $0 \leq x \leq 1$, and suppose that $P(X > \frac{1}{2}) = \frac{1}{3}$. Then:

(i) Determine the constants c and d.
(ii) Find the d.f. F of X.

Hint. One of the two relations needed follows from the use of Remark 2(ii), and the other from the given probability.

3.7 Show that the function $f(x) = (\frac{1}{2})^x$, $x = 1, 2, \ldots$ is a p.d.f.

Hint. See #4 in Table 6 in the Appendix in conjunction with Remark 2(ii).

3.8 For what value of c is the function $f(x) = c\alpha^x$, $x = 0, 1, \ldots$ a p.d.f.? The quantity α is a number such that $0 < \alpha < 1$, and c is to be expressed in terms of α.

Hint. As in Exercise 3.7.

3.9 For what value of the positive constant c is the function $f(x) = c^x$, $x = 1, 2, \ldots$ a p.d.f.?

Hint. As in Exercise 3.7.

3.10 The p.d.f. of an r.v. X is $f(x) = c(\frac{1}{3})^x$, for $x = 0, 1, \ldots$, where c is a positive constant.

(i) Determine the value of c.
(ii) Calculate the probability $P(X \geq 3)$.

Hint. As in Exercise 3.7.

3.11 The r.v. X has p.d.f. f given by: $f(x) = c(1 - x^2)$, $-1 \leq x \leq 1$.

(i) Determine the constant c.
(ii) Calculate the probability $P(-0.9 < X < 0.9)$.

3.12 Let X be an r.v. denoting the lifetime of a piece of electrical equipment, and suppose that the p.d.f. of X is: $f(x) = ce^{-cx}$, for $x > 0$ (for some constant $c > 0$).

(i) Determine the constant c.
(ii) Calculate the probability (in terms of c) that X is at least equal to 10 (time units).
(iii) If the probability in part (ii) is 0.5, what is the value of c?

3.13 The r.v. X has the so-called Pareto p.d.f. given by: $f(x) = \frac{1+\alpha}{x^{2+\alpha}}$, for $x > 1$, where α is a positive constant.

(i) Verify that f is, indeed, a p.d.f.
(ii) Calculate the probability $P(X > c)$ (in terms of c and α), for some $c > 1$.

3.14 Suppose that the r.v. X takes on the values $0, 1, \ldots$ with the respective probabilities $P(X = j) = f(j) = \frac{c}{3^j}$, $j = 0, 1, \ldots$. Then:

(i) Determine the constant c.

Compute the probabilities:

(ii) $P(X \geq 3)$.
(iii) $P(X = 2k + 1, \quad k = 0, 1, \ldots)$.
(iv) $P(X = 3k + 1, \quad k = 0, 1, \ldots)$.

Hint. As in Exercise 3.7.

3.15 Let X be an r.v. with p.d.f. f whose graph is given above (Figure 3.4). Without calculating f and by using geometric arguments, compute the following probabilities:

$$P(X \leq 3), \qquad P(1 \leq X \leq 2), \qquad P(X > 2), \qquad P(X > 5).$$

3.16 Let X be the r.v. denoting the number of a certain item sold by a merchant in a given day, and suppose that its p.d.f. is given by:

$$f(x) = \left(\frac{1}{2}\right)^{x+1}, \quad x = 0, 1, \ldots$$

Calculate the following probabilities:

(i) No items are sold.
(ii) More than three items are sold.
(iii) An odd number of items is sold.

Hint. As in Exercise 3.7.

3.17 Suppose an r.v. X has p.d.f. given by: $f(x) = \lambda e^{-\lambda x}$, $x > 0$, $(\lambda > 0)$, and you are invited to bet whether X would be $\geq c$ or $< c$ for some positive constant c. In terms of the probabilities $P(X \geq c)$ and $P(X < c)$:

(i) For what c (expressed in terms of λ) would you bet in favor of $X \geq c$?
(ii) What is the answer in part (i) if $\lambda = 4 \log 2$? (log, as always, is the natural logarithm.)

3.18 The lifetime in hours of electric tubes is an r.v. X with p.d.f. $f(x) = c^2 x e^{-cx}$, for $x \geq 0$, where c is a positive constant.

(i) Determine the constant c for which f is, indeed, a p.d.f.
(ii) Calculate the probability (in terms of c) that the lifetime will be at least t hours.
(iii) Find the numerical value in part (ii) for $c = 0.2$ and $t = 10$.

3.19 Let X be the r.v. denoting the number of forms required to be filled out by a contractor for participation in contract bids, where the values of X are 1, 2, 3, 4, and 5, and suppose that the respective probabilities are proportional to x; that is, $P(X = x) = f(x) = cx$, $x = 1, \ldots, 5$.

(i) Determine the constant c.
(ii) Calculate the probabilities:

$$P(X \leq 3), \qquad P(2 \leq X \leq 4).$$

3.20 The recorded temperature in an engine is an r.v. X whose p.d.f. is given by: $f(x) = n(1-x)^{n-1}$, $0 < x < 1$, $n \geq 1$, known integer. (See also Example 13.) The engine is equipped with a thermostat that is activated when the temperature

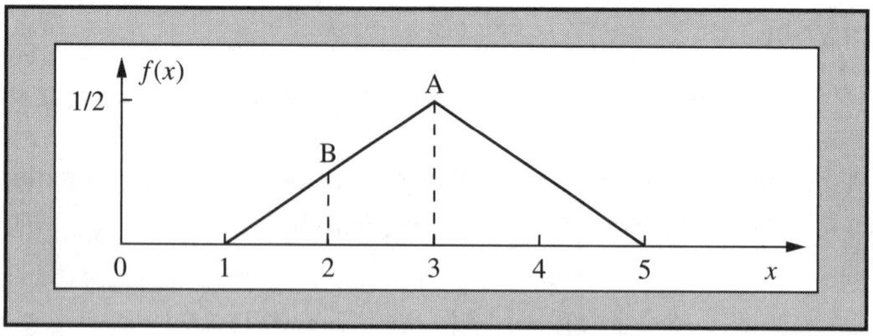

FIGURE 3.4
Triangular distribution.

exceeds a specified level x_0. If the probability of the thermostat being activated is $1/10^{2n}$, determine x_0.

3.21 Let X be an r.v. defined on a sample space $\mathcal{S}$ into the real line $\Re$. Then show that if B_1 and B_2 are any two disjoint subsets of $\Re$, so are the events $A_1 = (X \in B_1)$ and $A_2 = (X \in B_2)$.

3.22 Let X be an r.v. defined on a sample space $\mathcal{S}$ into the real line $\Re$, and let B_i, $i \geq 1$, be any subsets of $\Re$. Then show that $\cup_i (X \in B_i) = \big(X \in \cup_i (B_i)\big)$.

Hint. Show that the event on either side is contained in the event in the other side.

3.23 Consider the set function P_X (the probability distribution function of the r.v. X) defined in relation (3.1), and show that P_X is, actually, a probability function defined on subsets of $\Re$; that is, show that P_X satisfies properties (P1)–(P3) in the Axiomatic Definition of Probability.

Hint. Use Exercises 3.21 and 3.22.

3.24 Provide the details of the justification of Proposition 3.

Hint. For parts (iii) and (iv), use Theorem 2.

3.25 Provide the details of the justification of Proposition 4.

3.26 Let X be an r.v. of the continuous type with p.d.f f_X. Then by Remark 2(iii), $P(X = x) = 0$ for all $x \in \Re$, whereas by Proposition 5(ii), $\int_{\Re} f_X(x)dx = 1$. Reconcile these two seemingly contradictory facts.

Hint. Focus on property (P3) of the Axiomatic Definition of Probability.

3.27 An r.v. X takes on the values 0, 1, ... with respective probabilities given by:

$$\begin{aligned} p_0 = P(X = 0), \;\; p_k = P(X = k) &= cP(X = k - 1) \\ &= cp_{k-1} (0 < c < 1), \;\; k \geq 1. \end{aligned}$$

(i) Show that $p_k = c^k p_0$, and determine p_0 in terms of c.

(ii) Compute the probability $P(X > n)$ in terms of c, and find its numertical value for $n = 5$ and $c = 0.8$.

CHAPTER

Conditional Probability and Independence

4

This chapter consists of two sections. In the first section, the concept of the conditional probability of an event, given another event, is taken up. Its definition is given and its significance is demonstrated through a number of examples. The section is concluded with three theorems, formulated in terms of conditional probabilities. Through these theorems, conditional probabilities greatly simplify calculation of otherwise complicated probabilities. In the second section, the independence of two events is defined, and we also indicate how it carries over to any finite number of events. A result (Theorem 4) is stated which is often used by many authors without its use even being acknowledged. The section is concluded with an indication of how independence extends to random experiments. The definition of independence of r.v.'s is deferred to another chapter (Chapter 10).

4.1 Conditional Probability and Related Results

Conditional probability is a probability in its own right, as will be seen, and it is an extremely useful tool in calculating probabilities. Essentially, it amounts to suitably modifying a sample space $\mathcal{S}$, associated with a random experiment, on the evidence that a certain event has occurred. Consider the following examples, by way of motivation, before a formal definition is given.

Example 1. In tossing three distinct coins once (Example 1 in Chapter 2), consider the events B = "exactly 2 heads occur" = $\{HHT, HTH, THH\}$, A = "2 specified coins (e.g., coins #1 and #2) show heads" = $\{HHH, HHT\}$. Then $P(B) = \frac{3}{8}$ and $P(A) = \frac{2}{8} = \frac{1}{4}$. Now, suppose we are told that event B has occurred and we are asked to evaluate the probability of A on the basis of this evidence. Clearly, what really matters here is the event B, and given that B has occurred, the event A occurs only if the sample point HHT appears; that is, the event $\{HHT\} = A \cap B$ occurs. The required probability is then $\frac{1}{3} = \frac{1/8}{3/8} = \frac{P(A\cap B)}{P(B)}$, and the notation employed is $P(A \mid B)$ (probability of A, given that B has occurred or, just, given B). Thus, $P(A \mid B) = \frac{P(A\cap B)}{P(B)}$. Observe that $P(A \mid B) = \frac{1}{3} > \frac{1}{4} = P(A)$.

Example 2. In rolling two distinct dice once (Example 2 in Chapter 2), consider the event B defined by: B = "the sum of numbers on the upper face is ≤ 5", so that $B = \{(1,1), (1,2), (1,3), (1,4), (2,1), (2,2), (2,3), (3,1), (3,2), (4,1)\}$, and

Introduction to Probability, Second Edition. http://dx.doi.org/10.1016/B978-0-12-800041-0.00004-3

let A = "the sum of numbers on the upper faces is ≥ 4." Then A^c = "the sum of numbers on the upper faces is ≤ 3" = $\{(1,1),(1,2),(2,1)\}$, so that $P(B) = \frac{10}{36} = \frac{5}{18}$ and $P(A) = 1 - P(A^c) = 1 - \frac{3}{36} = \frac{33}{36} = \frac{11}{12}$. Next, if we are told that B has occurred, then the only way that A occurs is if $A \cap B$ occurs, where $A \cap B$ = "the sum of numbers on the upper faces is both ≥ 4 and ≤ 5 (i.e., either 4 or 5)" = $\{(1,3),(1,4),(2,2),(2,3),(3,1),(3,2),(4,1)\}$. Thus, $P(A \mid B) = \frac{7}{10} = \frac{7/36}{10/36} = \frac{P(A \cap B)}{P(B)}$, and observe that $P(A \mid B) = \frac{7}{10} < \frac{11}{12} = P(A)$.

Example 3. In recording the gender of children in a two-children family (Example 4 in Chapter 2), let B = "children of both genders" = $\{bg, gb\}$ and let A = "older child is a boy" = $\{bb, bg\}$, so that $A \cap B = \{bg\}$. Then $P(B) = \frac{1}{2} = P(A)$, and $P(A \mid B) = \frac{1}{2} = P(A)$.

These examples motivate the following definition of conditional probability.

Definition 1. The *conditional* probability of an event A, given the event B with $P(B) > 0$, is denoted by $P(A \mid B)$ and is defined by: $P(A \mid B) = P(A \cap B)/P(B)$.

Replacing B by the entire sample space $\mathcal{S}$, we are led back to the (*unconditional*) probability of A, as $P(A \mid \mathcal{S}) \frac{P(A \cap \mathcal{S})}{P(\mathcal{S})} = \frac{P(A)}{1} = P(A)$. Thus, the conditional probability is a generalization of the concept of probability where $\mathcal{S}$ is restricted to an event B.

The conditional probability is a full-fledged probability function, as the following proposition states.

Proposition 1. The conditional probability satisfies properties (P1)–(P3) in the Axiomatic Definition of Probability (in Chapter 3).

Proof. That the conditional probability is, indeed, a probability is seen formally as follows: $P(A \mid B) \geq 0$ for every A by definition;

$$P(\mathcal{S} \mid B) = \frac{P(\mathcal{S} \cap B)}{P(B)} = \frac{P(B)}{P(B)} = 1;$$

and if $A_1, A_2, \ldots$ are pairwise disjoint, then:

$$P\left(\bigcup_{j=1}^{\infty} A_j \mid B\right) = \frac{P[(\cup_{j=1}^{\infty} A_j) \cap B]}{P(B)} = \frac{P[\cup_{j=1}^{\infty}(A_j \cap B)]}{P(B)}$$

$$= \frac{\sum_{j=1}^{\infty} P(A_j \cap B)}{P(B)} = \sum_{j=1}^{\infty} \frac{P(A_j \cap B)}{P(B)} = \sum_{j=1}^{\infty} P(A_j \mid B). \quad \blacktriangle$$

It is to be noticed, furthermore, that the $P(A \mid B)$ can be smaller or larger than the $P(A)$, or equal to the $P(A)$. The case that $P(A \mid B) = P(A)$ is of special interest and will be discussed more extensively in the next section. This point is made by Examples 1, 2 and 3.

Here are another three examples pertaining to conditional probabilities.

Example 4. When we are recording the number of particles emitted by a certain radioactive source within a specified period of time (Example 8 in Chapter 2), we are going to see that if X is the number of particles emitted, then X is an r.v. taking on the values 0,1,…and that a suitable p.d.f. for it is $f_X(x) = e^{-\lambda}\frac{\lambda^x}{x!}$, $x = 0, 1, \ldots,$ for some constant $\lambda > 0$. Next, let B and A be the events defined by: $B = (X \geq 10)$, $A = (X \leq 11)$, so that $A \cap B = (10 \leq X \leq 11) = (X = 10 \text{ or } X = 11)$. Then

$$P(B) = \sum_{x=10}^{\infty} e^{-\lambda}\frac{\lambda^x}{x!} = e^{-\lambda}\sum_{x=10}^{\infty}\frac{\lambda^x}{x!},$$

$$P(A) = \sum_{x=0}^{11} e^{-\lambda}\frac{\lambda^x}{x!} = e^{-\lambda}\sum_{x=0}^{11}\frac{\lambda^x}{x!}, \quad \text{and}$$

$$P(A \mid B) = \left(e^{-\lambda}\frac{\lambda^{10}}{10!} + e^{-\lambda}\frac{\lambda^{11}}{11!}\right)\Big/ e^{-\lambda}\sum_{x=10}^{\infty}\frac{\lambda^x}{x!}.$$

For a numerical example, take $\lambda = 10$. Then we have (by means of the Poisson tables):

$$P(B) \simeq 0.5421, \quad P(A) \simeq 0.6968, \quad \text{and} \quad P(A \mid B) \simeq 0.441.$$

Example 5. When recording the lifetime of an electronic device, an electrical appliance, etc. (Example 9 in Chapter 2), if X is the lifetime under consideration, then X is an r.v. taking values in $(0, \infty)$, and a suitable p.d.f. for it is seen to be the function $f_X(x) = \lambda e^{-\lambda x}$, $x > 0$, for some constant $\lambda > 0$. Let B and A be the events: $B =$ "at the end of 5 time units, the equipment was still operating" $= (X \geq 5)$, $A =$ "the equipment lasts for no more than 2 additional time units" $= (X \leq 7)$. Then $A \cap B = (5 \leq X \leq 7)$, and:

$$P(B) = \int_5^{\infty} \lambda e^{-\lambda x}\,dx = e^{-5\lambda}, \quad P(A) = \int_0^7 \lambda e^{-\lambda x}\,dx = 1 - e^{-7\lambda},$$

$$P(A \cap B) = \int_5^7 \lambda e^{-\lambda x}\,dx = e^{-5\lambda} - e^{-7\lambda}, \quad \text{so that}$$

$$P(A \mid B) = \frac{P(A \cap B)}{P(B)} = \frac{e^{-5\lambda} - e^{-7\lambda}}{e^{-5\lambda}} = 1 - e^{-2\lambda}.$$

Take, for instance, $\lambda = \frac{1}{10}$. Then, given that $e^{-1} \simeq 0.36788$, the preceding probabilities are:

$$P(B) \simeq 0.607, \quad P(A) \simeq 0.503, \quad \text{and} \quad P(A \mid B) \simeq 0.181.$$

Example 6. If for the events A and B, $P(A)P(B) > 0$, then show that: $P(A \mid B) > P(A)$ if and only if $P(B \mid A) > P(B)$. Likewise, $P(A \mid B) < P(A)$ if and only if $P(B \mid A) < P(B)$.

Discussion. Indeed, $P(A \mid B) > P(A)$ is equivalent to $\frac{P(A\cap B)}{P(B)} > P(A)$ or $\frac{P(A\cap B)}{P(A)} > P(B)$ or $P(B \mid A) > P(B)$. Likewise, $P(A \mid B) < P(A)$ is equivalent to $\frac{P(A\cap B)}{P(B)} < P(A)$ or $\frac{P(A\cap B)}{P(A)} < P(B)$ or $P(B \mid A) < P(B)$.

This section is concluded with three simple but very useful results. They are the so-called multiplicative theorem, the total probability theorem, and the Bayes formula.

Theorem 1 (Multiplicative Theorem). For any n events $A_1, \ldots, A_n$ with $P\left(\bigcap_{j=1}^{n-1} A_j\right) > 0$, it holds:

$$P\left(\bigcap_{j=1}^{n} A_j\right) = P(A_n \mid A_1 \cap \ldots \cap A_{n-1})P(A_{n-1} \mid A_1 \cap \ldots \cap A_{n-2}) \\ \ldots P(A_2 \mid A_1)P(A_1).$$

Its justification is simple, is done by induction, and is left as an exercise (see Exercise 1.8). Its significance is that we can calculate the probability of the intersection of n events, step by step, by means of conditional probabilities. The calculation of these conditional probabilities is far easier. Here is a simple example which amply illustrates the point.

Example 7. An urn contains 10 identical balls, of which 5 are black, 3 are red, and 2 are white. Four balls are drawn one at a time and without replacement. Find the probability that the first ball is black, the second red, the third white, and the fourth black.

Discussion. Denoting by B_1 the event that the first ball is black, and likewise for R_2, W_3, and B_4, the required probability is:

$$P(B_1 \cap R_2 \cap W_3 \cap B_4) = P(B_4 \mid B_1 \cap R_2 \cap W_3)P(W_3 \mid B_1 \cap R_2)P(R_2 \mid B_1)P(B_1).$$

Assuming equally likely outcomes at each step, we have:

$$P(B_1) = \frac{5}{10}, \quad P(R_2 \mid B_1) = \frac{3}{9}, \quad P(W_3 \mid B_1 \cap R_2) = \frac{2}{8}, \\ P(B_4 \mid B_1 \cap R_2 \cap W_3) = \frac{4}{7}.$$

Therefore,

$$P(B_1 \cap R_2 \cap W_3 \cap B_4) = \frac{4}{7} \times \frac{2}{8} \times \frac{3}{9} \times \frac{5}{10} = \frac{1}{42} \simeq 0.024.$$

For the formulation of the next result, the concept of a partition of $\mathcal{S}$ is required.

Definition 2. The events $\{A_1, A_2, \ldots, A_n\}$ form a *partition* of $\mathcal{S}$, if these events are pairwise disjoint, $A_i \cap A_j = \varnothing, i \neq j$, and their union is $\mathcal{S}$, $\bigcup_{j=1}^{n} A_j = \mathcal{S}$; and similarly for countably infinite many events $\{A_1, A_2, \ldots\}$.

Then it is obvious that any event B in $\mathcal{S}$ may be expressed as follows, in terms of a partition of $\mathcal{S}$; namely, $B = \bigcup_{j=1}^{n}(A_j \cap B)$. Furthermore,

$$P(B) = \sum_{j=1}^{n} P(A_j \cap B) = \sum_{j=1}^{n} P(B \mid A_j)P(A_j), \quad \text{provided } P(A_j) > 0 \text{ for all } j;$$

and similarly for countably infinite many events. In the sequel, by writing $j = 1, 2, \ldots$ and $\sum_j$ we mean to include both cases, finitely many indices, and countably infinite many indices.

Thus, we have the following result.

Theorem 2 (Total Probability Theorem). Let $\{A_1, A_2, \ldots\}$ be a partition of $\mathcal{S}$, and let $P(A_j) > 0$ for all j. Then, for any event B,

$$P(B) = \sum_j P(B \mid A_j)P(A_j).$$

The significance of this result is that if it happens that we know the probabilities of the partitioning events, $P(A_j)$, as well as the conditional probabilities of B, given A_j, then these quantities may be combined, according to the preceding formula, to produce the probability $P(B)$. The probabilities $P(A_j)$, $j = 1, 2, \ldots$ are referred to as *a priori* or *prior* probabilities. The following examples illustrate the theorem and demonstrate its usefulness.

Example 8. In reference to Example 2 in Chapter 1, calculate the probability $P(+)$.

Discussion. Without having to refer specifically to a sample space, it is clear that the events D and N form a partition. Then,

$$P(+) = P(+ \text{ and } D) + P(+ \text{ and } N) = P(+ \mid D)P(D) + P(+ \mid N)P(N).$$

Here the a priori probabilities are $P(D) = p_1$, $P(N) = 1 - p_1$, and

$$P(+ \mid D) = 1 - P(- \mid D) = 1 - p_3, \quad P(+ \mid N) = p_2.$$

Therefore, $P(+) = (1 - p_3)p_1 + p_2(1 - p_1)$.

For a numerical application, take $p_1 = 0.02$ and $p_2 = p_3 = 0.01$. Then $P(+) = 0.0296$. So, on the basis of this testing procedure, about 2.96% of the population would test positive.

Example 9. The proportions of motorists at a given gas station using regular unleaded gasoline, extra unleaded, and premium unleaded over a specified period of time are 40%, 35%, and 25%, respectively. The respective proportions of filling their tanks are 30%, 50%, and 60%. What is the probability that a motorist selected at random from among the patrons of the gas station under consideration and for the specified period of time will fill his/her tank?

Discussion. Denote by R, E, and P the events of a motorist using unleaded gasoline which is regular, extra unleaded, and premium, respectively, and by F the event of having the tank filled. Then the translation into terms of probabilities of the proportions given above is:

$$P(R) = 0.40, \quad P(E) = 0.35, \quad P(P) = 0.25,$$
$$P(F \mid R) = 0.30, \quad P(F \mid E) = 0.50, \quad P(F \mid P) = 0.60.$$

Then the required probability is:

$$\begin{aligned}
P(F) &= P((F \cap R) \cup (F \cap E) \cup (F \cap P)) \\
&= P(F \cap R) + P(F \cap E) + P(F \cap P) \\
&= P(F \mid R)P(R) + P(F \mid E)P(E) + P(F \mid P)P(P) \\
&= 0.30 \times 0.40 + 0.50 \times 0.35 + 0.60 \times 0.25 \\
&= 0.445.
\end{aligned}$$

In reference to Theorem 2, stipulating the prior probabilities $P(B \mid A_j)$, $j = 1, 2, \ldots$, is often a precarious thing and guesswork. This being the case, the question then arises as to whether experimentation may lead to reevaluation of the prior probabilities on the basis of new evidence. To put it more formally, is it possible to use $P(A_j)$ and $P(B \mid A_j)$, $j = 1, 2, \ldots$ in order to calculate $P(A_j \mid B)$? The answer to this question is in the affirmative, is quite simple, and is the content of the next result.

Theorem 3 (Bayes' Formula). (See Theorem 2) Let $\{A_1, A_2, \ldots\}$ and B be as in the previous theorem. Then, for any $j = 1, 2, \ldots$:

$$P(A_j \mid B) = \frac{P(B \mid A_j)P(A_j)}{\sum_i P(B \mid A_i)P(A_i)}.$$

Proof. Indeed, $P(A_j \mid B) = P(A_j \cap B)/P(B) = P(B \mid A_j)P(A_j)/P(B)$, and then the previous theorem completes the proof. ▲

The probabilities $P(A_j \mid B)$, $j = 1, 2, \ldots$, are referred to as *posterior* probabilities in that they are reevaluations of the respective prior $P(A_j)$ probabilities after the event B has occurred.

Example 10. Referring to Example 8, a question of much importance is this: Given that the test shows positive, what is the probability that the patient actually has the disease? In terms of the notation adopted, this question becomes: $P(D \mid +) =$? Bayes' formula gives:

$$P(D \mid +) = \frac{P(+ \mid D)P(D)}{P(+ \mid D)P(D) + P(+ \mid N)P(N)} = \frac{p_1(1 - p_3)}{p_1(1 - p_3) + p_2(1 - p_1)}.$$

For the numerical values used above, we get:

$$P(D \mid +) = \frac{0.02 \times 0.99}{0.0296} = \frac{0.0198}{0.0296} = \frac{198}{296} \simeq 0.669.$$

So $P(D \mid +) \simeq 66.9\%$. This result is both reassuring and surprising—reassuring in that only 66.9% of those testing positive actually have the disease; surprising in that this proportion looks rather low, given that the test is quite good: it identifies correctly 99% of those having the disease. A reconciliation between these two seemingly contradictory aspects is as follows: The fact that $P(D) = 0.02$ means that on the average, 2 out of 100 persons have the disease. So, in 100 persons, 2 will have the disease and 98 will not. When 100 such persons are tested, $2 \times 0.99 = 1.98$ will be correctly confirmed as positive (because 0.99 is the probability of a correct positive), and $98 \times 0.01 = 0.98$ will be incorrectly diagnosed as positive (because 0.01 is the probability of an incorrect positive). Thus, the proportion of correct positives is equal to:

$$\begin{aligned}&(\text{correct positives})/(\text{correct positives} + \text{incorrect positives})\\ &\quad = 1.98/(1.98 + 0.98) = 1.98/2.96 = 198/296 \simeq 0.669.\end{aligned}$$

Remark 1. The fact that the probability $P(D \mid +)$ is less than 1 simply reflects the fact that the test, no matter how good, is imperfect. Should the test be perfect ($P(+ \mid D) = P(- \mid D^c) = 1$), then $P(D \mid +) = 1$, as follows from the preceding calculations, no matter what $P(D)$ is. The same, of course, is true for $P(D^c \mid -)$.

Example 11. Refer to Example 9 and calculate the probabilities: $P(R \mid F)$, $P(E \mid F)$, and $P(P \mid F)$.

Discussion. By Bayes' formula and Example 9,

$$P(R \mid F) = \frac{P(R \cap F)}{P(F)} = \frac{P(F \mid R)P(R)}{P(F)} = \frac{0.30 \times 0.40}{0.445} \simeq 0.270,$$

and likewise,

$$P(E \mid F) = \frac{0.50 \times 0.35}{0.445} \simeq 0.393, \quad P(P \mid F) = \frac{0.60 \times 0.25}{0.445} \simeq 0.337.$$

Exercises.

1.1 If $P(A) = 0.5$, $P(B) = 0.6$, and $P(A \cap B^c) = 0.4$, compute:

(i) $P(A \cap B)$; (ii) $P(A \mid B)$; (iii) $P(A \cup B^c)$; (iv) $P(B \mid A \cup B^c)$.

1.2 If $A \cap B = \varnothing$ and $P(A \cup B) > 0$, express the probabilities $P(A \mid A \cup B)$ and $P(B \mid A \cup B)$ in terms of $P(A)$ and $P(B)$.

1.3 A girls' club has in its membership rolls the names of 50 girls with the following descriptions:
20 blondes, 15 with blue eyes and 5 with brown eyes;
25 brunettes, 5 with blue eyes and 20 with brown eyes;
5 redheads, 1 with blue eyes and 4 with green eyes.
If one arranges a blind date with a club member, what is the probability that:

(i) The girl is blonde?
(ii) The girl is blonde, if it was revealed only that she has blue eyes?

1.4 Suppose that the probability that both of a pair of twins are boys is 0.30 and that the probability that they are both girls is 0.26. Given that the probability of the first child being a boy is 0.52, what is the probability that:

(i) The second twin is a boy, given that the first is a boy?
(ii) The second twin is a girl, given that the first is a girl?
(iii) The second twin is a boy?
(iv) The first is a boy and the second is a girl?

Hint. Denote by b_i and g_i the events that the ith child is a boy or a girl, respectively, $i = 1, 2$.

1.5 A shipment of 20 TV tubes contains 16 good tubes and 4 defective tubes. Three tubes are chosen successively and at random each time and are also tested successively. What is the probability that:

(i) The third tube is good if the first two were found to be good?
(ii) The third tube is defective if the first was found to be good and the second defective?
(iii) The third tube is defective if the first was found to be defective and the second was found to be good?
(iv) The third tube is defective if one of the other two was found to be good and the other was found to be defective?

Hint. Denote by D_i and G_i the events that the ith tube is defective or good, respectively, $i = 1, 2, 3$.

1.6 For any three events A, B, and C with $P(A)P(B)P(C) > 0$, show that:

(i) $P(A^c \mid B) = 1 - P(A \mid B)$.
(ii) $P(A \cup B \mid C) = P(A \mid C) + P(B \mid C) - P(A \cap B \mid C)$.
Also, by means of counterexamples, show that the following equations need not be true:
(iii) $P(A \mid B^c) = 1 - P(A \mid B)$.
(iv) $P(C \mid A \cup B) = P(C \mid A) + P(C \mid B)$, where $A \cap B = \varnothing$.

1.7 If A, B, and C are any events in the sample space $\mathcal{S}$, show that $\{A, A^c \cap B, A^c \cap B^c \cap C, (A \cup B \cup C)^c\}$ is a partition of $\mathcal{S}$.

Hint. Show that $A \cup (A^c \cap B) \cup (A^c \cap B^c \cap C) = A \cup B \cup C$.

1.8 Use induction to prove Theorem 1.

1.9 Let $\{A_j,\ j = 1, \ldots, 5\}$ be a partition of the sample space $\mathcal{S}$ and suppose that:

$$P(A_j) = \frac{j}{15} \quad \text{and} \quad P(A \mid A_j) = \frac{5-j}{15}, \quad j = 1, \ldots, 5.$$

Compute the probabilities $P(A_j \mid A),\ j = 1, \ldots, 5$.

1.10 A box contains 15 identical balls except that 10 are red and 5 are black. Four balls are drawn successively and without replacement.
Calculate the probability that the first and fourth balls are red.

Hint. Denote by R_i and B_i the events that the ith ball is red or black, respectively, $i = 1, \ldots, 4$, and use Theorem 1.

1.11 A box contains $m + n$ identical balls except that m of them are red and n are black. A ball is drawn at random, its color is noticed, and then the ball is returned to the box along with r balls of the same color. Finally, a ball is drawn also at random. (All probabilities in parts (i)–(iii) are to be expressed as functions of m and n.)

- **(i)** What is the probability that the first ball is red?
- **(ii)** What is the probability that the second ball is red?
- **(iii)** Compare the probabilities in parts (i) and (ii) and comment on them.
- **(iv)** What is the probability that the first ball is black if the second is red?
- **(v)** Find the numerical values in parts (i), (ii), and (iv) if $m = 9$, $n = 6$, and $r = 5$.

Hint. Denote by R_i and B_i the events that the ith ball is red or black, respectively, $i = 1, 2$, and use Theorems 1 and 2.

1.12 A test correctly identifies a disease D with probability 0.95 and wrongly diagnoses D with probability 0.01. From past experience, it is known that disease D occurs in a targeted population with frequency 0.2%. An individual is chosen at random from said population and is given the test.
Calculate the probability that:

- **(i)** The test is $+$, $P(+)$.
- **(ii)** The individual actually suffers from disease D if the test turns out to be positive, $P(D \mid +)$.

Hint. Use Theorems 2 and 3.

1.13 Suppose that the probability of correct diagnosis (either positive or negative) of cervical cancer in the Pap test is 0.95 and that the proportion of women in a given population suffering from this disease is 0.01%. A woman is chosen at random from the target population and the test is administered.
What is the probability that:

- **(i)** The test is positive?
- **(ii)** The subject actually has the disease, given that the diagnosis is positive?

Hint. Use Theorems 2 and 3.

1.14 A signal S is sent from point A to point B and is received at B if both switches I and II are closed (see Figure 4.1). It is assumed that the probabilities of I and II being closed are 0.8 and 0.6, respectively, and that $P(II \text{ is closed} \mid I \text{is closed}) = P(II \text{ is closed})$. Calculate the following probabilities:

- **(i)** The signal is received at B.
- **(ii)** The (conditional) probability that switch I was open, given that the signal was not received at B.
- **(iii)** The (conditional) probability that switch II was open, given that the signal was not received at B.

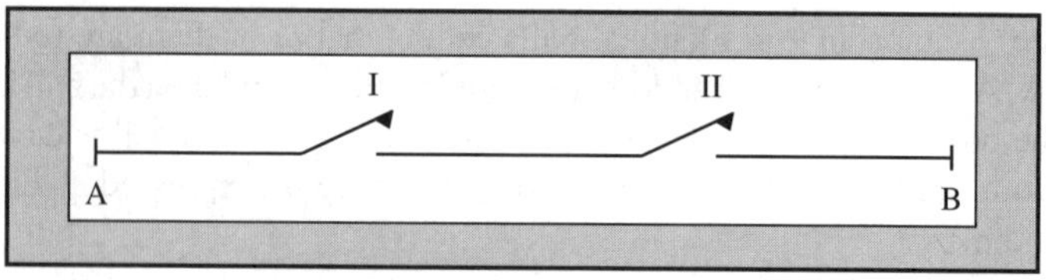

FIGURE 4.1

Sending a signal from point A to point B.

Hint. Use Theorems 2 and 3.

1.15 The student body at a certain college consists of 55% women and 45% men. Women and men smoke cigarettes in the proportions of 20% and 25%, respectively. If a student is chosen at random, calculate the probability that:

(i) The student is a smoker.
(ii) The student is a man, given that he/she is a smoker.

Hint. Use Theorems 2 and 3.

1.16 From a population consisting of 52% females and 48% males, an individual, drawn at random, is found to be color blind. If we assume that the proportions of color-blind females and males are 25% and 5%, respectively, what is the probability that the individual drawn is a male?

Hint. Use Theorems 2 and 3.

1.17 Drawers I and II contain black and red pencils as follows:
Drawer I: b_1 black pencils and r_1 red pencils; Drawer II: b_2 black pencils and r_2 red pencils. A drawer is chosen at random and then a pencil is also chosen at random from that drawer.

(i) What is the probability that the pencil is black?
(ii) If it is announced that the pencil is black, what is the probability it was chosen from drawer I?
(iii) Give numerical values in parts (i) and (ii) for $b_1 = 36, r_1 = 12, b_2 = 60, r_2 = 24$.

Hint. For parts (i) and (ii), use Theorems 2 and 3.

1.18 Three machines I, II, and III manufacture 30%, 30%, and 40%, respectively, of the total output of certain items. Of them, 4%, 3%, and 2%, respectively, are defective. One item is drawn at random from the total output and is tested.

(i) What is the probability that the item is defective?
(ii) If it is found to be defective, what is the probability the item was produced by machine I?
(iii) Same question as in part (ii) for each one of the machines II and III.

Hint. Use Theorems 2 and 3; probabilities are to be expressed as functions of b_1, b_2, r_1, and r_2.

1.19 Suppose that a multiple-choice test lists N alternative answers, of which only one is correct. If a student has done the homework, he/she is certain to identify the correct answer; otherwise the student chooses an answer at random. Denote by A the event that the student does the homework, set $p = P(A)$, and let B be the event that he/she answers the question correctly.

(i) Express the probability $P(A \mid B)$ in terms of p and n.
(ii) If $0 < p < 1$ and fixed, show that the probability $P(A \mid B)$, as a function of N, is increasing.
(iii) Does the result in part (ii) seem reasonable?

Hint. Use Theorem 2 for the computation of the P(B).

1.20 If the p.d.f. of the r.v. X is: $f(x) = \lambda e^{-\lambda x}$, for $x > 0$ ($\lambda > 0$), calculate:

(i) $P(X > t)$ (for some $t > 0$).
(ii) $P(X > s + t \mid X > s)$ (for some $s, t > 0$).
(iii) Compare the probabilities in parts (i) and (ii), and draw your conclusion.

1.21 A person is allowed to take a driver's exam up to 3 times. The probability that he/she passes the exam the first time is 0.7. If the first attempt results in a failure, the probability of passing the exam in the second attempt is 0.8. If both the first two attempts resulted in failures, the probability of passing the exam in the third (and final) attempt is 0.9. Compute the probability of:

(i) Passing the exam the second time.
(ii) Passing the exam the third time.
(iii) Passing the exam.

1.22 Three cards are drawn at random, without replacement, from a standard deck of 52 playing cards. Compute the (conditional) probability that the first card drawn is a spade, given that the second and the third cards drawn are spades.

Hint. Denote by S_i the event that the ith card is a spade, and use Theorem 1.

1.23 Four identical balls, numbered from 1 through 4, are placed at random into 4 urns, also numbered from 1 through 4, one ball in each urn. If the ith ball is placed in the ith urn, we say that a *match* occurred.

(i) Show that the probability that at least one match occurs is equal to:

$$1 - \frac{1}{2!} + \frac{1}{3!} - \frac{1}{4!} \quad \left(= \frac{5}{8} \simeq 0.625 \right).$$

(ii) Can you guess the probability of at least one match if the 4 balls and urns are replaced by n balls and urns?

(iii) In part (ii), suppose that n is large, and show that the computed probability is $\simeq 1 - e^{-1} \simeq 0.630$.

Hint. For $i = 1, 2, 3, 4$, set A_i for the event A_i = "a match occurs with the ith ball/urn". Then, clearly, the required probability is: $P(A_1 \cup A_2 \cup A_3 \cup A_4)$. Observe that $P(A_i) = \frac{3!}{4!}$, $i = 1, \ldots, 4$; $P(A_i \cap A_j) = \frac{2!}{4!}, 1 \leq i < j \leq 4$; $P(A_i \cap A_j \cap A_k) = \frac{1}{4!}, 1 \leq i < j < k \leq 4$; and $P(A_1 \cap A_2 \cap A_3 \cap A_4) = \frac{1}{4!}$. Then the expression in part (i) follows. For part (iii), use #6 in Table 6 in the Appendix.

Reamrk: For a more general treatment of the matching problem, see Theorem 10 in Chapter 2 in the book *A Course in Mathematical Statistics*, 2nd edition (1997), Academic Press, by G. G. Roussas.

1.24 It is known that 1% of a population suffers from a certain disease (event D), that a test employed for screening purposes identifies the disease correctly at the rate of 95%, and that it produces false positives at the rate of 0.5%. If a person is drawn at random from the target population, calculate the probability that the person:

(i) Tests positive (event $+$).
(ii) Actually has the disease, given that the person tests positive.

Hint. Use Theorems 2 and 3.

1.25 The totality of automobiles is divided into four groups A_1, A_2, A_3, and A_0 according to their makes (ith make, $i = 1, 2, 3$, and other makes), respectively. It is known that their respective proportions are 25%, 20%, 15%, and 40%. Next, let A be the event that an automobile accident occurs, and suppose that the respective probabilities of involvement in the accident when the automobile is from group A_i, $i = 0, 1, 2, 3$, are 0.06, 0.05, 0.04, and 0.03.

(i) Calculate the probability of an accident occurring, $P(A)$.
(ii) Also, calculate the (conditional) probabilities that the accident is due to an automobile from group $A_i, i = 0, 1, 2, 3$.

Hint. Use Theorems 2 and 3.

4.2 Independent Events and Related Results

In Example 3, it was seen that $P(A \mid B) = P(A)$. Thus, the fact that the event B occurred provides no information in reevaluating the probability of A. Under such a circumstance, it is only fitting to say that A is independent of B. For any two events A and B with $P(B) > 0$, we say that A is *independent* of B, if $P(A \mid B) = P(A)$. If, in addition, $P(A) > 0$, then B is also *independent* of A because

$$P(B \mid A) = \frac{P(B \cap A)}{P(A)} = \frac{P(A \cap B)}{P(A)} = \frac{P(A \mid B)P(B)}{P(A)} = \frac{P(A)P(B)}{P(A)} = P(B).$$

Because of this symmetry, we then say that A and B are independent. From the definition of either $P(A \mid B)$ or $P(B \mid A)$, it follows then that $P(A \cap B) = P(A)P(B)$. We further observe that this relation is true even if one or both of $P(A)$, $P(B)$ are equal to 0. We take this relation as the defining relation of independence.

Definition 3. Two events A_1 and A_2 are said to be *independent (statistically* or *stochastically* or in the *probability sense*) if $P(A_1 \cap A_2) = P(A_1)P(A_2)$. When $P(A_1 \cap A_2) \neq P(A_1)P(A_2)$ they are said to be *dependent*.

Remark 2. At this point, it should be emphasized that disjointness and independence of two events are two distinct concepts; the former does not even require the concept of probability. Nevertheless, they are related in that if $A_1 \cap A_2 = \varnothing$, then they are independent if and only if at least one of $P(A_1)$, $P(A_2)$ is equal to 0. Thus (subject to $A_1 \cap A_2 = \varnothing$), $P(A_1)P(A_2) > 0$ implies that A_1 and A_2 are definitely dependent.

The definition of independence extends to three events A_1, A_2, A_3, as well as to any number n of events $A_1, \ldots, A_n$. Thus, three events A_1, A_2, A_3 for which $P(A_1 \cap A_2 \cap A_3) > 0$ are said to be independent, if all conditional probabilities coincide with the respective (unconditional) probabilities:

$$\left.\begin{array}{c} P(A_1 \mid A_2) = P(A_1 \mid A_3) = P(A_1 \mid A_2 \cap A_3) = P(A_1) \\ P(A_2 \mid A_1) = P(A_2 \mid A_3) = P(A_2 \mid A_1 \cap A_3) = P(A_2) \\ P(A_3 \mid A_1) = P(A_3 \mid A_2) = P(A_3 \mid A_1 \cap A_2) = P(A_3) \\ P(A_1 \cap A_2 \mid A_3) = P(A_1 \cap A_2),\ P(A_1 \cap A_3 \mid A_2) \\ = P(A_1 \cap A_3),\ P(A_2 \cap A_3 \mid A_1) = P(A_2 \cap A_3). \end{array}\right\} \tag{4.1}$$

From the definition of conditional probability, relations (4.1) are equivalent to:

$$\left.\begin{array}{c} P(A_1 \cap A_2) = P(A_1)P(A_2),\ P(A_1 \cap A_3) = P(A_1)P(A_3), \\ P(A_2 \cap A_3) = P(A_2)P(A_3),\ P(A_1 \cap A_2 \cap A_3) = P(A_1)P(A_2)P(A_3). \end{array}\right\} \tag{4.2}$$

Furthermore, it is to be observed that relations (4.2) hold even if any of $P(A_1)$, $P(A_2)$, or $P(A_3)$ is equal to 0. These relations are taken as defining relations of independence of three events A_1, A_2, A_3.

As one would expect, all four relations (4.2) are needed for independence (that is, in order for them to imply relations (4.1)). That this is, indeed, the case is illustrated by the following examples.

Example 12. Let $\mathcal{S} = \{1, 2, 3, 4\}$ and let $P(\{1\}) = P(\{2\}) = P(\{3\}) = P(\{4\}) = 1/4$. Define the events A_1, A_2, A_3 by: $A_1 = \{1, 2\}$, $A_2 = \{1, 3\}$, $A_3 = \{1, 4\}$. Then it is easily verified that: $P(A_1 \cap A_2) = P(A_1)P(A_2)$, $P(A_1 \cap A_3) = P(A_1)P(A_3)$, $P(A_2 \cap A_3) = P(A_2)P(A_3)$. However, $P(A_1 \cap A_2 \cap A_3) \neq P(A_1)P(A_2)P(A_3)$.

Example 13. Let $\mathcal{S} = \{1, 2, 3, 4, 5\}$ and let $P(\{1\}) = \frac{2}{16}$, $P(\{2\}) = P(\{3\}) = P(\{4\}) = \frac{3}{16}$, $P(\{5\}) = \frac{5}{16}$. Define the events A_1, A_2, A_3 by: $A_1 = \{1, 2, 3\}$, $A_2 = \{1, 2, 4\}$, $A_3 = \{1, 3, 4\}$. Then it is easily verified that $P(A_1 \cap A_2 \cap A_3) = P(A_1) P(A_2)P(A_3)$, but none of the other three relations in (4.2) is satisfied.

Relations (4.2) provide the pattern of the definition of independence of n events. Thus:

Definition 4. The events $A_1, \ldots, A_n$ are said to be *independent* (*statistically* or *stochastically* or in the *probability sense*) if, for all possible choices of k out of n events $(2 \leq k \leq n)$, the probability of their intersection equals the product of their probabilities. More formally, for any k with $2 \leq k \leq n$ and any integers $j_1, \ldots, j_k$ with $1 \leq j_1 < \cdots < j_k \leq n$, we have:

$$P\left(\bigcap_{i=1}^{k} A_{j_i}\right) = \prod_{i=1}^{k} P(A_{j_i}). \tag{4.3}$$

If at least one of the relations in (4.3) is violated, the events are said to be *dependent.* The number of relations of the form (4.3) required to express independence of n events is (see also Exercise 4.11(ii) in Chapter 2):

$$\binom{n}{2} + \binom{n}{3} + \cdots + \binom{n}{n} = 2^n - \binom{n}{1} - \binom{n}{0} = 2^n - n - 1.$$

For example, for $n = 2, 3$, these relations are: $2^2 - 2 - 1 = 1$ and $2^3 - 3 - 1 = 4$, respectively.

Typical cases in which independent events occur are whenever we are sampling with replacement from finite populations, such as selecting successively and with replacement balls from an urn containing balls of several colors, pulling successively and with replacement playing cards out of a standard deck of such cards, and the like.

The following property of independence of events is often used without even being acknowledged; it is stated here as a theorem.

Theorem 4.

(i) If the events A_1, A_2 are independent, then so are all three sets of events: A_1, A_2^c; A_1^c, A_2; A_1^c, A_2^c.

(ii) More generally, if the events $A_1, \ldots, A_n$ are independent, then so are the events $A_1', \ldots, A_n'$, where A_i' stands either for A_i or A_i^c, $i = 1, \ldots, n$.

For illustrative purposes, we present the proof of part (i) only.

Proof of Part (i). Clearly, $A_1 \cap A_2^c = A_1 - A_1 \cap A_2$. Thus,

$$\begin{aligned} P\left(A_1 \cap A_2^c\right) &= P(A_1 - A_1 \cap A_2) = P(A_1) - P(A_1 \cap A_2) \text{ (since } A_1 \cap A_2 \subseteq A_1) \\ &= P(A_1) - P(A_1)P(A_2) \quad \text{(by independence of } A_1,\ A_2) \\ &= P(A_1)[1 - P(A_2)] = P(A_1)P\left(A_2^c\right). \end{aligned}$$

The proof of $P(A_1^c \cap A_2) = P(A_1^c)P(A_2)$ is entirely symmetric. Finally,

$$\begin{aligned}
P(A_1^c \cap A_2^c) &= P((A_1 \cup A_2)^c) \quad \text{(by DeMorgan's laws)} \\
&= 1 - P(A_1 \cup A_2) \\
&= 1 - P(A_1) - P(A_2) + P(A_1 \cap A_2) \\
&= 1 - P(A_1) - P(A_2) + P(A_1)P(A_2) \text{ (by independence of } A_1,\ A_2) \\
&= [1 - P(A_1)] - P(A_2)[1 - P(A_1)] \\
&= P(A_1^c)P(A_2^c). \qquad \blacktriangle
\end{aligned}$$

Remark 3. The interested reader may find the proof of part (ii) in Theorem 6 of Chapter 2 of the book *A Course in Mathematical Statistics*, 2nd edition (1997), Academic Press, by G. G. Roussas.

The following examples will help illustrate concepts and results discussed in this section.

Example 14. Suppose that $P(B)P(B^c) > 0$. Then the events A and B are independent if and only if $P(A \mid B) = P(A \mid B^c)$.

Discussion. First, if A and B are independent, then A and B^c are also independent, by Theorem 4. Thus, $P(A \mid B^c) = \frac{P(A \cap B^c)}{P(B^c)} = \frac{P(A)P(B^c)}{P(B^c)} = P(A)$. Since also $P(A \mid B) = P(A)$, the equality $P(A \mid B) = P(A \mid B^c)$ holds. Next, $P(A \mid B) = P(A \mid B^c)$ is equivalent to $\frac{P(A \cap B)}{P(B)} = \frac{P(A \cap B^c)}{P(B^c)}$ or $P(A \cap B)P(B^c) = P(A \cap B^c)P(B)$ or $P(A \cap B)[1 - P(B)] = P(A \cap B^c)P(B)$ or $P(A \cap B) - P(A \cap B)P(B) = P(A \cap B^c)P(B)$ or $P(A \cap B) = [P(A \cap B) + P(A \cap B^c)]P(B) = P(A)P(B)$, since $(A \cap B) \cup (A \cap B^c) = A$. Thus, A and B are independent.

Remark 4. It is to be pointed out that the condition $P(A \mid B) = P(A \mid B^c)$ for independence of the events A and B is quite natural, intuitively. It says that the (conditional) probability of A remains the same no matter which one of B or B^c is given.

Example 15. Let $P(C)P(C^c) > 0$. Then the inequalities $P(A \mid C) > P(B \mid C)$ and $P(A \mid C^c) > P(B \mid C^c)$ imply $P(A) > P(B)$.

Discussion. The inequalities $P(A \mid C) > P(B \mid C)$ and $P(A \mid C^c) > P(B \mid C^c)$ are equivalent to $P(A \cap C) > P(B \cap C)$ and $P(A \cap C^c) > P(B \cap C^c)$. Adding up these inequalities, we obtain $P(A \cap C) + P(A \cap C^c) > P(B \cap C) + P(B \cap C^c)$ or $P(A) > P(B)$, since $A = (A \cap C) \cup (A \cap C^c)$ and $B = (B \cap C) \cup (B \cap C^c)$.

Remark 5. Once again, that the inequalities of the two conditional probabilities should imply the same inequality for the unconditional probabilities is quite obvious on intuitive grounds. The justification given above simply makes it rigorous.

Example 16. If the events A, B, and C are independent, then $P(A \cup B \cup C) = 1 - [1 - P(A)][1 - P(B)][1 - P(C)]$.

Discussion. Clearly,

$$\begin{aligned}P(A \cup B \cup C) &= P[(A^c \cap B^c \cap C^c)^c] && \text{(by DeMorgan's laws, Proposition 4 in Chapter 2)}\\ &= 1 - P(A^c \cap B^c \cap C^c) && \text{(by Proposition 1(iii) in Chapter 3)}\\ &= 1 - P(A^c)P(B^c)P(C^c) && \text{(by Theorem 4(ii) applied with n=3)}\\ &= 1 - [1 - P(A)][1 - P(B)][1 - P(C)].\end{aligned}$$

Example 17. A mouse caught in a maze has to maneuver through three successive escape hatches in order to escape. If the hatches operate independently and the probabilities for the mouse to maneuver successfully through them are 0.6, 0.4, and 0.2, respectively, calculate the probabilities that the mouse: (i) will be able to escape, (ii) will not be able to escape.

Discussion. Denote by H_1, H_2, and H_3 the events that the mouse successfully maneuvers through the three hatches, and by E the event that the mouse is able to escape. We have that H_1, H_2, and H_3 are independent, $P(H_1) = 0.6$, $P(H_2) = 0.4$, and $P(H_3) = 0.2$, and $E = H_1 \cap H_2 \cap H_3$. Then: (i) $P(E) = P(H_1 \cap H_2 \cap H_3) = P(H_1)P(H_2)P(H_3) = 0.6 \times 0.4 \times 0.2 = 0.048$, and (ii) $P(E^c) = 1 - P(E) = 1 - 0.048 = 0.952$.

Example 18. Out of a set of 3 keys, only 1 opens a certain door. Someone tries the keys successively and independently, and let A_k be the event that the right key appears the kth time. Calculate the probabilities $P(A_k)$:

(i) If the keys tried are not replaced, $k = 1, 2, 3$.
(ii) If the keys tried are replaced, $k = 1, 2, \ldots$.

Discussion.

(i) By enumeration:

$$P(A_1) = \frac{1}{3};\ P(A_2) = \frac{2 \times 1}{3 \times 2} = \frac{1}{3};\ P(A_3) = \frac{2 \times 1 \times 1}{3 \times 2 \times 1} = \frac{1}{3}.$$

$$\text{So, } P(A_1) = P(A_2) = P(A_3) = \frac{1}{3} \simeq 0.333.$$

To calculate the probabilities in terms of conditional probabilities, let R and W stand for the events that the right and a wrong key are selected, respectively. Then:

$$P(A_1) = P(R) = \frac{1}{3};\ P(A_2) = P(W \cap R) = P(R \mid W)P(W) = \frac{1}{2} \times \frac{2}{3} = \frac{1}{3};$$

$$\begin{aligned}P(A_3) = P(W \cap W \cap R) &= P(R \mid W \cap W)P(W \mid W)\\ &= \times P(W)\frac{1}{1} \times \frac{1}{2} \times \frac{2}{3} = \frac{1}{3}.\end{aligned}$$

(ii) For $k = 1$, $P(A_1) = \frac{1}{3}$, whereas for $k \geq 2$, clearly,

$$\begin{aligned} P(A_k) &= P(\underbrace{W \cap W \cap \cdots \cap W}_{k-1} \cap R) \\ &= P(W)P(W)\dots P(W)P(R) \quad \text{(by independence)} \\ &= \left(\frac{2}{3}\right)^{k-1} \times \frac{1}{3} \quad \text{(and this also holds for } k = 1\text{).} \end{aligned}$$

Example 19. Each of the $2n$ members of a committee flips a fair coin independently in deciding whether or not to attend a meeting of the committee; a committee member attends the meeting if an H appears. What is the probability that a majority will show up for the meeting?

Discussion. There will be majority if there are at least $n + 1$ committee members present, which amounts to having at least $n + 1$ H's in $2n$ independent throws of a fair coin. If X is the r.v. denoting the number of H's in the $2n$ throws, then the required probability is: $P(X \geq n + 1) = \sum_{x=n+1}^{2n} P(X = x)$. However,

$$P(X = x) = \binom{2n}{x}\left(\frac{1}{2}\right)^x\left(\frac{1}{2}\right)^{2n-x} = \frac{1}{2^{2n}}\binom{2n}{x},$$

since there are $\binom{2n}{x}$ ways of having x H's in $2n$ throws. Therefore:

$$\begin{aligned} P(X \geq n + 1) &= \frac{1}{2^{2n}} \sum_{x=n+1}^{2n} \binom{2n}{x} = \frac{1}{2^{2n}}\left[\sum_{x=0}^{2n}\binom{2n}{x} - \sum_{x=0}^{n}\binom{2n}{x}\right] \\ &= \frac{1}{2^{2n}}\left[2^{2n} - \sum_{x=0}^{n}\binom{2n}{x}\right] = 1 - \frac{1}{2^{2n}}\sum_{x=0}^{n}\binom{2n}{x}. \end{aligned}$$

For example, for $2n = 10$, $P(X \geq 6) = 1 - 0.6230 = 0.377$ (from the binomial tables).

Example 20. Refer to Example 3 in Chapter 1, and let S_i, $i = 1, \dots, 5$ be events defined as follows: S_i = "switch i works," $i = 1, \dots, 5$. Also, set $C_1 = S_1 \cap S_2$, $C_2 = S_5$, and $C_3 = S_3 \cap S_4$, and let C be the event defined by: C = " current is transmitted from point A to point B." Then:

$$\begin{aligned} P(C) &= P(C_1 \cup C_2 \cup C_3) \text{(obviously)} \\ &= P(C_1) + P(C_2) + P(C_3) - P(C_1 \cap C_2) - P(C_1 \cap C_3) \\ &\quad - P(C_2 \cap C_3) + P(C_1 \cap C_2 \cap C_3) \quad \text{(by Proposition 2(ii) in Chapter 3).} \end{aligned}$$

However, $C_1 \cap C_2 = S_1 \cap S_2 \cap S_5$, $C_1 \cap C_3 = S_1 \cap S_2 \cap S_3 \cap S_4$, $C_2 \cap C_3 = S_3 \cap S_4 \cap S_5$, and $C_1 \cap C_2 \cap C_3 = S_1 \cap S_2 \cap S_3 \cap S_4 \cap S_5$. Then, by assuming that the switches work independently of each other, so that the events S_i, $i = 1, \ldots, 5$ are independent, we have:

$$\begin{aligned} P(C) = {} & P(S_1)P(S_2) + P(S_5) + P(S_3)P(S_4) \\ & -P(S_1)P(S_2)P(S_5) - P(S_1)P(S_2)P(S_3)P(S_4) \\ & -P(S_3)P(S_4)P(S_5) + P(S_1)P(S_2)P(S_3)P(S_4)P(S_5). \end{aligned}$$

By setting $p_i = P(S_i)$, $i = 1, \ldots, 5$, we get:

$$\begin{aligned} P(C) = {} & p_5 + p_1p_2 + p_3p_4 - p_1p_2p_5 - p_3p_4p_5 \\ & -p_1p_2p_3p_4 + p_1p_2p_3p_4p_5. \end{aligned}$$

For $p_1 = \cdots = p_5 = p$, $P(C) = p + 2p^2 - 2p^3 - p^4 + p^5$, and for $p = 0.9$, we obtain

$$P(C) = 0.9 + 2(0.9)^2 - 2(0.9)^3 - (0.9)^4 + (0.9)^5 \simeq 0.99639.$$

The concept of independence carries over to random experiments. Although a technical definition of independence of random experiments is available, we are not going to indulge in it. The concept of independence of random experiments will be taken in its intuitive sense, and somewhat more technically, in the sense that random experiments are independent if they give rise to independent events associated with them.

Finally, independence is also defined for r.v.'s. This topic will be taken up in Chapter 10 (see Definition 1 there). Actually, independence of r.v.'s is one of the founding blocks of most discussions taking place in this book.

Exercises.

2.1 If $P(A) = 0.4$, $P(B) = 0.2$, and $P(C) = 0.3$, calculate the probability $P(A \cup B \cup C)$, if the events A, B, and C are:

(i) Pairwise disjoint.
(ii) Independent.

2.2 Show that the event A is independent of itself if and only if $P(A) = 0$ or $P(A) = 1$.

2.3

(i) For any two events A and B, show that $P(A \cap B) \geq P(A) + P(B) - 1$.
(ii) If A and B are disjoint, then show that they are independent if and only if at least one of $P(A)$ and $P(B)$ is zero.
(iii) If the events A, B, and C are pairwise disjoint, under what conditions are they independent?

2.4 Suppose that the events A_1, A_2, and B_1 are independent, the events A_1, A_2, and B_2 are independent, and $B_1 \cap B_2 = \varnothing$. Then show that the events A_1, A_2, $B_1 \cup B_2$ are independent.

Hint. Just check condition (4.2) or (4.3) (for $k = 3$) for the events: A_1, A_2, $B_1 \cup B_2$.

2.5

(i) If for the events A, B, and C, it so happens that $P(A) = P(B) = P(C) = \frac{1}{2}$, $P(A \cap B) = P(A \cap C) = P(B \cap C) = \frac{1}{4}$, and $P(A \cap B \cap C) = \frac{1}{6}$, determine whether or not these events are independent. Justify your answer.

(ii) For the values given in part (i), calculate the probabilities: $P(A^c)$, $P(A \cup B)$, $P(A^c \cap B^c)$, $P(A \cup B \cup C)$, and $P(A^c \cap B^c \cap C^c)$.

2.6 For the events A, B, C and their complements, suppose that:

$$P(A \cap B \cap C) = \tfrac{1}{16},\quad P(A^c \cap B \cap C) = \tfrac{2}{16},\quad P(A \cap B^c \cap C) = \tfrac{5}{16},$$
$$P(A \cap B \cap C^c) = \tfrac{3}{16},\quad P(A^c \cap B^c \cap C) = \tfrac{1}{16},\quad P(A^c \cap B \cap C^c) = \tfrac{1}{16},$$
$$P(A \cap B^c \cap C^c) = \tfrac{2}{16},\quad P(A^c \cap B^c \cap C^c) = \tfrac{1}{16}.$$

(i) Calculate the probabilities: $P(A)$, $P(B)$, $P(C)$.

(ii) Determine whether or not the events A, B, and C are independent.

(iii) Calculate the (conditional) probability $P(A \mid B)$.

(iv) Determine whether or not the events A and B are independent.

Hint. Use Theorem 2 for part (i) and for the calculation of the $P(A \cap B)$ in part (iii).

2.7 If the events $A_1, \ldots, A_n$ are independent, show that

$$P\left(\bigcup_{j=1}^{n} A_j\right) = 1 - \prod_{j=1}^{n} P(A_j^c).$$

Hint. Use Proposition 4 in Chapter 2, Proposition 1(iii) in Chapter 3, and Theorem 4(ii) here.

2.8

(i) Three coins, with probability of falling heads being p, are tossed once, and you win if all three coins show the same face (either all H or all T). What is the probability of winning?

(ii) What are the numerical answers in part (i) for $p = 0.5$ and $p = 0.4$?

2.9 Suppose that men and women are distributed in the freshman and sophomore classes of a college according to the proportions listed in the following table.

Class\Gender	M	W	Totals
F	4	6	10
S	6	x	$6 + x$
Totals	10	$6 + x$	$16 + x$

A student is chosen at random and let M, W, F, and S be the events, respectively, that the student is a man, a woman, a freshman, or a sophomore. Then, being a man or a woman and being a freshman or a sophomore are independent, if:

$$P(M \cap F) = P(M)P(F), \qquad P(W \cap F) = P(W)P(F),$$
$$P(M \cap S) = P(M)P(S), \qquad P(W \cap S) = P(W)P(S).$$

Determine the number x so that the preceding independence relations hold.

Hint. Determine x by using any one of the above four relations (and check that this value of x also satisfies the remaining three relations).

2.10 The r.v. X has p.d.f. given by:

$$f(x) = \begin{cases} cx, & 0 \le x < 5 \\ c(10 - x), & 5 \le x < 10 \\ 0, & \text{elsewhere.} \end{cases}$$

(i) Determine the constant c.
(ii) Draw the graph of f. Define the events A and B by: $A = (X > 5)$, $B = (5 < X < 7.5)$.
(iii) Calculate the probabilities $P(A)$ and $P(B)$.
(iv) Calculate the conditional probability $P(B \mid A)$.
(v) Are the events A and B independent or not? Justify your answer.

2.11 A student is given a test consisting of 30 questions. For each question, 5 different answers (of which only one is correct) are supplied. The student is required to answer correctly at least 25 questions in order to pass the test. If he/she knows the right answers to the first 20 questions and chooses an answer to the remaining questions at random and independently of each other, what is the probability that the student will pass the test?

Hint. The student passes the test if he/she answers correctly at least 5 of the last 10 questions.

2.12 From an urn containing n_R red balls, n_B black balls, and n_W white balls (all identical except for color), 3 balls are drawn at random. Calculate the following probabilities:

(i) All 3 balls are red.
(ii) At least one ball is red.
(iii) One ball is red, 1 is black, and 1 is white.

Do this when the balls are drawn:

(a) Successively and with replacement;
(b) Without replacement.

(All probabilities are to be expressed as functions of n_R, n_B, and n_W.)

2.13 Two people toss independently n times each a coin whose probability of falling heads is p. What is the probability (expressed in terms of n and p) that they have the same number of heads? How does this probability become for $p = \frac{1}{2}$ and any n? Also, for $p = \frac{1}{2}$ and $n = 5$?

Hint. If A_m and B_m are the events that the two people toss m heads each, then the required event is $\cup_{m=0}^{n}(A_m \cap B_m)$.

2.14 Consider two urns U_1 and U_2 such that urn U_1 contains m_1 white balls and n_1 black balls, and urn U_2 contains m_2 white balls and n_2 black balls. All balls are identical except for color. One ball is drawn at random from each of the urns U_1 and U_2 independently and is placed into a third urn. Then a ball is drawn at random from the third urn. Compute the probability that the ball is:

(i) Black;
(ii) White.
(iii) Give numerical answers to parts (i) and (ii) for: $m_1 = 10, n_1 = 15; m_2 = 35, n_2 = 25$.
(Probabilities in parts (i) and (ii) are to be expressed in terms of m_1, m_2, and n_1, n_2.)

Hint. For parts (i) and (ii), denote by B_i and W_i the events that the ball drawn from the ith urn, $i = 1, 2$, is black or white, respectively, and by B and $W$the events that the ball drawn from the third urn is black or white, respectively, and then use Theorem 2.

2.15 The probability that a missile fired against a target is not intercepted by an antimissile is $\frac{2}{3}$. If the missile is not intercepted, then the probability of a successful hit is $\frac{3}{4}$.

(i) What is the probability that the missile hits the target?
If four missiles are fired independently, what is the probability that:
(ii) All four will successfully hit the target?
(iii) At least one will do so?
(iv) What is the minimum number of missiles to be fired so that at least one is not intercepted with probability at least 0.95?
(v) What is the minimum number of missiles to be fired so that at least one hits the target with probability at least 0.99?

2.16 Electric current is transmitted from point A to point B, provided at least one of the circuits #1 through #n here is closed (Figure 4.2). It is assumed that the n circuits close independently of each other and with respective probabilities $p_1, \ldots, p_n$.
Determine the following probabilities:

(i) No circuit is closed.
(ii) At least one circuit is closed.

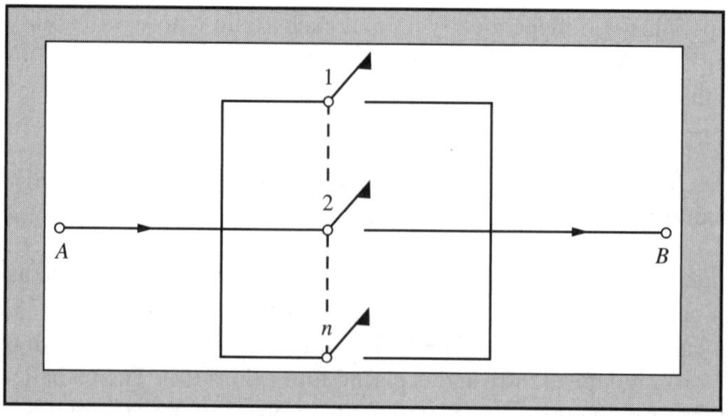

FIGURE 4.2

Transmitting electric current from point A to point B.

- **(iii)** Exactly one circuit is closed.
- **(iv)** How do the expressions in parts (i)–(iii) simplify if $p_1 = \cdots = p_n = p$?
- **(v)** What are the numerical values in part (iv) for $n = 5$ and $p = 0.6$?

Hint. Use Theorem 4(ii). Probabilities in parts (i)–(iii) are to be expressed in terms of $p_1, \ldots, p_n$ and p, respectively.

2.17 Jim takes the written and road driver's license tests repeatedly until he passes them. It is given that the probability that he passes the written test is 0.9, that he passes the road test is 0.6, and that the tests are independent of each other. Furthermore, it is assumed that the road test cannot be taken unless he passes the written test, and that once he passes the written test, he does not have to take it again ever, no matter whether he passes or fails his road tests. Also, it is assumed that the written and road tests are distinct attempts.

- **(i)** What is the probability that he will pass the road test on his nth attempt? (Just write down the correct formula.)
- **(ii)** What is the numerical value in part (i) for $n = 5$?

Hint. Denote by W_i and R_j the events that Jim passes the written test and the road test the ith and jth time, respectively. Then the required event is expressed as follows:

$$\left(W_1 \cap R_1^c \cap \ldots \cap R_{n-2}^c \cap R_{n-1}\right) \cup \left(W_1^c \cap W_2 \cap R_1^c \cap \ldots \cap R_{n-3}^c \cap R_{n-2}\right)$$
$$\cup \ldots \cup \left(W_1^c \cap \ldots \cap W_{n-2}^c \cap W_{n-1} \cap R_n\right).$$

Also, use Theorem 4(ii).

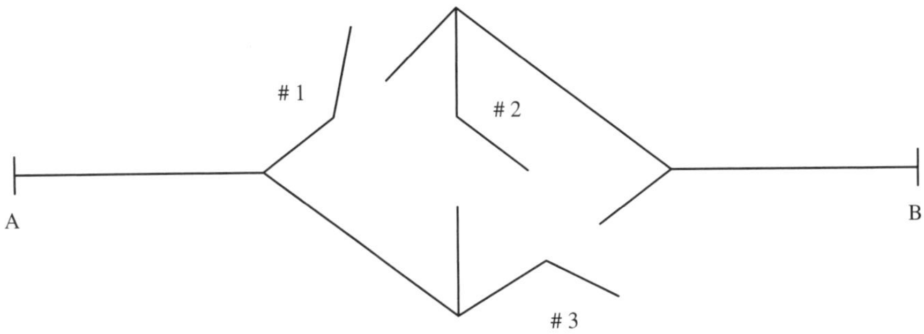

FIGURE 4.3

Transmitting electric current from point A to point B.

2.18 Three players *I*, *II*, and *III* throw simultaneously three coins (one coin each) with probability of falling heads (H) 0.5. A sample space describing this experiment is:

$$\mathcal{S} = \{HHH, HHT, HTH, THH, HTT, THT, TTH, TTT\}.$$

Define the events $A_i, i = 1, 2, 3$, and B by:

$$A_1 = \{HTT, THH\} \quad A_2 = \{THT, HTH\}, \quad A_3 = \{TTH, HHT\}$$

(i.e., A_i is the event that the outcome for the ith player, $i = 1, 2, 3$, is different from those for the other two players),

$$B = \{HHH, TTT\}.$$

If any one of the events $A_i, i = 1, 2, 3$ occurs, the ith player wins and the game ends. If event B occurs, the game is repeated independently as many times as needed until one of the events A_1, A_2, A_3 occurs.
Calculate the probabilities: $P(A_i), i = 1, 2, 3$.

Hint. By symmetry, it suffices to calculate $P(A_1)$. Let A_{1j} = "event A_1 occurs the jth time", B_j = "event B occurs the jth time". Then (with slight abuse of notation)

$$A_1 = A_{11} \cup (B_1 \cap A_{12}) \cup (B_1 \cap B_2 \cap A_{13}) \cup \ldots.$$

At this point, also refer to #4 in Table 6 in the Appendix.

2.19 In the circuit diagram depicted below (Figure 4.3), assume that the three switches turn on and off independent of each other with respective probabilities $P(\text{Switch } \#i \text{ turns on}) = p_i, i = 1, 2, 3$.

(i) Compute the probability that current flows from A to B.
(ii) What is the probability in part (i) if $p_1 = p_2 = p_3 = p$, say?
(iii) What is the numerical value of the probabilities in parts (i) and (ii), respectively, if $p_1 = 0.90$, $p_2 = 0.95$, $p_3 = 0.99$; $p = 0.96$?

Hint. For part (i), use Proposition 2(ii) in Chapter 3, and in parts (i) and (ii), express probabilities in terms of p_1, p_2, p_3, and p, respectively.

2.20 If the events A, B, and C are independent, show that:

(i) The events A and $B \cup C$ are also independent.
(ii) The events A^c and $B \cap C^c$ are also independent.
(iii) Evaluate the probability $P(B \cup C \mid A)$ (assuming that $P(A) > 0$) in terms of the probabilities $P(B)$, $P(C)$.

Hint. For part (ii), use Theorem 4.

2.21 An investor buys three stocks A, B, and C which perform independently of each other. On the basis of market analysis, the stocks are expected to rise with probabilities 0.4, 0.6, and 0.7, respectively. Compute the probability that:

(i) All three stocks rise.
(ii) None of the stocks rise.
(iii) At least one stock rises.
(iv) Exactly one stock rises.

Hint. For parts (ii) and (iv), use Theorem 4.

2.22 A coin with probability $p(0 < p < 1)$ of showing H is tossed repeatedly and independently, and let A_n and A be events defined as follows:

$$\begin{aligned} A_n &= \text{“}H\text{'s the first } n \text{ times, and whatever (i.e., either } H\text{'s or } T\text{'s) thereafter”} \\ &= \{\underbrace{H \cdots H}_{n\text{times}} \text{ whatever thereafter}\}, \quad n \geq 1, \\ A &= \text{“only}H\text{'s”} = \{HH \cdots\}. \end{aligned}$$

(i) Show that the events $\{A_n\}$, $n \geq 1$, form a decreasing sequence, $A_n \supset A_{n+1}$ for $n \geq 1$.
(ii) Also, $\bigcap_{n=1}^{\infty} A_n \left(= \lim_{n \to \infty} A_n\right) = A$.
(iii) Use parts (i)-(ii), Theorem 2 in Chapter 3 (and independence) to show that $P(A) = 0$.

CHAPTER

5 Numerical Characteristics of a Random Variable

In this chapter, we discuss the following material. In Section 5.1, the concepts of expectation and variance of an r.v. are introduced and interpretations are provided. Higher order moments are also defined and their significance is pointed out. Also, the moment-generating function of an r.v. is defined, and its usefulness as a mathematical tool is commented upon. In Section 5.2, the Markov and Tchebichev inequalities are introduced and their role in estimating probabilities is explained. The chapter is concluded by Section 5.3 with a discussion of the concepts of median and mode, which are illustrated by concrete examples.

5.1 Expectation, Variance, and Moment-Generating Function of a Random Variable

The ideal situation in life would be to know with certainty what is going to happen next. This being almost never the case, the element of chance enters in all aspects of our lives. An r.v. is a mathematical formulation of a random environment. Given that we have to deal with an r.v. X, the best thing to expect is to know the values of X and the probabilities with which these values are taken on, for the case that X is discrete, or the probabilities with which X takes values in various subsets of the real line $\Re$ when X is of the continuous type. That is, we would like to know the *probability distribution function* of X. In real life, often, even this is not feasible. Instead, we are forced to settle for some numerical characteristics of the distribution of X. This line of argument leads us to the concepts of the mathematical expectation and variance of an r.v., as well as to moments of higher order.

Definition 1. Let X be a (discrete) r.v. taking on the values x_i with corresponding probabilities $f(x_i), i = 1, \ldots, n$. Then the *mathematical expectation* of X (or just *expectation* or *mean value* of X or just *mean* of X) is denoted by EX and is defined by:

$$EX = \sum_{i=1}^{n} x_i f(x_i). \tag{5.1}$$

Introduction to Probability, Second Edition. http://dx.doi.org/10.1016/B978-0-12-800041-0.00005-5

If the r.v. X takes on (countably) infinite many values x_i with corresponding probabilities $f(x_i), i = 1, 2, \ldots$, then the expectation of X is defined by:

$$EX = \sum_{i=1}^{\infty} x_i f(x_i), \quad \text{provided} \sum_{i=1}^{\infty} |x_i| f(x_i) < \infty. \tag{5.2}$$

Finally, if the r.v. X is continuous with p.d.f. f, its expectation is defined by:

$$EX = \int_{-\infty}^{\infty} x f(x) dx, \quad \text{provided this integral exists and is finite.} \tag{5.3}$$

The alternative notations $\mu(X)$ or μ_X are also often used.

Remark 1.

(i) The condition $\sum_{i=1}^{\infty} |x_i| f(x_i) < \infty$ is needed because, if it is violated, it is known that $\sum_{i=1}^{\infty} x_i f(x_i)$ may take on different values, depending on the order in which the terms involved are summed up. This, of course, would render the definition of EX meaningless.

(ii) An example will be presented later on (see Exercise 1.16) where the integral $\int_{-\infty}^{\infty} x f(x) dx = \infty - \infty$, so that the integral does not exist.

The expectation has several interpretations, some of which are illustrated by the following Examples 1 and 2. One basic interpretation, however, is that of center of gravity. Namely, if one considers the material system where mass $f(x_i)$ is placed at the point $x_i, i = 1, \ldots, n$, then EX is the *center of gravity* (point of equilibrium) of this system. In this sense, EX is referred to as a *measure of location of the center* of the distribution of X. The same interpretation holds when X takes on (countably) infinitely many values or is of the continuous type.

Example 1. Suppose an insurance company pays the amount of \$1,000 for lost luggage on an airplane trip. From past experience, it is known that the company pays this amount in 1 out of 200 policies it sells. What premium should the company charge?

Discussion. Define the r.v. X as follows: $X = 0$ if no loss occurs, which happens with probability $1 - (1/200) = 0.995$, and $X = -1,000$ with probability $\frac{1}{200} = 0.005$. Then the expected loss to the company is: $EX = -1,000 \times 0.005 + 0 \times 0.995 = -5$. Thus, the company must charge \$5 to break even. To this, it will normally add a reasonable amount for administrative expenses and a profit.

Even in this simple example, but most certainly so in more complicated cases, it is convenient to present the values of a (discrete) r.v. and the corresponding probabilities in a tabular form as follows.

x	0	−1,000	Total
$f(x)$	199/200	1/200	1

Example 2. A roulette wheel consists of 18 black slots, 18 red slots, and 2 green slots. If a gambler bets \$10 on red, what is the gambler's expected gain or loss?

Discussion. Define the r.v. X by: $X = 10$ with probability $18/38$ and $X = -10$ with probability $20/38$, or in a tabular form

x	10	−10	Total
$f(x)$	18/38	20/38	1

Then $EX = 10 \times \frac{18}{38} - 10 \times \frac{20}{38} = -\frac{10}{19} \simeq -0.526$. Thus, the gambler's expected loss is about 53 cents.

From the definition of the expectation and familiar properties of summations or integrals, it follows that:

Proposition 1.

(i) $E(cX) = cEX, \quad E(cX + d) = cEX + d$, where c and d are constants. (5.4)

(ii) $X \geq c$ constant, implies $EX \geq c$, and, in particular, $X \geq 0$ implies $EX \geq 0$. (5.5)

The verification of relations (5.4) and (5.5) is left as an exercise (see Exercise 1.18).

Now suppose the r.v. X represents the lifetime of a new piece of equipment put in service, and let $g(x)$ be the cost of operating this equipment up to time x. Then $Eg(X)$ is, clearly, the expected cost of operating said equipment over its lifetime. This example points to the need for defining the expectation of a function of an r.v., $Eg(X)$. In principle, one may be able to determine the p.d.f. of $Y = g(X)$ and proceed to defining its expectation by the appropriate version of formulas (5.1–5.3). It can be shown, however, that this is not necessary. Instead, the expectation of Y is defined by using the p.d.f. of X; namely:

Definition 2. Consider an r.v. X with p.d.f. $f(x)$ and let $g : \Re \to \Re$. Then the *expectation* of the r.v. $Y = g(X)$, EY, is defined by

$$EY = \sum_{i=1}^{n} g(x_i)f(x_i) \quad \text{or} \quad EY = \sum_{i=1}^{\infty} g(x_i)f(x_i) \quad \text{or} \quad EY = \int_{-\infty}^{\infty} g(x)f(x)dx, \tag{5.6}$$

under provisions similar to the ones mentioned in connection with (5.2) and (5.3). By taking $g(x) = x^k$, where k is a positive integer, we obtain the kth *moment* of X:

$$EX^k = \sum_{i=1}^{n} x_i^k f(x_i) \quad \text{or} \quad EX^k = \sum_{i=1}^{\infty} x_i^k f(x_i) \quad \text{or} \quad EX^k = \int_{-\infty}^{\infty} x^k f(x)dx. \tag{5.7}$$

For $k = 1$, we revert to the expectation of X, and for $k = 2$, we get its *second moment*.

Moments are important, among other things, in that, in certain circumstances, a number of them completely determine the distribution of X. See Theorem 2 and Exercise 1.17 here, and the special distributions discussed in Chapter 6.

The following simple example illustrates that the expectation, as a measure of location of the distribution, may reveal very little about the entire distribution.

$$X = \begin{cases} -1, & 5/8 \\ 1, & 1/8 \\ 2, & 2/8 \end{cases}, \quad Y = \begin{cases} -10, & 5/8 \\ 10, & 1/8 \\ 20, & 2/8 \end{cases}; \quad EX = EY = 0.$$

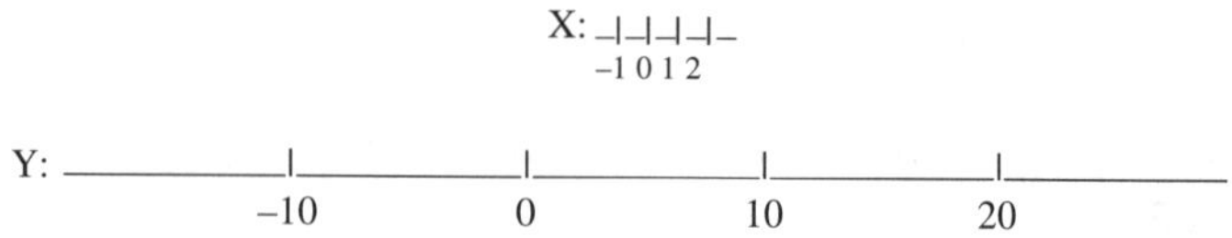

FIGURE 5.1
Spread of the r.v.'s X and Y.

Example 3. Consider the r.v.'s X and Y defined above (Figure 5.1), and compute their expectations.

Clearly, $EX = EY = 0$.
The distribution of X is over an interval of length 3, whereas the distribution of Y is over an interval of length 10 times as large. Yet, they have the same center of location. This simple example clearly indicates that the expectation by itself is not an adequate measure of description of a distribution, and an additional measure is needed to be associated with the spread of a distribution. Such a measure exists and is the variance of an r.v. or of its distribution. Its definition follows.

Definition 3. The *variance* of an r.v. X is denoted by $Var(X)$ and is defined by:

$$Var(X) = E(X - EX)^2. \tag{5.8}$$

The explicit expression of the right-hand side in (5.8) is taken from (5.6) for $g(x) = (x - EX)^2$. The alternative notations $\sigma^2(X)$ and σ_X^2 are also often used for the $Var(X)$.

The positive square root of the $Var(X)$ is called the *standard deviation* (s.d.) of X denoted by $\sigma(X)$ of σ_X. Unlike the variance, the s.d. is measured in the same units as X (and EX) and serves as a yardstick of measuring deviations of X from EX.

For the r.v.'s X and Y in Example 3, we have $Var(X) = 1.75$ and $Var(Y) = 175$. Thus, the variance does convey adequately the difference in size of the range of the distributions of the r.v.'s X and Y. Also, $\sigma(X) = \sqrt{1.75} \simeq 1.32$ and $\sigma(Y) = \sqrt{175} \simeq 13.23$.

In reference to Examples 1 and 2, the variances and the s.d.'s of the r.v.'s involved are: $\sigma^2(X) = 4,975$, $\sigma(X) \simeq 70.534$,and $\sigma^2(X) = \frac{36,000}{361} \simeq 99.723$, $\sigma(X) \simeq 9.986$, respectively.

More generally, for an r.v. X taking on finitely many values $x_1, \ldots, x_n$ with respective probabilities $f(x_1), \ldots, f(x_n)$, the variance is: $Var(X) = \sum_{i=1}^{n}(x_i - EX)^2 f(x_i)$ and represents the sum of the weighted squared distances of the points x_i, $i = 1, \ldots, n$ from the center of location of the distribution, EX. Thus, the farther from EX the x_i's are located, the larger the variance, and vice versa. The same interpretation holds for the case that X takes on (countably) infinitely many values or is of the continuous type. Because of this characteristic property of the variance, the variance is referred to as a measure of *dispersion* of the underlying distribution. In mechanics, the variance is referred to as the moment of *inertia*.

From (5.8), (5.6), and familiar properties of summations and integrals, one obtains formula (5.9) below, which often facilitates the actual calculation of the variance. From (5.8), formula (5.10) below also follows immediately.

For an r.v. Y which is a function of X, $Y = g(X)$, the calculation of the $Var[g(X)]$ reduces to calculating expectations as in (5.6) because, by means of (5.8) and (5.9), one has formula (5.11) below. Actually, formulas (5.8) and (5.9) are special cases of (5.11). In summary, we have:

Proposition 2. For an r.v. X:

(i)
$$Var(X) = EX^2 - (EX)^2. \tag{5.9}$$

(ii)
$$Var(cX) = c^2\,Var(X), \quad Var(cX + d) = c^2\,Var(X), \tag{5.10}$$

where c and d are constants.

(iii)
$$Var[g(X)] = Eg^2(X) - [Eg(X)]^2. \tag{5.11}$$

The justification of this proposition is left as an exercise (see Exercises 1.6(i) and 1.19). Some of the last relations are illustrated by the following example.

Example 4. Let X be an r.v. with p.d.f. $f(x) = 3x^2$, $0 < x < 1$. Then:

(i) Calculate the quantities EX, EX^2, and $Var(X)$.
(ii) If the r.v. Y is defined by $Y = 3X - 2$, calculate the EY and the $Var(Y)$.

Discussion.

(i) By (5.3), $EX = \int_0^1 x \times 3x^2\,dx = \frac{3}{4}x^4\big|_0^1 = \frac{3}{4} = 0.75$, whereas by (5.7), applied with $k = 2$, $EX^2 = \int_0^1 x^2 \times 3x^2\,dx = \frac{3}{5} = 0.60$, so that, by (5.9), $Var(X) = 0.60 - (0.75)^2 = 0.0375$.
(ii) By (5.4) and (5.6), $EY = E(3X-2) = 3EX-2 = 3\times 0.75-2 = 0.25$, whereas by (5.10), $Var(Y) = Var(3X - 2) = 9\,Var(X) = 9 \times 0.0375 = 0.3375$.

In (5.6), the EY was defined for $Y = g(X)$, some function of X. In particular, we may take $Y = e^{tX}$ for an arbitrary but fixed $t \in \Re$. Assuming that there exist t's in $\Re$ for which Ee^{tX} is finite, then this expectation defines a function in t. This function is denoted by $M(t)$ and is called the moment-generating function of X. That is,

Definition 4. The function $M(t) = Ee^{tX}$, defined for all those t in $\Re$ for which Ee^{tX} is finite, is called the *moment-generating function* (m.g.f.) of X.

Sometimes the notation $M_X(t)$ is also used to emphasize the fact that the m.g.f. under discussion is that of the r.v. X. The m.g.f. of any r.v. always exists for $t = 0$, since $Ee^{0X} = E1 = 1$; it may exist only for $t = 0$, or for t's in a proper subset (interval) in $\Re$, or for every t in $\Re$. All these points will be demonstrated by concrete examples later on (see, for example, the m.g.f.'s of the special distributions discussed in Chapter 6, relations (6.2), (6.7), (6.9), (6.17), (6.19), (6.20), (6.23), (6.31), and (6.33)). The following properties of $M(t)$ follow immediately from its definition.

Proposition 3. For an r.v. X:

(i) $M_X(0) = 1$.

(ii)
$$M_{cX}(t) = M_X(ct), \quad M_{cX+d}(t) = e^{dt} M_X(ct), \tag{5.12}$$
where c and d are constants.

(iii) Under certain conditions,
$$\frac{d}{dt} M_X(t)\big|_{t=0} = EX, \quad \text{and} \quad \frac{d^n}{dt^n} M_X(t)\big|_{t=0} = EX^n, \quad n = 2, 3, \ldots \tag{5.13}$$

Proof.

(i) It has already been justified.

(ii) Indeed,
$$M_{cX}(t) = Ee^{t(cX)} = Ee^{(ct)X} = M_X(ct),$$
and
$$M_{cX+d}(t) = Ee^{t(cX+d)} = E\big[e^{dt} \times e^{(ct)X}\big] = e^{dt} Ee^{(ct)X} = e^{dt} M_X(ct).$$

(iii) For example, for the first property, we have:
$$\frac{d}{dt} M_X(t)\Big|_{t=0} = \left(\frac{d}{dt} Ee^{tX}\right)\Big|_{t=0} = E\left(\frac{\partial}{\partial t} e^{tX}\Big|_{t=0}\right)$$
$$= E(Xe^{tX}|_{t=0}) = EX.$$

What is required for the derivation in part (iii) to be legitimate is that the order in which the operators d/dt and E operate on e^{tX} can be interchanged. The justification of the property in (5.13) for $n \geq 2$ is quite similar. ▲

On account of property (5.13), $M_X(t)$ *generates* the moments of X through differentiation and evaluation of the derivatives at $t = 0$. It is from this property that the m.g.f. derives its name.

For Examples 1 and 2, the m.g.f.'s of the r.v.'s involved are: $M_X(t) = 0.995 + 0.005e^{-1{,}000t}$, $t \in \Re$, and $M_X(t) = \frac{1}{19}(9e^{10t} + 10e^{-10t})$, $t \in \Re$. Then, by differentiation, we get: $\frac{d}{dt} M_X(t)|_{t=0} = -5 = EX$, $\frac{d^2}{dt^2} M_X(t)|_{t=0} = 5{,}000 = EX^2$, so that $\sigma^2(X) = 4{,}975$; and $\frac{d}{dt} M_X(t)|_{t=0} = -\frac{10}{19} = EX$, $\frac{d^2}{dt^2} M_X(t)|_{t=0} = 100 = EX^2$, so that $\sigma^2(X) = \frac{36{,}000}{361} \simeq 99.723$.

Here is a further example which helps illustrate the last proposition.

Example 5. Let X be an r.v. with p.d.f. $f(x) = e^{-x}$, $x > 0$. Then:

(i) Find the m.g.f. $M_X(t)$ for the t's for which it is finite.

(ii) Using M_X, obtain the quantities EX, EX^2, and $Var(X)$.
(iii) If the r.v. Y is defined by $Y = 2 - 3X$, determine $M_Y(t)$ for the t's for which it is finite.

Discussion.

(i) By Definition 4,

$$\begin{aligned} M_X(t) = Ee^{tX} &= \int_0^\infty e^{tx} \times e^{-x} dx = \int_0^\infty e^{-(1-t)x}\, dx \\ &= -\frac{1}{1-t} e^{-(1-t)x}\Big|_0^\infty \quad \text{(provided } t \neq 1) \\ &= -\frac{1}{1-t}(0-1) = \frac{1}{1-t} \quad \text{(provided } 1-t > 0 \text{ or } t < 1). \end{aligned}$$

Thus, $M_X(t) = \frac{1}{1-t}$, $t < 1$.

(ii) By (5.13),

$$\begin{aligned} \frac{d}{dt} M_X(t)|_{t=0} &= \frac{d}{dt}\left(\frac{1}{1-t}\right)|_{t=0} = \frac{1}{(1-t)^2}|_{t=0} = 1 = EX, \\ \frac{d^2}{dt^2} M_X(t)|_{t=0} &= \frac{d}{dt}\left(\frac{1}{(1-t)^2}\right)|_{t=0} = \frac{2}{(1-t)^3}|_{t=0} = 2 = EX^2, \end{aligned}$$

so that, by (5.9), $Var(X) = 2 - 1^2 = 1$.

(iii) By (5.12), $M_Y(t) = M_{2-3X}(t) = M_{-3X+2}(t) = e^{2t} M_X(-3t) = e^{2t} \frac{1}{1-(-3t)} = \frac{e^{2t}}{1+3t}$, provided $t > -\frac{1}{3}$.

It is to be emphasized that the m.g.f. does not have an intuitive interpretation, as the expectation and the variance do. It is to be looked upon as a valuable technical mathematical tool at our disposal. Relation (5.13) provides an example of using profitably an m.g.f. It is used in many other cases, some of which will be dealt with in subsequent chapters. Presently, it suffices only to state one fundamental property of the m.g.f. in the form of a theorem.

Theorem 1. Under certain conditions, the m.g.f. M_X of an r.v. X uniquely determines the distribution of X.

This theorem is, actually, a rather deep probability result, referred to as the *inversion formula.*

Some forms of such a formula for characteristic functions which are a version of an m.g.f. may be found, e.g., in Section 6.2 in the book *A Course in Mathematical Statistics*, 2nd edition (1997), Academic Press, by G. G. Roussas.

Still another important result associated with m.g.f.'s is stated (but not proved) in the following theorem.

Theorem 2. If for the r.v. X all moments $EX^n, n = 1, 2, \ldots$ are finite, then, under certain conditions, these moments uniquely determine the m.g.f. M_X of X, and hence (by Theorem 1) the distribution of X.

Exercise 1.17 provides an example of an application of the theorem just stated.

Exercises.

Remark: In several calculations required in solving some exercises in this section, the formulas #4 and #5 in Table 6 in the Appendix may prove useful.

1.1 Refer to Exercise 3.1 in Chapter 3 and calculate the quantities EX and $Var(X)$, and the s.d. of X.

1.2 For the r.v. X for which $P(X = -c) = P(X = c) = 1/2$ (for some $c > 0$):

(i) Calculate the EX and the $Var(X)$ in terms of c.
(ii) Show that $P(|X - EX| \leq c) = Var(X)/c^2$.

1.3 A chemical company currently has in stock 100 lb of a certain chemical, which it sells to customers in 5-lb packages. Let X be the r.v. denoting the number of packages ordered by a randomly chosen customer, and suppose that the p.d.f. of X is given by: $f(1)=0.2$, $f(2)=0.4$, $f(3)=0.3$, $f(4)=0.1$.

x	1	2	3	4
$f(x)$	0.2	0.4	0.3	0.1

(i) Compute the following quantities: EX, EX^2, and $Var(X)$.
(ii) Compute the expected number of pounds left after the order of the customer in question has been shipped, as well as the s.d. of the number of pounds around the expected value.

Hint. For part (ii), observe that the leftover number of pounds is the r.v. $Y = 100 - 5X$.

1.4 Let X be an r.v. denoting the damage incurred (in $) in a certain type of accident during a given year, and suppose that the distribution of X is given by the following table:

x	0	1,000	5,000	10,000
$f(x)$	0.8	0.1	0.08	0.02

A particular insurance company offers a $500 deductible policy. If the company's expected profit on a given policy is $100, what premium amount should it charge?

Hint. If the r.v. Y denotes the net loss to the insurance company, then Y makes on the value: 0, 500, 4,500, and 9,500 with probabilities: 0.8, 0.1, 0.08, and 0.02. So, if P is the premium, there must be: $P - EY = 100$.

1.5 Let X be the r.v. denoting the number in the uppermost side of a fair die when rolled once.

(i) Determine the m.g.f. of X.
(ii) Use the m.g.f. to calculate EX, EX^2, $Var(X)$, and the s.d. of X.

1.6 For any r.v. X, for which the EX and the EX^2 are finite, show that:

(i) $Var(X) = EX^2 - (EX)^2$;
(ii) $Var(X) = E[X(X-1)] + EX - (EX)^2$.

1.7 Suppose that for an r.v. X it is given that: $EX = 5$ and $E[X(X-1)] = 27.5$. Calculate:

(i) EX^2.
(ii) $Var(X)$ and s.d. of X.

Hint. Refer to Exercise 1.6.

1.8 For the r.v. X with p.d.f. $f(x) = (1/2)^x$, $x = 1, 2, \ldots$:

(i) Calculate the EX and the $E[X(X-1)]$.
(ii) Use part (i) and Exercise 1.6(ii) to compute the $Var(X)$.

Hint. See #5 in Table 6 in the Appendix.

1.9 The p.d.f. f of an r.v. X is given by: $f(x) = (2/3)(1/3)^x$, for $x = 0, 1, \ldots$.

(i) Calculate the EX.
(ii) Determine the m.g.f. M_X of X and specify the range of its argument.
(iii) Employ the m.g.f. in order to derive the EX.

Hint. For part (i), see #5 in Table 6 in the Appendix, and for part (ii), use #4 in Table 6 in the Appendix.

1.10 For the r.v. X with p.d.f. $f(x) = 0.5x$, for $0 \leq x \leq 2$, calculate EX, $Var(X)$, and the s.d. of X.

1.11 If the r.v. X has p.d.f. $f(x) = 3x^2 - 2x + 1$, for $0 < x < 1$, compute the expectation and variance of X.

1.12 If the r.v. X has p.d.f. f given by:

$$f(x) = \begin{cases} c_1 x, & -2 < x < 0 \\ c_2 x, & 0 \leq x < 1 \\ 0, & \text{otherwise}, \end{cases}$$

and if we suppose that $EX = \frac{1}{3}$, determine the constants c_1 and c_2.

Hint. Two relations are needed for the determination of c_1 and c_2.

1.13 The lifetime in hours of electric tubes is an r.v. X with p.d.f. f given by: $f(x) = \lambda^2 x e^{-\lambda x}$, for $x > 0$ ($\lambda > 0$). Calculate the expected life of such tubes in terms of λ.

1.14 Let X be an r.v. whosc $EX = \mu \in \Re$. Then:

(i) For any constant c, show that:

$$E(X-c)^2 = E(X-\mu)^2 + (\mu - c)^2 = Var(X) + (\mu - c)^2.$$

(ii) Use part (i) to show that $E(X-c)^2$, as a function of c, is minimized for $c = \mu$.

Hint. In part (i), add and subtract μ.

1.15 Let X be an r.v. with p.d.f. $f(x) = \frac{|x|}{c^2}$, for $-c < x < c,\ c > 0$.

(i) Verify that $f(x)$ is, indeed, a p.d.f.
(ii) For any $n = 1, 2, \ldots$, calculate the EX^n, and as a special case, derive the EX and the *Var* (X).

Hint. Split the integral from $-c$ to 0 and from 0 to c.

1.16 Let X be an r.v. with p.d.f. given by: $f(x) = \frac{1}{\pi} \times \frac{1}{1+x^2}$, $x \in \Re$. Show that:

(i) f is, indeed, a p.d.f. (called the *Cauchy* p.d.f.).
(ii) $\int_{-\infty}^{\infty} x f(x)\, dx = \infty - \infty$, so that the EX *does not* exist.

Hint. For part (i), use the transformation $x = \tan u$.

1.17 If X is an r.v. for which all moments EX^n, $n = 0, 1, \ldots$ are finite, show that:

$$M_X(t) = \sum_{n=0}^{\infty} (EX^n) \frac{t^n}{n!}.$$

Hint. Use the expansion $e^x = \sum_{n=0}^{\infty} \frac{x^n}{n!}$ (see #6 in Table 6 in the Appendix).

Remark. The result in this exercise says that the moments of an r.v. determine (under certain conditions) the m.g.f. of the r.v., and hence its distribution, as Theorems 2 and 1 state.

1.18 Establish the inequalities stated in relations (5.4) and (5.5), for both the discrete and the continuous case.

1.19 Establish relations (5.10) and (5.11) in Proposition 2.

1.20 The claim sizes of an auto insurance company are the values x of an r.v. X with respective probabilities given below:

x	$f(x)$	x	$f(x)$
10	0.13	60	0.09
20	0.12	70	0.11
30	0.10	80	0.08
40	0.13	90	0.08
50	0.10	100	0.06

(i) What is the probability $P(40 \le X \le 60)$?
(ii) Compute the $EX = \mu$ and the s.d. of X, $\sigma = \sqrt{Var(X)}$.

(iii) What is the proportion of claims lying within σ, 1.5σ, and 2σ from μ?

1.21 The claims submitted to an insurance company over a specified period of time t is an r.v. X with p.d.f. $f(x) = \frac{c}{(1+x)^4}$, $x > 0$ $(c > 0)$.

(i) Determine the constant c.
(ii) Compute the probability $P(1 \leq X \leq 4)$.
(iii) Compute the expected number of claims over the period of time t.

5.2 Some Probability Inequalities

If the r.v. X has a known p.d.f. f, then, in principle, we can calculate probabilities $P(X \in B)$ for $B \subseteq \Re$. This, however, is easier said than done in practice. What one would be willing to settle for would be some suitable and computable bounds for such probabilities. This line of thought leads us to the inequalities discussed here.

Theorem 3.

(i) For any nonnegative r.v. X and for any constant $c > 0$, it holds:

$$P(X \geq c) \leq EX/c.$$

(ii) More generally, for any nonnegative function of any r.v. X, $g(X)$, and for any constant $c > 0$, it holds:

$$P[g(X) \geq c] \leq Eg(X)/c. \tag{5.14}$$

(iii) By taking $g(X) = |X - EX|$ in part (ii), the inequality reduces to the *Markov* inequality; namely,

$$P(|X-EX| \geq c) = P(|X-EX|^r \geq c^r) \leq E|X-EX|^r/c^r, \quad r>0. \tag{5.15}$$

(iv) In particular, for $r = 2$ in (5.15), we get the *Tchebichev* inequality; namely,

$$P(|X - EX| \geq c) \leq \frac{E(X - EX)^2}{c^2} = \frac{\sigma^2}{c^2} \quad \text{or}$$

$$P(|X - EX| < c) \geq 1 - \frac{\sigma^2}{c^2}, \tag{5.16}$$

where σ^2 stands for the $Var(X)$. Furthermore, if $c = k\sigma$, where σ is the s.d. of X, then:

$$P(|X - EX| \geq k\sigma) \leq \frac{1}{k^2} \quad \text{or} \quad P(|X - EX| < k\sigma) \geq 1 - \frac{1}{k^2}. \tag{5.17}$$

Remark 2. From the last expression, it follows that X lies within k s.d.'s from its mean with probability at least $1 - \frac{1}{k^2}$, regardless of the distribution of X. It is in this sense that the s.d. is used as a yardstick of deviations of X from its mean, as already mentioned elsewhere.

Thus, for example, for $k = 2, 3$, we obtain, respectively:

$$P(|X - EX| < 2\sigma) \geq 0.75, \qquad P(|X - EX| < 3\sigma) \geq \frac{8}{9} \simeq 0.889. \tag{5.18}$$

Proof of Theorem 3. Clearly, all one has to do is to justify (5.14) and this only for the case that X is continuous with p.d.f. f, because the discrete case is entirely analogous.

Indeed, let $A = \{x \in \Re;\ g(x) \geq c\}$, so that $A^c = \{x \in \Re;\ g(x) < c\}$. Then, clearly:

$$\begin{aligned} Eg(X) &= \int_{-\infty}^{\infty} g(x)f(x)\,dx = \int_A g(x)f(x)\,dx + \int_{A^c} g(x)f(x)\,dx \\ &\geq \int_A g(x)f(x)\,dx \quad (\text{since } g(x) \geq 0) \\ &\geq c\int_A f(x)\,dx \quad (\text{since } g(x) \geq c \text{ on } A) \\ &= cP(A) = cP[g(X) \geq c]. \end{aligned}$$

Solving for $P[g(X) \geq c]$, we obtain the desired result. ▲

Example 6. Let the r.v. X take on the values $-2, -1/2, 1/2$, and 2 with respective probabilities 0.05, 0.45, 0.45, and 0.05. Then $EX = 0$ and $\sigma^2 = Var(X) = 0.625$, so that $2\sigma \simeq 1.582$. Then: $P(|X| < 2\sigma) = P(-1.582 < X < 1.582) = P(X = -\frac{1}{2}) + P(X = \frac{1}{2}) = 0.90$, compared with the lower bound of 0.75.

The following example provides for another use of Tchebichev's inequality.

Example 7.

(i) If $EX = \mu$ and $Var(X) = \sigma^2$, determine the smallest value of the (positive) constant c for which $P(|X - \mu| < c) \geq p$, where p is in the interval (0, 1); express c in terms of σ and p.

(ii) Determine the numerical value of c if $\sigma = 0.5$ and $p = 0.95$.

Discussion.

(i) Since $P(|X - \mu| < c) \geq 1 - \frac{\sigma^2}{c^2}$, it suffices to determine c so that $1 - \frac{\sigma^2}{c^2} \geq p$, from where we obtain $c \geq \frac{\sigma}{\sqrt{1-p}}$, and hence $c = \frac{\sigma}{\sqrt{1-p}}$.

(ii) Here $c = \frac{0.5}{\sqrt{0.05}} \simeq 0.5 \times 4.47 = 2.235$.

Exercises.

2.1 Suppose the distribution of the r.v. X is given by the following table:

x	-1	0	1
$f(x)$	$1/18$	$8/9$	$1/18$

(i) Calculate the EX (call it μ), the $Var(X)$, and the s.d. of X (call it σ).
(ii) Compute the probability: $P(|X - \mu| \geq k\sigma)$ for $k = 2, 3$.
(iii) By the Tchebichev inequality: $P(|X - \mu| \geq k\sigma) \leq 1/k^2$. Compare the exact probabilities computed in part (ii) with the respective upper bounds.

2.2 If X is an r.v. with expectation μ and s.d. σ, use the Tchebichev inequality:

(i) To determine c in terms of σ and α, so that:

$$P(|X - \mu| < c) \geq \alpha \qquad (0 < \alpha < 1).$$

(ii) Give the numerical value of c for $\sigma = 1$ and $\alpha = 0.95$.

2.3 Let X be an r.v. with p.d.f. $f(x) = c(1 - x^2)$, for $-1 \leq x \leq 1$. Then, by Exercise 3.11(i) in Chapter 3, $c = 3/4$. Do the following:

(i) Calculate the EX and $Var(X)$.
(ii) Use the Tchebichev inequality to find a lower bound for the probability $P(-0.9 < X < 0.9)$, and compare it with the exact probability calculated in Exercise 3.11(ii) in Chapter 3.

2.4 Let X be an r.v. with (finite) mean μ and variance 0. Then:

(i) Use the Tchebichev inequality to show that $P(|X - \mu| \geq c) = 0$ for all $c > 0$.
(ii) Use part (i) and Theorem 2 in Chapter 3 in order to conclude that $P(X = \mu) = 1$.

5.3 Median and Mode of a Random Variable

Although the mean of an r.v. X does specify the center of location of the distribution of X, sometimes this is not what we actually wish to know. A case in point is the distribution of yearly income in a community (e.g., in a state or in a country). For the sake of illustration, consider the following (rather) extreme example. A community consisting of 10 households comprises 1 household with yearly income \$500,000 and 9 households with respective yearly incomes $x_i = \$20{,}000 + \$1{,}000(i - 1)$, $i = 1, \ldots, 9$, $x_0 = \$500{,}000$. Defining the r.v. X to take the values x_i, $i = 1, \ldots, 9$ and $x_0 = \$500{,}000$ with respective probabilities 0.10, we obtain: $EX = \$71{,}600$. Thus, the average yearly income in this community would be \$71,600, significantly above the national average yearly income, which would indicate a rather prosperous community. The reality, however, is that this community is highly stratified, and the expectation does not reveal this characteristic. What is more appropriate for cases like this are numerical characteristics of a distribution known as median or, more generally, percentiles or quantiles.

The median of the distribution of an r.v. X is usually defined as a point, denoted by $x_{0.50}$, for which

$$P(X \leq x_{0.50}) \geq 0.50 \quad \text{and} \quad P(X \geq x_{0.50}) \geq 0.50, \tag{5.19}$$

or, equivalently,

$$P(X < x_{0.50}) \leq 0.50 \quad \text{and} \quad P(X \leq x_{0.50}) \geq 0.50. \tag{5.20}$$

If the underlying distribution is continuous, the median is (essentially) unique and may be simply defined by:

$$P(X \leq x_{0.50}) = P(X \geq x_{0.50}) = 0.50. \tag{5.21}$$

However, in the discrete case, relation (5.19) (or (5.20)) may not define the median in a unique manner, as the following example shows.

Example 8. Examine the median of the r.v. X distributed as follows.

x	1	2	3	4	5	6	7	8	9	10
$f(x)$	2/32	1/32	5/32	3/32	4/32	1/32	2/32	6/32	2/32	6/32

Discussion. We have $P(X \leq 6) = 16/32 = 0.50 \geq 0.50$ and $P(X \geq 6) = 17/32 > 0.05 \geq 0.50$, so that (5.19) is satisfied. Also,

$$P(X \leq 7) = 18/32 > 0.50 \geq 0.50 \quad \text{and} \quad P(X \geq 7) = 16/32 = 0.50 \geq 0.50,$$

so that (5.19) is satisfied again. However, if we define the median as the point $(6 + 7)/2 = 6.5$, then $P(X \leq 6.5) = P(X \geq 6.5) = 0.50$, as (5.19) requires, and the median is uniquely defined.

Relations (5.19) and (5.20) and Example 8 suggest the following definition of the median.

Definition 5. The *median* of the distribution of a continuous r.v. X is the (essentially) unique point $x_{0.50}$ defined by (5.21). For the discrete case, consider two cases: Let x_k be the value for which $P(X \leq x_k) = 0.50$, if such a value exists. Then the *median* is defined to be the midpoint between x_k and x_{k+1}; that is, $x_{0.50} = (x_k + x_{k+1})/2$. If there is no such value, the *median* is defined by the relations: $P(X < x_{0.50}) < 0.50$ and $P(X \leq x_{0.50}) > 0.50$ (or $P(X \leq x_{0.50}) > 0.50$ and $P(X \geq x_{0.50}) > 0.50$).

Thus, in Example 9 below, $x_{0.50} = 6$, because $P(X < 6) = P(X \leq 5) = 15/32 < 0.50$ and $P(X \leq 6) = 17/32 > 0.50$.

Example 9. Determine the median of the r.v. X distributed as follows.

x	1	2	3	4	5	6	7	8	9	10
f(x)	2/32	1/32	2/32	6/32	4/32	2/32	1/32	7/32	1/32	6/32

For the yearly incomes of the 10 households considered at the beginning of this section, the median is \$24,500 by the first part of Definition 5 regarding the discrete case.

More generally, the pth quantile is defined as follows.

Definition 6. For any p with $0 < p < 1$, the pth *quantile* of the distribution of a r.v. X, denoted by x_p, is defined as follows: If X is continuous, then the (essentially) unique x_p is defined by:

$$P(X \leq x_p) = p \quad \text{and} \quad P(X \geq x_p) = 1 - p.$$

For the discrete case, consider two cases: Let x_k be the value for which $P(X \leq x_k) = p$, if such a value exists. Then the pth *quantile* is defined to be the midpoint between x_k and x_{k+1}; that is, $x_p = (x_k + x_{k+1})/2$. If there is no such value, the (unique) pth *quantile* is defined by the relation: $P(X < x_p) < p$ and $P(X \leq x_p) > p$ (or $P(X \leq x_p) > p$ and $P(X \geq x_p) > 1 - p$).

For $p = 0.25$, $x_{0.25}$ is called the first *quartile*, and for $p = 0.75$, $x_{0.75}$ is the third *quartile*. For $p = 0.50$, we revert to the median.

Thus, the pth quantile is a point x_p, which divides the distribution of X into two parts, and $(-\infty, x_p]$ contains exactly $100p\%$ (or at least $100p\%$) of the distribution, and $[x_p, \infty)$ contains exactly $100(1-p)\%$ (or at least $100(1-p)\%$) of the distribution of X.

Another numerical characteristic which helps shed some light on the distribution of an r.v. X is the so-called mode.

Definition 7. A *mode* of the distribution of an r.v. X is any point, if such points exist, which maximizes the p.d.f. of X, f.

A mode, being defined as a maximizing point, is subject to all the shortcomings of maximization: It may not exist at all; it may exist but is not obtainable in closed form; there may be more than one mode (the distribution is a *multimodal* one). It may also happen that there is a unique mode (*unimodal* distribution). Clearly, if a mode exists, it will be of particular importance for discrete distributions, as the modes provide the values of the r.v. X that occur with the largest probability. In the continuous case, the interpretation is that if a (small) interval I is centered at the mode of a p.d.f., then the probability that X takes values in this I attains its maximum value as it compares with any other location of the interval I. See also Figures 5.2 to 5.4 below.

Exercises.

3.1 Let X be an r.v. with p.d.f. $f(x) = 3x^2$, for $0 \leq x \leq 1$.

(i) Calculate the EX and the median of X and compare them.

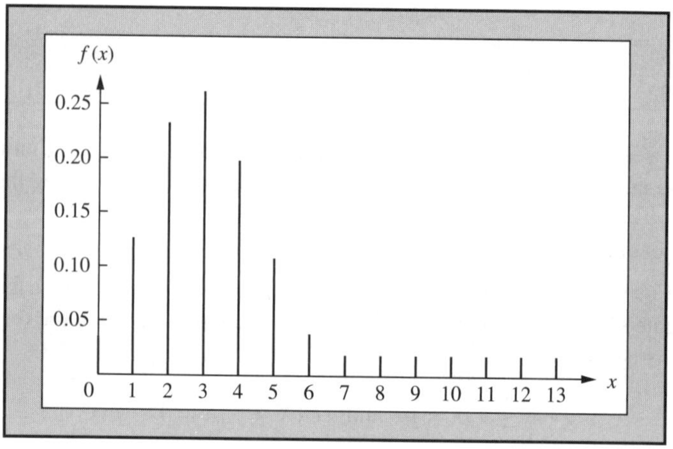

FIGURE 5.2
Graph of a unimodal (discrete) p.d.f.

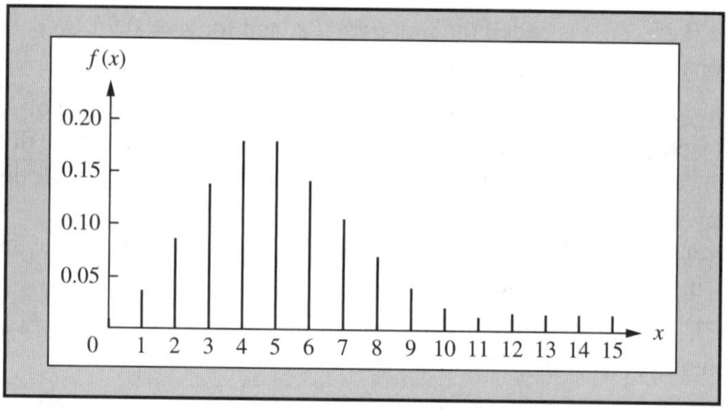

FIGURE 5.3
Graph of a bimodal (discrete) p.d.f.

(ii) Determine the 0.125-quantile of X.

3.2 Let X be an r.v. with p.d.f. $f(x) = x^n$, for $0 \leq x \leq c$ (n positive integer), and let $0 < p < 1$. Determine:

(i) The pth quantile x_p of X in terms of n and p.
(ii) The median $x_{0.50}$ for $n = 3$.

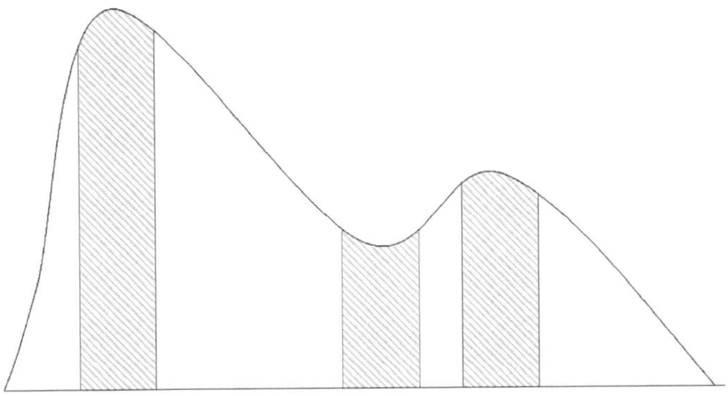

FIGURE 5.4
Graph of a unimodal (continuous) p.d.f. Probabilities corresponding to an interval I centered at the mode and another two locations.

3.3 **(i)** If the r.v. X has p.d.f. $f(x) = \lambda e^{-\lambda x}$, for $x > 0$ ($\lambda > 0$), determine the pth quantile x_p in terms of λ and p.
(ii) What is the numerical value of x_p for $\lambda = \frac{1}{10}$ and $p = 0.25$?

3.4 Let X be an r.v. with p.d.f. f given by:

$$f(x) = \begin{cases} c_1 x^2, & -1 \leq x \leq 0 \\ c_2(1 - x^2), & 0 < x \leq 1 \\ 0, & \text{otherwise.} \end{cases}$$

(i) If it is also given that $EX = 0$, determine the constants c_1 and c_2.
(ii) Determine the $\frac{1}{3}$-quantile of the distribution.

Hint. In part (i), two relations are needed for the determination of c_1, c_2.

3.5 Let X be an r.v. with d.f. F given by:

$$F(x) = \begin{cases} 0, & x \leq 0 \\ \frac{3}{4}(x^2 - \frac{1}{3}x^3), & 0 < x \leq 2 \\ 1, & x > 2. \end{cases}$$

(i) Find the corresponding p.d.f.
(ii) Determine the mode of the p.d.f.
(iii) Show that $\frac{1}{2}$ is the $\frac{5}{32} = 0.15625$-quantile of the distribution.

Hint. For part (i), use Proposition 5(iii) in Chapter 3.

3.6 Two fair and distinct dice are rolled once, and let X be the r.v. denoting the sum of the numbers shown on the uppermost sides, so that the possible values of X are: $2, 3, \ldots, 12$.

(i) Derive the p.d.f. f of the r.v. X.
(ii) Compute the EX.
(iii) Find the median of f, as well as its mode.

3.7 Determine the modes of the following p.d.f.'s:

(i) $f(x) = (\frac{1}{2})^x$, $x = 1, 2, \ldots$
(ii) $f(x) = (1 - \alpha)^x$, $x = 1, 2, \ldots$ $(0 < \alpha < 1)$. Also, what is the value of α?
(iii) $f(x) = \frac{2}{3^{x+1}}$, $x = 0, 1, \ldots$

3.8 Let X be an r.v. of the continuous type with p.d.f. f symmetric about a constant c (i.e., $f(c-x) = f(c+x)$ for all x; in particular, if $c = 0$, then $f(-x) = f(x)$ for all x). Then show that c is the median of X.

Hint. Start with $P(X \leq c) = \int_{-\infty}^{c} f(x)dx$ and, by making a change of the variable x, show that this last integral equals $\int_0^\infty f(c - y)dy$. Likewise, $P(X \geq c) = \int_c^\infty f(x)dx$, and a change of the variable x leads to the integral $\int_0^\infty f(c + y)dy$. Then the use of symmetry completes the proof.

3.9 Let X be an r.v. of the continuous type with p.d.f. f, with finite expectation, and median m, and let c be any constant. Then:

(i) Show that:

$$E|X - c| = E|X - m| + 2\int_m^c (c - x)f(x)dx.$$

(ii) Use part (i) to conclude that the constant c which minimizes the $E|X - c|$ is $c = m$.

Hint. For $m < c$, show that:

$$|x - c| - |x - m| = \begin{cases} c - m, & x < m \\ c + m - 2x, & m \leq x \leq c \\ m - c, & x > c. \end{cases}$$

Then

$$\begin{aligned} E|X - c| - E|X - m| &= \int_{-\infty}^{m} (c - m)f(x)\,dx + \int_m^c (c + m - 2x)f(x)\,dx \\ &\quad + \int_c^\infty (m - c)f(x)\,dx \\ &= \frac{c - m}{2} + (c + m)\int_m^c f(x)\,dx - 2\int_m^c xf(x)\,dx \\ &\quad + (m - c)\int_m^\infty f(x)\,dx - (m - c)\int_m^c f(x)\,dx \end{aligned}$$

$$= \frac{c-m}{2} + \frac{m-c}{2} + 2c\int_m^c f(x)\,dx - 2\int_m^c xf(x)\,dx$$
$$= 2\int_m^c (c-x)f(x)\,dx.$$

For $m \geq c$, show that:

$$|x-c| - |x-m| = \begin{cases} c-m, & x < c \\ -c-m+2x, & c \leq x \leq m \\ m-c, & x > m. \end{cases}$$

Then

$$\begin{aligned} E|X-c| - E|X-m| &= \int_{-\infty}^c (c-m)f(x)\,dx + \int_c^m (-c-m+2x)f(x)\,dx \\ &\quad + \int_m^\infty (m-c)f(x)\,dx \\ &= (c-m)\int_{-\infty}^m f(x)\,dx - (c-m)\int_c^m f(x)\,dx \\ &\quad - (c+m)\int_c^m f(x)\,dx + 2\int_c^m xf(x)\,dx \\ &\quad + (m-c)\int_m^\infty f(x)\,dx \\ &= \frac{c-m}{2} + \frac{m-c}{2} - 2c\int_c^m f(x)\,dx + 2\int_c^m xf(x)\,dx \\ &= -2\int_c^m (c-x)f(x)\,dx = 2\int_m^c (c-x)f(x)\,dx. \end{aligned}$$

Combining the two results, we get

$$E|X-c| = E|X-m| + 2\int_m^c (c-x)f(x)dx.$$

3.10 Let X be a continuous r.v. with pth quantile x_p, and let $Y = g(X)$, where g is a strictly increasing function, so that the inverse g^{-1} exists (and is also strictly increasing). Let y_p be the pth quantile of the r.v. Y.

(i) Show that the pth quantile of Y is given by $g(x_p)$; i.e., $y_p = g(x_p)$.
(ii) If the p.d.f. of X is $f(x) = e^{-x}$, $x > 0$, determine x_p in terms of p.
(iii) Find the corresponding y_p of part (i).

CHAPTER

Some Special Distributions 6

This chapter is devoted to discussing some of the most commonly occurring distributions: They are the binomial, geometric, Poisson, hypergeometric, gamma (negative exponential and chi-square), normal, and uniform distributions. In all cases, the mathematical expectation, variance, and moment-generating function (m.g.f.) involved are presented. In the case of the binomial and the Poisson distributions, the mode(s) are also established.

The chapter consists of two sections. In the first section, the four discrete distributions are discussed, whereas the second section is devoted to the continuous distributions.

6.1 Some Special Discrete Distributions

In this section, we discuss four discrete distributions, which occur often. These are the binomial, the geometric, the Poisson, and the hypergeometric. At this point, it should be mentioned that a p.d.f. is 0 for all the values of its argument not figuring in its definition.

6.1.1 Binomial Distribution

We first introduced the concept of *a binomial experiment*, which is meant to result in two possible outcomes, one termed a *success*, denoted by S and occurring with probability p, and the other termed a *failure*, denoted by F and occurring with probability $q = 1 - p$. A binomial experiment is performed n independent times (with p remaining the same), and let X be the r.v. denoting the number of successes. Then, clearly, X takes on the values $0, 1, \ldots, n$, with the respective probabilities:

$$P(X = x) = f(x) = \binom{n}{x} p^x q^{n-x}, \quad x = 0, 1, \ldots, n, 0 < p < 1, \; q = 1 - p. \tag{6.1}$$

The r.v. X is said to be *binomially* distributed, its distribution is called *binomial* with *parameters* n and p, and the fact that X is so distributed is denoted by $X \sim B(n, p)$. That $\sum_{x=0}^{n} \binom{n}{x} p^x q^{n-x} = 1$ is immediate, since the left-hand side is the expansion of $(p + q)^n = 1^n = 1$ (see #2 in Table 6 in the Appendix).

Introduction to Probability, Second Edition. http://dx.doi.org/10.1016/B978-0-12-800041-0.00006-7

Remark 1. Simple examples of a binomially distributed r.v. X is the case where X represents the number of H's of a coin tossed n independent times; or X represents the number of times a 6 appears when a die is rolled independently n times; or X represents the number of times an ace appears when a card is drawn at random and with replacement from a standard deck of 52 playing cards; or X represents the number of defective items produced by a manufacturing process during a day's work; or X represents the number of those voters out of n interviewed who favor a certain legislative proposal, etc.

It is to be observed that the outcomes of an experiment do not have to be only two literally for the binomial model to apply. They can be as many as they may, but we can group them into two disjoint groups. Then the occurrence of a member in one of the groups is termed a success arbitrarily, and the occurrence of a member in the other group is characterized as a failure. Thus, in reference to Example 16 in Chapter 1, the selection of a person of blood type A or B may be called a success, so that the selection of a person of blood type AB or O will be a failure.

The graph of f depends on n and p; two typical cases, for $n = 12$, $p = \frac{1}{4}$, and $n = 10$, $p = \frac{1}{2}$ are given in Figures 6.1 and 6.2.

Values of the p.d.f. f of the $B(12, \frac{1}{4})$ distribution	
$f(0) = 0.0317$	$f(7) = 0.0115$
$f(1) = 0.1267$	$f(8) = 0.0024$
$f(2) = 0.2323$	$f(9) = 0.0004$
$f(3) = 0.2581$	$f(10) = 0.0000$
$f(4) = 0.1936$	$f(11) = 0.0000$
$f(5) = 0.1032$	$f(12) = 0.0000$
$f(6) = 0.0401$	

Values of the p.d.f. f of the $B(10, \frac{1}{2})$ distribution	
$f(0) = 0.0010$	$f(6) = 0.2051$
$f(1) = 0.0097$	$f(7) = 0.1172$
$f(2) = 0.0440$	$f(8) = 0.0440$
$f(3) = 0.1172$	$f(9) = 0.0097$
$f(4) = 0.2051$	$f(10) = 0.0010$
$f(5) = 0.2460$	

For selected n and p, the d.f. $F(k) = \sum_{j=0}^{k} \binom{n}{j} p^j q^{n-j}$ is given by the binomial tables (see, however, Exercise 1.1). The individual probabilities $\binom{n}{j} p^j q^{n-j}$ may be found by subtraction. Alternatively, such probabilities can be calculated recursively (see Exercise 1.9).

For $n = 1$, the corresponding r.v. is known as the *Bernoulli* r.v. It is then clear that a $B(n, p)$ r.v. X is the sum of $nB(1, p)$ r.v.'s. More precisely, in n independent binomial experiments, associate with the ith performance of the experiment the r.v. X_i defined by: $X_i = 1$ if the outcome is S (a success) and $X_i = 0$ otherwise, $i = 1, \dots, n$.

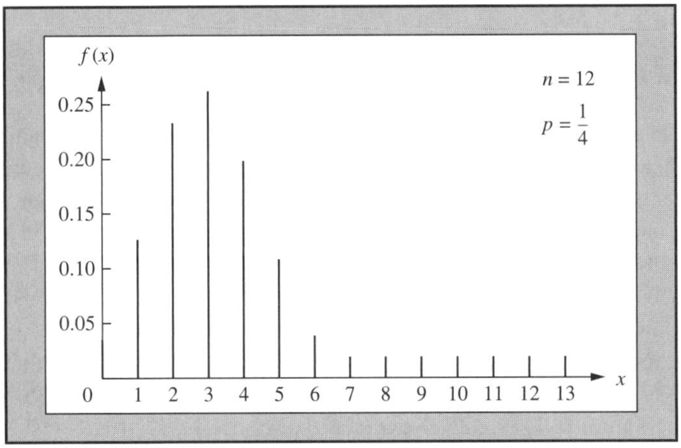

FIGURE 6.1

Graph of the p.d.f. of the binomial distribution for $n = 12$, $p = \frac{1}{4}$.

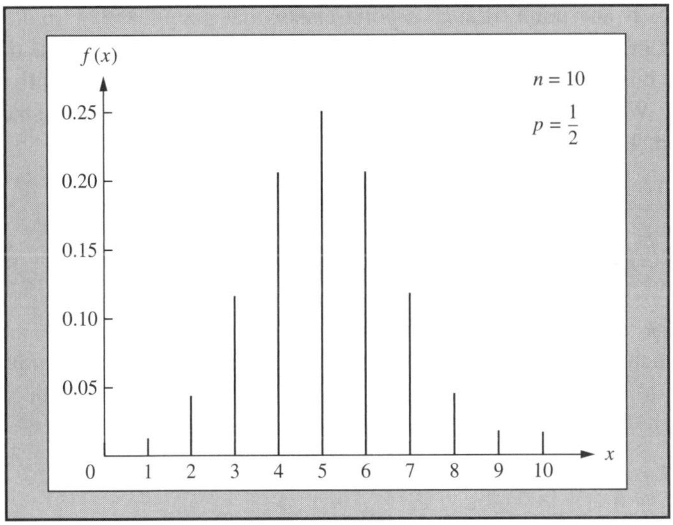

FIGURE 6.2

Graph of the p.d.f. of the binomial distribution for $n = 10$, $p = \frac{1}{2}$.

Then, clearly, $\sum_{i=1}^{n} X_i$ is the number of 1's in the n trials, or, equivalently, the number of S's, which is exactly what the r.v. X stands for. Thus, $X = \sum_{i=1}^{n} X_i$. Finally, it is mentioned here that if $X \sim B(n, p)$, then:

Proposition 1.

$$EX = np, \quad Var(X) = npq, \quad \text{and} \quad M_X(t) = (pe^t + q)^n, \quad t \in \Re. \tag{6.2}$$

The relevant derivations are left as exercises (see Exercises 1.10 and 1.11). A brief justification of formula (6.1) is as follows: Think of the n outcomes of the n experiments as n points on a straight line segment, where an S or an F is to be placed. By independence, the probability that there will be exactly x S's in x *specified* positions (and therefore $n - x$ F's in the remaining positions) is $p^x q^{n-x}$, and this probability is independent of the locations where the x S's occur. Because there are $\binom{n}{x}$ ways of selected x points for the S's, the conclusion follows.

Finally, for illustrative purposes, refer to Example 7 in Chapter 1. In that example, clearly, $X \sim B(n, 0.75)$, and for the sake of specificity take $n = 25$, so that X takes on the values $0, 1, \ldots, 25$. Next (see Exercise 1.1), $\binom{25}{x}(0.75)^x(0.25)^{25-x} = \binom{25}{y}(0.25)^y(0.75)^{25-y}$, where $y = 25 - x$. Therefore, for $a = 15$ and $b = 20$, for example, $P(15 \le X \le 20) = \sum_{y=5}^{10}\binom{25}{y}(0.25)^y(0.75)^{25-y} = 0.9703 - 0.2137 = 0.7566$. Finally, $EX = 25 \times 0.75 = 18.75$, $Var(X) = 25 \times 0.75 \times 0.25 = 4.6875$, so that $\sigma(X) \simeq 2.165$. Examples 8 and 9 in Chapter 1 fit into the same framework.

At this point, recall (Definition 7 in Chapter 5) that a mode of the distribution of an r.v. X is any point, if such a point exists, that maximizes the p.d.f. of X, f. Clearly, a mode, if it exists, will be of particular importance for a discrete distribution such as the binomial, as the modes provide the values of X that occur with the largest probability. With this in mind, let us proceed in determining the modes of the binomial distribution. To this effect, we have:

Theorem 1. Let X be $B(n, p)$; that is,

$$f(x) = \binom{n}{x} p^x q^{n-x}, \quad x = 0, 1, \ldots, n, \; 0 < p < 1, \; q = 1 - p.$$

Consider the number $(n+1)p$ and set $m = [(n+1)p]$, where $[y]$ denotes the largest integer, which is $\le y$. Then, if $(n+1)p$ is *not* an integer, $f(x)$ has a unique mode at $x = m$. If $(n+1)p$ *is* an integer, then $f(x)$ has two modes obtained for $x = m$ and $x = m - 1$.

Proof. For $x \ge 1$, we have:

$$\frac{f(x)}{f(x-1)} = \frac{\binom{n}{x} p^x q^{n-x}}{\binom{n}{x-1} p^{x-1} q^{n-x+1}}$$

$$= \frac{\frac{n!}{x!(n-x)!} p^x q^{n-x}}{\frac{n!}{(x-1)!(n-x+1)!} p^{x-1} q^{n-x+1}} = \frac{n-x+1}{x} \times \frac{p}{q}.$$

That is,

$$\frac{f(x)}{f(x-1)} = \frac{n-x+1}{x} \times \frac{p}{q}.$$

Hence,

$$f(x) > f(x-1) \text{ if and only if } \frac{f(x)}{f(x-1)} > 1 \text{ if and only if}$$
$$\frac{n-x+1}{x} \times \frac{p}{q} > 1 \text{ if and only if } x < (n+1)p. \tag{6.3}$$

Also,

$$f(x) < f(x-1) \text{ if and only if } \frac{f(x)}{f(x-1)} < 1 \text{ if and only if}$$
$$\frac{n-x+1}{x} \times \frac{p}{q} < 1 \text{ if and only if } x > (n+1)p. \tag{6.4}$$

Finally,

$$f(x) = f(x-1) \text{ if and only if } \frac{f(x)}{f(x-1)} = 1 \text{ if and only if}$$
$$\frac{n-x+1}{x} \times \frac{p}{q} = 1 \text{ if and only if } x = (n+1)p. \tag{6.5}$$

First, consider the case that $(n+1)p$ is *not* an integer. Then we have the following diagram.

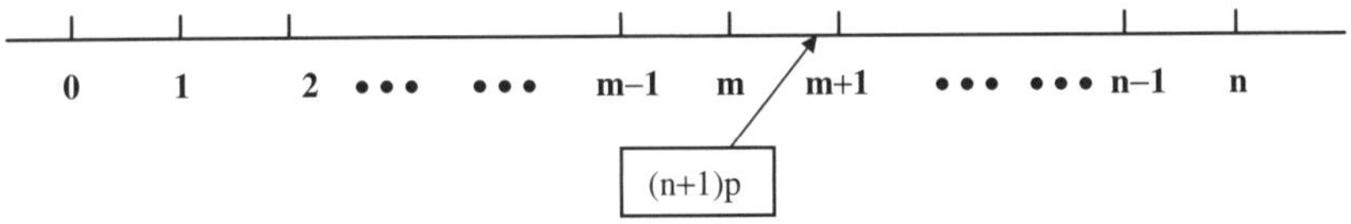

From relations (6.3) and (6.4), it follows that:

$$f(x) < f(m), x = 0, 1, \ldots, m-1; \quad f(x) < f(m), x = m+1, \ldots, n,$$

so that $f(x)$ attains its unique maximum at $x = m$.

Now, consider the case that $(n+1)p$ *is* an integer and look at the following diagram.

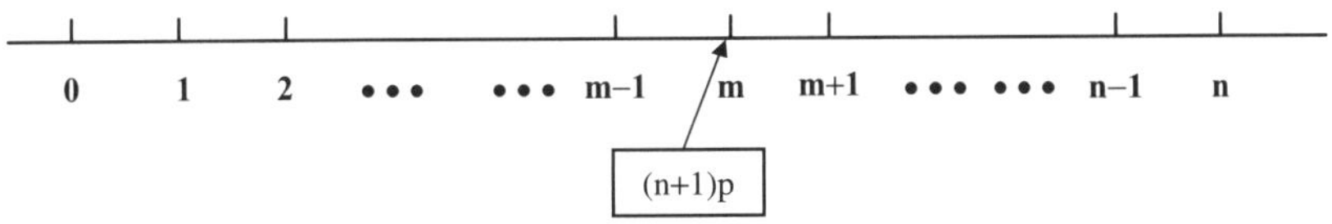

From relations (6.5) we have $f(m) = f(m-1)$, whereas from relations (6.3) and (6.4), we have that:

$$f(x) < f(m-1), x = 0, 1, \ldots, m-2; \quad f(x) < f(m), x = m+1, \ldots, n,$$

so that $f(x)$ has two maxima at $x = m$ and $x = m-1$. ▲

As an illustration, consider the following simple example.

Example 1. Let $X \sim B(n, p)$ with $n = 20$ and $p = \frac{1}{4}$. Then $(n+1)p = \frac{21}{4}$ is not an integer and therefore there is a unique mode. Since $\frac{21}{4} = 5.25$, the mode is $[5.25] = 5$. The maximum probability is $\binom{20}{5}(0.25)^5(0.75)^{15} = 0.2024$. If $n = 15$ and $p = \frac{1}{4}$, then $(n+1)p = \frac{16}{4} = 4$ and therefore there are two modes; they are 4 and 3. The respective maximum probability is $\binom{15}{4}(0.25)^4(0.75)^{11} = 0.2252$.

The discussion of binomial distribution concludes with an example of determination of the median and the quartiles of the distribution (see Definitions 5 and 6 in Chapter 5).

Example 2. Refer to Figure 6.1 ($B(12, 1/4)$) and determine $x_{0.25}$, $x_{0.50}$, and $x_{0.75}$.

Discussion. Here $x_{0.25} = 2$, since $P(X < 2) = P(X = 0) + P(X = 1) = 0.1584 < 0.25$ and $P(X \leq 2) = 0.1584 + P(X = 2) = 0.3907 > 0.25$. Likewise, $x_{0.50} = 3$, since $P(X < 3) = 0.3907 \leq 0.50$ and $P(X \leq 3) = 0.6488 \geq 0.50$. Finally, $x_{0.75} = 4$, since $P(X < 4) = 0.6488 \leq 0.75$ and $P(X \leq 4) = 0.8424 > 0.75$.

6.1.2 Geometric Distribution

This discrete distribution arises in a binomial-experiment situation when trials are carried out independently (with constant probability p of an S) until the *first* S occurs. The r.v. X denoting the number of required trials is a *geometrically* distributed r.v. with *parameter* p and its distribution is *geometric* with *parameter* p. It is clear that X takes on the values $1, 2, \ldots$ with the respective probabilities:

$$P(X = x) = f(x) = pq^{x-1}, \quad x = 1, 2, \ldots, \quad 0 < p < 1, \ q = 1 - p. \tag{6.6}$$

The justification of this formula is immediate because, if the first S is to appear in the xth position, the overall outcome is $\underbrace{FF \cdots F}_{x-1} S$, whose probability (by independence) is $q^{x-1}p$.

The algebraic verification that $\sum_{x=1}^{\infty} pq^{x-1} = 1$ is immediate, since $\sum_{x=1}^{\infty} pq^{x-1} = p\sum_{x=1}^{\infty} q^{x-1} = p \times \frac{1}{1-q} = \frac{p}{p} = 1$ (see also #4 in Table 6 in the Appendix).

Values of $f(x) = (0.25)(0.75)^{x-1}$, $x = 1, 2, \ldots$	Values of $f(x) = (0.5)^x$, $x = 1, 2, \ldots$
$f(1) = 0.2500$	$f(1) = 0.5000$
$f(2) = 0.1875$	$f(2) = 0.2500$
$f(3) = 0.1406$	$f(3) = 0.1250$
$f(4) = 0.1055$	$f(4) = 0.0625$
$f(5) = 0.0791$	$f(5) = 0.0313$
$f(6) = 0.0593$	$f(6) = 0.0156$
$f(7) = 0.0445$	$f(7) = 0.0078$
$f(8) = 0.0334$	
$f(9) = 0.0250$	
$f(10) = 0.0188$	

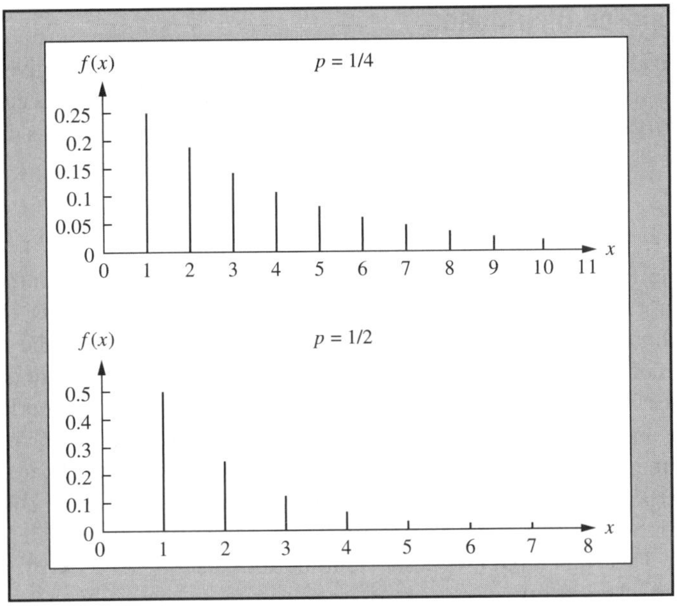

FIGURE 6.3

Graphs of the p.d.f.'s of geometric distribution with $p = \frac{1}{4}$ and $p = \frac{1}{2}$.

For some concrete examples, refer to Remark 1, and let X be the number of (independent) trials until the first success occurs.

The graph of f depends on p; two typical cases for $p = \frac{1}{4}$ and $p = \frac{1}{2}$ are given in Figure 6.3.

If the r.v. X is geometrically distributed with parameter p, then:

Proposition 2.

$$EX = \frac{1}{p}, \quad Var(X) = \frac{q}{p^2}, \quad M_X(t) = \frac{pe^t}{1 - qe^t}, \quad t < -\log q. \tag{6.7}$$

The verification of (6.7) is left as an exercise (see Exercises 1.17 and 1.18).

Remark 2. Sometimes the p.d.f. of X is given in the form: $f(x) = pq^x, x = 0, 1, \ldots$; then $EX = \frac{q}{p}$, $Var(X) = \frac{p}{q^2}$ and $M_X(t) = \frac{p}{1-qe^t}, t < -\log q$.

In reference to Example 10 in Chapter 1, assume for mathematical convenience that the number of cars passing by may be infinite. Then the r.v. X described there has geometric distribution with some p. Here probabilities are easily calculated. For example, $P(X \geq 20) = \sum_{x=20}^{\infty} pq^{x-1} = pq^{19} \sum_{x=0}^{\infty} q^x = pq^{19} \frac{1}{1-q} = q^{19}$; that is, $p(X \geq 20) = q^{19}$. For instance, if $p = 0.01$, then $q = 0.99$ and $P(X \geq 20) \simeq 0.826$. (See also #4 in Table 6 in the Appendix.)

6.1.3 Poisson Distribution

An r.v. X taking on the values $0, 1, \ldots$ with respective probabilities given in (6.8) is said to have *Poisson distribution* with *parameter* λ; its distribution is called *Poisson* with *parameter* λ. That X is Poisson distributed with parameter λ is denoted by $X \sim P(\lambda)$.

$$P(X = x) = f(x) = e^{-\lambda}\frac{\lambda^x}{x!}, \quad x = 0, 1, \ldots, \lambda > 0. \tag{6.8}$$

Example 11 in Chapter 1 may serve as an illustration of the usage of the Poisson distribution. Assuming, for mathematical convenience, that the number of bacteria may be infinite, then the Poisson distribution may be used to describe the actual distribution of bacteria (for a suitable value of λ) quite accurately. There is a host of similar cases for the description of which the Poisson distribution is appropriate. These include the number of telephone calls served by a certain telephone exchange center within a certain period of time, the number of particles emitted by a radioactive source within a certain period of time, the number of typographical errors in a book, etc.

The graph of f depends on λ; for example, for $\lambda = 5$, the graph looks like Figure 6.4. That f is a p.d.f. follows from the formula $\sum_{x=0}^{\infty} \frac{\lambda^x}{x!} = e^{\lambda}$ (see #6 in Table 6 in the Appendix).

For selected values of λ, the d.f. $F(k) = \sum_{j=0}^{k} e^{-\lambda}\frac{\lambda^j}{j!}$ is given by the Poisson tables. The individual values $e^{-\lambda}\frac{\lambda^j}{j!}$ are found by subtraction. Alternatively, such probabilities can be calculated recursively (see Exercise 1.21). It is not hard to see (see Exercises 1.22 and 1.23) that if $X \sim P(\lambda)$, then:

Proposition 3.

$$EX = \lambda, \quad Var(X) = \lambda, \quad \text{and} \quad M_X(t) = e^{\lambda e^t - \lambda}, \quad t \in \Re. \tag{6.9}$$

(See Exercises 1.22 and 1.23.)

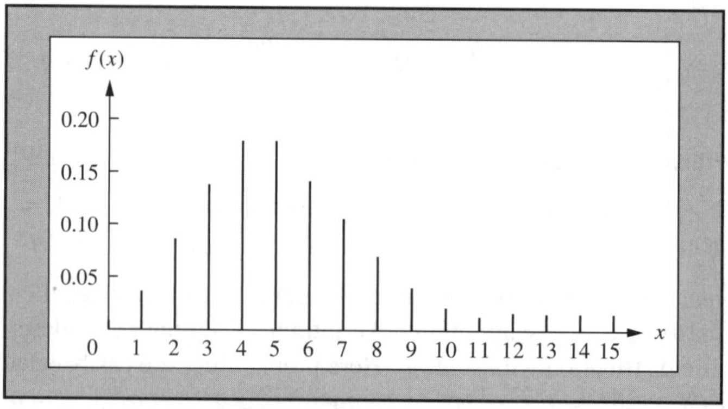

FIGURE 6.4
Graph of the p.d.f. of Poisson distribution with $\lambda = 5$.

Values of the p.d.f. f of the $P(5)$ distribution	
$f(0) = 0.0067$	$f(9) = 0.0363$
$f(1) = 0.0337$	$f(10) = 0.0181$
$f(2) = 0.0843$	$f(11) = 0.0082$
$f(3) = 0.1403$	$f(12) = 0.0035$
$f(4) = 0.1755$	$f(13) = 0.0013$
$f(5) = 0.1755$	$f(14) = 0.0005$
$f(6) = 0.1462$	$f(15) = 0.0001$
$f(7) = 0.1044$	
$f(8) = 0.0653$	$f(n)$ is negligible for $n \geq 16$.

From these expressions, the parameter λ acquires a special meaning: it is both the mean and the variance of the r.v. X.

Regarding the modes of a Poisson distribution, there is a clear-cut way of determining them. This is the content of the theorem below.

Theorem 2. Let X be $P(\lambda)$; that is,

$$f(x) = e^{-\lambda}\frac{\lambda^x}{x!}, \quad x = 0, 1, 2, \ldots, \quad \lambda > 0.$$

Then, if λ is *not* an integer, $f(x)$ has a unique mode at $x = [\lambda]$. If λ *is* an integer, then $f(x)$ has two modes obtained for $x = \lambda$ and $x = \lambda - 1$.

Proof. It goes along the same lines as that of Theorem 1 about the binomial distribution. Briefly, for $x \geq 1$, we have:

$$\frac{f(x)}{f(x-1)} = \frac{e^{-\lambda}\lambda^x/x!}{e^{-\lambda}\lambda^{x-1}/(x-1)!} = \frac{\lambda}{x}.$$

Hence, $f(x) > f(x-1)$ if and only if $\lambda > x$, $f(x) < f(x-1)$ if and only if $\lambda < x$, and $f(x) = f(x-1)$ if and only if $x = \lambda$ in case λ *is* an integer. Thus, if λ is *not* an integer, $f(x)$ keeps increasing for $x \leq [\lambda]$ and then decreases. Thus the maximum of $f(x)$ occurs at $x = [\lambda]$. If λ *is* an integer, then the maximum occurs at $x = \lambda$. But in this case $f(x) = f(x-1)$, which implies that $x = \lambda - 1$ is a second point which gives the maximum value to the p.d.f. ▲

The following simple example serves as an illustration of Theorem 2.

Example 3. Let $X \sim P(\lambda)$ and let $\lambda = 4.5$. Then there is a unique mode that is $[4.5] = 4$. The respective maximum probability is 0.1898. If, on the other hand, $\lambda = 7$, then there are two modes 7 and 6. The respective maximum probability is 0.149.

Example 4. Refer to Figure 6.4 ($P(5)$) and determine $x_{0.25}$, $x_{0.50}$, and $x_{0.75}$.

As in Example 2, $x_{0.25} = 3$, $x_{0.50} = 5$, and $x_{0.75} = 6$.

There is an intimate relationship between Poisson and binomial distributions: the former may be obtained as the limit of the latter, as explained below. Namely, it is seen (see Exercise 1.25) that in the binomial, $B(n, p)$, situation, if n is large and p is small, then the binomial probabilities $\binom{n}{x}p^x(1-p)^{n-x}$ are close to the Poisson probabilities $e^{-np}\frac{(np)^x}{x!}$. More precisely, $\binom{n}{x}p_n^x(1-p_n)^{n-x} \to e^{-\lambda}\frac{\lambda^x}{x!}$, provided $n \to \infty$

and $p_n \to 0$ so that $np_n \to \lambda \in (0, \infty)$. Here p_n is the probability of a success in the nth trial. Thus, for large values of n, $\binom{n}{x} p_n^x (1-p_n)^{n-x} \simeq e^{-\lambda}\frac{\lambda^x}{x!}$; or, upon replacing λ by np_n, we obtain the approximation mentioned before.

A rough explanation as to why binomial probabilities are approximated by Poisson probabilities is given next. To this end, suppose an event A occurs once in a small time interval of length h with approximate probability proportional to h and coefficient of proportionally λ; that is, A occurs once in an interval of length h with approximate probability λh. It occurs two or more times with probability approximately 0, so that it occurs zero times with probability approximately $1 - \lambda h$. Finally, occurrences in nonoverlapping intervals of length h are independent. Next, consider the unit interval $(0, 1]$ and divide it into a large number n of nonoverlapping subintervals of equal length $h : (t_{i-1}, t_i], i = 1, \ldots, n, t_0 = 0, t_n = 1, h = \frac{1}{n}$. With the ith interval $(t_{i-1}, t_i]$, associate the r.v. X_i defined by: $X_i = 1$ with approximate probability λh and 0 with approximate probability $1 - \lambda h$. Then the r.v. $X = \sum_{i=1}^n X_i$ denotes the number of occurrences of A over the unit $(0, 1]$ interval with approximate probabilities $\binom{n}{x}(\lambda h)^x (1-\lambda h)^{n-x}$. The exact probabilities are found by letting $n \to \infty$ (which implies $h \to 0$). Because here $p_n = \lambda h$ and $np_n = n\lambda h = n\lambda\frac{1}{n} = \lambda$, we have that $\binom{n}{x}(\lambda h)^x (1-\lambda h)^{n-x} \to e^{-\lambda}\frac{\lambda^x}{x!}$, as $n \to \infty$ (by Exercise 1.25), so that the exact probabilities are $e^{-\lambda}\frac{\lambda^x}{x!}$. So, the exact probability that A occurs x times in $(0, 1]$ is the Poisson probability $e^{-\lambda}\frac{\lambda^x}{x!}$, and the approximate probability that A occurs the same number of times is the binomial probability $\binom{n}{x}(\lambda h)^x (1-\lambda h)^{n-x} = \binom{n}{x} p_n^x (1-p_n)^{n-x}$; these two probabilities are close to each other for large n.

The following example sheds some light on the approximation just discussed.

Example 5. If X is an r.v. distributed as $B(25, \frac{1}{16})$, we find from the binomial tables that $P(X = 2) = 0.2836$. Next, considering an r.v. Y distributed as $P(\lambda)$ with $\lambda = \frac{25}{16} = 1.5625$, we have that $P(Y = 2) = e^{-1.5625}\frac{(1.5625)^2}{2!} \simeq 0.2556$. Thus, the exact probability is underestimated by the amount 0.028. The error committed is of the order of 9.87%. Given the small value of $n = 25$, the approximate value is not bad at all.

6.1.4 Hypergeometric Distribution

This discrete distribution occurs quite often and is suitable in describing situations of the following type: m identical objects (e.g., balls) are thoroughly mixed with n identical objects (e.g., balls) that are distinct from the m objects. From these $m+n$ objects, r are drawn at random and *without replacement*, and let X be the number among the r that come from the m objects. Then the r.v. X takes on the values $0, 1, \ldots, \min(r, m)$ with respective probabilities given below. Actually, by defining $\binom{m}{x} = 0$ for $x > m$, we have:

$$P(X = x) = f(x) = \frac{\binom{m}{x}\binom{n}{r-x}}{\binom{m+n}{r}}, \quad x = 0, \ldots, r; \tag{6.10}$$

m and n may be referred to as the *parameters* of the distribution. By assuming that the selections of r objects out of the $m+n$ are all equally likely, there are $\binom{m+n}{r}$ ways of selecting these r objects, whereas there are $\binom{m}{x}$ ways of selecting x out of the m objects, and $\binom{n}{r-x}$ ways of selecting the remaining $r-x$ objects out of n objects. Thus, the probability that $X=x$ is as given in the preceding formula. The simple justification that these probabilities actually sum to 1 follows from Exercise 4.10 in Chapter 2. For large values of any one of m, n, and r, actual calculation of the probabilities in (6.10) may be quite involved. A recursive formula (see Exercise 1.28) facilitates significantly these calculations. The calculation of the expectation and of the variance of X is based on the same ideas as those used in Exercise 1.10 in calculating the EX and $Var(X)$ when $X \sim B(n, p)$. We omit the details and give the relevant formulas; namely,

$$EX = \frac{mr}{m+n}, \quad Var(X) = \frac{mnr(m+n-r)}{(m+n)^2(m+n-1)}. \tag{6.11}$$

Finally, by utilizing ideas and arguments similar to those employed in Exercise 1.25, it is shown that as m and $n \to \infty$ so that $\frac{m}{m+n} \to p \in (0,1)$, then $\binom{m}{x}\binom{n}{r-x}/\binom{m+n}{r}$ tends to $\binom{r}{x}p^x(1-p)^{r-x}$. Thus, for large values of m and n, hypergeometric probabilities $\binom{m}{x}\binom{n}{r-x}/\binom{m+n}{r}$ may be approximated by simpler binomial probabilities $\binom{r}{x}p_{m,n}^x(1-p_{m,n})^{r-x}$, where $p_{m,n} = \frac{m}{m+n}$. (See also Exercise 1.29.)

Example 6. As an application of formula (6.10) and the approximation discussed, take $m=70$, $n=90$, $r=25$, and $x=10$. Then:

$$f(10) = \frac{\binom{70}{10}\binom{90}{25-10}}{\binom{70+90}{25}} = \frac{\binom{70}{10}\binom{90}{15}}{\binom{160}{25}} \simeq 0.166,$$

after quite a few calculations. On the other hand, since $\frac{m}{m+n} = \frac{70}{160} = \frac{7}{16}$, the binomial tables give for the $B(25, \frac{7}{16})$ distribution: $\binom{25}{10}\left(\frac{7}{16}\right)^{10}\left(\frac{9}{16}\right)^{15} = 0.15$. Therefore, the approximation overestimates the exact probability by the amount 0.016. The error committed is of the order of 10.7%.

Remark 3. By Theorem 2 in Chapter 5, the moments of an r.v. completely determine its distribution (under certain conditions). This is certainly true in all four distributions discussed here. In the binomial, geometric, and Poisson cases, the expectation alone determines the distribution. In hypergeometric distribution, the first two moments do so (through the determination of m and n).

Exercises.

1.1 If $X \sim B(n, p)$ with $p > 0.5$, the binomial tables (in this book) cannot be used, even if n is suitable. This problem is resolved by the following result.

(i) If $X \sim B(n, p)$, show that $P(X = x) = P(Y = n - x)$, where $Y \sim B(n, q)(q = 1 - p)$.

(ii) Apply part (i) for $n = 20$, $p = 0.625$, and $x = 8$.

1.2 Let X be an r.v. distributed as $B(n, p)$, and recall that $P(X = x) = f(x) = \binom{n}{x} p^x q^{n-x}$, $x = 0, 1, \ldots, n$ $(q = 1 - p)$. Set $B(n, p; x) = f(x)$.

(i) By using the relationship $\binom{m+1}{y} = \binom{m}{y} + \binom{m}{y-1}$ (see Exercise 4.9 in Chapter 2), show that:

$$B(n + 1, p; x) = pB(n, p; x - 1) + qB(n, p; x).$$

(ii) By using this recursive relation of $B(n+1, p; .)$, calculate the probabilities $B(n, p; x)$ for $n = 26$, $p = 0.25$, and $x = 10$.

1.3 Someone buys one ticket in each of 50 lotteries, and suppose that each ticket has probability $1/100$ of winning a prize. Compute the probability that the person in question will win a prize:

(i) Exactly once.
(ii) At least once.

1.4 Suppose that 15 people, chosen at random from a (large) target population, are asked if they favor a certain proposal. If 43.75% of the target population favor the proposal, calculate the probability that:

(i) At least 5 of the 15 polled favor the proposal.
(ii) A majority of those polled favor the proposal.
(iii) Compute the expected number of those favoring the proposal, and the s.d. around this number.

1.5 A fair die is tossed independently 18 times, and the appearance of a 6 is called a success. Find the probability that:

(i) The number of successes is greater than the number of failures.
(ii) The number of successes is twice as large as the number of failures.
(iii) The number of failures is 3 times the number of successes.

1.6 Suppose you are throwing darts at a target and you hit the bull's eye with probability p. It is assumed that the trials are independent and that p remains constant throughout.

(i) If you throw darts 100 times, what is the probability that you hit the bull's eye at least 40 times?
(ii) How does this expression become for $p = 0.25$?
(iii) What is the expected number of hits, and what is the s.d. around this expected number?

(For parts (i) and (ii), just write down the correct formula.)

1.7 If $X \sim B(100, 1/4)$, use the Tchebichev inequality to determine a lower bound for the probability $P(|X - 25| < 10)$.

1.8 A manufacturing process produces defective items at the constant (but unknown to us) proportion p. Suppose that n items are sampled independently, and let X be the r.v. denoting the number of defective items among the n, so that $X \sim B(n, p)$. Use the Tchebichev inequality in order to determine the smallest value of the sample size n, so that: $P(|\frac{X}{n} - p| < 0.05\sqrt{pq}) \geq 0.95$ $(q = 1-p)$.

1.9 If $X \sim B(n, p)$ show that $f(x+1) = \frac{p}{q} \times \frac{n-x}{x+1} f(x), x = 0, 1, \ldots, n-1$, $(q = 1 - p)$ (so that probabilities can be computed recursively).

Hint. Write the combinations in terms of factorials, and make the appropriate grouping of terms.

1.10 If $X \sim B(n, p)$:

(i) Calculate the EX and the $E[X(X-1)]$.

(ii) Use part (i) and Exercise 1.6(ii) in Chapter 5 to calculate the $\lambda \times \lambda^{x-1} Var(X)$.

Hint. For part (i), observe that:

$$EX = \sum_{x=1}^{n} x \frac{n(n-1)!}{x(x-1)!(n-x)!} p^x q^{n-x} = np \sum_{x=1}^{n} \binom{n-1}{x-1} p^{x-1} q^{(n-1)-x}$$

$$= np \sum_{y=0}^{n-1} \binom{n-1}{y} p^y q^{(n-1)-y} = np,$$

$$\text{and } E[X(X-1)] = \sum_{x=2}^{n} x(x-1) \frac{n(n-1)(n-2)!}{x(x-1)(x-2)!(n-x)!} p^x q^{n-x}$$

$$= n(n-1)p^2 \sum_{x=2}^{n} \binom{n-2}{x-2} p^{x-2} q^{(n-2)-(x-2)}$$

$$= n(n-1)p^2 \sum_{y=0}^{n-2} \binom{n-2}{y} p^y q^{(n-2)-y} = n(n-1)p^2.$$

Also, use Exercise 1.6(ii) in Chapter 5.

1.11 If $X \sim B(n, p)$:

(i) Show that $M_X(t) = (pe^t + q)^n, t \in \Re$ $(q = 1 - p)$.

(ii) Use part (i) to rederive the EX and the $Var(X)$.

1.12 Let $X \sim B(100, 1/4)$ and suppose you were to bet on the observed value of X. On which value would you bet in terms of probabilities?

1.13 Let the r.v. X have the geometric p.d.f. $f(x) = pq^{x-1}, x = 1, 2, \ldots$ $(q = 1 - p)$.

(i) What is the probability that the first success will occur by the 10th trial? (Express it in terms of $q = 1 - p$.)

(ii) What is the numerical value of this probability for $p = 0.2$?

Hint. For part (i), see #4 in Table 6 in the Appendix.

1.14 A manufacturing process produces defective items at the rate of 1%. Let X be the r.v. denoting the number of (independent) trials required until the first defective item is produced. Then calculate the probability that X is not larger than 10.

Hint. See #4 in Table 6 in the Appendix.

1.15 A fair die is tossed repeatedly (and independently) until a 6 appears for the first time. Calculate the probability that:

(i) This happens on the 3rd toss.
(ii) At least 5 tosses will be needed.

Hint. For part (ii), see #4 in Table 6 in the Appendix.

1.16 A coin with probability p of falling heads is tossed repeatedly and independently until the first head appears.

(i) Determine the smallest number of tosses, n, required to have the first head appear by the nth time with prescribed probability α. (Determine n in terms of α and $q = 1 - p$.)
(ii) Determine the value of n for $\alpha = 0.95$ and $p = 0.25$ $(q = 0.75)$ and $p = 0.50(=q)$.

Hint. For part (ii), see #4 in Table 6 in the Appendix.

1.17 If X has geometric distribution; that is, $f(x) = pq^{x-1}$, for $x = 1, 2, \ldots$ $(q = 1 - p)$:

(i) Calculate the EX and the $E[X(X-1)]$.
(ii) Use part (i) and Exercise 1.6(ii) in Chapter 5 to calculate the $Var(X)$.

Hint. For part (i), refer to #5 in Table 6 in the Appendix.

1.18 If X has geometric distribution, then:

(i) Derive the m.g.f. of X and specify the range of its argument. (See #4 in Table 6 in the Appendix.)
(ii) Specify the m.g.f. for $p = 0.01$.
(iii) Employ the m.g.f. in order to derive: EX, EX^2, and $Var(X)$, and compute their values for $p = 0.01$.

1.19 Suppose that r.v. X is distributed as $P(\lambda)$; i.e., $f(x) = e^{-\lambda}\frac{\lambda^x}{x!}$, for $x = 0, 1, \ldots$, and that $f(2) = 2f(0)$. Determine the value of the parameter λ.

1.20 Let X be a Poisson-distributed r.v. with parameter λ, and suppose that $P(X = 0) = 0.1$. Calculate the probability $P(X = 5)$.

1.21 If $X \sim P(\lambda)$, show that $f(x+1) = \frac{\lambda}{x+1} f(x)$, $x = 0, 1, \ldots$ (so that probabilities can be computed recursively).

1.22 If $X \sim P(\lambda)$:

(i) Calculate the EX and the $E[X(X-1)]$.
(ii) Use part (i) and Exercise 1.6(ii) in Chapter 5 to calculate the $Var(X)$.

Hint. For part (i), observe that (by #6 in Table 6 in the Appendix):

$$EX = e^{-\lambda}\sum_{x=1}^{\infty} x \frac{\lambda \times \lambda^{x-1}}{x(x-1)!} = \lambda e^{-\lambda}\sum_{y=0}^{\infty}\frac{\lambda^y}{y!} = \lambda e^{-\lambda}e^{\lambda} = \lambda, \text{ and}$$

$$E[X(X-1)] = \lambda^2 e^{-\lambda}\sum_{x=2}^{\infty} x(x-1)\frac{\lambda^{x-2}}{x(x-1)(x-2)!} = \lambda^2 e^{-\lambda}\sum_{y=0}^{\infty}\frac{\lambda^y}{y!} = \lambda^2.$$

1.23 If $X \sim P(\lambda)$:

(i) Show that $M_X(t) = e^{\lambda(e^t-1)}$, $t \in \Re$. (See #6 in Table 6 in the Appendix.)
(ii) Use the m.g.f. to rederive the EX and the $Var(X)$.

1.24 Let X be the r.v. denoting the number of particles arriving independently at a detector at the average rate of 3 per second.

(i) Specify a probability model for the r.v. X.
(ii) State which number(s) of particles arrive within 1 second with the maximum probability, and specify the value of this probability.

1.25 For $n = 1, 2, \ldots$, let the r.v. $X_n \sim B(n, p_n)$ where, as $n \to \infty$, $0 < p_n \to 0$, and $np_n \to \lambda \in (0, \infty)$. Then show that:

$$\binom{n}{x} p_n^x q_n^{n-x} \xrightarrow[n \to \infty]{} e^{-\lambda}\frac{\lambda^x}{x!} \quad (q_n = 1 - p_n).$$

Hint. Write $\binom{n}{x}$ as $n(n-1)\cdots(n-x+1)/x!$, set $np_n = \lambda_n$, so that $p_n = \frac{\lambda_n}{n} \underset{n\to\infty}{\longrightarrow} 0$ and $q_n = 1 - p_n = 1 - \frac{\lambda_n}{n} \underset{n\to\infty}{\to} 1$. Group terms suitably, take the limit as $n \to \infty$, and use the calculus fact that $\left(1 + \frac{x_n}{n}\right)^n \to e^x$ when $x_n \to x$ as $n \to \infty$, as mentioned in #6 of Table 6 in the Appendix.

1.26 In an undergraduate statistics class of 80, 10 of the students are actually graduate students. If 5 students are chosen at random from the class, what is the probability that:

(i) No graduate students are included?
(ii) At least 3 undergraduate students are included?

1.27 Suppose a geologist has collected 15 specimens of a certain rock, call it R_1, and 10 specimens of another rock, call it R_2. A laboratory assistant selects randomly 15 specimens for analysis, and let X be the r.v. denoting the number of specimens of rock R_1 selected for analysis.

(i) Specify the p.d.f. of the r.v. X.
(ii) What is the probability that at least 10 specimens of the rock R_1 are included in the analysis?
(iii) What is the probability that all specimens come from the rock R_2?

Hint. For part (ii), just write down the right formula.

1.28 If the r.v. X has hypergeometric distribution; i.e.,

$$P(X = x) = f(x) = \frac{\binom{m}{x}\binom{n}{r-x}}{\binom{m+n}{r}},$$

$x = 0, 1, \ldots, r$, then show that:

$$f(x+1) = \frac{(m-x)(r-x)}{(n-r+x+1)(x+1)} f(x)$$

(so that probabilities can be computed recursively).

Hint. Start with $f(x+1)$ and write the numerator in terms of factorials. Then modify suitably some terms, and regroup them to arrive at the expression on the right-hand side.

The following exercise, Exercise 1.29, is recorded here mostly for reference purposes; its solution goes along the same lines as that of Exercise 1.25, but it is somewhat more involved. The interested reader is referred for details to Theorem 2, Chapter 3, in the book *A Course in Mathematical Statistics*, 2nd edition (1997), Academic Press, by G. G. Roussas.

1.29 Let X be an r.v. having hypergeometric distribution with parameters m and n, so that its p.d.f. is given by $f(x) = \binom{m}{x}\binom{n}{r-x}/\binom{m+n}{r}$, $x = 0, 1, \ldots, r$. Suppose that m and $n \to \infty$, so that $\frac{m}{m+n} \to p \in (0, 1)$. Then:

$$\frac{\binom{m}{x}\binom{n}{r-x}}{\binom{m+n}{r}} \to \binom{r}{x} p^x (1-p)^{r-x}, \quad x = 0, 1, \ldots, r.$$

Thus, for large m and n, the hypergeometric probabilities may be approximated by the (simpler) binomial probabilities, as follows:

$$\frac{\binom{m}{x}\binom{n}{r-x}}{\binom{m+n}{r}} \simeq \binom{r}{x}\left(\frac{m}{m+n}\right)^x \left(1 - \frac{m}{m+n}\right)^{r-x}, \quad x = 0, 1, \ldots, r.$$

1.30 If $X \sim B(n, 0.15)$ determine the smallest value of n for which $P(X = 1) > P(X = 0)$.

1.31 Suppose that in a preelection campaign, propositions #1 and #2 are favored by $100p_1\%$ and $100p_2\%$ of voters, respectively, whereas the remaining $100(1 - p_1 - p_2)\%$ of the voters are either undecided or refuse to respond ($0 < p_1 < 1, 0 < p_2 < 1, p_1 + p_2 \leq 1$). A random sample of size $2n$ is taken. Then:

- **(i)** What are the expected numbers of voters favoring each one of the two propositions?
- **(ii)** What is the probability that the number of voters favoring proposition #1 is at most n?
- **(iii)** What is the probability that the number of voters favoring proposition #2 is at least n?
- **(iv)** Give the numerical values in parts (i)–(iii) for $p_1 = 31.25\%$, $p_2 = 43.75\%$, and $2n = 24$.

Hint. For parts (i)–(iii), just write down the right formulas.

1.32 A quality control engineer suggests that for each shipment of 1,000 integrated circuit components, a sample of 100 be selected from the 1,000 for testing. If either 0 or 1 defective is found in the sample, the entire shipment is declared acceptable.

- **(i)** Compute the probability of accepting the entire shipment if there are 3 defectives among the 1,000 items.
- **(ii)** What is the expected number of defective items?
- **(iii)** What is the variance and the s.d. of the defective items?

1.33 Refer to Exercise 1.32, and use the binomial approximation to hypergeometric distribution (see Exercise 1.29 in this chapter) to find an approximate value for the probability computed in part (i) of Exercise 1.32.

1.34 Refer to Exercise 1.33, and use the Poisson approximation to binomial distribution (see Exercise 1.25 here) to find an approximate value for the probability computed in part (i) of Exercise 1.32.

1.35 Suppose that 125 out of 1,000 California homeowners have earthquake insurance. If 25 such homeowners are chosen at random:

- **(i)** What is the expected number of earthquake insurance holders, and what is the s.d. around this expected number?
- **(ii)** What is the probability that the number of earthquake insurance holders is within one s.d. of the mean, inclusive?
- **(iii)** Use both the binomial and the Poisson approximation in part (ii).

Hint. For part(ii), just write down the correct formula.

1.36 Let the number of claims filed by policyholders over a period of time be an r.v. X which has Poisson distribution; i.e., $X \sim P(\lambda)$. If $P(X = 2) = 3P(X = 4)$, determine:

- **(i)** The EX and $Var(X)$.
- **(ii)** The probabilities $P(2 \leq X \leq 4)$ and $P(X \geq 5)$.

1.37 Refer to the geometric distribution with parameter p (see relation (6.6)), but now suppose that the predetermined number of successes $r \geq 1$. Then the number of (independent) trials required in order to obtain exactly r successes is an r.v. X which takes on the values $r, r+1, \ldots$ with respective probabilities

$$P(X = x) = f(x) = \binom{x-1}{r-1} q^{x-r} p^r = \binom{x-1}{x-r} q^{x-r} p^r \quad (q = 1 - p).$$

This distribution is called *negative binomial* (for obvious reasons) with parameter p.

(i) Show that the p.d.f. of X is, indeed, given by the above formula.

(ii) Use the identity

$$\frac{1}{(1-x)^n} = 1 + \binom{n}{1} x + \binom{n+1}{2} x^2 + \cdots \quad (|x| < 1, n \geq 1 \text{integer})$$

in order to show that the m.g.f. of X is given by

$$M_X(t) = \left(\frac{pe^t}{1 - qe^t}\right)^r, \quad t < -\log q.$$

(iii) For $t < -\log q$, show that

$$\frac{d}{dt} M_X(t) = \frac{r(pe^t)^r}{(1 - qe^t)^{r+1}}, \quad \frac{d^2}{dt^2} M_X(t) = \frac{r(pe^t)^r (r + qe^t)}{(1 - qe^t)^{r+2}}.$$

(iv) Use part (iii) in order to find that $EX = \frac{r}{p}$, $EX^2 = \frac{r(r+q)}{p^2}$, and $\sigma^2(X) = \frac{rq}{p^2}$.

(v) For $r = 5$ and $p = 0.25, 0.50, 0.75$, compute the EX, $\sigma^2(X)$, and $\sigma(X)$.

Hint.

(i) For part (i), derive the expression $f(x) = \binom{x-1}{r-1} q^{x-r} p^r$; the second expression follows from it and the identity $\binom{n}{m} = \binom{n}{n-m}$.

(ii) For part (ii), we have

$$M_X(t) = \sum_{x=r}^{\infty} e^{tx} \binom{x-1}{x-r} q^{x-r} p^r$$
$$= \frac{p^r (qe^t)^r}{q^r} \left[1 + \binom{r}{1}(qe^t) + \binom{r+1}{2}(qe^t)^2 + \cdots\right].$$

6.2 Some Special Continuous Distributions

In this section, we discuss basically three continuous distributions, which occur often. They are the gamma—and its special cases, the negative exponential and the chi-square—the normal, and the uniform distributions. In all cases, the p.d.f. is 0 for all values of its argument not figuring in the definition.

6.2.1 Gamma Distribution

As an introduction, the so-called gamma function, will be defined first. It is shown that the integral $\int_0^\infty y^{\alpha-1}e^{-y}\,dy$ is finite for $\alpha > 0$ and thus defines a function (in α); namely,

$$\Gamma(\alpha) = \int_0^\infty y^{\alpha-1}e^{-y}\,dy, \quad \alpha > 0. \tag{6.12}$$

This is the *gamma function*. By means of the gamma function, the *gamma distribution* is defined by its p.d.f. as follows:

$$f(x) = \frac{1}{\Gamma(\alpha)\beta^\alpha}x^{\alpha-1}e^{-x/\beta}, \quad x > 0,\ \alpha > 0,\ \beta > 0; \tag{6.13}$$

α and β are the *parameters* of the distribution. That the function f integrates to 1 is an immediate consequence of the definition of $\Gamma(\alpha)$. An r.v. X taking on values in $\Re$ and having p.d.f. f, given in (6.13), is said to be *gamma-distributed* with *parameters* α and β; one may choose the notation $X \sim \Gamma(\alpha, \beta)$ to express this fact.

The graph of f depends on α and β but is, of course, always concentrated on $(0, \infty)$. Typical cases for several values of the pair (α, β) are given in Figures 6.5 and 6.6.

The gamma distribution derives its usefulness primarily from the fact that it is widely used as a survival distribution, both for living organisms (including humans) and for equipment. It is also successfully employed to characterize the distribution of the size of claims submitted to an insurance company by claimants, as well as waiting times between successive occurrences of events following a Poisson distribution. (See also Exercise 2.6 below.)

In all cases, a gamma distribution provides great flexibility through its two parameters α and β. For specific values of the pair (α, β), we obtain the negative exponential and the chi-square distributions to be studied subsequently. By integration by parts, one may derive the following useful recursive relation for the gamma function (see Exercise 2.1):

$$\Gamma(\alpha) = (\alpha - 1)\Gamma(\alpha - 1). \tag{6.14}$$

In particular, if α is an integer, repeated applications of the recursive relation (6.14) produce:

$$\Gamma(\alpha) = (\alpha - 1)(\alpha - 2)\cdots\Gamma(1).$$

But $\Gamma(1) = \int_0^\infty e^{-y}\,dy = 1$, so that:

$$\Gamma(\alpha) = (\alpha - 1)(\alpha - 2)\cdots 1 = (\alpha - 1)! \tag{6.15}$$

For later reference, we mention here (see also Exercise 2.22) that, by integration, we obtain:

$$\Gamma\left(\frac{1}{2}\right) = \sqrt{\pi}, \tag{6.16}$$

and then, by means of this and the recursive formula (6.14), we can calculate $\Gamma\left(\frac{3}{2}\right)$, $\Gamma\left(\frac{5}{2}\right)$, etc. Finally, by integration (see Exercises 2.2 and 2.3), it is seen that:

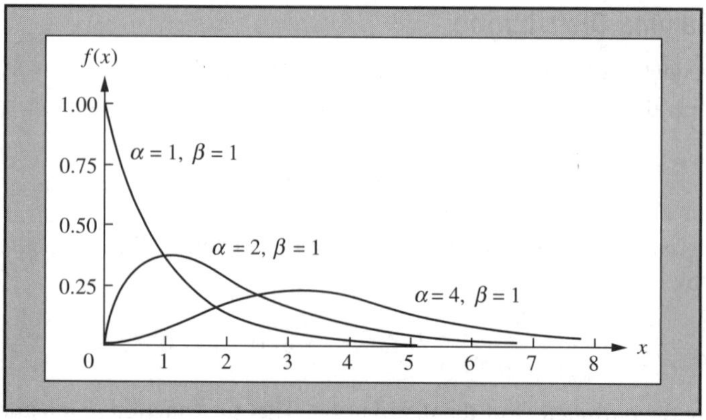

FIGURE 6.5
Graphs of the p.d.f. of the gamma distribution for several values of α, β.

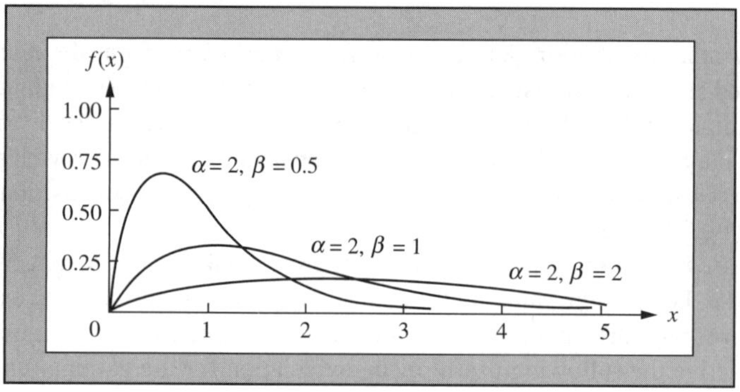

FIGURE 6.6
Graphs of the p.d.f. of the gamma distribution for several values of α, β.

Proposition 4.

$$EX = \alpha\beta, \quad Var(X) = \alpha\beta^2 \text{ and } M_X(t) = \frac{1}{(1-\beta t)^\alpha}, \quad t < \frac{1}{\beta}. \tag{6.17}$$

Example 7. The lifetime of a certain equipment is described by an r.v. X whose distribution is gamma with parameters $\alpha = 2$ and $\beta = \frac{1}{3}$, so that the corresponding p.d.f. is $f(x) = 9xe^{-3x}$, for $x > 0$. Determine the expected lifetime, the variation around it, and the probability that the lifetime is at least 1 unit of time.

Discussion. Since $EX = \alpha\beta$ and $Var(X) = \alpha\beta^2$, we have here: $EX = \frac{2}{3}$ and $Var(X) = \frac{2}{9}$ so that s.d. of $X = \frac{\sqrt{2}}{3} \simeq 0.471$. Also,

$$P(X > 1) = \int_1^\infty 9xe^{-3x}\,dx = \frac{4}{e^3} \simeq 0.199.$$

6.2.2 Negative Exponential Distribution

In (6.13), set $\alpha = 1$ and $\beta = \frac{1}{\lambda}$ $(\lambda > 0)$ to obtain:

$$f(x) = \lambda e^{-\lambda x}, \quad x > 0, \lambda > 0. \tag{6.18}$$

This is the so-called *negative exponential* distribution with *parameter* λ. The graph of f depends on λ but, typically, looks as in Figure 6.7.

For an r.v. X having negative exponential distribution with parameter λ, formula (6.17) gives:

$$EX = \frac{1}{\lambda}, \quad Var(X) = \frac{1}{\lambda^2}, \quad \text{and} \quad M_X(t) = \frac{1}{1 - \frac{t}{\lambda}}, \quad t < \lambda. \tag{6.19}$$

The expression $EX = \frac{1}{\lambda}$ provides special significance for the parameter λ: its inverse value is the mean of X. This fact also suggests the *reparameterization* of f; namely, set $\frac{1}{\lambda} = \mu$, in which case:

$$f(x) = \frac{1}{\mu}e^{-\frac{x}{\mu}}, \quad x > 0, \quad EX = \mu, \quad Var(X) = \mu^2, \quad \text{and}$$

$$M_X(t) = \frac{1}{1 - \mu t}, \quad t < \frac{1}{\mu}. \tag{6.20}$$

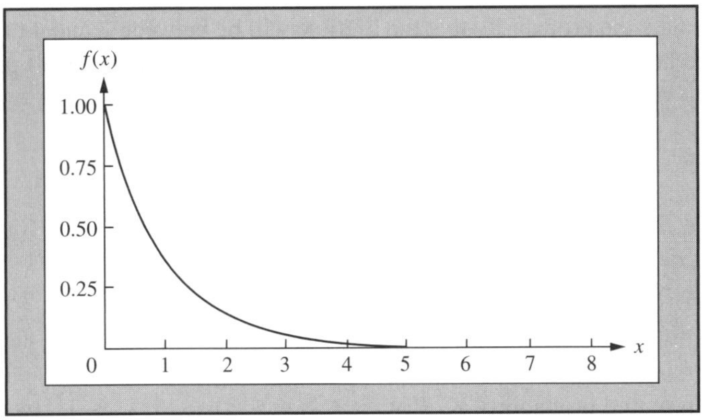

FIGURE 6.7
Graph of the negative exponential p.d.f. with $\lambda = 1$.

From (6.18), one finds by a simple integration:

$$F(x) = 1 - e^{-\lambda x}, \quad x > 0, \quad \text{so that } P(X > x) = e^{-\lambda x}, \quad x > 0. \tag{6.21}$$

The negative exponential distribution is used routinely as a survival distribution, describing the lifetime of a piece of equipment, etc., put in service at what may be termed time zero. As such, it exhibits a *lack of memory* property, which may not be desirable in this context. Namely, if one poses the question: "What is the probability that a piece of equipment will last for t additional units of time, given that it has already survived s units of time?" the answer (by means of the negative exponential distribution) is, by (6.21):

$$\begin{aligned} P(X > s+t \mid X > s) &= \frac{P(X > s+t, X > s)}{P(X > s)} = \frac{P(X > s+t)}{P(X > s)} = \frac{e^{-\lambda(s+t)}}{e^{-\lambda s}} \\ &= e^{-\lambda t} = P(X > t); \end{aligned}$$

that is, $P(X > s+t \mid X > s) = P(X > t)$ independent of s! Well, in real life, used pieces of equipment do not exactly behave as brand-new ones! Finally, it is to be mentioned that the negative exponential distribution is the waiting time distribution between the occurrence of any two successive events, which occur according to a Poisson distribution (see also Exercise 2.6 below).

By the fact that the negative exponential distribution involves one parameter only is easier to handle than the gamma distribution, which involves two parameters. On the other hand, the latter provides more flexibility and a better fit.

Example 8. The lifetime of an automobile battery is described by an r.v. X having the negative exponential distribution with parameter $\lambda = \frac{1}{3}$. Then:

- **(i)** Determine the expected lifetime of the battery and the variation around this mean.
- **(ii)** Calculate the probability that the lifetime will be between 2 and 4 time units.
- **(iii)** If the battery has lasted for 3 time units, what is the (conditional) probability that it will last for at least an additional time unit?

Discussion.

- **(i)** Since $EX = \frac{1}{\lambda}$ and $Var(X) = \frac{1}{\lambda^2}$, we have here: $EX = 3$, $Var(X) = 9$, and the $s.d.$ of X is equal to 3.
- **(ii)** Since, by (6.21), $F(x) = 1 - e^{-\frac{x}{3}}$ for $x > 0$, we have $P(2 < X < 4) = P(2 < X \leq 4) = P(X \leq 4) - P(X \leq 2) = F(4) - F(2) = (1 - e^{-\frac{4}{3}}) - (1 - e^{-\frac{2}{3}}) = e^{-\frac{2}{3}} - e^{-\frac{4}{3}} \simeq 0.252$.
- **(iii)** The required probability is: $P(X > 4 \mid X > 3) = P(X > 1)$, by the memoryless property of this distribution, and $P(X > 1) = 1 - P(X \leq 1) = 1 - F(1) = e^{-\frac{1}{3}} \simeq 0.716$.

6.2.3 Chi-Square Distribution

In formula (6.13), set $\alpha = \frac{r}{2}$ for a positive integer r and $\beta = 2$ to obtain:

$$f(x) = \frac{1}{\Gamma(\frac{r}{2})2^{r/2}} x^{(r/2)-1} e^{-x/2}, \quad x > 0, \ r > 0 \text{ integer.} \tag{6.22}$$

The resulting distribution is known as *chi-square* distribution with r *degrees of freedom* (d.f.). This distribution is used in certain statistical inference problems involving point estimation of variances, confidence intervals for variances, and testing hypotheses about variances. The notation used for an r.v. X having chi-square distribution with r d.f. is $X \sim \chi_r^2$. For such an r.v., formulas (6.17) then become:

$$\begin{aligned} &EX = r, \quad Var(X) = 2r \quad \text{(both easy to remember) and} \\ &M_X(t) = \frac{1}{(1-2t)^{r/2}}, \quad t < \frac{1}{2}. \end{aligned} \tag{6.23}$$

The shape of the graph of f depends on r and typically looks like Figure 6.8.

Later on (see Remark 5 in Chapter 10), it will be seen why r is referred to as the number of d.f. of the distribution.

6.2.4 Normal Distribution

This is by far the most important distribution, in both probability and statistics. The reason for this is twofold: First, many observations follow to a very satisfactory degree a normal distribution (see, for instance, Examples 12–14 in Chapter 1). The distribution of weights or heights of a large human population, the distribution of yearly income of households in a large (not stratified) community, the distribution of grades in a test in a large class are some additional examples where normal distribution is a good

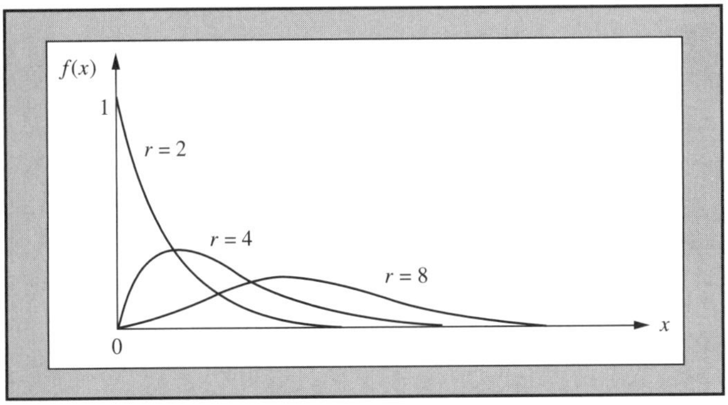

FIGURE 6.8

Graph of the p.d.f. of the chi-square distribution for several values of r.

approximation.Second, (almost) no matter what the underlying distribution of observations is, the sum of sufficiently many observations behaves pretty much as if it were normally distributed, under very mild conditions. This second property is referred to as *normal approximation* or as the *Central Limit Theorem* and will be revisited later (see Section 12.2 in Chapter 12). The p.d.f. of a normal distribution is given by:

$$f(x) = \frac{1}{\sqrt{2\pi}\sigma} e^{-(x-\mu)^2/2\sigma^2}, \quad x \in \Re, \ \mu \in \Re, \ \sigma > 0; \tag{6.24}$$

μ and σ^2 (or σ) are referred to as the *parameters* of the distribution. The graph of f depends on μ and σ; typical cases for $\mu = 1.5$ and various values of σ are given in Figure 6.9.

No matter what μ and σ are, the curve representing f attains its maximum at $x = \mu$ and this maximum is equal to $1/\sqrt{2\pi}\sigma$; it is symmetric around μ (i.e., $f(\mu - x) = f(\mu + x)$); and $f(x)$ tends to 0 as $x \to \infty$ or $x \to -\infty$. All these observations follow immediately from formula (6.24). Also, they lead to the general form of f depicted in Figure 6.9 for several values of the pair (μ, σ).

For $\mu = 0$ and $\sigma = 1$, formula (6.24) is reduced to:

$$f(x) = \frac{1}{\sqrt{2\pi}} e^{-x^2/2}, \quad x \in \Re, \tag{6.25}$$

and this is referred to as the *standard normal* distribution (see Figure 6.10 for its graph).

The fact that an r.v. X is *normally distributed* with *parameters* μ and σ^2 (or σ) is conveniently denoted by: $X \sim N(\mu, \sigma^2)$. In particular, $X \sim N(0, 1)$ for $\mu = 0, \sigma = 1$. We often use the notation Z for an $N(0, 1)$-distributed r.v.

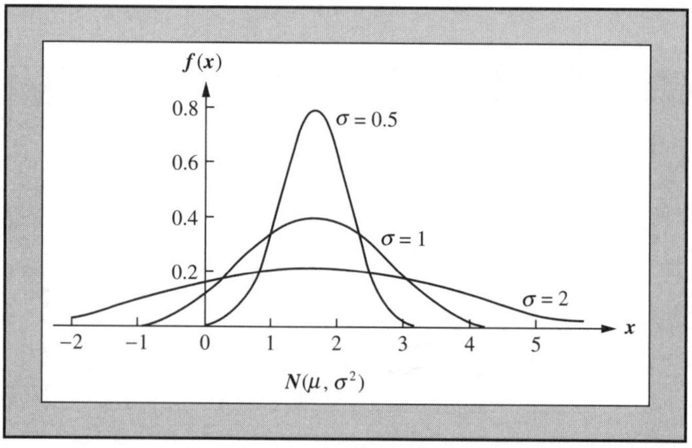

FIGURE 6.9
Graph of the p.d.f. of the normal distribution with $\mu = 1.5$ and several values of σ.

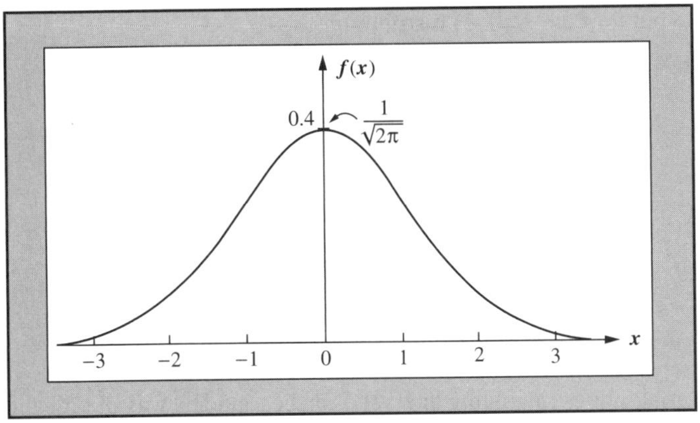

FIGURE 6.10
Graph of the p.d.f. of the standard normal distribution.

The d.f. of the $N(0, 1)$ distribution is usually denoted by Φ; that is, if $Z \sim N(0, 1)$, then:

$$P(Z \leq x) = \Phi(x) = \int_{-\infty}^{x} \frac{1}{\sqrt{2\pi}} e^{-t^2/2}\, dt, \quad x \in \Re. \tag{6.26}$$

Calculations of probabilities of the form $P(a < X < b)$ for $-\infty \leq a < b \leq \infty$ are done through two steps: First, turn the r.v. $X \sim N(\mu, \sigma^2)$ into an $N(0, 1)$-distributed r.v., or, as we say, *standardize* it, as indicated in Proposition 5 below, and then use available normal tables. Finally, that $f(x)$ integrates to 1 is seen through a technique involving a double integral and polar coordinates (see Exercise 2.21).

Proposition 5. If $X \sim N(\mu, \sigma^2)$, then $Z = \frac{X-\mu}{\sigma}$ is $\sim N(0, 1)$.

Proof. Indeed, for $y \in \Re$,

$$\begin{aligned} F_Z(y) &= P(Z \leq y) = P\left(\frac{X-\mu}{\sigma} \leq y\right) \\ &= P(X \leq \mu + \sigma y) \\ &= \int_{-\infty}^{\mu+\sigma y} \frac{1}{\sqrt{2\pi}\,\sigma} e^{-(t-\mu)^2/2\sigma^2}\, dt. \end{aligned}$$

Set $\frac{t-\mu}{\sigma} = z$, so that $t = \mu + \sigma z$ with range from $-\infty$ to y, and $dt = \sigma\, dz$, to obtain:

$$\begin{aligned} F_Z(y) &= \int_{-\infty}^{y} \frac{1}{\sqrt{2\pi}\,\sigma} e^{-z^2/2} \sigma\, dz \\ &= \int_{-\infty}^{y} \frac{1}{\sqrt{2\pi}} e^{-z^2/2}\, dz, \quad \text{so that} \\ f_Z(y) &= \frac{d}{dy} F_Z(y) = \frac{1}{\sqrt{2\pi}} e^{-y^2/2}, \end{aligned}$$

which is the p.d.f. of the $N(0, 1)$ distribution.

Thus, if $X \sim N(\mu, \sigma^2)$ and a, b are as above, then:

$$P(a < X < b) = P\left(\frac{a-\mu}{\sigma} < \frac{X-\mu}{\sigma} < \frac{b-\mu}{\sigma}\right) = P\left(\frac{a-\mu}{\sigma} < Z < \frac{b-\mu}{\sigma}\right)$$
$$= \Phi\left(\frac{b-\mu}{\sigma}\right) - \Phi\left(\frac{a-\mu}{\sigma}\right).$$

That is,

$$P(a < X < b) = \Phi\left(\frac{b-\mu}{\sigma}\right) - \Phi\left(\frac{a-\mu}{\sigma}\right). \tag{6.27}$$

Any other probabilities (involving intervals) can be found by way of probability (6.26) by exploiting the symmetry (around 0) of the $N(0, 1)$ curve.

Now, if $Z \sim N(0, 1)$, it is clear that $EZ^{2n+1} = 0$ for $n = 0, 1, \ldots$; by integration by parts, the following recursive relation is also easily established:

$$m_{2n} = (2n-1)m_{2n-2}, \quad \text{where } m_k = \int_{-\infty}^{\infty} x^k \frac{1}{\sqrt{2\pi}} e^{-x^2/2}\, dx, \tag{6.28}$$

from which it follows that $EZ = 0$ and $EZ^2 = 1$, so that $Var(Z) = 1$. (For details, see Exercise 2.25.)

If $X \sim N(\mu, \sigma^2)$, then $Z = \frac{X-\mu}{\sigma} \sim N(0, 1)$, so that (by Propositions 1 and 2 in Chapter 5):

$$0 = EZ = \frac{EX}{\sigma} - \frac{\mu}{\sigma}, \quad 1 = Var(Z) = \frac{1}{\sigma^2} Var(X), \quad \text{or} \quad EX = \mu \quad \text{and}$$
$$Var(X) = \sigma^2.$$

In other words:

Proposition 6.

$$\text{If } X \sim N(\mu, \sigma^2), \quad \text{then } EX = \mu \quad \text{and} \quad Var(X) = \sigma^2. \tag{6.29}$$

In particular, $EZ = 0$ and $Var(Z) = 1$, where $Z = \frac{X-\mu}{\sigma} \sim N(0, 1)$.

Thus, the parameters μ and σ^2 have specific interpretations: μ is the mean of X, and σ^2 is its variance (so that σ is its s.d.).

If $Z \sim N(0, 1)$, it is seen from the normal tables that:

$$P(-1 < Z < 1) = 0.68269, \quad P(-2 < Z < 2) = 0.95450,$$
$$P(-3 < Z < 3) = 0.99730, \tag{6.30}$$

so that almost all of the probability mass lies within 3 standard deviations from the mean. The same is true, by means of formula (6.27), applied with $a = \mu - k\sigma$ and

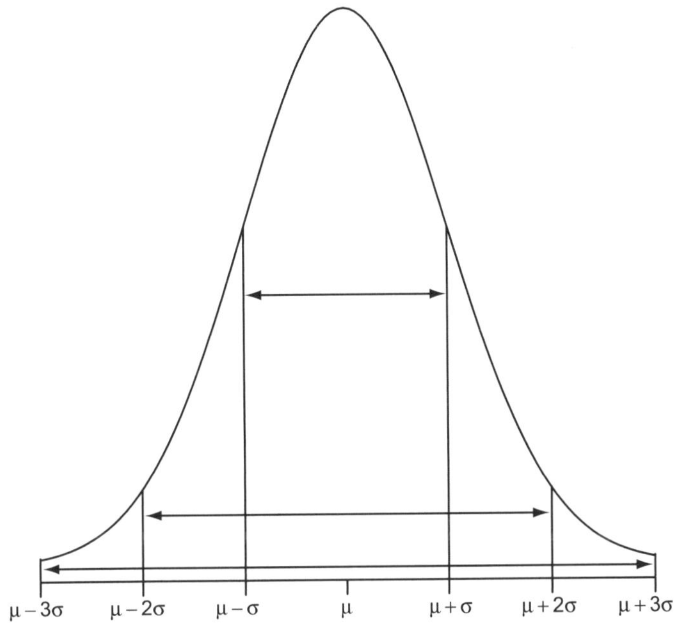

FIGURE 6.11

Probabilities within one, two, and three s.d.'s σ from the mean μ (approximately 0.68269, 0.95450, and 0.99730, respectively).

$b = \mu + k\sigma$ with $k = 1, 2, 3$ in case $X \sim N(\mu, \sigma^2)$. That is:

$$P(\mu - \sigma < X < \mu + \sigma) = 0.68269, \quad P(\mu - 2\sigma < X < \mu + 2\sigma) = 0.95450,$$
$$P(\mu - 3\sigma < X < \mu + 3\sigma) = 0.99730.$$

(See also Figure 6.11 above.)

Finally, simple integration produces the m.g.f. of X (see also Exercise 2.23); namely,

Proposition 7.

$$\begin{aligned} M_X(t) &= e^{\mu t + \sigma^2 t^2/2}, \quad t \in \Re, \text{ for } X \sim N(\mu, \sigma^2); \\ M_Z(t) &= e^{t^2/2}, \quad t \in \Re, \text{ for } Z \sim N(0, 1). \end{aligned} \tag{6.31}$$

The normal distribution is widely used in problems of statistical inference, involving point estimation, interval estimation, and testing hypotheses. Some instances where a normal distribution is assumed as an appropriate (approximate) underlying distribution are described in Examples 12–14 in Chapter 1, as mentioned already.

Example 9. Suppose that numerical grades in a statistics class are values of an r.v. X which is (approximately) normally-distributed with mean $\mu = 65$ and s.d. $\sigma = 15$. Furthermore, suppose that letter grades are assigned according to the following rule: the student receives an A if $X \geq 85$; B if $70 \leq X < 85$; C if $55 \leq X < 70$; D if $45 \leq X < 55$; and F if $X < 45$.

(i) If a student is chosen at random from that class, calculate the probability that the student will earn a given letter grade.
(ii) Identify the expected proportions of letter grades to be assigned.

Discussion.

(i) The student earns an A with probability $P(X \geq 85) = 1 - P(X < 85) = 1 - P(X \leq 85) = 1 - P\left(\frac{X-\mu}{\sigma} < \frac{85-65}{15}\right) \simeq 1 - P(Z \leq 1.34) \simeq 1 - \Phi(1.34) = 1 - 0.909877 = 0.090123 \simeq 0.090$. Likewise, the student earns a B with probability $P(70 \leq X < 85) = P\left(\frac{70-65}{15} \leq \frac{X-\mu}{\sigma} < \frac{85-65}{15}\right) \simeq P(0.34 \leq Z < 1.34) \simeq \Phi(1.34) - \Phi(0.34) = 0.909877 - 0.633072 = 0.276805 \simeq 0.277$. Similarly, the student earns a C with probability $P(55 \leq X < 70) \simeq \Phi(0.34) + \Phi(0.67) - 1 = 0.381643 \simeq 0.382$. The student earns a D with probability $P(45 \leq X < 55) \simeq \Phi(1.34) - \Phi(0.67) = 0.161306 \simeq 0.161$, and the student is assigned an F with probability $P(X < 45) \simeq \Phi(-1.34) = 1 - \Phi(1.34) = 0.09123 \simeq 0.091$.

(ii) The respective expected proportions for A, B, C, D, and F are about: 9%, 28%, 38%, 16%, and 9%.

Indeed, suppose there are n students, and let X_A be the number of those whose numerical grades are $\geq$85. By assuming that the n events that the numerical grade of each one of the n students is $\geq$85 are independent, we have that $X_A \sim B(n, 0.09)$. Then, $\frac{X_A}{n}$ is the proportion of A grades, and $E\left(\frac{X_A}{n}\right) = \frac{1}{n} \times n \times 0.090 = 0.090 = 9\%$ is the expected proportion of A's. Likewise for the other grades.

Finally, here is an example of determination of the pth quantile of the $N(0, 1)$ distribution for some selected values of p.

Example 10. If $Z \sim N(0, 1)$, take $p = 0.10, 0.20, 0.30, 0.40, 0.50, 0.60, 0.70, 0.80$, and 0.90 and determine the corresponding x_p.
From the normal tables, we obtain: $x_{0.10} = -x_{0.90} = -1.282$, $x_{0.20} = -x_{0.80} = -0.842$, $x_{0.30} = -x_{0.70} = -0.524$, $x_{0.40} = -x_{0.60} = -0.253$, and $x_{0.50} = 0$.

This section is concluded with a simple distribution, uniform (or rectangular) distribution.

6.2.5 Uniform (or Rectangular) Distribution

Such a distribution is restricted to finite intervals between the *parameters* α and β with $-\infty < \alpha < \beta < \infty$, and its p.d.f. is given by:

$$f(x) = \frac{1}{\beta - \alpha}, \quad \alpha \leq x \leq \beta \quad (-\infty < \alpha < \beta < \infty). \tag{6.32}$$

Its graph is given in Figure 6.12, and it also justifies its name as rectangular.

The term "uniform" is justified by the fact that intervals of equal length in (α, β) are assigned the same probability regardless of their location. The notation used for such a distribution is $U(\alpha, \beta)$ (or $R(\alpha, \beta)$), and the fact that the r.v. X is distributed as such is denoted by $X \sim U(\alpha, \beta)$ (or $X \sim R(\alpha, \beta)$). Simple integrations give (see also Exercise 2.28 for the EX and the Var(X)):

Proposition 8.

$$EX = \frac{\alpha+\beta}{2}, \quad Var(X) = \frac{(\alpha-\beta)^2}{12}, \quad \text{and} \quad M_X(t) = \frac{e^{\beta t} - e^{\alpha t}}{(\beta-\alpha)t}, \quad t \in \Re. \quad (6.33)$$

Example 11. A bus is supposed to arrive at a given bus stop at 10:00 a.m., but the actual time of arrival is an r.v. X which is uniformly distributed over the 16-minute interval from 9:52 to 10:08. If a passenger arrives at the bus stop at exactly 9:50, what is the probability that the passenger will board the bus no later than 10 minutes from the time of his or her arrival?

Discussion. The p.d.f. of X is $f(x) = 1/16$ for x ranging between 9:52 and 10:08, and 0 otherwise. The passenger will board the bus no later than 10 minutes from the time of his or her arrival at the bus stop if the bus arrives at the bus stop between 9:52 and 10:00 (as the passenger will necessarily have to wait for 2 minutes, between 9:50 and 9:52). The probability for the bus to arrive between 9:52 and 10:00 is $8/16 = 0.5$. This is the required probability.

For any $0 < p < 1$, the pth quantile of the distribution is immediately computed, as the following example indicates.

Example 12. If $X \sim U(0, 1)$, take $p = 0.10, 0.20, 0.30, 0.40, 0.50, 0.60, 0.70, 0.80$, and 0.90 and determine the corresponding x_p.

Here $F(x) = \int_0^x dt = x, 0 \le x \le 1$. Therefore $F(x_p) = p$ gives $x_p = p$.

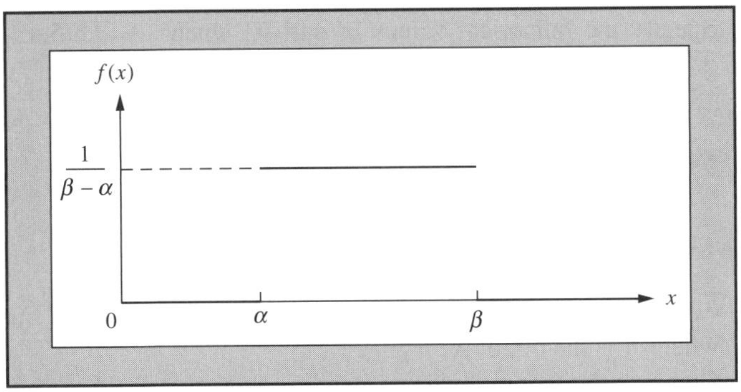

FIGURE 6.12
Graph of the p.d.f. of the $U(\alpha, \beta)$ distribution.

Remark 4. As was the case in Remark 3, the same is true here; namely, the first one or two moments of the distributions completely determine the distribution itself.

Exercises.

2.1 By using the definition of $\Gamma(\alpha)$ by (6.12) and integrating by parts, show that: $\Gamma(\alpha) = (\alpha - 1)\Gamma(\alpha - 1), \alpha > 1$.

Hint. Use #13 in Table 6 in the Appendix, and formula (6.12).

2.2 Let the r.v. X have gamma distribution with parameters α and β. Then:

(i) Show that $EX = \alpha\beta$, $Var(X) = \alpha\beta^2$.
(ii) As a special case of part (i), show that: If X has negative exponential distribution with parameter λ, then $EX = \frac{1}{\lambda}$, $Var(X) = \frac{1}{\lambda^2}$.
(iii) If $X \sim \chi^2_r$, then $EX = r$, $Var(X) = 2r$.

Hint. For part (i), use a suitable transformation first, and then formula (6.12) and Exercise 2.1.

2.3 If the r.v. X is distributed as gamma with parameters α and β, then:

(i) Show that $M_X(t) = 1/(1 - \beta t)^\alpha$, provided $t < 1/\beta$.
(ii) Use the m.g.f. to rederive the EX and the $Var(X)$.

Hint. For part (i), use formula (6.13).

2.4 Let X be an r.v. denoting the lifetime of a certain component of a system, and suppose that X has negative exponential distribution with parameter λ. Also, let $g(x)$ be the cost of operating this piece of equipment to time $X = x$.

(i) Compute the expected cost of operation over the lifetime of the component under consideration, when:

(a) $g(x) = cx$, where c is a positive constant,
(b) $g(x) = c(1 - 0.5e^{-\alpha x})$, where α is a positive constant.

(ii) Specify the numerical values in part (i) when $\lambda = 1/5, c = 2$, and $\alpha = 0.2$.

Hint. For part (i), use Definition 2 in Chapter 5, and compute the required expressions in terms of c and λ.

2.5 If the r.v. X has negative exponential p.d.f. with parameter λ:

(i) Calculate the *failure rate* $r(x)$ defined by $r(x) = \frac{f(x)}{1-F(x)}$, for $x > 0$, where F is the d.f. of X.
(ii) Compute the (conditional) probability $P(X > s + t | X > s) (s, t > 0)$, and comment on the result (see also the discussion following relation (6.21)).

2.6 Suppose that certain events occur in time intervals of length t according to Poisson distribution with parameter λt. Then show that the waiting time between any two such successive occurrences of events is an r.v. T which has negative exponential distribution with parameter λ, by showing that $P(T > t) = e^{-\lambda t}, t > 0$.

2.7 Let X be the r.v. denoting the number of particles arriving independently at a detector at the average rate of 3 per second (so that, by Exercise 2.6, we may assume that $X \sim P(\lambda t)$ with $\lambda = 3$ and $t = 1$), and let Y be the r.v. denoting the waiting time between two successive arrivals.
Then compute:

- **(i)** The probability that the first particle will arrive within 1 second.
- **(ii)** Given that we have waited for 1 second since the arrival of the last particle without a new arrival, what is the probability that we have to wait for at least another second?

2.8 Let X be an r.v. with p.d.f. $f(x) = \alpha\beta x^{\beta-1}e^{-\alpha x^{\beta}}$, for $x > 0$ (where the parameters α and β are > 0). This is the so-called *Weibull* distribution employed in describing the lifetime of living organisms or of mechanical systems.

- **(i)** Show that f is, indeed, a p.d.f.
- **(ii)** For what values of the parameters does f become a negative exponential p.d.f.?
- **(iii)** Calculate the quantities EX, EX^2, and $Var(X)$.

Hint. For part (i), observe that $\int_0^{\infty} \alpha\beta x^{\beta-1}e^{-\alpha x^{\beta}}dx = \int_0^{\infty} e^{-\alpha x^{\beta}} \times (\alpha\beta x^{\beta-1}) dx = -\int_0^{\infty} de^{-\alpha x^{\beta}} = -e^{-\alpha x^{\beta}} |_0^{\infty} = 1$.
For part (iii), set $\alpha x^{\beta} = t$, so that $x = t^{1/\beta}/\alpha^{1/\beta}$, $dx = (t^{\frac{1}{\beta}-1}/\beta\alpha^{1/\beta})dt$ and $0 < t < \infty$. Then:

$$EX^n = \frac{1}{\alpha^{n/\beta}} \int_0^{\infty} t^{\left(\frac{n}{\beta}+1\right)-1} e^{-t} dt.$$

Then multiply and divide by the constant $\Gamma\left(\frac{n}{\beta}+1\right)$ and observe that

$$\frac{1}{\Gamma\left(\frac{n}{\beta}+1\right)} t^{\left(\frac{n}{\beta}+1\right)-1} e^{-t} (t > 0)$$

is the gamma p.d.f. with parameters $\frac{n}{\beta}+1$ and 1.

2.9 In reference to Exercise 2.8, calculate:

- **(i)** The *failure rate* $r(x) = \frac{f(x)}{1-F(x)}$, $x > 0$, where F is the d.f. of the r.v. X.
- **(ii)** The conditional probability $P(X > s + t \mid X > s)$, $s > 0, t > 0$.
- **(iii)** Compare the results in parts (i) and (ii) with the respective results in Exercise 2.5.

2.10

- **(i)** If X has negative exponential distribution with $\lambda = 1$, calculate the pth quantile x_p.

(ii) Use part (i) and Exercise 3.10 in Chapter 5 to determine y_p without calculations, where $Y = e^X$.
(iii) How do parts (i) and (ii) become for $p = 0.5$?

2.11 If Φ is the d.f. of the r.v. $Z \sim N(0, 1)$, use symmetry of the $N(0, 1)$ p.d.f. in order to show that:

(i) For $0 \le a < b$, $P(a < Z < b) = \Phi(b) - \Phi(a)$.
(ii) For $a \le 0 < b$, $P(a < Z < b) = \Phi(-a) + \Phi(b) - 1$.
(iii) For $a \le b < 0$, $P(a < Z < b) = \Phi(-a) - \Phi(-b)$.
(iv) For $c > 0$, $P(-c < Z < c) = 2\Phi(c) - 1$.

2.12 If the r.v. $Z \sim N(0, 1)$, use Exercise 2.11(iv) and the normal tables in the appendix to verify that:

(i) $P(-1 < Z < 1) = 0.68269$.
(ii) $P(-2 < Z < 2) = 0.9545$.
(iii) $P(-3 < Z < 3) = 0.9973$.

2.13

(i) If the r.v. X is distributed as $N(\mu, \sigma^2)$, identify the constant c, in terms of μ and σ, for which:

$$P(X < c) = 2 - 9P(X > c).$$

(ii) What is the numerical value of c for $\mu = 5$ and $\sigma = 2$?

2.14 For any r.v. X with expectation μ and variance σ^2 (both finite), use the Tchebichev inequality to determine a lower bound for the probabilities $P(|X - \mu| < k\sigma)$, for $k = 1, 2, 3$. Compare these bounds with the respective probabilities when $X \sim N(\mu, \sigma^2)$ (see Exercise 2.12 and comment following relation (6.30)).

2.15 The distribution of IQs of the people in a given group is approximated well by normal distribution with $\mu = 105$ and $\sigma = 20$. What proportion of the individuals in the group in question has an IQ:

(i) At least 50?
(ii) At most 80?
(iii) Between 95 and 125?

2.16 A certain manufacturing process produces light bulbs whose lifetime (in hours) is an r.v. X distributed as normal with $\mu = 2{,}000$ and $\sigma = 200$. A light bulb is supposed to be defective if its lifetime is less than 1,800. If 25 light bulbs are tested, what is the probability that at most 15 of them are defective?

Hint. Use the required independence and the binomial distribution suitably. Just write down the correct formula.

2.17 A manufacturing process produces 1/2-inch ball bearings, which are assumed to be satisfactory if their diameter lies in the interval 0.5 ± 0.0006 and defective otherwise. A day's production is examined, and it is found that the distribution of the actual diameters of the ball bearings is approximately normal with $\mu = 0.5007$ inch and $\sigma = 0.0005$ inch. What would you expect the proportion of defective ball bearings to be equal to?

Hint. Use the required independence and the binomial distribution suitably. Also, refer to the concluding part of the discussion in Example 9.

2.18 Let f be the p.d.f. of the $N(\mu, \sigma^2)$ distribution. Then show that:

(i) f is symmetric about μ.
(ii) $\max_{x \in \Re} f(x) = 1/\sqrt{2\pi}\sigma$.

2.19

(i) If $X \sim N(\mu, \sigma^2)$ and $0 < p < 1$, show that the pth quantile x_p of X is given by: $x_p = \mu + \sigma\Phi^{-1}(p)$.
(ii) Determine x_p in terms of μ and σ for $p = 0.25, 0.50, 0.75$.

2.20 If f is the p.d.f. of the r.v. $X \sim N(\mu, \sigma^2)$, show that the points $x = \mu \pm \sigma$ are inflection points of $f(x)$; i.e., $f''(\mu \pm \sigma) = 0$.

2.21

(i) Show that $f(x) = \frac{1}{\sqrt{2\pi}} e^{-\frac{x^2}{2}}, x \in \Re$, is a p.d.f.
(ii) Use part (i) in order to show that $f(x) = \frac{1}{\sqrt{2\pi}\sigma} e^{-\frac{(x-\mu)^2}{2\sigma^2}}, x \in \Re\ (\mu \in \Re, \sigma > 0)$ is also a p.d.f.

Hint. Set $I = \frac{1}{\sqrt{2\pi}} \int_{-\infty}^{\infty} e^{-\frac{x^2}{2}} dx$ and show that $I^2 = 1$, by writing I^2 as a product of two integrals and then as a double integral; at this point, use polar coordinates: $x = r\cos\theta$, $y = r\sin\theta$, $0 < r < \infty$, $0 \le \theta < 2\pi$. Part (ii) is reduced to part (i) by letting $\frac{x-\mu}{\sigma} = y$.

2.22 Refer to the definition of $\Gamma(\alpha)$ by (6.12) and show that $\Gamma(\frac{1}{2}) = \sqrt{\pi}$.

Hint. Use the transformation $y^{1/2} = t/\sqrt{2}$, and observe that the outcome is a multiple of the $N(0, 1)$ p.d.f.

2.23

(i) If $Z \sim N(0, 1)$, show that $M_Z(t) = e^{t^2/2}, t \in \Re$.
(ii) If $X \sim N(\mu, \sigma^2)$, use part (i) to show that $M_X(t) = e^{\mu t + \frac{\sigma^2 t^2}{2}}, t \in \Re$.
(iii) Employ the m.g.f. in part (ii) in order to show that $EX = \mu$ and $Var(X) = \sigma^2$.

Hint. For part (i), complete the square in the exponent, and for part (ii), set $Z = (X - \mu)/\sigma$ and apply property (5.12) in Chapter 5.

2.24 If the r.v. X has m.g.f. $M_X(t) = e^{\alpha t + \beta t^2}$, where $\alpha \in \Re$ and $\beta > 0$, identify the distribution of X.

Hint. Just go through the list of m.g.f.'s given in Table 5 in the Appendix, and then use Theorem 1 in Chapter 5.

2.25 If $Z \sim N(0, 1)$, show that:

- **(i)** $EZ^{2n+1} = 0$ and $EZ^{2n} = \frac{(2n)!}{2^n(n!)}$, $n = 0, 1, \ldots$
- **(ii)** From part (i), derive that $EZ = 0$ and $Var(Z) = 1$.
- **(iii)** Employ part (ii) in order to show that, if $X \sim N(\mu, \sigma^2)$, then $EX = \mu$ and $Var(X) = \sigma^2$.

Hint. For part (i), that $EZ^{2n+1} = 0$ follows by the fact that the integrand is an odd function. For EZ^{2n}, establish a recursive relation, integrating by parts, and then multiply out the resulting recursive relations to find an expression for EZ^{2n}. The final form follows by simple manipulations. For part (iii), recall that $X \sim N(\mu, \sigma^2)$ implies $\frac{X-\mu}{\sigma} \sim N(0, 1)$.

2.26 Let X be an r.v. with moments given by:

$$EX^{2n+1} = 0, \quad EX^{2n} = \frac{(2n)!}{2^n(n!)}, \quad n = 0, 1, \ldots$$

- **(i)** Use Exercise 1.17 in Chapter 5 in order to express the m.g.f. of X in terms of the moments given.
- **(ii)** From part (i) and Exercise 2.23(i) here, conclude that $X \sim N(0, 1)$.

2.27 If the r.v. X is distributed as $U(-\alpha, \alpha) (\alpha > 0)$, determine the parameter α, so that each of the following equalities holds:

- **(i)** $P(-1 < X < 2) = 0.75$.
- **(ii)** $P(|X| < 1) = P(|X| > 2)$.

2.28 If $X \sim U(\alpha, \beta)$, show that $EX = \frac{\alpha+\beta}{2}$, $Var(X) = \frac{(\alpha-\beta)^2}{12}$.

2.29 If the r.v. X is distributed as $U(0, 1)$, compute the expectations:

- **(i)** $E(3X^2 - 7X + 2)$.
- **(ii)** $E(2e^X)$.

Hint. Use Definition 2 and Proposition 1(i) in Chapter 5.

2.30 The number of customers arriving at a service counter in a supermarket in a 5-minute period is an r.v. X which has Poisson distribution with mean 0.2.

(i) What is the probability that the number of arrivals in a given 5-minute period will not exceed 1?

(ii) It is known that the waiting time between two successive arrivals is a r.v. Y having negative exponential distribution with parameter $\lambda = \frac{0.2}{5} = 0.04$ (see Exercise 2.6 above). On the basis of this, what is the probability that the waiting time between two successive arrivals is more than 5 minutes?

2.31 The amount paid by an insurance company to a policyholder is an r.v. X uniformly distributed over the interval $(0, \alpha)(\alpha > 0)$, so that $EX = \alpha/2$. With the introduction of a deductible d $(0 < d < \alpha)$, the amount paid is an r.v. Y defined by: $Y = 0$ if $0 \leq X \leq d$, and $Y = X - d$ if $d < X \leq \alpha$.

(i) Determine the EY in terms of α and d.

(ii) If we wish that $EY = cEX$ for some $0 < c < 1$, express d in terms of c and α.

(iii) For $c = 1/4$, express d in terms of α.

Hint. It may be helpful to write Y as an indicator function of the suitable interval of values of X. Also, for part (i), use Proposition 1(i) in Chapter 5.

2.32 The lifetime of a certain new piece of equipment is an r.v. X having negative exponential distribution with mean μ (i.e., its p.d.f. is given by: $f(x) = \mu^{-1}e^{-x/\mu}, x > 0$), and it costs \$C. The accompanying guarantee provides for a full refund, if the equipment fails within the $(0, \mu/2]$ time interval, and one-half refund if it fails within the $(\mu/2, \mu]$ time interval.

(i) In terms of μ and C, compute the expected amount to be paid as a refund by selling one piece of equipment.

(ii) What is the numerical value in part (i) for $C = \$2,400$?

Hint. Introduce an r.v. Y expressing the refund provided, depending on the behavior of X, and then compute the EY.

2.33 The lifetime of a certain new equipment is an r.v. X with p.d.f. given by: $f(x) = c/x^3, x > c$ $(c > 0)$.

(i) Determine the constant c.

(ii) Determine the median m of X.
Suppose the company selling the equipment provides a reimbursement should the equipment fail, described by the r.v. Y defined as follows: $Y = m - X$ when $c < X < m$, and $Y = 0$ otherwise (i.e., when $X \leq c$ or $X \geq m$), or $Y = (m - X)I_{(c,m)}(X)$.

(iii) Determine the p.d.f. of Y, f_Y.

Hint. Observe that $P(Y = 0) = P(X < c, \text{or } X > m) = P(X > m) = 1/2$, and that $c < X < m$ if and only if $0 < Y < m - c$. Then, for $0 < y < m - c$, determine the d.f. F_Y, and by differentiation, the p.d.f. f_Y.

(iv) Check that $\int_0^{m-c} f_Y(y)dy = 1/2$.
(v) Determine the EY.

Hint. Observe that $EY = (0 \times \frac{1}{2}) + \int_0^{m-c} y f_Y(y)dy = \int_0^{m-c} y f_Y(y)dy$, with $f_Y(y)$ as given in part (iii) for $0 < y < m - c$.

2.34 The size of certain claims submitted to an insurance company is an r.v. X having negative exponential distribution; that is, $f(x) = \lambda e^{-\lambda x}$, $x > 0$, $\lambda > 0$.

(i) Find the median m of the claim sizes in terms of λ, and its numerical value for $\lambda = 0.005$.
(ii) If the maximum amount reimbursed by the insurance company is M, determine the actual distribution of the claim sizes in terms of λ and M.
(iii) Find the precise expression of the p.d.f. in part (ii) for $\lambda = 0.005$ and $M = 200$.

Hint. The actual claim is an r.v. Y, where $Y = X$ if $X < M$, and $Y = M$ if $X \geq M$.

2.35 Refer to Exercise 2.34, and consider the r.v. Y given in the Hint of that exercise.

(i) By means of the result in part (ii) of Exercise 2.34, compute the EY in terms of M.
(ii) Find the numerical value of EY for $\lambda = 0.005$ and $M = 200$.

6.2.6 The basics of the Central Limit Theorem (CLT)

As it has already been mentioned, the CLT is discussed extensively in Section 12.2 of Chapter 12. Because of its great significance and broad applicability, it might be appropriate, however, to have a brief look at it at an earlier part of the book; this may just be the suitable place for such a discussion. It would be necessary, however, for some concepts to be treated with less than a rigorous manner.

To start with, consider n independent repetitions of a random experiment, giving rise to n independent and identically distributed r.v.'s $X_1, \ldots, X_n$ (see Section 10.1 in Chapter 10 for the precise definition of independence of r.v.'s). Since the r.v.'s $X_1, \ldots, X_n$ are identically distributed, they have the same expectation (assuming it exists), call it μ, and the same variance, call it σ^2. It is assume that μ is finite and that $0 < \sigma^2 < \infty$.

Set $\bar{X}_n = n^{-1}\sum_{i=1}^n X_i$ for the *sample mean* of the r.v.'s $X_1, \ldots, X_n$, and S_n for their sum, $S_n = \sum_{i=1}^n X_i$. Then, it is seen that the r.v.'s $\bar{X}_n$ and S_n have means and variances expressed in terms of μ and σ^2 as follows:

$$E\bar{X}_n = \mu, \quad \sigma^2(\bar{X}_n) = \frac{\sigma^2}{n}, \quad ES_n = n\mu, \quad \sigma^2(S_n) = n\sigma^2.$$

(The justification of it is found in the discussion in Section 9.1 of Chapter 9.)

Then what the CLT does is to make the striking statement that, regardless of the (common) distribution of the r.v.'s $X_1, \ldots, X_n$, and subject only to the restrictions that $\mu \in \Re$ and $0 < \sigma^2 < \infty$, the "summary" r.v.'s $\bar{X}_n$ and S_n have approximately normal distributions, provided that n is large enough. More precisely, $\bar{X}_n$ is approximately distributed as $N(\mu, \sigma^2/n)$, and S_n is approximately distributed as $N(n\mu, n\sigma^2)$. In brief notation,

$$\bar{X}_n \simeq N(\mu, \sigma^2/n), \quad S_n \simeq N(n\mu, n\sigma^2). \tag{6.34}$$

Now, if the r.v. $X \sim N(\mu, \sigma^2)$, then the "normalized" (or "standardized") r.v. $(X - \mu)/\sigma$ is distributed as $N(0, 1)$ (see Proposition 5 in this chapter). Then by (6.34), one would expect (and this, actually, does happen) that, for sufficient large n:

$$\frac{\bar{X}_n - \mu}{\sqrt{\sigma^2/n}} = \frac{\sqrt{n}(\bar{X}_n - \mu)}{\sigma} \simeq N(0, 1), \quad \frac{S_n - n\mu}{\sqrt{n\sigma^2}} = \frac{S_n - n\mu}{\sigma\sqrt{n}} \simeq N(0, 1). \tag{6.35}$$

An immediate consequence of the approximations in (6.35) ought to be that the d.f. of the r.v. $\frac{\sqrt{n}(\bar{X}_n - \mu)}{\sigma}$, evaluated at any $x \in \Re$, should be close to $\Phi(x)$, the d.f. of $Z \sim N(0, 1)$, evaluated at the same x; i.e.,

$$P\left[\frac{\sqrt{n}(\bar{X}_n - \mu)}{\sigma} \le x\right] \simeq \Phi(x), \quad x \in \Re. \tag{6.36}$$

Likewise,

$$P\left(\frac{S_n - n\mu}{\sigma\sqrt{n}} \le x\right) \simeq \Phi(x), \quad x \in \Re. \tag{6.37}$$

The utilitarian aspect of the CLT derives form the fact that probabilities on the left-hand side of (6.36) and (6.37) are not easy (or even possible) to compute but they can always be approximated by the probabilities $\Phi(x)$ read out of the normal tables.

Questions which are often posed are of the following kind.

(i) For given $a, b \in \Re$ with $a < b$, find the approximate value to the probability $P(a < S_n \le b)$. This question is dealt with as follows:

$$\begin{aligned} P(a < S_n \le b) &= P(S_n \le b) - P(S_n \le a) \\ &= P\left(\frac{S_n - n\mu}{\sigma\sqrt{n}} \le \frac{b - n\mu}{\sigma\sqrt{n}}\right) - P\left(\frac{S_n - n\mu}{\sigma\sqrt{n}} \le \frac{a - n\mu}{\sigma\sqrt{n}}\right) \\ &\simeq \Phi\left(\frac{b - n\mu}{\sigma\sqrt{n}}\right) - \Phi\left(\frac{a - n\mu}{\sigma\sqrt{n}}\right) \text{ (by means of (6.37)).} \end{aligned} \tag{6.38}$$

(However, be aware of Remark 4(ii) in Chapter 12.)

(ii) With a and b as above, find an approximate value to the probability that the difference between the sample mean $\bar{X}_n$ and the population mean μ is larger than a and smaller or equal to b; i.e., $P(a < \bar{X}_n - \mu \le b)$. This question is resolved as follows:

$$P(a < \bar{X}_n - \mu \le b) = P\left[\frac{\sqrt{n}a}{\sigma} < \frac{\sqrt{n}(\bar{X}_n - \mu)}{\sigma} \le \frac{\sqrt{n}b}{\sigma}\right]$$

(so that the middle expression above will look like the left-hand side of (6.36))

$$= P\left[\frac{\sqrt{n}(\bar{X}_n - \mu)}{\sigma} \leq \frac{\sqrt{n}b}{\sigma}\right] - P\left[\frac{\sqrt{n}(\bar{X}_n - \mu)}{\sigma} \leq \frac{\sqrt{n}a}{\sigma}\right]$$

$$\simeq \Phi\left(\frac{\sqrt{n}b}{\sigma}\right) - \Phi\left(\frac{\sqrt{n}a}{\sigma}\right) \text{ (by means of (6.36)).} \tag{6.39}$$

(Again, be aware of Remark 4(ii) in Chapter 12.)

For the sake of illustration, consider the following example.

Example 13. Suppose that the X_i's are distributed as $B(1, p)$, so that S_n represents the number of 1's (successes) in n independent tossings of a coin with probability p of a success. Take $n = 100, p = 1/2, a = 45, b = 60$. In the present case, $\mu = p, \sigma^2 = pq$ (with $q = 1 - p) = 1/4$, so that $\sigma = 1/2$, and therefore $\frac{a-n\mu}{\sigma\sqrt{n}} = \frac{45-50}{5} = -1$, $\frac{b-n\mu}{\sigma\sqrt{n}} = \frac{60-50}{5} = 2$. From (6.38), it follows that

$$P(40 < S_{100} \leq 60) \simeq \Phi(2) - \Phi(-1) = \Phi(2) + \Phi(1) - 1$$
$$= 0.977250 + 0.841345 - 1 = 0.818595.$$

Now, with n and p as before, take $a = -\frac{1}{16}$, and $b = 2/16$. Then $\frac{\sqrt{n}a}{\sigma} = -1.25$, $\frac{\sqrt{n}b}{\sigma} = 2.5$, so that, by (6.39),

$$P(-\frac{1}{16} < \bar{X}_{100} - \frac{1}{2} \leq \frac{2}{16}) \simeq \Phi(2.5) - \Phi(-1.25) = \Phi(2.5) + \Phi(1.25) - 1$$
$$= 0.993790 + 0.894350 - 1 = 0.88814.$$

Once again, the reader is referred to Section 12.2 of Chapter 12 for rigorous justification of results employed here, as well as numerous and interesting applications.

CHAPTER

7

Joint Probability Density Function of Two Random Variables and Related Quantities

A brief description of the material discussed in this chapter is as follows. In the first section, two r.v.'s are considered and the concepts of their joint probability distribution, joint d.f., and joint p.d.f. are defined. The basic properties of the joint d.f. are given, and a number of illustrative examples are provided. On the basis of a joint d.f., marginal d.f.'s are defined. Also, through a joint p.d.f., marginal and conditional p.d.f.'s are defined, and illustrative examples are supplied. By means of conditional p.d.f.'s, conditional expectations and conditional variances are defined and are applied to some examples. These things are done in the second section of the chapter.

7.1 Joint d.f. and Joint p.d.f. of Two Random Variables

In carrying out a random experiment, we are often interested simultaneously in two outcomes rather than one. Then with each one of these outcomes an r.v. is associated, and thus we are furnished with two r.v.'s, or a *2-dimensional random vector.* Let us denote by (X, Y) the two relevant r.v.'s or the 2-dimensional random vector. Here are some examples in which two r.v.'s arise in a natural way. The pair of r.v.'s (X, Y) denote, respectively, the SAT and GPA scores of a student chosen at random from a specified student population; the number of customers waiting for service in two lines in your local favorite bank; the days of a given year that the Dow Jones averages closed with a gain and the corresponding gains; the number of hours a student spends daily for studying and for other activities; the weight and the height of an individual chosen at random from a targeted population; the amount of fertilizer used and the yield of a certain agricultural commodity; the lifetimes of two components used in an electronic system; the dosage of a drug used for treating a certain allergy and the number of days a patient enjoys relief.

We are going to restrict ourselves to the case where both X and Y are either discrete or of the continuous type. The concepts of probability distribution, distribution function, and probability density function are defined by a straightforward generalization of the definition of these concepts in Section 3.3 of Chapter 3. Thus, the *joint probability distribution* of (X, Y), to be denoted by $P_{X,Y}$, is defined by $P_{X,Y}(B) = P[(X, Y) \in B]$, $B \subseteq \Re^2 = \Re \times \Re$, the 2-dimensional Euclidean space, the plane. (See Figure 7.1.)

Introduction to Probability, Second Edition. http://dx.doi.org/10.1016/B978-0-12-800041-0.00007-9

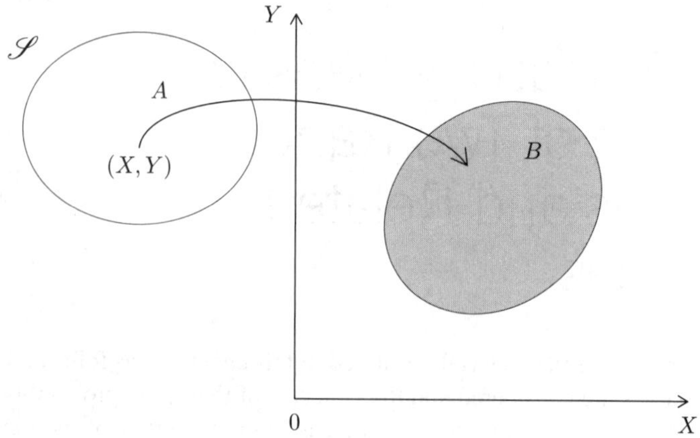

FIGURE 7.1

Event A is mapped onto B under (X, Y); i.e., $A = \{s \in S; (X(s), Y(s)) \in B\}$ and $P_{X,Y}(B) = P(A)$.

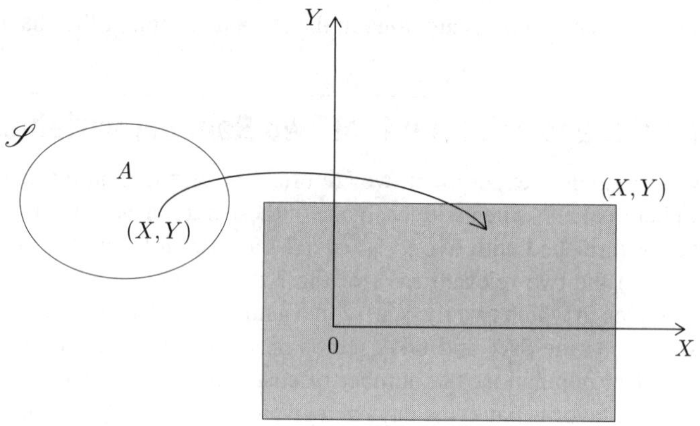

FIGURE 7.2

Event A is mapped onto $[-\infty, x] \times [-\infty, y]$; that is, $A = \{s \in S; X(s) \leq x, Y(s) \leq y\}$, and $F_{X,Y}(x, y) = P(A)$.

In particular, by taking $B = (-\infty, x] \times (-\infty, y]$, we obtain the *joint d.f.* of X, Y, to be denoted by $F_{X,Y}$; namely, $F_{X,Y}(x, y) = P(X \leq x, Y \leq y), x, y \in \Re$. (See Figure 7.2.)

More formally, we have the following definition.

Definition 1.

(i) The *joint probability distribution* (or just *joint distribution*) of the pair of r.v.'s (X, Y) is a set function which assigns values to subsets B of $\Re^2 = \Re \times \Re$

according to the formula

$$P[(X,Y) \in B] = P(\{s \in \mathcal{S}; \quad (X(s), Y(s)) \in B\}), \quad B \subseteq \Re^2; \qquad (7.1)$$

the value assigned to B is denoted by $P_{X,Y}(B)$.

(ii) By taking B to be a rectangle of the form $(-\infty, x] \times (-\infty, y]$; that is, $B = (-\infty, x] \times (-\infty, y]$, relation (7.1) becomes

$$\begin{aligned} P[(X,Y) \in (-\infty, x] \times (-\infty, y]] &= P(\{s \in \mathcal{S}; \quad X(s) \le x, Y(s) \le y\}) \\ &= P(X \le x, Y \le y), \end{aligned}$$

and it defines a function on the plane $\Re^2$ denoted by $F_{X,Y}$ and called the *joint distribution function* (d.f.) of X and Y.

Remark 1.

(i) The joint probability distribution of the r.v.'s X, Y, $P_{X,Y}$, is a set function defined on subsets of the plane $\Re^2$, and it is seen to be a probability function (see Exercise 3.23 in Chapter 3).

(ii) From Definition 1, it follows that if we know $P_{X,Y}$, then we can determine $F_{X,Y}$. The converse is also true, but its proof is beyond the scope of this book. However, it does provide a motivation for our being preoccupied with it, since what we are really interested in is $P_{X,Y}$.

The d.f. $F_{X,Y}$ has properties similar to the ones mentioned in the case of a single r.v.; namely:

Proposition 1. The joint d.f. of the r.v.'s X, Y, $F_{X,Y}$, has the following properties:

(i) $0 \le F_{X,Y}(x,y) \le 1$ for all $x, y \in \Re$.
Whereas it is clearly still true that $x_1 \le x_2$ and $y_1 \le y_2$ imply $F_{X,Y}(x_1, y_1) \le F_{X,Y}(x_2, y_2)$, property (ii) in the case of a single r.v. may be restated as follows:

(ii) The *variation* of $F_{X,Y}$ over rectangles with sides parallel to the axes, given in Figure 7.3, is ≥ 0.

(iii) $F_{X,Y}$ is continuous from the right (right-continuous); that is, if $x_n \downarrow x$ and $y_n \downarrow y$, then $F_{X,Y}(x_n, y_n) \to F_{X,Y}(x,y)$ as $n \to \infty$.

(iv) $F_{X,Y}(+\infty, +\infty) = 1$ and $F_{X,Y}(-\infty, -\infty) = F_{X,Y}(-\infty, y) = F_{X,Y}(x, -\infty) = 0$ for any $x, y \in \Re$, where, of course, $F_{X,Y}(+\infty, +\infty)$ is defined to be $\lim_{n\to\infty} F_{X,Y}(x_n, y_n)$ as $x_n \uparrow \infty$ and $y_n \uparrow \infty$, and similarly for the remaining cases.

Property (i) is immediate, and property (ii) follows by the fact that the variation of $F_{X,Y}$ as described is simply the probability that the pair (X, Y) lies in the rectangle of Figure 7.3, or, more precisely, the probability $P(x_1 < X \le x_2, y_1 < Y \le y_2)$, which, of course, is ≥ 0; the justification of properties (iii) and (iv) is based on Theorem 2 in Chapter 3.

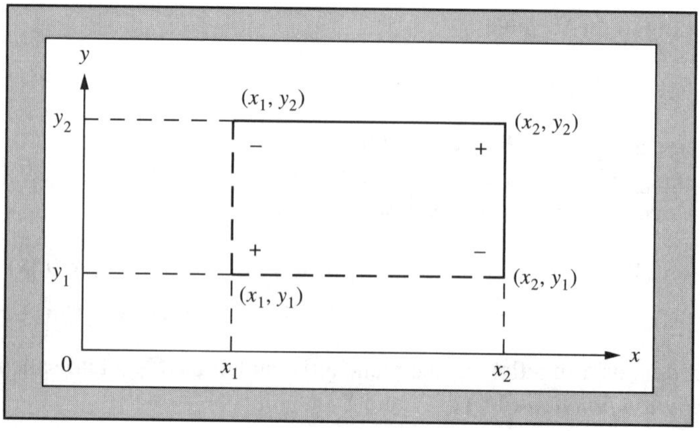

FIGURE 7.3

The variation of $F_{X,Y}$ over the rectangle is $F_{X,Y}(x_1, y_1) + F_{X,Y}(x_2, y_2) - F_{X,Y}(x_1, y_2) - F_{X,Y}(x_2, y_1)$.

Now, suppose that the r.v.'s X and Y are discrete and take on the values x_i and y_j, $i, j \geq 1$, respectively. Then the *joint p.d.f.* of X and Y, to be denoted by: $f_{X,Y}$, is defined by: $f_{X,Y}(x_i, y_j) = P(X = x_i, Y = y_j)$ and $f_{X,Y}(x, y) = 0$ when $(x, y) \neq (x_i, y_j)$ (i.e., at least one of x or y is not equal to x_i or y_j, respectively). It is then immediate that for $B \subseteq \Re^2$, $P[(X, Y) \in B] = \sum_{(x_i, y_j) \in B} f_{X,Y}(x_i, y_j)$, and, in particular, $\sum_{(x_i, y_j) \in \Re^2} f_{X,Y}(x_i, y_j) = 1$, and $F_{X,Y}(x, y) = \sum_{x_i \leq x, y_j \leq y} f_{X,Y}(x_i, y_j)$. More formally, we have the following definition and result.

Definition 2. Let X and Y be two (discrete) r.v.'s taking on the values x_i and y_j (finitely or infinitely many) with respective probabilities $P(X = x_i, Y = y_j)$, $i, j \geq 1$. Define the function $f_{X,Y}(x, y)$ as follows:

$$f_{X,Y}(x, y) = \begin{cases} P(X = x_i, Y = y_j) & \text{if } x = x_i \text{ and } y = y_j, \quad i,\ j \geq 1 \\ 0 & \text{otherwise.} \end{cases} \tag{7.2}$$

The function $f_{X,Y}$ is called the *joint probability density function* (*joint p.d.f.*) of the r.v.'s X and Y.

The following properties are immediate from the definition.

Proposition 2. Let $f_{X,Y}$ be as in Definition 2. Then:

(i) $f_{X,Y}(x, y) \geq 0$ for all $x, y \in \Re$.

(ii) For any $B \subseteq \Re^2$, $P[(X, Y) \in B] = \sum_{(x_i, y_j) \in B} f_{X,Y}(x_i, y_j)$.

(iii) In particular,

$$F_{X,Y}(x, y) = \sum_{x_i \leq x} \sum_{y_j \leq y} f_{X,Y}(x_i, y_j), \quad \text{and} \quad \sum_{x_i \in \Re} \sum_{y_j \in \Re} f_{X,Y}(x_i, y_j) = 1.$$

Table 7.1 Joint distribution of the r.v.'s X and Y.

y/x	0	1	2	3	Totals
0	0.05	0.21	0	0	0.26
1	0.20	0.26	0.08	0	0.54
2	0	0.06	0.07	0.02	0.15
3	0	0	0.03	0.02	0.05
Totals	0.25	0.53	0.18	0.04	1

Consider the following illustrative example.

Example 1. The number of customers lining up for service in front of two windows in your local bank are r.v.'s X and Y, and suppose that the r.v.'s X and Y take on four values only, 0, 1, 2, 3, with joint probabilities $f_{X,Y}$ expressed best in a matrix form as in Table 7.1.

There are many questions which can be posed. Some of them are:

(i) For $x = 2$ and $y = 1$, compute the $F_{X,Y}(2, 1)$.
(ii) Compute the probability $P(2 \leq X \leq 3, 0 \leq Y \leq 2)$.

Discussion.

(i) $F_{X,Y}(x, y) = F_{X,Y}(2, 1) = \sum_{u \leq 2, v \leq 1} f_{X,Y}(u, v) = f_{X,Y}(0, 0) + f_{X,Y}(0, 1) + f_{X,Y}(1, 0) + f_{X,Y}(1, 1) + f_{X,Y}(2, 0) + f_{X,Y}(2, 1) = 0.05 + 0.20 + 0.21 + 0.26 + 0 + 0.08 = 0.80$; also,
(ii) $P(2 \leq X \leq 3, 0 \leq Y \leq 2) = f_{X,Y}(2, 0) + f_{X,Y}(2, 1) + f_{X,Y}(2, 2) + f_{X,Y}(3, 0) + f_{X,Y}(3, 1) + f_{X,Y}(3, 2) = 0 + 0.08 + 0.07 + 0 + 0 + 0.02 = 0.17.$

Now, suppose that both X and Y are of the continuous type, and, indeed, a little bit more; namely, there exists a nonnegative function $f_{X,Y}$ defined on $\Re^2$ such that, for all x and y in $\Re$: $F_{X,Y}(x, y) = \int_{-\infty}^{y} \int_{-\infty}^{x} f_{X,Y}(s, t) ds\, dt$. Then for $B \subseteq \Re^2$ (interpret B as a familiar geometric figure in $\Re^2$): $P[(X, Y) \in B] = \int_B \int f_{X,Y}(x, y) dx\, dy$, and, in particular, $\int_{-\infty}^{\infty} \int_{-\infty}^{\infty} f_{X,Y}(x, y) dx\, dy = 1$. The function $f_{X,Y}$ is called the *joint p.d.f.* of X and Y. Analogously to the case of a single r.v., the relationship $\frac{\partial^2}{\partial x\, \partial y} F_{X,Y}(x, y) = f_{X,Y}(x, y)$ holds true (for continuity points (x, y) of $f_{X,Y}$), so that not only does the joint p.d.f. determine the joint d.f. through an integration process, but the converse is also true; that is, the joint d.f. determines the joint p.d.f. through differentiation. Again, as in the case of a single r.v., $P(X = x, Y = y) = 0$ for all $x, y \in \Re$.

Summarizing these things in the form of a definition and a proposition, we have

Definition 3. Let X and Y be two r.v.'s of the continuous type, and suppose there exists a function $f_{X,Y}$ such that:

$$f_{X,Y}(x, y) \geq 0 \text{ for all } x, y \in \Re,$$

and

$$P[(X,Y)\in B]=\int_B\int f_{X,Y}(x,y)dx\,dy,\quad B\subseteq\Re^2,\quad \int_{\Re^2}\int f_{X,Y}(x,y)dx\,dy=1. \tag{7.3}$$

The function f_{XY} is called the *joint probability density function (joint p.d.f.)* of the r.v.'s X and Y.

From the above definition and familiar results from calculus, we have:

Proposition 3. Let f_{XY} be as in Definition 3. Then:

(i) $F_{X,Y}(x,y)=\int_{-\infty}^{y}\int_{-\infty}^{x}f_{X,Y}(u,v)du\,dv$ for all $x,y\in\Re$ (by taking $B=(-\infty,x]\times(-\infty,x]$ in (7.3)).
(ii) $\frac{\partial^2}{\partial x\partial y}F_{X,Y}(x,y)=f_{X,Y}(x,y)$ (for all x and y *in* $\Re$ which are continuity points of $f_{X,Y}$).

Example 10 in Chapter 2 provides an example of two r.v.'s of the continuous type, where X and Y are the coordinates of the point of impact.

Remark 2.

(i) As was the case in a single r.v., it happens here also that calculation of probabilities is reduced either to a summation (for the discrete case) (see Proposition 2(ii)) or to an integration (for the continuous case) (see relation (7.3)).
(ii) Also, given a function $F_{X,Y}$ which satisfies properties (i)–(iv) in Proposition 1, one can always construct a pair if r.v.'s X, Y whose joint d.f. $F_{XY}(x,y)=F(x,y)$ for all $x,y\in\Re$.
(iii) The above discussion raises the question:
When is a function $f_{X,Y}$ the joint p.d.f. of two r.v.'s X and Y?
The answer is this: First, $f_{X,Y}(x,y)$ must be ≥ 0 for all $x,y\in\Re$; and second,

$$\sum_{x_i}\sum_{y_j}f(x_i,y_j)=1$$

for the discrete case, and $\int_{-\infty}^{\infty}\int_{-\infty}^{\infty}f(x,y)dx\,dy=1$ for the continuous case.

This section concludes with three illustrative examples.

Example 2. Let the r.v.'s X and Y have the joint p.d.f. $f_{X,Y}(x,y)=\lambda_1\lambda_2e^{-\lambda_1x-\lambda_2y}$, $x,y>0$, $\lambda_1,\lambda_2>0$. For example, X and Y may represent the lifetimes of two components in an electronic system. Derive the joint d.f. $F_{X,Y}$.

Discussion. The corresponding joint d.f. is:

$$F_{X,Y}(x,y)=\int_0^y\int_0^x\lambda_1\lambda_2\times e^{-\lambda_1s-\lambda_2t}ds\,dt=\int_0^y\lambda_2e^{-\lambda_2t}\left(\int_0^x\lambda_1e^{-\lambda_1s}ds\right)dt$$
$$=\int_0^y\lambda_2e^{-\lambda_2t}(1-e^{-\lambda_1x})dt=(1-e^{-\lambda_1x})(1-e^{-\lambda_2y})$$

for $x > 0$, $y > 0$, and 0 otherwise. That is,

$$F_{X,Y}(x, y) = (1 - e^{-\lambda_1 x})(1 - e^{-\lambda_2 y}), \qquad x > 0, y > 0, \tag{7.4}$$
and $F_{X,Y}(x, y) = 0$ otherwise.

Example 3. If the function $F_{X,Y}$ given by:

$$F_{X,Y}(x, y) = \frac{1}{16}xy(x + y), \quad 0 \leq x \leq 2, 0 \leq y \leq 2,$$

is the joint d.f. of the r.v.'s X and Y, then:

(i) Determine the corresponding joint p.d.f. $f_{X,Y}$.
(ii) Verify that $f_{X,Y}$ found in part (ii) is, indeed, a p.d.f.
(iii) Calculate the probability $P(0 \leq X \leq 1, 1 \leq Y \leq 2)$.

Discussion.

(i) For $0 \leq x \leq 2$ and $0 \leq y \leq 2$,

$$\begin{aligned} f_{X,Y}(x, y) &= \frac{\partial^2}{\partial x \partial y}\left(\frac{1}{16}xy(x + y)\right) = \frac{1}{16}\frac{\partial^2}{\partial x \partial y}(x^2 y + xy^2) \\ &= \frac{1}{16}\frac{\partial}{\partial y}\frac{\partial}{\partial x}(x^2 y + xy^2) = \frac{1}{16}\frac{\partial}{\partial y}(2xy + y^2) \\ &= \frac{1}{16}(2x + 2y) = \frac{1}{8}(x + y); \end{aligned}$$

that is, $f_{X,Y}(x, y) = \frac{1}{8}(x + y)$, $0 \leq x \leq 2$, $0 \leq y \leq 2$. For (x, y) outside the rectangle $[0, 2] \times [0, 2]$, $f_{X,Y}$ is 0, since $F_{X,Y}$ is constantly either 0 or 1.

(ii) Since $f_{X,Y}$ is nonnegative, all we have to show is that it integrates to 1. In fact,

$$\begin{aligned} \int_{-\infty}^{\infty}\int_{-\infty}^{\infty} f_{X,Y}(x, y)dx\,dy &= \int_0^2\int_0^2 \frac{1}{8}(x + y)dx\,dy \\ &= \frac{1}{8}\left(\int_0^2\int_0^2 x\,dx\,dy + \int_0^2\int_0^2 y\,dx\,dy\right) \\ &= \frac{1}{8}(2 \times 2 + 2 \times 2) \\ &= 1. \end{aligned}$$

(iii) Here, $P(0 \leqslant X \leq 1, 1 \leq Y \leqslant 2) = \int_1^2\int_0^1 \frac{1}{8}(x+y)dx\,dy = \frac{1}{8}\left[\int_1^2\left(\int_0^1 x\,dx\right)dy + \int_1^2\left(y\int_0^1 dx\right)dy\right] = \frac{1}{8}\left(\frac{1}{2} \times 1 + 1 \times \frac{3}{2}\right) = \frac{1}{4}$.

Example 4. It is known that tire pressure in an automobile improves (up to a certain point) the mileage efficiency in terms of fuel. For an automobile chosen at random, let the r.v.'s X and Y stand for tire pressure and mileage rate, respectively, and suppose that their joint p.d.f. $f_{X,Y}$ is given by $f_{X,Y}(x, y) = cx^2y$ for $0 < x^2 \leq y < 1, x > 0$ (and 0 otherwise):

(i) Determine the constant c, so that $f_{X,Y}$ is a p.d.f.

(ii) Calculate the probability $P(0 < X < \frac{3}{4}, \frac{1}{4} \leq Y < 1)$.

Discussion.

(i) Clearly, for the function to be nonnegative, c must be > 0. The actual value of c will be determined through the relationship below for $x > 0$:

$$\iint_{\{(x,y);0<x^2\leq y<1\}} cx^2y\,dx\,dy = 1.$$

The region over which the p.d.f. is positive is the shaded region in Figure 7.4, determined by a branch of the parabola $y = x^2$, the y-axis, and the line segment connecting the points (0, 1) and (1, 1). Since for each fixed x with $0 < x < 1$, y ranges from x^2 to 1, we have:

$$\iint_{\{x^2\leq y<1\}} cx^2y\,dx\,dy = c\int_0^1\left(x^2\int_{x^2}^1 y\,dy\right)dx = \frac{c}{2}\int_0^1 x^2(1-x^4)dx$$

$$= \frac{c}{2}\left(\frac{1}{3}-\frac{1}{7}\right) = \frac{2c}{21} = 1$$

and $c = \frac{21}{2}$.

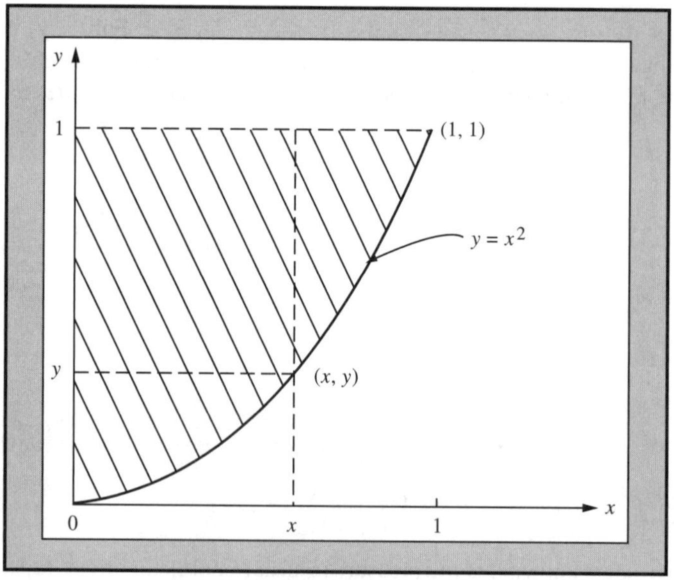

FIGURE 7.4

Range of the pair (x, y).

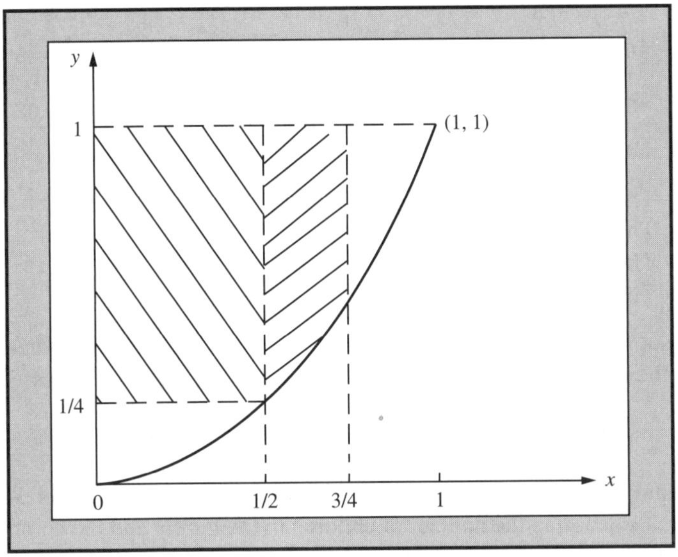

FIGURE 7.5
Diagram facilitating integration.

(ii) Since $y = x^2 = \frac{1}{4}$ for $x = \frac{1}{2}$, it follows that for each x with $0 < x \leq \frac{1}{2}$, the range of y is from $\frac{1}{4}$ to 1; on the other hand, for each x with $\frac{1}{2} < x \leq \frac{3}{4}$, the range of y is from x^2 to 1 (see Figure 7.5).
Thus,

$$\begin{aligned}
P\left(0 < X \leq \frac{3}{4}, \frac{1}{4} \leq Y < 1\right) &= c\int_0^{\frac{1}{2}}\int_{\frac{1}{4}}^1 x^2 y\,dy\,dx + c\int_{\frac{1}{2}}^{\frac{3}{4}}\int_{x^2}^1 x^2 y\,dy\,dx \\
&= c\int_0^{\frac{1}{2}}\left(x^2\int_{\frac{1}{4}}^1 y\,dy\right)dx + c\int_{\frac{1}{2}}^{\frac{3}{4}}\left(x^2\int_{x^2}^1 y\,dy\right)dx \\
&= \frac{c}{2}\int_0^{\frac{1}{2}} x^2\left(1 - \frac{1}{16}\right)dx + \frac{c}{2}\int_{\frac{1}{2}}^{\frac{3}{4}} x^2(1 - x^4)dx \\
&= \frac{15c}{3 \times 2^8} + \frac{38c}{3 \times 2^8} - \frac{2,059c}{7 \times 2^{15}} = c \times \frac{41,311}{21 \times 2^{15}} \\
&= \frac{21}{2} \times \frac{41,311}{21 \times 2^{15}} = \frac{41,311}{2^{16}} = \frac{41,311}{65,536} \simeq 0.63.
\end{aligned}$$

Exercises.

1.1 Let X and Y be r.v.'s denoting the number of cars and buses, respectively, lined up at a stoplight at a given point in time, and suppose their joint p.d.f. is given by the following table:

y\x	0	1	2	3	4	5
0	0.025	0.050	0.125	0.150	0.100	0.050
1	0.015	0.030	0.075	0.090	0.060	0.030
2	0.010	0.020	0.050	0.060	0.040	0.020

Calculate the following probabilities:

(i) There are exactly 4 cars and no buses.
(ii) There are exactly 5 cars.
(iii) There is exactly 1 bus.
(iv) There are at most 3 cars and at least 1 bus.

1.2 In a sociological project, families with 0, 1, and 2 children are studied. Suppose that the numbers of children occur with the following frequencies:

$$0 \text{ children: } 30\%; \quad 1 \text{ child : } 40\%; \quad 2 \text{ children: } 30\%.$$

A family is chosen at random from the target population, and let X and Y be the r.v.'s denoting the number of children in the family and the number of boys among those children, respectively. Finally, suppose that $P(\text{observing a boy}) = P(\text{observing a girl}) = 0.5$.

Calculate the joint p.d.f. $f_{X,Y}(x, y) = P(X = x, Y = y), 0 \leq y \leq x$, $x = 0, 1, 2$.

Hint. Tabulate the joint probabilities as indicated below by utilizing the formula:

$$P(X = x, Y = y) = P(Y = y \mid X = x)P(X = x).$$

y\x	0	1	2
0			
1			
2			

1.3 If the r.v.'s X and Y have the joint p.d.f. given by:

$$f_{X,Y}(x, y) = x + y, \quad 0 < x < 1, \quad 0 < y < 1,$$

calculate the probability $P(X < Y)$.

Hint. Can you guess the answer without doing any calculations?

1.4 The r.v.'s X and Y have the joint p.d.f. $f_{X,Y}$ given by:

$$f_{X,Y}(x, y) = \frac{6}{7}\left(x^2 + \frac{xy}{2}\right), \quad 0 < x \leq 1, \quad 0 < y \leq 2.$$

(i) Show that $f_{X,Y}$ is, indeed, a p.d.f.
(ii) Calculate the probability $P(X > Y)$.

1.5 The r.v.'s X and Y have the joint p.d.f. $f_{X,Y}(x, y) = e^{-x-y}$, $x > 0$, $y > 0$.

(i) Calculate the probability $P(X \leq Y \leq c)$, in terms of c, for some $c > 0$.
(ii) Find the numerical value in part (i) for $c = \log 2$, where log is, as always, the natural logarithm.

Hint. The integration in part (i) may be facilitated by drawing the picture of the set for which $x \leq y$.

1.6 If the r.v.'s X and Y have the joint p.d.f. $f_{X,Y}(x, y) = e^{-x-y}$, for $x > 0$ and $y > 0$, compute the following probabilities:

(i) $P(X \leq x)$; (ii) $P(Y \leq y)$; (iii) $P(X < Y)$; (iv) $P(X + Y \leq 3)$.

Hint. For part (iii), draw the picture of the set for which $0 < x < y$, and for part (iv), draw the picture of the set for which $0 < x + y \leq 3$.

1.7 Let X and Y be r.v.'s jointly distributed with p.d.f. $f_{X,Y}(x, y) = 2/c^2$, for $0 < x \leq y < c$.
Determine the constant c.

Hint. Draw the picture of the set for which $0 < x \leq y < c$.

1.8 The r.v.'s X and Y have the joint p.d.f. $f_{X,Y}$ given by:

$$f_{X,Y}(x, y) = cye^{-xy/2}, \quad 0 < y < x.$$

Determine the constant c.

Hint. Draw the picture of the set for which $x > y > 0$.

1.9 The joint p.d.f. of the r.v.'s X and Y is given by:

$$f_{X,Y}(x, y) = xy^2, \quad 0 < x \leq c_1, \quad 0 < y \leq c_2.$$

Determine the condition that c_1 and c_2 must satisfy so that $f_{X,Y}$ is, indeed, a p.d.f.

Hint. All that can be done here is to find a relation that c_1 and c_2 satisfy; c_1 and c_2 cannot be determined separately.

1.10 The joint p.d.f. of the r.v.'s X and Y is given by:

$$f_{X,Y}(x, y) = cx, x > 0, y > 0, 1 \leq x + y < 2 \quad (c > 0).$$

Determine the constant c.

Hint. The following diagram, Figure 7.6, should facilitate the calculations. The range of the pair (x, y) is the shadowed area.

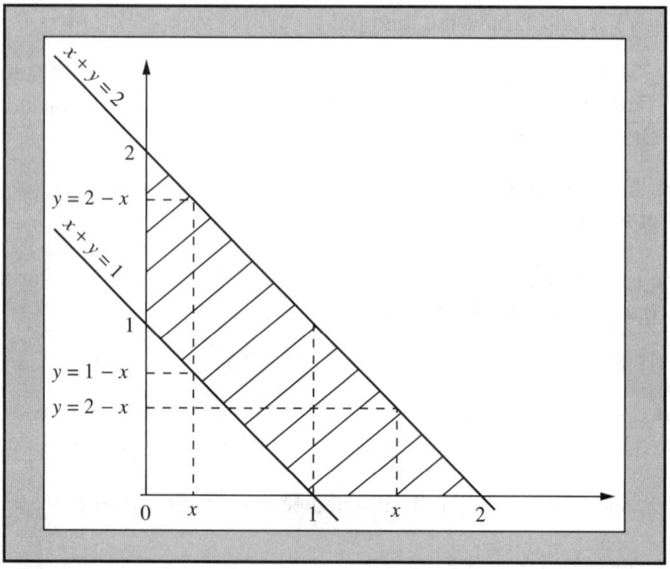

FIGURE 7.6
The shaded area is the set of pairs (x, y) for which $x > 0, y > 0$ and $1 < x + y < 2$.

1.11 The r.v.'s X and Y have joint p.d.f. $f_{X,Y}$ given by:

$$f_{X,Y}(x, y) = c(y^2 - x^2)e^{-y}, \quad -y < x < y, \quad 0 < y < \infty.$$

Determine the constant c.

Hint. We have that $\int_0^\infty y\, e^{-y} dy = 1$, and the remaining integrals are computed recursively.

1.12 Let X and Y be r.v.'s jointly uniformly distributed over the triangle OAC (see Figure 7.7), and let $U = X^2 + Y^2$ be the square distance of the point (X, Y) from the origin in said triangle.

(i) Compute the EU in terms of C (without finding the p.d.f. of U).
(ii) Find the numerical value of EU for $C = 1$.

1.13 Let X and Y be two r.v.'s for which $P(X > x, Y > y) = P(X > x)P(Y > y)$ for all x, y in $\Re$. Then $P(X \leq x, Y \leq y) = P(X \leq x)P(Y \leq y)$.

Remark: The same is true for any finite number of r.v.'s, although its justification is not pursued here.

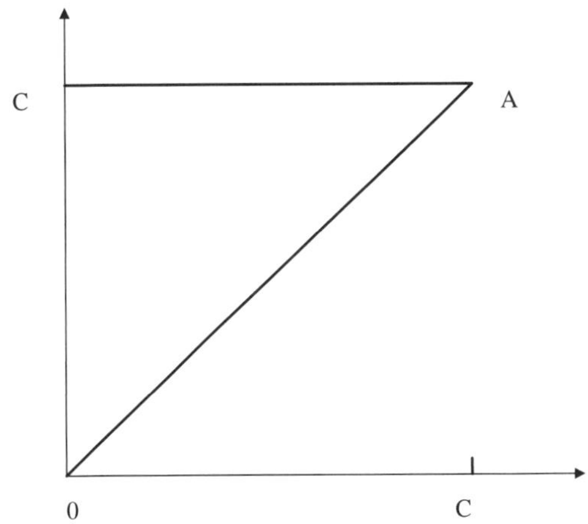

FIGURE 7.7
The triangle OAC is the area over which the pair of r.v.'s (X, Y) is distributed uniformly.

7.2 Marginal and Conditional p.d.f.'s, Conditional Expectation, and Variance

In the case of two r.v.'s with joint d.f. $F_{X,Y}$ and joint p.d.f. $f_{X,Y}$, we may define quantities that were not available in the case of a single r.v. These quantities are marginal d.f.'s and p.d.f.'s, conditional p.d.f.'s, and conditional expectations and variances. To this end, consider the joint d.f. $F_{X,Y}(x, y) = P(X \leq x, Y \leq y)$, and let $y \to \infty$. Then we obtain $F_{X,Y}(x, \infty) = P(X \leq x, Y < \infty) = P(X \leq x) = F_X(x)$; thus, $F_X(x) = F_{X,Y}(x, \infty)$, and likewise, $F_Y(y) = F_{X,Y}(\infty, y)$. That is, the d.f.'s of the r.v.'s X and Y are obtained from their joint d.f. by eliminating one of the variables x or y through a limiting process. The d.f.'s F_X and F_Y are referred to as *marginal* d.f.'s. If the r.v.'s X and Y are discrete with joint p.d.f. $f_{X,Y}$, then $P(X = x_i) = P(X = x_i, -\infty < Y < \infty) = \sum_{y_j \in \Re} f_{X,Y}(x_i, y_j)$; that is, $f_X(x_i) = \sum_{y_j \in \Re} f_{X,Y}(x_i, y_j)$, and likewise, $f_Y(y_j) = \sum_{x_i \in \Re} f_{X,Y}(x_i, y_j)$. Because of this marginalization process, the p.d.f.'s of the r.v.'s. X and Y, f_X and f_Y, are referred to as *marginal* p.d.f.'s. In the continuous case, f_X and f_Y are obtained by integrating out the "superfluous" variables; i.e., $f_X(x) = \int_{-\infty}^{\infty} f_{X,Y}(x, y)dy$ and $f_Y(y) = \int_{-\infty}^{\infty} f_{X,Y}(x, y)dx$. The marginal f_X is, indeed, the p.d.f. of X because $P(X \leq x) = P(X \leq x, -\infty < Y < \infty) = \int_{-\infty}^{x} \int_{-\infty}^{\infty} f_{X,Y}(s, t)dt\, ds = \int_{-\infty}^{x} [\int_{-\infty}^{\infty} f_{X,Y}(s, t)dt]\, ds = \int_{-\infty}^{x} f_X(s)ds$; that is, $F_X(x) = P(X \leq x) = \int_{-\infty}^{x} f_X(s)ds$, so that $\frac{d}{dx} F_X(x) = f_X(x)$, and likewise, $\frac{d}{dy} F_Y(y) = f_Y(y)$ (for continuity points x and y of f_X and f_Y, respectively).

Summarizing these things in the form of a definition, we have then:

Definition 4.

(i) Let $F_{X,Y}$ be the joint d.f. of the r.v.'s X and Y. Then:

$$F_X(x) = F_{X,Y}(x, \infty) = \lim_{y \to \infty} F_{X,Y}(x, y),$$
$$F_Y(y) = F_{X,Y}(\infty, y) = \lim_{x \to \infty} F_{X,Y}(x, y) \tag{7.5}$$

are called *marginal d.f.*'s of $F_{X,Y}$, and they are the d.f.'s of the r.v.'s X and Y, respectively.

(ii) Let $f_{X,Y}$ be the joint p.d.f. of the r.v.'s X and Y. For the discrete and the continuous case, set, respectively,

$$f_X(x_i) = \sum_{y_j \in \Re} f_{X,Y}(x_i, y_j), \quad f_Y(y_j) = \sum_{x_i \in \Re} f_{X,Y}(x_i, y_j),$$
$$f_X(x) = \int_{-\infty}^{\infty} f_{X,Y}(x, y)dy, \quad f_Y(y) = \int_{-\infty}^{\infty} f_{X,Y}(x, y)dx. \tag{7.6}$$

The functions f_X, f_Y, are called *marginal p.d.f.*'s of $f_{X,Y}$ and they are the p.d.f.'s of the r.v.'s X and Y, respectively.

In terms of the joint and the marginal p.d.f.'s, one may define formally the functions:

$$f_{X|Y}(x|y) = f_{X,Y}(x, y)/f_Y(y) \quad \text{for fixed } y \text{ with } f_Y(y) > 0, \tag{7.7}$$

and

$$f_{Y|X}(y|x) = f_{X,Y}(x, y)/f_X(x) \quad \text{for fixed } x \text{ with } f_X(x) > 0. \tag{7.8}$$

These nonnegative functions are, actually, p.d.f.'s. For example, for the continuous case:

$$\int_{-\infty}^{\infty} f_{X|Y}(x|y)dx = \frac{1}{f_Y(y)} \int_{-\infty}^{\infty} f_{X,Y}(x, y)dx = \frac{f_Y(y)}{f_Y(y)} = 1,$$

and similarly for $f_{Y|X}(\cdot \mid x)$; in the discrete case, integrals are replaced by summation signs. The p.d.f. $f_{X|Y}(\cdot|y)$ is called the *conditional* p.d.f. of X, given $Y = y$, and $f_{Y|X}(\cdot \mid x)$ is the *conditional* p.d.f. of Y, given $X = x$. The motivation for this terminology is as follows: For the discrete case, $f_{X|Y}(x \mid y) = \frac{f_{X,Y}(x,y)}{f_Y(y)} = \frac{P(X=x, Y=y)}{P(Y=y)} = P(X = x \mid Y = y)$; i.e., $f_{X|Y}(x \mid y)$ does, indeed, stand for the conditional probability that $X = x$, given that $Y = y$. Likewise for $f_{Y|X}(\cdot \mid x)$. In the continuous case, the points x and y are to be replaced by "small" intervals around them.

Definition 5. The quantities defined in relations (7.7) and (7.8) are p.d.f.'s and they are called the *conditional p.d.f.* of X, given $Y = y$, and the *conditional p.d.f.* of Y, given $X = x$, respectively.

The concepts introduced so far are now illustrated by means of examples.

Example 5. Refer to Example 1 and derive the marginal and conditional p.d.f.'s involved.

Discussion. From Table 7.1, we have: $f_X(0) = 0.25$, $f_X(1) = 0.53$, $f_X(2) = 0.18$, and $f_X(3) = 0.04$; also, $f_Y(0) = 0.26$, $f_Y(1) = 0.54$, $f_Y(2) = 0.15$, and $f_Y(3) = 0.05$. Thus, the probability that there are 2 people in the x-line, for instance, regardless of how many people are in the y-line, is: $P(X = 2) = f_X(2) = 0.18$.

Next, all values of conditional p.d.f.'s are tabulated below for easy reading and later reference.

x	0	1	2	3
$f_{X\|Y}(x\|0)$	$\frac{0.05}{0.26} = \frac{5}{26} \simeq 0.192$	$\frac{0.21}{0.26} = \frac{21}{26} \simeq 0.808$	0	0
$f_{X\|Y}(x\|1)$	$\frac{0.20}{0.54} = \frac{20}{54} \simeq 0.37$	$\frac{0.26}{0.54} = \frac{26}{54} \simeq 0.482$	$\frac{0.08}{0.54} = \frac{8}{54} \simeq 0.148$	0
$f_{X\|Y}(x\|2)$	0	$\frac{0.06}{0.15} = \frac{6}{15} = 0.40$	$\frac{0.07}{0.15} = \frac{7}{15} \simeq 0.467$	$\frac{0.02}{0.15} = \frac{2}{15} \simeq 0.133$
$f_{X\|Y}(x\|3)$	0	0	$\frac{0.03}{0.05} = \frac{3}{5} = 0.60$	$\frac{0.02}{0.05} = \frac{2}{5} = 0.40$

Likewise,

y	0	1	2	3
$f_{Y\|X}(y\|0)$	$\frac{0.05}{0.25} = \frac{5}{25} = 0.2$	$\frac{0.20}{0.25} = \frac{20}{25} = 0.8$	0	0
$f_{Y\|X}(y\|1)$	$\frac{0.21}{0.53} = \frac{21}{53} \simeq 0.396$	$\frac{0.26}{0.53} = \frac{26}{53} \simeq 0.491$	$\frac{0.06}{0.53} = \frac{6}{53} \simeq 0.113$	0
$f_{Y\|X}(y\|2)$	0	$\frac{0.08}{0.18} = \frac{8}{18} \simeq 0.444$	$\frac{0.07}{0.18} = \frac{7}{18} \simeq 0.389$	$\frac{0.03}{0.18} = \frac{3}{18} \simeq 0.167$
$f_{Y\|X}(y\|3)$	0	0	$\frac{0.02}{0.04} = \frac{2}{4} = 0.50$	$\frac{0.02}{0.04} = \frac{2}{4} = 0.50$

Thus, for example, the (conditional) probability that there will be 2 customers in the x-line, given that there is 1 customer in the y-line, is approximately 0.148; and the (conditional) probability that there will be 1 customer in the y-line, given that there is 1 customer in the x-line, is approximately 0.491.

Example 6. Refer to Example 2 and derive the marginal d.f.'s and p.d.f.'s, as well as the conditional p.d.f.'s, involved.

Discussion. In (7.1), let $y \to \infty$ to obtain $F_X(x) = 1 - e^{-\lambda_1 x}, x > 0$, and likewise $F_Y(y) = 1 - e^{-\lambda_2 y}, y > 0$, by letting $x \to \infty$. Next, by differentiation, $f_X(x) = \lambda_1 e^{-\lambda_1 x}, x > 0$, and $f_Y(y) = \lambda_2 e^{-\lambda_2 y}, y > 0$, so that the r.v.'s X and Y have negative exponential distribution with parameters λ_1 and λ_2, respectively. Finally, for $x > 0$ and $y > 0$:

$$f_{X|Y}(x|y) = \frac{\lambda_1 \lambda_2 e^{-\lambda_1 x - \lambda_2 y}}{\lambda_2 e^{-\lambda_2 y}} = \lambda_1 e^{-\lambda_1 x} = f_X(x), \text{ and likewise}$$

$$f_{Y|X}(y|x) = f_Y(y).$$

Example 7. Refer to Example 4 and determine the marginal and conditional p.d.f.'s f_X, f_Y, $f_{X|Y}$, and $f_{Y|X}$. Also, compute the $P(a < Y < b|X = x)$, $x^2 < a < b < 1$, and its numerical value for $a = \frac{3}{8}$, $b = \frac{3}{4}$, $x = \frac{1}{2}$.

Discussion. We have:

$$f_X(x) = \int_{x^2}^{1} cx^2 y\,dy = cx^2 \int_{x^2}^{1} y\,dy = \frac{21}{4}x^2(1-x^4), \qquad 0 < x < 1,$$

$$f_Y(y) = \int_{0}^{\sqrt{y}} cx^2 y\,dx = cy \int_{0}^{\sqrt{y}} x^2\,dx = \frac{7}{2}y^2\sqrt{y}, \qquad 0 < y < 1,$$

and therefore

$$f_{X|Y}(x|y) = \frac{\frac{21}{2}x^2 y}{\frac{21}{6}y^2\sqrt{y}} = \frac{3x^2}{y\sqrt{y}}, \qquad 0 < x \le \sqrt{y},\ 0 < y < 1,$$

$$f_{Y|X}(y|x) = \frac{\frac{21}{2}x^2 y}{\frac{21}{4}x^2(1-x^4)} = \frac{2y}{1-x^4}, \qquad x^2 \le y < 1,\ 0 < x < 1.$$

Finally,

$$P(a < Y < b|X = x) = \int_a^b \frac{2y}{1-x^4}dy = \frac{b^2-a^2}{1-x^4},$$

and its numerical value is 0.45. Thus, when the tire pressure is equal to 1/2 units, then the probability that the mileage rate will be between 3/8 and 3/4 units is equal to 0.45.

Example 8. Refer to Examples 5 and 7 and for later reference, also compute: EX, EY, *Var*(X), and *Var*(Y).

Discussion. For Example 5, we have:

$$EX = 1.01,\ EX^2 = 1.61, \text{ so that } Var(X) = 0.5899;$$
$$EY = 0.99,\ EY^2 = 1.59, \text{ so that } Var(Y) = 0.6099.$$

For Example 7, we have:

$$EX = \frac{21}{4}\int_0^1 x^3(1-x^4)dx = \frac{21}{4}\left(\frac{x^4}{4}\Big|_0^1 - \frac{x^8}{8}\Big|_0^1\right) = \frac{21}{32},$$

$$EX^2 = \frac{21}{4}\int_0^1 x^4(1-x^4)dx = \frac{21}{4}\left(\frac{x^5}{5}\Big|_0^1 - \frac{x^9}{9}\Big|_0^1\right) = \frac{7}{15},$$

$$\text{so that } Var(X) = \frac{7}{15} - \left(\frac{21}{32}\right)^2 = \frac{553}{15{,}360} \simeq 0.036;$$

$$EY = \frac{7}{2}\int_0^1 y^3\sqrt{y}\,dy = \frac{7}{2} \times \frac{2}{9}y^{9/2}\Big|_0^1 = \frac{7}{9},$$

$$EY^2 = \frac{7}{2}\int_0^1 y^4\sqrt{y}\,dy = \frac{7}{2} \times \frac{2}{11}y^{11/2}\Big|_0^1 = \frac{7}{11},$$

$$\text{so that } Var(Y) = \frac{7}{11} - \left(\frac{7}{9}\right)^2 = \frac{28}{891} \simeq 0.031.$$

Once a conditional p.d.f. is at hand, an expectation can be defined as previously done in relations (5.1), (5.2), and (5.3) of Chapter 5. However, a modified notation will be needed to reveal the fact that the expectation is calculated with respect to a *conditional* p.d.f. The resulting expectation is the *conditional expectation* of one r.v., given another r.v., as specified below.

Definition 6. For two r.v.'s X and Y, of either the discrete or the continuous type, the *conditional expectation* of one of these r.v.'s, given the other, is defined by:

$$E(X \mid Y = y_j) = \sum_{x_i \in \Re} x_i f_{X|Y}(x_i \mid y_j) \quad \text{or} \quad E(X \mid Y = y) = \int_{-\infty}^{\infty} x f_{X|Y}(x \mid y)dx, \tag{7.9}$$

for the discrete and continuous case, respectively; similarly:

$$E(Y \mid X = x_i) = \sum_{y_j \in \Re} y_j f_{Y|X}(y_j \mid x_i) \quad \text{or}$$

$$E(Y \mid X = x) = \int_{-\infty}^{\infty} y f_{Y|X}(y \mid x)dy. \tag{7.10}$$

Of course, it is understood that the preceding expectations exist as explained right after relations (5.2) and (5.3) were defined.

Remark 3. It is to be emphasized that unlike the results in (5.1)–(5.3), which are numbers, in relations (7.9) and (7.10) above the outcomes depend on y_j or y, and x_i or x, respectively, which reflect the values that the "conditioning" r.v.'s assume.

For illustrative purposes, let us calculate some conditional expectations.

Example 9. In reference to Example 1 (see also Example 5), calculate: $E(X \mid Y = 0)$ and $E(Y \mid X = 2)$.

Discussion. In Example 5, we have calculated the conditional p.d.f.'s $f_{X|Y}(\cdot \mid 0)$ and $f_{Y|X}(\cdot \mid 2)$. Therefore:

$$E(X \mid Y = 0) = 0 \times \frac{5}{26} + 1 \times \frac{21}{26} + 2 \times 0 + 3 \times 0 = \frac{21}{26} \simeq 0.808, \quad \text{and}$$

$$E(Y \mid X = 2) = 0 \times 0 + 1 \times \frac{8}{18} + 2 \times \frac{7}{18} + 3 \times \frac{3}{18} = \frac{31}{18} \simeq 1.722.$$

So, if in the y-line there are no customers waiting, the expected number of those waiting in the x-line will be about 0.81; likewise, if there are 2 customers waiting in the x-line, the expected number of those waiting in the y-line will be about 1.72.

Example 10. In reference to Example 2 (see also Example 6), calculate: $E(X \mid Y = y)$ and $E(Y \mid X = x)$.

Discussion. In Example 6, we have found that $f_{X|Y}(x \mid y) = f_X(x) = \lambda_1 e^{-\lambda_1 x}$ $(x > 0)$, and $f_{Y|X}(y \mid x) = f_Y(y) - \lambda_2 e^{-\lambda_2 y} (y > 0)$, so that: $E(X \mid Y = y) =$

$\int_0^\infty x\,\lambda_1 e^{-\lambda_1 x}dx = 1/\lambda_1$, and $E(Y \mid X = x) = \int_0^\infty y\lambda_2 e^{-\lambda_2 y}\,dy = 1/\lambda_2$, by integration by parts, or simply by utilizing known results.

Example 11. In reference to Example 4 (see also Example 7), calculate: $E(X \mid Y = y)$ and $E(Y \mid X = x)$.

Discussion. In Example 7, we have found that $f_{X|Y}(x \mid y) = \frac{3x^2}{y\sqrt{y}}, 0 < x \leq \sqrt{y} < 1$, so that:

$$E(X \mid Y = y) = \int_0^{\sqrt{y}} x \times \frac{3x^2}{y\sqrt{y}}dx = \frac{3}{y\sqrt{y}}\int_0^{\sqrt{y}} x^3\,dx = \frac{3\sqrt{y}}{4}, \qquad 0 < y < 1.$$

Also, $f_{Y|X}(y \mid x) = \frac{2y}{1-x^4}, x^2 \leq y < 1, 0 < x < 1$, so that

$$E(Y \mid X = x) = \int_{x^2}^1 y \times \frac{2y}{1-x^4}dy = \frac{2}{1-x^4}\int_{x^2}^1 y^2\,dy = \frac{2(1-x^6)}{3(1-x^4)}, \quad 0 < x < 1.$$

Following up the interpretation of the r.v.'s X and Y (given in Example 4), by taking $x = 0.5$, we have then that $E(Y|X = \frac{1}{2}) = 0.7$. That is, the expected mileage rate is 0.7 when the tire pressure is 0.5.

Now, for the discrete case, set $g(y_j) = E(X \mid Y = y_j)$ and proceed to replace y_j by the r.v. Y. We obtain the r.v. $g(Y) = E(X \mid Y)$, and then it makes sense to talk about its expectation $Eg(Y) = E[E(X \mid Y)]$. Although the $E(X \mid Y = y_j)$ depends on the particular values of Y, it turns out that its average does not, and, indeed, is the same as the EX. More precisely, it holds:

Proposition 4.

$$E[E(X \mid Y)] = EX \quad \text{and} \quad E[E(Y \mid X)] = EY. \tag{7.11}$$

That is, the expectation of the conditional expectation of X is equal to its expectation, and likewise for Y. Relation (7.11) is true for both the discrete and the continuous case.

Proof. The justification of (7.11) for the continuous case, for instance, is as follows: We have $g(Y) = E(X \mid Y)$ and therefore

$$\begin{aligned}
Eg(Y) &= \int_{-\infty}^{\infty} g(y) f_Y(y)dy = \int_{-\infty}^{\infty} E(X \mid y) f_Y(y)dy \\
&= \int_{-\infty}^{\infty}\left[\int_{-\infty}^{\infty} x f_{X|Y}(x \mid y)dx\right] f_Y(y)dy \\
&= \int_{-\infty}^{\infty}\int_{-\infty}^{\infty} [x f_{X|Y}(x \mid y) f_Y(y)dx]dy = \int_{-\infty}^{\infty}\int_{-\infty}^{\infty} x f_{X,Y}(x, y)dx\,dy \\
&= \int_{-\infty}^{\infty} x\left[\int_{-\infty}^{\infty} f_{X,Y}(x, y)dy\right]dx = \int_{-\infty}^{\infty} x f_X(x)dx = EX; \text{ that is,}
\end{aligned}$$

$Eg(Y) = E[E(X \mid Y)] = EX$. ▲

Remark 4. However, $Var[E(X \mid Y)] \leqslant) Var(X)$ with equality holding, if and only if Y is a function of X (with probability 1). A proof of this fact may be found in Section 5.3.1 in *A Course in Mathematical Statistics*, 2nd edition (1997), Academic Press, by G. G. Roussas.

Example 12. Verify the first relation $E[E(X \mid Y)] = EX$, in (7.11) for Example 4 (see also Examples 7 and 11).

Discussion. By Example 7, $f_X(x) = \frac{21}{4}x^2(1-x^4), 0 < x < 1$, so that:

$$EX = \int_0^1 x \times \frac{21}{4}x^2(1-x^4)dx = \frac{21}{4}\left(\int_0^1 x^3 dx - \int_0^1 x^7 dx\right) = \frac{21}{32}.$$

From Example 11, $E(X \mid Y) = \frac{3\sqrt{Y}}{4}, 0 < Y < 1$, whereas from Example 7, $f_Y(y) = \frac{21}{6}y^2\sqrt{y}, 0 < y < 1$, so that:

$$E[E(X \mid Y)] = \int_0^1 \frac{3\sqrt{y}}{4} \times \frac{21}{6}y^2\sqrt{y}\,dy = \frac{21}{8}\int_0^1 y^3\,dy = \frac{21}{32} = EX.$$

Remark 5. The point made in Remark 4 is ascertained below by way of Example 12. Indeed, in Example 12, $Var[E(X \mid Y)] = Var\left(\frac{3\sqrt{Y}}{4}\right) = \frac{9}{16}Var(\sqrt{Y}) = \frac{9}{16} \times [EY - (E\sqrt{Y})^2] = \frac{9}{16}\left(\frac{4}{5} - \frac{49}{64}\right) = \frac{99}{5{,}120} < \frac{2}{75} = Var(Y)$.

In addition to the conditional expectation of X, given Y, one may define the *conditional variance* of X, given Y, by utilizing the conditional p.d.f. and formula (5.8) (in Chapter 5); the notation to be used is $Var(X \mid Y = y_j)$ or $Var(X \mid Y = y)$ for the discrete and continuous case, respectively. Thus:

Definition 7. For two r.v.'s X and Y, either of the discrete or of the continuous type, the *conditional variance* of one of these r.v.'s, given the other, is defined by:

$$Var(X \mid Y = y_j) = \sum_{x_i \in \Re} [x_i - E(X \mid Y = y_j)]^2 f_{X|Y}(x_i \mid y_j), \tag{7.12}$$

and

$$Var(X \mid Y = y) = \int_{-\infty}^{\infty} [x - E(X \mid Y = y)]^2 f_{X|Y}(x \mid y)dx, \tag{7.13}$$

for the discrete and the continuous case, respectively.

Remark 6. The conditional variances depend on the values of the conditioning r.v., as was the case for the conditional expectations.

From formulas (7.12) and (7.13) (see also Exercise 2.20), it is not hard to see that:

Proposition 5.

$$\begin{aligned} Var(X \mid Y = y_j) &= E(X^2 \mid Y = y_j) - [E(X \mid Y = y_j)]^2 \quad \text{or} \\ Var(X \mid Y = y) &= E(X^2 \mid Y = y) - [E(X \mid Y = y)]^2, \end{aligned} \tag{7.14}$$

for the discrete and the continuous case, respectively.

Here is an illustration of (7.14).

Example 13. In reference to Example 11 and relation (7.14), we find: $E(Y^2|X = x) = \frac{1-x^8}{2(1-x^4)}$, which for $x = \frac{1}{2}$ becomes $E(Y^2|X = \frac{1}{2}) = \frac{17}{32}$. Therefore $Var(Y|X = \frac{1}{2}) = \frac{17}{32} - \left(\frac{7}{10}\right)^2 = \frac{33}{880}$, and the conditional s.d. of Y, given $X = \frac{1}{2}$ is $\left(\frac{33}{880}\right)^{1/2} \simeq 0.203$. So, in the tire pressure/mileage rate interpretation of the r.v.'s X and Y (see Example 4), the conditional s.d. around the expected mean, given $X = 0.5$, is about 0.203.

Exercises.

2.1 Refer to Exercise 1.1 and calculate the marginal p.d.f.'s f_X and f_Y.

2.2 Refer to Exercise 1.2 and calculate the marginal p.d.f.'s f_X and f_Y.

2.3 If the joint p.d.f. of the r.v.'s X and Y is given by the following table, determine the marginal p.d.f.'s f_X and f_Y.

y\x	−4	−2	2	4
−2	0	0.25	0	0
−1	0	0	0	0.25
1	0.25	0	0	0
2	0	0	0.25	0

2.4 The r.v.'s X and Y take on the values 1, 2, and 3, as indicated in the following table:

y\x	1	2	3
1	2/36	2/36	3/36
2	1/36	10/36	3/36
3	4/36	5/36	6/36

(i) Determine the marginal p.d.f.'s f_X and f_Y.

(ii) Determine the conditional p.d.f.'s $f_{X|Y}(\cdot \mid y)$ and $f_{Y|X}(\cdot \mid x)$.

2.5 The r.v.'s X and Y have joint p.d.f. $f_{X,Y}$ given by the entries of the following table:

y\x	0	1	2	3
1	1/8	1/16	3/16	1/8
2	1/16	1/16	1/8	1/4

(i) Determine the marginal p.d.f.'s f_X and f_Y, and the conditional p.d.f. $f_{X|Y}(\cdot \mid y)$, $y = 1, 2$.

(ii) Calculate: EX, EY, $E(X \mid Y = y)$, $y = 1, 2$, and $E[E(X \mid Y)]$.

(iii) Compare EX and $E[E(X \mid Y)]$.

(iv) Calculate: $Var(X)$ and $Var(Y)$.

2.6 Let the r.v.'s X and Y have the joint p.d.f.:

$$f_{X,Y}(x, y) = \frac{2}{n(n+1)}, \qquad y = 1, \ldots, x; \; x = 1, \ldots, n.$$

Then compute:

(i) The marginal p.d.f.'s f_X and f_Y.
(ii) The conditional p.d.f.'s $f_{X|Y}(\cdot \mid y)$ and $f_{Y|X}(\cdot \mid x)$.
(iii) The conditional expectations $E(X \mid Y = y)$ and $E(Y \mid X = x)$.

Hint. For part (iii), use the appropriate part of #1 in Table 6 in the Appendix.

2.7 In reference to Exercise 1.3, calculate the marginal p.d.f.'s f_X and f_Y.

2.8 Show that the marginal p.d.f.'s of the r.v.'s X and Y whose joint p.d.f. is given by:

$$f_{X,Y}(x, y) = \frac{6}{5}(x + y^2), \quad 0 \le x \le 1, \quad 0 \le y \le 1,$$

are as follows:

$$f_X(x) = \frac{2}{5}(3x + 1), 0 \le x \le 1; \; f_Y(y) = \frac{3}{5}(2y^2 + 1), 0 \le y \le 1.$$

2.9 Let X and Y be two r.v.'s with joint p.d.f. given by:

$$f_{X,Y}(x, y) = ye^{-x}, \qquad 0 < y \le x < \infty.$$

(i) Determine the marginal p.d.f.'s f_X and f_Y, and specify the range of the arguments involved.
(ii) Determine the conditional p.d.f.'s $f_{X|Y}(\cdot \mid y)$ and $f_{Y|X}(\cdot \mid x)$, and specify the range of the arguments involved.
(iii) Calculate the (conditional) probability $P(X > 2\log 2 \mid Y = \log 2)$, where, as always, log stands for the natural logarithm.

2.10 The joint p.d.f. of the r.v.'s X and Y is given by:

$$f_{X,Y}(x, y) = xe^{-(x+y)}, \quad x > 0, \quad y > 0.$$

(i) Determine the marginal p.d.f.'s f_X and f_Y.
(ii) Determine the conditional p.d.f. $f_{Y|X}(\cdot \mid x)$.
(iii) Calculate the probability $P(X > \log 4)$, where, as always, log stands for the natural logarithm.

2.11 The joint p.d.f. of the r.v.'s X and Y is given by:

$$f_{X,Y}(x, y) = \frac{1}{2}ye^{-xy}, \quad 0 < x < \infty, \quad 0 < y < 2.$$

(i) Determine the marginal p.d.f. f_Y.
(ii) Find the conditional p.d.f. $f_{X|Y}(\cdot \mid y)$, and evaluate it at $y = 1/2$.
(iii) Compute the conditional expectation $E(X \mid Y = y)$, and evaluate it at $y = 1/2$.

2.12 In reference to Exercise 1.4, calculate:

(i) The marginal p.d.f.'s f_X, f_Y, and the conditional p.d.f. $f_{Y|X}(\cdot \mid x)$; in all cases, specify the range of the variables involved.

(ii) EY and $E(Y \mid X = x)$.
(iii) $E[E(Y \mid X)]$ and observe that it is equal to EY.
(iv) The probability $P(Y > \frac{1}{2} \mid X < \frac{1}{2})$.

2.13 In reference to Exercise 1.7, calculate in terms of c:

(i) The marginal p.d.f.'s f_X and f_Y.
(ii) The conditional p.d.f.'s $f_{X|Y}(\cdot \mid y)$ and $f_{Y|X}(\cdot \mid x)$.
(iii) The probability $P(X \leq 1)$.

2.14 In reference to Exercise 1.8, determine the marginal p.d.f. f_Y and the conditional p.d.f. $f_{X|Y}(\cdot \mid y)$.

2.15 Refer to Exercise 1.9, and in terms of c_1, c_2:

(i) Determine the marginal p.d.f.'s f_X and f_Y.
(ii) Determine the conditional p.d.f. $f_{X|Y}(\cdot \mid y)$.
(iii) Calculate the EX and $E(X \mid Y = y)$.
(iv) Show that $E[E(X \mid Y)] = EX$.

2.16 In reference to Exercise 1.10, determine:

(i) The marginal p.d.f. f_X.
(ii) The conditional p.d.f. $f_{Y|X}(\cdot \mid x)$.

Hint. Consider separately the cases: $0 < x \leq 1$, $1 < x \leq 2$, x whatever else.

2.17 In reference to Exercise 1.11, determine:

(i) The marginal p.d.f. f_Y.
(ii) The conditional p.d.f. $f_{X|Y}(\cdot \mid y)$.
(iii) The marginal p.d.f. f_X.

Hint. For part (iii), consider separately the case that $x < 0$ (so that $-x < y$) and $x \geq 0$ (so that $x < y$).

2.18

(i) For a fixed $y > 0$, consider the function $f(x, y) = e^{-y}\frac{y^x}{x!}$, $x = 0, 1, \ldots$ and show that it is the conditional p.d.f. of a r.v. X, given that another r.v. $Y = y$.
(ii) Now, suppose that the marginal p.d.f. of Y is negative exponential with parameter $\lambda = 1$. Determine the joint p.d.f. of the r.v.'s X and Y.
(iii) Show that the marginal p.d.f. f_X is given by:

$$f_X(x) = \left(\frac{1}{2}\right)^{x+1}, \qquad x = 0, 1, \ldots.$$

Hint. ;For part (iii), observe that the integrand is essentially the p.d.f. of a gamma distribution (except for constants). Also, use the fact that $\Gamma(x+1) = x!$ (for $x \geq 0$ integer).

2.19 Suppose the r.v. Y is distributed as $P(\lambda)$ and that the conditional p.d.f. of an r.v. X, given $Y = y$, is $B(y, p)$. Then show that:

(i) The marginal p.d.f. f_X is Poisson with parameter λp.
(ii) The conditional p.d.f. $f_{Y|X}(\cdot \mid x)$ is Poisson with parameter λq (with $q = 1 - p$) over the set: $x, x + 1, \ldots$.

Hint. For part (i), form first the joint p.d.f. of X and Y. Also, use the appropriate part of #6 in Table 6 in the Appendix.

2.20

(i) Let X and Y be two discrete r.v.'s with joint p.d.f. $f_{X,Y}$. Then show that the conditional variance of X, given Y, satisfies the following relation:

$$Var(X \mid Y = y_j) = E(X^2 \mid Y = y_j) - [E(X \mid Y = y_j)]^2.$$

(ii) Establish the same relation, if the r.v.'s X and Y are of the continuous type.

2.21 Consider the function $f_{X,Y}$ defined by:

$$f_{X,Y}(x, y) = 8xy, 0 < x \leq y < 1.$$

(i) Verify that $f_{X,Y}$ is, indeed, a p.d.f.
(ii) Show that the marginal p.d.f.'s are given by:

$$f_X(x) = 4x(1 - x^2), 0 < x < 1; \; f_Y(y) = 4y^3, 0 < y < 1.$$

(iii) Show that $EX = \frac{8}{15}$, $EX^2 = \frac{1}{3}$, so that $Var(X) = \frac{11}{225}$.
(iv) Also, show that $EY = \frac{4}{5}$, $EY^2 = \frac{2}{3}$, so that $Var(Y) = \frac{6}{225}$.

2.22 In reference to Exercise 2.21, show that:

(i) $f_{X|Y}(x|y) = \frac{2x}{y^2}, 0 < x \leq y < 1; \; f_{Y|X}(y|x), 0 < x \leq y < 1.$
(ii) $E(X|Y = y) = \frac{2y}{3}, 0 < y < 1; \; E(Y|X = x) = \frac{2(1-x^3)}{3(1-x^2)}, 0 < x < 1.$

2.23 From Exercise 2.22(ii), we have that $E(X|Y) = \frac{2Y}{3}, 0 < Y < 1$.

(i) Use this expression and the p.d.f. of Y found in Exercise 2.21(ii) in order to show that $E[E(X|Y)] = EX(= 8/15)$.
(ii) Observe, however, that:

$$Var[E(X|Y)] = Var\left(\frac{2Y}{3}\right) = \frac{4}{9}Var(Y) < Var(Y)\left(= \frac{6}{225}\right).$$

2.24 By using the following expression (see relation (7.14)),

$$Var(X|Y = y) = E(X^2|Y = y) - [E(X|Y = y)]^2,$$

and the conditional p.d.f. $f_{X|Y}(\cdot|y)$ found in Exercise 2.22(i), compute $Var(X|Y = y), 0 < y < 1$.

2.25 The joint p.d.f. of the r.v.'s X and Y is given by the formula $f_{X,Y}(x, y) = 2, 0 < y < x < 1$ (and 0 elsewhere).

(i) Show that $f_{X,Y}$ is, indeed, a p.d.f.
(ii) In the region represented by $0 < y < x < 1$, draw the diagram represented by $y > x^2$.
(iii) Use part (ii) to compute the probability $P(Y > X^2)$.
(iv) Determine the marginal p.d.f.'s f_X and f_Y, including the range of the variables involved, and compute the EY.
(v) Determine $f_{Y|X}(\cdot|x)$ and compute the $E(Y|X = x)$.
(vi) Verify that $E[E(Y|X)] = EY$.

2.26 The number of domestically made and imported automobiles sold by a dealership daily are r.v.'s X and Y, respectively, whose joint distribution is given below.

y\x	0	1	2	3
0	0.05	0.15	0.12	0.08
1	0.07	0.11	0.10	0.07
2	0.01	0.08	0.07	0.05
3	0.01	0.02	0.01	0

(i) Compute the marginal p.d.f.'s f_X and f_Y.
(ii) Compute the probabilities $P(X \geq 2)$, $P(Y \leq 2)$, $P(X \geq 2, Y \leq 2)$.
(iii) Compute the expected numbers EX and EY of automobiles sold.
(iv) Also, compute the s.d.'s of X and Y.

2.27 Let the r.v.'s X and Y represent the proportions of customers of a computer store who buy computers only and who buy computers and printers, respectively. Suppose that the joint p.d.f. of X and Y is given by $f_{X,Y}(x, y) = c(x + y)$, $0 < y < x < 1$, $(c > 0)$.

(i) Determine the value of the constant c.
(ii) Find the marginal and the conditional p.d.f.'s f_X and $f_{Y|X}(\cdot|x)$.
(iii) Compute the conditional probability $P(Y \leq y_0|X = x_0)$.
(iv) Evaluate the probability in part (iii) for $x_0 = 0.15$, $y_0 = 0.10$.

Hint. For part (i), you may wish to draw the picture of the set for which $(0 <)y < x(< 1)$.

2.28 The loss covered by an insurance policy is an r.v. X with p.d.f. given by $f_X(x) = cx^2, 0 < x < 1/\sqrt{c}$ $(c > 0)$.

(i) Determine the constant c, so that f_X is, indeed, a p.d.f.
Given that $X = x$, the time of processing the claim is an r.v. T with (conditional) p.d.f. uniform over the interval $(0.5x, 2.5x)$; that is, $f_{T|X}(t|x) = \frac{1}{2x}$, $0.5x < t < 2.5x$.

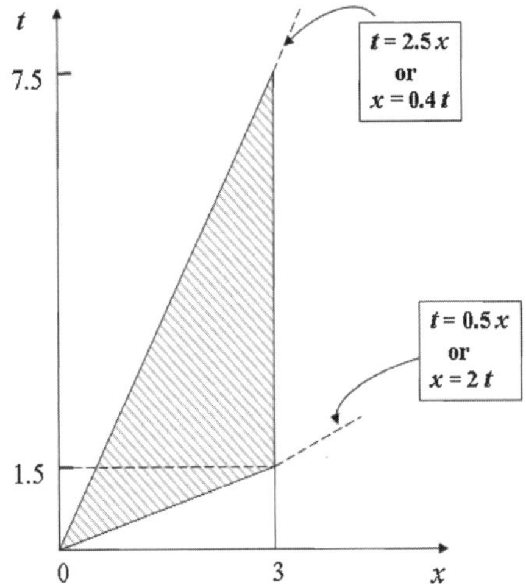

FIGURE 7.8
The shaded area is the region over which the joint p.d.f. of the r.v.'s X and T is positive.

(ii) Determine the joint p.d.f. $f_{X,T}$ of the r.v.'s X and T.
(iii) Determine the marginal p.d.f. f_T of the r.v. T.
(iv) Use f_T found in part (iii) in order to compute the probability $P(2 < T < 4)$ (see Figure 7.8).

Hint. From the inequalities $0.5x < t < 2.5x$ and $0 < x < 3$, we have that (x, y) lies in the part of the plane defined by: $2.5x = t, 0.5x = t, x = 3$; or $x = 0.4t, x = 2t, x = 3$ (see the lined area in the figure above). It follows that for $0 < t < 1.5$, x varies from $0.4t$ to $2t$; and for $1.5 \leq t < 7.5$, x varies from $0.4t$ to 3.

2.29 The annual numbers of snowfalls in two adjacent counties A and B are r.v.'s X and Y, respectively, with joint p.d.f. given below:

	Annual number of snowfalls in county *B*				
	$y \backslash x$	0	1	2	3
Annual number of snowfalls in county *A*	0	0.15	0.07	0.06	0.02
	1	0.13	0.11	0.09	0.05
	2	0.04	0.12	0.10	0.06

Find:

(i) The marginal p.d.f.'s f_X, f_Y, and the conditional p.d.f. $f_{Y|X}(\cdot|x)$.
(ii) The expectations EX, EY, and the conditional expectation $E(Y|X = x)$.
(iii) The variances $Var(X)$, $Var(Y)$, and the conditional variance $Var(Y|X = x)$.

CHAPTER

8 Joint Moment-Generating Function, Covariance, and Correlation Coefficient of Two Random Variables

In this chapter, we pursue the study of two r.v.'s X and Y with joint p.d.f. $f_{X,Y}$. To this end, consider an r.v. which is a function of the r.v.'s X and Y, $g(X, Y)$, and define its expectation. A special choice of $g(X, Y)$ gives the joint m.g.f. of the r.v.'s X and Y, which is studied to some extent in the first section. Another choice of $g(X, Y)$ produces what is known as the covariance of the r.v.'s X and Y, as well as their correlation coefficient. Some properties of these quantities are investigated in the second section of this chapter. Proofs of some results and some further properties of the correlation coefficient are discussed in Section 8.3.

8.1 The Joint m.g.f. of Two Random Variables

In this section, a function of the r.v.'s X and Y is considered and its expectation and variance are defined. As a special case, one obtains the joint m.g.f. of X and Y. To this end, let g be a real-valued function defined on $\Re^2$, so that $g(X, Y)$ is an r.v. Then the expectation of $g(X, Y)$ is defined as in (5.6) in Chapter 5 except that the joint p.d.f. of X and Y is to be used. Thus:

Definition 1.

(i)

$$Eg(X,Y) = \sum_{x_i \in \Re, y_j \in \Re} g(x_i, y_j) f_{X,Y}(x_i, y_j)$$

$$\text{or} = \int_{-\infty}^{\infty}\int_{-\infty}^{\infty} g(x,y) f_{X,Y}(x,y)dxdy, \tag{8.1}$$

for the discrete and the continuous case, respectively, provided, of course, the quantities defined exist.

(ii)

$$Var[g(X,Y)] = E[g(X,Y)]^2 - [Eg(X,Y)]^2. \tag{8.2}$$

Introduction to Probability, Second Edition. http://dx.doi.org/10.1016/B978-0-12-800041-0.00008-0

Properties analogous to those in Propositions 1 and 2(ii) in Chapter 5 apply here, too. Namely, assuming all the expectations figuring below are finite, we have:

Proposition 1. For c and d constants:

(i)

$$E[cg(X,Y)] = cEg(X,Y), \qquad E[cg(X,Y)+d] = cEg(X,Y)+d. \tag{8.3}$$

(ii)

$$E[g_1(X,Y)+g_2(X,Y)] = Eg_1(X,Y)+Eg_2(X,Y), \tag{8.4}$$

and, in particular,

(ii′)

$$E(cX+dY) = cEX+dEY. \tag{8.5}$$

Also, if h is another real-valued function, then

(iii)

$$g(X,Y) \leq h(X,Y) \ \text{ implies } \ Eg(X,Y) \leq Eh(X,Y), \tag{8.6}$$

and, in particular,

(iii′)

$$g(X) \leq h(X) \ \text{ implies } \ Eg(X) \leq Eh(X). \tag{8.7}$$

Furthermore,

(iv)

$$\begin{aligned} Var[cg(X,Y)] &= c^2 Var[g(X,Y)], \\ Var[cg(X,Y)+d] &= c^2 Var[g(X,Y)]. \end{aligned} \tag{8.8}$$

The justification of the above assertions is left as an exercise (see Exercise 1.2). As an illustration of relation (8.1), consider the following examples.

Example 1. Consider the r.v.'s X and Y jointly distributed as in Example 1 of Chapter 7. Then set $g(X,Y) = XY$ and compute the $E(XY)$.

Discussion. From Table 7.1 (in Chapter 7), it follows that the r.v. XY takes on values from 0 to 9 with respective probabilities as shown below.

xy	0	1	2	3	4	6	9
$f_{XY(x,y)}$	0.46	0.26	0.14	0	0.07	0.05	0.02

Therefore:

$$\begin{aligned} E(XY) = {} & (0 \times 0.46) + (1 \times 0.26) + (2 \times 0.14) + (3 \times 0) + (4 \times 0.07) \\ & + (6 \times 0.05) + (9 \times 0.02) = 1.3. \end{aligned}$$

Example 2. In reference to Example 4 in Chapter 7, set $g(X,Y) = XY$ and compute the $E(XY)$.

Discussion. We have:

$$E(XY) = \int_0^1 \int_{x^2}^1 xy \frac{21}{2} x^2 y dy dx = \frac{21}{2} \int_0^1 x^3 \left(\int_{x^2}^1 y^2 dy \right)$$
$$= \frac{21}{2 \times 3} \int_0^1 x^3 \left(y^3 \Big|_{x^2}^1 \right) = \frac{7}{2} \int_0^1 x^3 (1 - x^6) dx = \frac{21}{40}.$$

For the special choice of the function $g(x, y) = e^{t_1 x + t_2 y}$, t_1, t_2 reals, the expectation $E \exp(t_1 X + t_2 Y)$ defines a function in t_1, t_2 for those t_1, t_2 for which this expectation is finite. That is:

Definition 2. For two r.v.'s X and Y, define $M_{X,Y}(t_1, t_2)$ as:

$$M_{X,Y}(t_1, t_2) = E e^{t_1 X + t_2 Y},$$
$$(t_1, t_2) \in C \subseteq \Re^2 \text{ (with } C \text{ containing the origin in } \Re^2). \tag{8.9}$$

Thus, for the discrete and the continuous case, we have, respectively,

$$M_{X,Y}(t_1, t_2) = \sum_{x_i \in \Re, y_j \in \Re} e^{t_1 x_i + t_2 y_j} f_{X,Y}(x_i, y_j), \tag{8.10}$$

and

$$M_{X,Y}(t_1, t_2) = \int_{-\infty}^{\infty} \int_{-\infty}^{\infty} e^{t_1 x + t_2 y} f_{X,Y}(x, y) dx\, dy. \tag{8.11}$$

The function $M_{X,Y}(t, t)$ so defined is called the *joint m.g.f.* of the r.v.'s X and Y.

Here are two examples of joint m.g.f.'s.

Example 3. Refer to Example 1 in Chapter 7 and calculate the joint m.g.f. of the r.v.'s involved.

Discussion. For any $t_1, t_2 \in \Re$, we have, by means of (8.10):

$$M_{X,Y}(t_1, t_2) = \sum_{x=0}^{3} \sum_{y=0}^{3} e^{t_1 x + t_2 y} f_{X,Y}(x, y)$$
$$= 0.05 + 0.20 e^{t_2} + 0.21 e^{t_1} + 0.26 e^{t_1 + t_2} + 0.06 e^{t_1 + 2t_2}$$
$$+ 0.08 e^{2t_1 + t_2} + 0.07 e^{2t_1 + 2t_2} + 0.03 e^{2t_1 + 3t_2}$$
$$+ 0.02 e^{3t_1 + 2t_2} + 0.02 e^{3t_1 + 3t_2}. \tag{8.12}$$

Example 4. Refer to Example 2 in Chapter 7 and calculate the joint m.g.f. of the r.v.'s involved.

Discussion. By means of (8.11), we have here:

$$M_{X,Y}(t_1, t_2) = \int_0^{\infty} \int_0^{\infty} e^{t_1 x + t_2 y} \lambda_1 \lambda_2 e^{-\lambda_1 x - \lambda_2 y} dx\, dy$$
$$= \int_0^{\infty} \lambda_1 e^{-(\lambda_1 - t_1) x} dx \times \int_0^{\infty} \lambda_2 e^{-(\lambda_2 - t_2) y} dy.$$

But $\int_0^\infty \lambda_1 e^{-(\lambda_1-t_1)x}\,dx = -\frac{\lambda_1}{\lambda_1-t_1}e^{-(\lambda_1-t_1)x}|_0^\infty = \frac{\lambda_1}{\lambda_1-t_1}$, provided $t_1 < \lambda_1$, and likewise $\int_0^\infty \lambda_2 e^{-(\lambda_2-t_2)y}\,dy = \frac{\lambda_2}{\lambda_2-t_2}$ for $t_2 < \lambda_2$. (We arrive at the same results without integration by recalling (Example 6 in Chapter 7) that the r.v.'s X and Y have negative exponential distributions with parameters λ_1 and λ_2, respectively.) Thus,

$$M_{X,Y}(t_1,t_2) = \frac{\lambda_1}{\lambda_1-t_1} \times \frac{\lambda_2}{\lambda_2-t_2}, \qquad t_1 < \lambda_1, \quad t_2 < \lambda_2. \tag{8.13}$$

From relation (8.9), we have the following properties of the joint m.g.f. of the r.v.'s X and Y.

Proposition 2.

(i) Clearly, $M_{X,Y}(0,0) = 1$ for any r.v.'s X and Y, and it may happen that $C = \{(0,0)\}$.

(ii) In (8.9), by setting successively $t_2 = 0$ and $t_1 = 0$, we obtain:

$$M_{X,Y}(t_1,0) = Ee^{t_1X} = M_X(t_1), \qquad M_{X,Y}(0,t_2) = Ee^{t_2Y} = M_Y(t_2). \tag{8.14}$$

Thus, the m.g.f.'s of the individual r.v.'s X and Y are taken as marginals from the joint m.g.f. of X and Y, and they are referred to as *marginal m.g.f.'s*.

(iii) For c_1, c_2 and d_1, d_2 constants:

$$M_{c_1X+d_1,c_2Y+d_2}(t_1,t_2) = e^{d_1t_1+d_2t_2}M_{X,Y}(c_1t_1,c_2t_2). \tag{8.15}$$

(iv)

$$\frac{\partial}{\partial t_1}M_{X,Y}(t_1,t_2)|_{t_1=t_2=0} = EX, \qquad \frac{\partial}{\partial t_2}M_{X,Y}(t_1,t_2)|_{t_1=t_2=0} = EY, \tag{8.16}$$

and

$$\frac{\partial^2}{\partial t_1 \partial t_2}M_{X,Y}(t_1,t_2)|_{t_1=t_2=0} = E(XY) \tag{8.17}$$

(provided one may interchange the order of differentiating and taking expectations).

The justification of parts (iii) and (iv) is left as exercises (see Exercises 1.3 and 1.4). For example, in reference to (8.12) and (8.13), we obtain:

$$M_X(t_1) = 0.25 + 0.53\,e^{t_1} + 0.18\,e^{2t_1} + 0.04e^{3t_1}, \quad t_1 \in \Re,$$
$$M_Y(t_2) = 0.26 + 0.54\,e^{t_2} + 0.15\,e^{2t_2} + 0.05e^{3t_2}, \quad t_2 \in \Re,$$

and

$$M_X(t_1) = \frac{\lambda_1}{\lambda_1-t_1}, \quad t_1 < \lambda_1, \qquad M_Y(t_2) = \frac{\lambda_2}{\lambda_2-t_2}, \quad t_2 < \lambda_2.$$

Remark 1. Although properties (8.16) and (8.17) allow us to obtain moments by means of the m.g.f.'s of the r.v.'s X and Y, the most significant property of the m.g.f. is that it allows (under certain conditions) retrieval of the distribution of the r.v.'s X and Y. This is done through the so-called inversion formula. (See also Theorem 1 in Chapter 5.)

Exercises.

1.1 Let X and Y be the r.v.'s denoting the number of sixes when two fair dice are rolled independently 15 times each. Determine the $E(X+Y)$.

1.2

(i) Justify the properties stated in relations (8.3), (8.4), and (8.6).
(ii) Justify the properties stated in relations (8.8).

1.3 Show that the joint m.g.f. of two r.v.'s X and Y satisfies the following property, where c_1, c_2, d_1, and d_2 are constants:

$$M_{c_1X+d_1, c_2Y+d_2}(t_1, t_2) = e^{d_1t_1+d_2t_2} M_{X,Y}(c_1t_1, c_2t_2).$$

1.4 Justify the properties stated in relations (8.16) and (8.17). (Assume that you can differentiate under the expectation sign.)

8.2 Covariance and Correlation Coefficient of Two Random Variables

In this section, we define the concepts of covariance and correlation coefficient of two r.v.'s for a specific selection of the function $g(X, Y)$ in relation (8.1). First, a basic result is established (Theorem 1 and its corollary), and then it is explained how the covariance and the correlation coefficient may be used as measures of degree of *linear* dependence between the r.v.'s X and Y involved.

Definition 3. Consider the r.v.'s X and Y with finite expectations EX and EY and finite second moments, and in relation (8.1) take $g(X, Y) = (X - EX)(Y - EY)$. Then the expectation $E[(X - EX)(Y - EY)]$ is denoted by $Cov(X, Y)$ and is called the *covariance* of the r.v.'s X and Y; that is,

$$Cov(X, Y) = E[(X - EX)(Y - EY)]. \tag{8.18}$$

Next, by setting $Var(X) = \sigma_X^2$ and $Var(Y) = \sigma_Y^2$, we can formulate the following fundamental result regarding the covariance.

Theorem 1. Consider the r.v.'s X and Y with finite expectations EX and EY, finite and positive variances σ_X^2 and σ_Y^2, and s.d.'s σ_X and σ_Y. Then:

$$-\sigma_X\sigma_Y \leq Cov(X, Y) \leq \sigma_X\sigma_Y, \tag{8.19}$$

and

$$Cov(X, Y) = \sigma_X\sigma_Y \text{ if and only if } P\left[Y = EY + \frac{\sigma_Y}{\sigma_X}(X - EX)\right] = 1, \tag{8.20}$$

$$Cov(X, Y) = -\sigma_X\sigma_Y \text{ if and only if } P\left[Y = EY - \frac{\sigma_Y}{\sigma_X}(X - EX)\right] = 1. \tag{8.21}$$

Definition 4. The quantity $Cov(X, Y)/\sigma_X \sigma_Y$ is denoted by $\rho(X, Y)$ and is called the *correlation coefficient* of the r.v.'s X and Y; that is,

$$\rho(X, Y) = \frac{Cov(X, Y)}{\sigma_X \sigma_Y}. \tag{8.22}$$

Then from relations 8.19–8.22, we have the following result.

Corollary (to Theorem 1). For two r.v.'s X and Y,

$$-1 \leq \rho(X, Y) \leq 1, \tag{8.23}$$

and

$$\rho(X, Y) = 1 \text{ if and only if } P\left[Y = EY + \frac{\sigma_Y}{\sigma_X}(X - EX)\right] = 1, \tag{8.24}$$

$$\rho(X, Y) = -1 \text{ if and only if } P\left[Y = EY - \frac{\sigma_Y}{\sigma_X}(X - EX)\right] = 1. \tag{8.25}$$

The proof of the theorem and of the corollary is deferred to Section 8.3.

The figure below, Figure 8.1, depicts the straight lines $y = EY + \frac{\sigma_Y}{\sigma_X}(x - EX)$ and $y = EY - \frac{\sigma_Y}{\sigma_X}(x - EX)$ appearing within the brackets in relations (8.20) and (8.21).

Relations 8.19–8.21 state that the $Cov(X, Y)$ is always between $-\sigma_X \sigma_Y$ and $\sigma_X \sigma_Y$ and that it takes on the boundary values $\sigma_X \sigma_Y$ and $-\sigma_X \sigma_Y$ if and only if the pair (X, Y) lies (with probability 1) on the (straight) line $y = EY + \frac{\sigma_Y}{\sigma_X}(x - EX)$ and $y = EY - \frac{\sigma_Y}{\sigma_X}(x - EX)$, respectively.

Relations 8.23–8.25 state likewise that the $\rho(X, Y)$ is always between -1 and 1 and that it takes the boundary values 1 and -1 if and only if the pair (X, Y) lies (with probability 1) on the straight line $y = EY + \frac{\sigma_Y}{\sigma_X}(x - EX)$ and $y = EY - \frac{\sigma_Y}{\sigma_X}(x - EX)$, respectively.

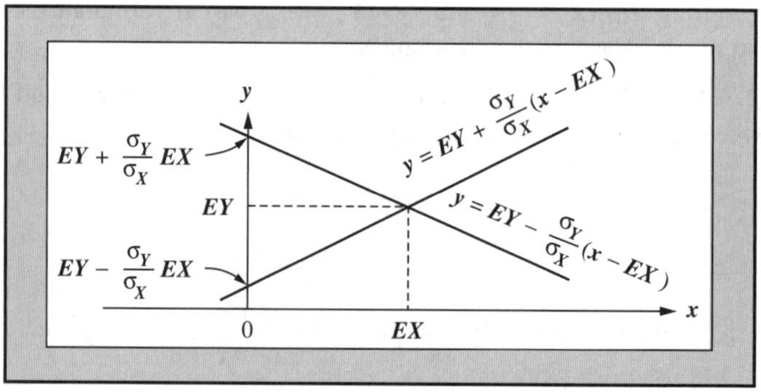

FIGURE 8.1
Lines of perfect linear relation of x and y.

Remark 2. From relations(8.18) and (8.19), it follows that the $Cov(X, Y)$ is measured in the same unit as the r.v.'s X and Y and that the range of its value depends on the magnitude of the product $\sigma_X \sigma_Y$. On the other hand, the $\rho(X, Y)$ is a pure number (dimensionless quantity), as it follows from relation (8.22), and it always lies in $[-1, 1]$. These are the reasons that we often focus on the correlation coefficient, rather than the covariance of two r.v.'s.

The following simple example, taken together with other arguments, reinforces the assertion that the correlation coefficient (or the covariance) may be used as a measure of *linear* dependence between two r.v.'s.

Example 5. Consider the events A and B with $P(A)P(B) > 0$ and set $X = I_A$ and $Y = I_B$ for the *indicator* functions, where, we recall that, $I_A(s) = 1$ if $s \in A$ and $I_A(s) = 0$ if $s \in A^c$. Then, clearly, $EX = P(A)$, $EY = P(B)$, and $XY = I_{A\cap B}$, so that $E(XY) = P(A \cap B)$. It follows that $Cov(X, Y) = P(A \cap B) - P(A)P(B)$. Next,

$$\begin{aligned} P(A)[P(Y = 1|X{=}1) - P(Y{=}1)] &= P(A \cap B) - P(A)P(B) \\ &= Cov(X, Y), \qquad (8.26) \\ P(A^c)[P(Y = 0|X = 0) - P(Y = 0)] &= P(A^c \cap B^c) - P(A^c)P(B^c) \\ &= P(A \cap B) - P(A)P(B) = Cov(X, Y), \qquad (8.27) \\ P(A^c)[P(Y = 1|X = 0) - P(Y = 1)] &= P(A^c \cap B) - P(A^c)P(B) \\ &= -[P(A \cap B) - P(A)P(B)] = -Cov(X, Y), \qquad (8.28) \\ P(A)[P(Y = 0|X = 1) - P(Y = 0)] &= P(A \cap B^c) - P(A)P(B^c) \\ &= -[P(A \cap B) - P(A)P(B)] = -Cov(X, Y), \qquad (8.29) \end{aligned}$$

(see also Exercise 2.1).

From (8.26) and (8.27), it follows that $Cov(X, Y) > 0$ if and only if $P(Y = 1 \mid X = 1) > P(Y = 1)$, or $P(Y = 0 \mid X = 0) > P(Y = 0)$. That is, $Cov(X, Y) > 0$ if and only if, given that X has taken a "large" value (namely, 1), it is more likely that Y does so as well than it otherwise would; also, given that X has taken a "small" value (namely, 0), it is more likely that Y does so too than it otherwise would. On the other hand, from relations (8.28) and (8.29), we see that $Cov(X, Y) < 0$ if and only if $P(Y = 1 \mid X = 0) > P(Y = 1)$, or $P(Y = 0 \mid X = 1) > P(Y = 0)$. That is, $Cov(X, Y) < 0$ if and only if, given that X has taken a "small" value, it is more likely for Y to take a "large" value than it otherwise would, and given that X has taken a "large" value, it is more likely for Y to take a "small" value than it otherwise would.

The above argument is consistent with the general interpretation of the expectation of an r.v. Thus, if EU is positive, then values of U tend to be positive, and values of U tend to be negative if $EU < 0$. Applying this rough argument to the $Cov(X, Y) = E[(X - EX)(Y - EY)]$, one concludes that values of X and Y tend to be simultaneously either both "large" or both "small" for $Cov(X, Y) > 0$, and they tend to go to opposite directions for $Cov(X, Y) < 0$. The same arguments apply to the correlation coefficient, since the $Cov(X, Y)$ and $\rho(X, Y)$ have the same sign.

Putting together arguments we have worked out above, we arrive at the following conclusions.

From relation (8.24), we have that $\rho(X, Y) = 1$ if and only if (X, Y) are linearly related (with probability 1). On the other hand, from Example 5 and the ensuing comments, we have that $Cov(X, Y) > 0$ if and only if X and Y tend to take simultaneously either "large" values or "small" values. Since $Cov(X, Y)$ and $\rho(X, Y)$ have the same sign, the same statement can be made about $\rho(X, Y)$, being positive if and only if X and Y tend to take simultaneously either "large" values or "small" values. The same arguments apply for the case that $Cov(X, Y) < 0$ (equivalently, $\rho(X, Y) < 0$). This reasoning indicates that $\rho(X, Y)$ may be looked upon as a measure of *linear* dependence between X and Y. The pair (X, Y) lies on the line $y = EY + \frac{\sigma_Y}{\sigma_X}(x - EX)$ if $\rho(X, Y) = 1$; pairs identical to (X, Y) tend to be arranged along this line, if $(0 <)$ $\rho(X, Y) < 1$, and they tend to move farther and farther away from this line as $\rho(X, Y)$ gets closer to 0; the pairs bear no sign of linear tendency whatever, if $\rho(X, Y) = 0$. Rough arguments also hold for the reverse assertions. For $0 < \rho(X, Y) \leq 1$, the r.v.'s X and Y are said to be *positively correlated*, and *uncorrelated* if $\rho(X, Y) = 0$. Likewise, the pair (X, Y) lies on the line $y = EY - \frac{\sigma_Y}{\sigma_X}(x - EX)$ if $\rho(X, Y) = -1$ from relation (8.25); pairs identical to (X, Y) tend to be arranged along this line if $-1 < \rho(X, Y) < 0$. Again, rough arguments can also be made for the reverse assertions. For $-1 \leq \rho(X, Y) < 0$, the r.v.'s X and Y are said to be *negatively correlated.*

The behavior of the pair (X, Y), as interpreted by means of the correlation coefficient $\rho(X, Y)$, is depicted in Figure 8.2 below. In (a), $\rho(X, Y) = 1$, the r.v.'s X and Y are perfectly positively linearly related. In (b), $\rho(X, Y) = -1$, the r.v.'s X and Y are perfectly negatively linearly related. In (c), $0 < \rho(X, Y) < 1$, the r.v.'s X and Y are positively correlated. In (d), $-1 < \rho(X, Y) < 0$, the r.v.'s X and Y are negatively correlated. In (e), $\rho(X, Y) = 0$, the r.v.'s X and Y are uncorrelated.

The following simple results facilitated the actual computation of a covariance.

Proposition 3. For the covariance of two r.v.'s X and Y, defined by relation (8.18), we have:

$$Cov(X, Y) = E[(X - EX)(Y - EY)] = E(XY) - (EX)(EY). \tag{8.30}$$

Proof. Indeed,

$$\begin{aligned} E[(X - EX)(Y - EY)] &= E[XY - X(EY) - (EX)Y + (EX)(EY)] \\ &= E(XY) - (EX)(EY) - (EX)(EY) + (EX)(EY) \\ &= E(XY) - (EX)(EY). \quad \blacktriangle \end{aligned}$$

As an illustration of actual calculation of a covariance and a correlation coefficient, let us do it for Examples 1 and 4 in Chapter 7.

Example 6.

(i) In reference to Example 1 (see also Example 8), in Chapter 7, compute the $Cov(X, Y)$ and the $\rho(X, Y)$.

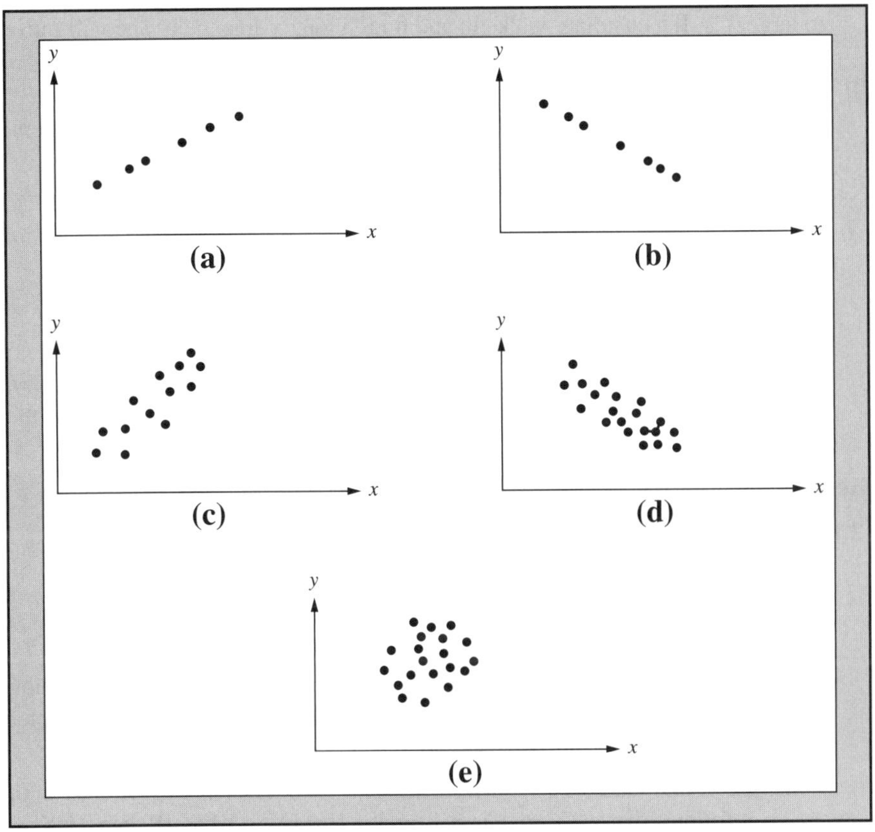

FIGURE 8.2
(a) Perfect linear relation (positive slope); (b) perfect linear relation (negative slope); (c) positive correlation; (d) negative correlation; (e) uncorrelated r.v.'s.

(ii) In reference to Example 4 (see also Example 8), in Chapter 7, compute the $Cov(X, Y)$ and the $\rho(X, Y)$.

Discussion.

(i) From Example 8 (in Chapter 7):
$EX = 1.01$, $EY = 0.99$, $\sigma_X^2 = 0.5899$, $\sigma_Y^2 = 0.6099$, whereas from Example 1 here, $E(XY) = 1.3$. Therefore, by relation (8.30), $Cov(X, Y) = E(XY) - (EX)(EY) = 1.3 - (1.01)(0.99) = 0.3001$, and $\rho(X, Y) = \frac{0.3001}{\sqrt{0.5899 \times 0.6099}} \simeq \frac{0.3001}{0.5998} \simeq 0.5$.

This result is consistent with what one would expect. Namely, although the X and Y are not linearly related, there is a tendency to have large Y's corresponding to large X's: If a customer walks in and finds a long x-line, he or she will move to the y-line, increasing its size.

(ii) From Example 8 (in Chapter 7):
$EX = \frac{21}{32}$, $EY = \frac{7}{9}$, $\sigma_X^2 = \frac{553}{15{,}360}$, $\sigma_Y^2 = \frac{28}{891}$, whereas from Example 2 here, $E(XY) = \frac{21}{40}$. Therefore, $Cov(X, Y) = \frac{21}{40} - \frac{21}{32} \times \frac{7}{9} = \frac{7}{480}$, and:

$$\rho(X, Y) = \frac{\frac{7}{480}}{\sqrt{\frac{553}{15{,}360} \times \frac{28}{891}}} = \frac{7}{480}\sqrt{\frac{891 \times 15,360}{28 \times 553}}$$

$$\simeq \frac{7}{480} \times 29.73 \simeq 0.434.$$

Again, the correlation coefficient says what one would expect to see. Namely, to larger values of tire pressure there is a tendency for larger values of the corresponding mileage rates.

Exercises.

2.1 Provide a justification of relations 8.27–8.29. That is:

(i) $P(A^c \cap B^c) - P(A^c)P(B^c) = P(A \cap B) - P(A)P(B)$.
(ii) $P(A^c \cap B) - P(A^c)P(B) = -P(A \cap B) + P(A)P(B)$.
(iii) $P(A \cap B^c) - P(A)P(B^c) = -P(A \cap B) + P(A)P(B)$.

Hint. For part (i), use Proposition 4 in Chapter 2 along with Proposition 1(iii) in Chapter 3. For Parts (ii) and (iii), use Proposition 1(iv) in Chapter 3 after rewriting suitably $A^c \cap B$ and $A \cap B^c$.

2.2 Let X and Y be two r.v.'s with $EX = EY = 0$. Then, if $Var(X - Y) = 0$, it follows that $P(X = Y) = 1$, and if $Var(X + Y) = 0$, then $P(X = -Y) = 1$.

Hint. Use Exercise 2.4 (ii) in Chapter 5.

2.3 In Exercise 1.1 in Chapter 7, two discrete r.v.'s X and Y are given with tabulated joint probabilities. In Exercise 2.1 of the same chapter, the marginal p.d.f.'s f_X and f_Y were calculated.
Use these results in order to:

(i) Calculate EX, EY, $Var(X)$, and $Var(Y)$.
(ii) Calculate $Cov(X, Y)$ and $\rho(X, Y)$.
(iii) Decide on the kind of correlation of the r.v.'s X and Y.

2.4 In Exercise 1.2 in Chapter 7, one is asked to compute the joint p.d.f. of two discrete r.v.'s X and Y; and in Exercise 2.2 of the same chapter, the marginal p.d.f.'s f_X and f_Y were derived. Use these results in order to calculate:

(i) $EX, EY, Var(X), Var(Y)$.
(ii) $E(XY), Cov(X, Y)$.
(iii) $\rho(X, Y)$.

(iv) Decide on the kind of correlation, if any, the r.v.'s X and Y exhibit.

2.5 In Exercise 2.3 in Chapter 7, the joint p.d.f. of two discrete r.v.'s X and Y is given in tabular form, and then the marginal p.d.f.'s f_X and f_Y were derived.
Use these results in order to:

(i) Calculate EX, EY, $Var(X)$, and $Var(Y)$.
(ii) Calculate $Cov(X, Y)$ and $\rho(X, Y)$.
(iii) Plot the points $(-4, 1)$, $(-2, -2)$, $(2, 2)$, and $(4, -1)$, and reconcile this graph with the value of $\rho(X, Y)$ found in part (ii).

2.6 In Exercise 2.4 in Chapter 7, the joint p.d.f. of two discrete r.v.'s X and Y is given in tabular form, and then the marginal p.d.f.'s f_X and f_Y were derived.
Use these results in order to compute:

(i) EX, EY, $Var(X)$, and $Var(Y)$.
(ii) $Cov(X, Y)$ and $\rho(X, Y)$.

2.7 In Exercise 2.5 in Chapter 7, the joint p.d.f. of two discrete r.v.'s X and Y is given in tabular form, and then the EX, EY, $Var(X)$, and $Var(Y)$ were computed.
Use these results in order to calculate the $Cov(X, Y)$ and $\rho(X, Y)$.

2.8 Let X be an r.v. taking on the values $-2, -1, 1, 2$, each with probability 1/4, and define the r.v. Y by: $Y = X^2$. Then calculate the quantities: EX, $Var(X)$, EY, $Var(Y)$, $E(XY)$, $Cov(X, Y)$, and $\rho(X, Y)$. Are you surprised by the value of $\rho(X, Y)$? Explain.

2.9 In Exercise 1.3 in Chapter 7, the joint p.d.f. $f_{X,Y}$ of two r.v.'s (of the continuous type) is given, and then the marginal p.d.f.'s f_X and f_Y were derived in Exercise 2.7 of the same chapter.
Use these results in order to compute:

(i) The expectations EX and EY.
(ii) The variances $Var(X)$ and $Var(Y)$.
(iii) The covariance $Cov(X, Y)$ and the correlation coefficient $\rho(X, Y)$.
(iv) On the basis of part (iii), decide on the kind of correlation of the r.v.'s X and Y.

2.10 In Exercise 2.8 in Chapter 7, the joint p.d.f. $f_{X,Y}$ of two r.v.'s (of the continuous type) is given, and then the marginal p.d.f.'s f_X and f_Y were derived.
Use these results in order to calculate:

(i) The expectations EX and EY.
(ii) The variances $Var(X)$ and $Var(Y)$.
(iii) The covariance $Cov(X, Y)$ and the correlation coefficient $\rho(X, Y)$.
(iv) On the basis of part (iii), decide on the kind of correlation of the r.v.'s X and Y.

2.11 Refer to Exercise 2.21 in Chapter 7 and compute the $E(XY)$ and $Cov(X, Y)$.

2.12 From relation (7.11) in Chapter 7, we have that, for two r.v.'s X and Y:

$$EX = E[E(X|Y)] \text{ and } EY = E[E(Y|X)].$$

In a similar fashion, show that:

(i) $EY^2 = E[E(Y^2|X)]$.
(ii) $E(XY) = E[E(XY|X)] = E[XE(Y|X)]$.

Hint. Restrict the proof to the continuous case only. In part (ii), show that $E[XE(Y|X)] = E(XY)$, and $E[E(XY|X)] = E(XY)$.

2.13 Let X be an r.v. distributed as $U(0, \alpha)(\alpha > 0)$, and let Y be an r.v. distributed as $U(0, x)$, given that $X = x$. From the fact that $X \sim U(0, \alpha)$, it follows that $EX = \frac{\alpha}{2}$ and $Var(X) = \frac{\alpha^2}{12}$.

(i) Since $Y|X = x$ is distributed as $U(0, x)$, use the relation $EY = E[E(Y|X)]$ to compute the EY.
(ii) Use the relation $EY^2 = E[E(Y^2|X)]$ to compute the EY^2 and then the $Var(Y)$.
(iii) Use the relation $E(XY) = E[E(XY|X)] = E[XE(Y|X)]$ to compute the $E(XY)$ and then the $Cov(X, Y)$.
(iv) Show that the $\rho(X, Y)$ is independent of α, and compute its value.

Hint. For parts (ii) and (iii), refer also to Exercise 2.12, parts (i) and (ii), respectively.

8.3 Proof of Theorem 1, Some Further Results

In this section, we present the proof of Theorem 1 and its corollary. Also, we point out a desirable property of the correlation coefficient (Theorem 2), and illustrate it by means of an example. Finally, the covariance and the correlation coefficient are used in expressing the variance of the sum of two r.v.'s (Theorem 3).

Proof (of Theorem 1). The proof of the theorem is split into two parts as follows:

(i) Assume first that $EX = EY = 0$ and $\sigma_X^2 = \sigma_Y^2 = 1$. Then, by means of (8.30), relation (8.19) becomes: $-1 \leq E(X, Y) \leq 1$, whereas relations (8.20) and (8.21) take the form: $E(XY) = 1$ if and only if $P(Y = X) = 1$, and $E(XY) = -1$ if and only if $P(Y = -X) = 1$.

Clearly, $0 \leq E(X - Y)^2 = EX^2 + EY^2 - 2E(XY) = 2 - 2E(XY)$, so that $E(XY) \leq 1$; also, $0 \leq E(X + Y)^2 = EX^2 + EY^2 + 2E(XY) = 2 + 2E(XY)$, so that $-1 \leq E(XY)$. Combining these results, we obtain $-1 \leq E(XY) \leq 1$. As for equalities, observe that if $P(X = Y) = 1$, then $E(XY) = EX^2 = 1$, and if $P(X = -Y) = 1$, then $E(XY) = -EX^2 = -1$. Next, $E(XY) = 1$ implies $E(X - Y)^2 = 0$ or $Var(X - Y) = 0$. But then $P(X - Y = 0) = 1$ or $P(X = Y) = 1$ (see Exercise 2.4(ii) in Chapter 5). Also, $E(XY) = -1$ implies

$E(X+Y)^2 = 0$ or $Var(X+Y) = 0$, so that $P(X = -Y) = 1$ (by the exercise just cited).

(ii) Remove the assumptions made in part (i), and replace the r.v.'s X and Y by the r.v.'s $X^* = \frac{X-EX}{\sigma_X}$ and $Y^* = \frac{Y-EY}{\sigma_Y}$, for which $EX^* = EY^* = 0$ and $Var(X^*) = Var(Y^*) = 1$. Then the inequalities $-1 \le E(X^*Y^*) \le 1$ become

$$-1 \le E\left[\left(\frac{X-EX}{\sigma_X}\right)\left(\frac{Y-EY}{\sigma_Y}\right)\right] \le 1$$

from which (8.19) follows. Also, $E(X^*Y^*) = 1$ if and only if $P(X^* = Y^*) = 1$ becomes $E[(X-EX)(Y-EY)] = \sigma_X\sigma_Y$ if and only if $P[Y = EY + \frac{\sigma_Y}{\sigma_X}(X - EX)] = 1$, and $E(X^*Y^*) = -1$ if and only if $P(X^* = -Y^*) = 1$ becomes $E[(X-EX)(Y-EY)] = -\sigma_X\sigma_Y$ if and only if $P[Y = EY - \frac{\sigma_Y}{\sigma_X}(X-EX)] = 1$. A restatement of the last two conclusions is: $Cov(X, Y) = \sigma_X\sigma_Y$ if and only if $P[Y = EY + \frac{\sigma_Y}{\sigma_X}(X - EX)] = 1$, and $Cov(X, Y) = -\sigma_X\sigma_Y$ if and only if $P[Y = EY - \frac{\sigma_Y}{\sigma_X}(X - EX)] = 1$. ▲

Proof (of the corollary to Theorem 1). In relation (8.19), divide all three sides by $\sigma_X\sigma_Y$ to obtain:

$$-1 \le \frac{Cov(X, Y)}{\sigma_X\sigma_Y} = \rho(X, Y) \le 1.$$

Then assertions (8.23)–(8.25) follow immediately from (8.19)–(8.21). ▲

The following result presents an interesting property of the correlation coefficient.

Theorem 2. Let X and Y be r.v.'s with finite first and second moments and positive variances, and let c_1, c_2, d_1, d_2 be constants with $c_1c_2 \neq 0$. Then:

$$\rho(c_1X + d_1, c_2Y + d_2) = \pm\rho(X, Y), \text{ with } + \text{ if } c_1c_2 > 0 \text{ and } - \text{ if } c_1c_2 < 0. \quad (8.31)$$

Proof. Indeed, $Var(c_1X + d_1) = c_1^2 Var(X)$, $Var(c_2Y + d_2) = c_2^2 Var(Y)$, and $Cov(c_1X + d_1, c_2Y + d_2) = E\{[(c_1X + d_1) - E(c_1X + d_1)][(c_2Y + d_2) - E(c_2Y + d_2)]\} = E[c_1(X - EX) \cdot c_2(Y - EY)] = c_1c_2E[(X - EX)(Y - EY)] = c_1c_2Cov(X, Y)$. Therefore

$$\rho(c_1X + d_1, c_2Y + d_2) = \frac{c_1c_2Cov(X, Y)}{|c_1c_2|\sqrt{Var(X)Var(Y)}},$$

and the conclusion follows. ▲

Here is an illustrative example of this theorem.

Example 7. Let X and Y be temperatures in two localities measured on the Celsius scale, and let U and V be the same temperatures measured on the Fahrenheit scale. Then $\rho(X, Y) = \rho(U, V)$, as it should be. This is so because $U = \frac{9}{5}X + 32$ and $V = \frac{9}{5}Y + 32$, so that (8.31) applies with the $+$ sign.

This section is concluded with the following result and two examples.

Theorem 3. For two r.v.'s X and Y with finite first and second moments and (positive) standard deviations σ_X and σ_Y, it holds:

$$Var(X+Y) = \sigma_X^2 + \sigma_Y^2 + 2Cov(X,Y) = \sigma_X^2 + \sigma_Y^2 + 2\sigma_X\sigma_Y\rho(X,Y), \quad (8.32)$$

and

$$Var(X+Y) = \sigma_X^2 + \sigma_Y^2 \text{ if } X \text{ and } Y \text{are uncorrelated.} \quad (8.33)$$

Proof. Since (8.33) follows immediately from (8.32), and $Cov(X,Y) = \sigma_X \times \sigma_Y \times \rho(X,Y)$, it suffices to establish only the first equality in (8.32). Indeed,

$$\begin{aligned} Var(X+Y) &= E[(X+Y) - E(X+Y)]^2 = E[(X-EX) + E(Y-EY)]^2 \\ &= E(X-EX)^2 + E(Y-EY)^2 + 2E[(X-EX)(Y-EY)] \\ &= \sigma_X^2 + \sigma_Y^2 + 2Cov(X,Y). \quad \blacktriangle \end{aligned}$$

Example 8. In reference to Examples 1 and 4 in Chapter 7, compute the $Var(X+Y)$.

Discussion. By (8.32),
$Var(X+Y) = Var(X) + Var(Y) + 2Cov(X,Y)$. For Example 1, we have:
$Var(X) = 0.5899$ and $Var(Y) = 0.6099$ from Example 8 in Chapter 7, and $Cov(X,Y) = 0.3001$ from Example 6(i) here. Therefore:

$$Var(X+Y) = 0.5899 + 0.6099 + 2 \times 0.3001 = 1.8.$$

For Example 4, we have:
$Var(X) = \frac{553}{15{,}360}$ and $Var(Y) = \frac{28}{891}$ from Example 8 in Chapter 7, and $Cov(X,Y) = \frac{7}{480}$ from Example 6(ii) here. Therefore $Var(X+Y) = \frac{553}{15{,}360} + \frac{28}{891} + 2 \times \frac{7}{480} \simeq 0.097$.

Example 9. Let X and Y be two r.v.'s with finite expectations and equal (finite) variances, and set $U = X+Y$ and $V = X-Y$. Then the r.v.'s U and V are uncorrelated, and $Var(U+V) = 4Var(X)$.

Discussion. Indeed,

$$\begin{aligned} E(UV) &= E[(X+Y)(X-Y)] = E(X^2 - Y^2) = EX^2 - EY^2, \\ (EU)(EV) &= [E(X+Y)][E(X-Y)] \\ &= (EX+EY)(EX-EY) = (EX)^2 - (EY)^2, \end{aligned}$$

so that

$$\begin{aligned} Cov(U,V) &= E(UV) - (EU)(EV) = [EX^2 - (EX)^2] - [EY^2 - (EY)^2] \\ &= Var(X) - Var(Y) = 0. \end{aligned}$$

Furthermore,

$$\begin{aligned} Var(U+V) &= Var(U) + Var(V) = Var(X+Y) + Var(X-Y) \\ &= Var(X) + Var(Y) + 2Cov(X,Y) + Var(X) \\ &\quad + Var(Y) - 2Cov(X,Y) \\ &= 2Var(X) + 2Var(Y) = 4Var(X). \end{aligned}$$

Alternatively, since $U + V = 2X$, we have again $Var(U + V) = Var(2X) = 4Var(X)$.

Exercises.

3.1 Refer to Exercises 2.21 in Chapter 7 and 2.11 in this chapter and compute:

(i) The covariance $Cov(X, Y)$ and the correlation coefficient $\rho(X, Y)$. Decide on the kind of correlation of the r.v.'s X and Y.

(ii) The $Var(X + Y)$.

Hint. For part (ii), refer to Theorem 3.

3.2 Let X be an r.v. with finite expectation and finite and positive variance, and set $Y = aX + b$, where a and b are constants and $a \neq 0$. Then compute $EY, Var(Y), E(XY), Cov(X, Y)$. Also, show that $|\rho(X, Y)| = 1$ and, indeed $\rho(X, Y) = 1$ if and only if $a > 0$, and $\rho(X, Y) = -1$ if and only if $a < 0$.

3.3 For any two r.v.'s X and Y, set $U = X + Y$ and $V = X - Y$. Then show that:

(i) $P(UV < 0) = P(|X| < |Y|)$.

(ii) If $EX^2 = EY^2 < \infty$, then $E(UV) = 0$.

(iii) If $EX^2 < \infty$, $EY^2 < \infty$ and $Var(X) = Var(Y)$, then the r.v.'s U and V are uncorrelated.

3.4 Let X and Y be r.v.'s with finite second moments EX^2, EY^2, and $Var(X) > 0$. Suppose we know X and we wish to *predict* Y in terms of X through the *linear* relationship $\alpha X + \beta$, where α and β are (unknown) constants. Further, suppose there exist values $\hat{\alpha}$ and $\hat{\beta}$ of α and β, respectively, for which the expectation of the square difference $[Y - (\hat{\alpha}X + \hat{\beta})]^2$ is minimum. Then $\hat{Y} = \hat{\alpha}X + \hat{\beta}$ is called the *best linear predictor* of Y in terms of X (when the criterion of optimality is that of minimizing $E[Y - (\alpha X + \beta)]^2$ over all α and β).
Then show that $\hat{\alpha}$ and $\hat{\beta}$ are given as follows:

$$\hat{\alpha} = \frac{\sigma_Y}{\sigma_X}\rho(X, Y), \qquad \hat{\beta} = EY - \hat{\alpha}EX,$$

where σ_X and σ_Y are the s.d.'s of the r.v.'s X and Y, respectively.

Hint. Set $g(\alpha, \beta) = E[Y - (\alpha X + \beta)]^2$, carry out the operations on the right-hand side in order to get: $g(\alpha, \beta) = EY^2 + \alpha^2 EX^2 + \beta^2 + 2\alpha\beta EX - 2\alpha E(XY) - 2\beta EY$, minimize $g(\alpha, \beta)$ by equating to 0 the two partial derivatives in order to find the values $\hat{\alpha}$ and $\hat{\beta}$ given above. Finally, show that these values $\hat{\alpha}$ and $\hat{\beta}$ do, indeed, minimize $g(\alpha, \beta)$ by showing that the 2×2 matrix of the second order derivatives has its 1×1 and 2×2 determinants positive. Or that the matrix of the second order derivatives is positive definite.

CHAPTER

9 Some Generalizations to k Random Variables, and Three Multivariate Distributions

In this chapter, some of the concepts and results discussed in previous chapters are carried over to the case, in which we are faced with $k(\geq 2)$ r.v.'s, rather than one or two r.v.'s. It consists of four sections. In the first section, the joint probability distribution of the r.v.'s involved is defined, as well as their joint d.f. and joint p.d.f. From the joint p.d.f., marginal (joint) p.d.f.'s are derived, and then conditional p.d.f.'s are defined. Next, the expectation of a function of the r.v.'s involved is defined, and for a specific choice of such a function, one obtains the joint m.g.f. of the underlying r.v.'s. The section is concluded with a formula providing the variance of the sum of r.v.'s. In the second section, the multinomial distribution is introduced, and its p.d.f. is derived. It is shown that all marginal and all conditional p.d.f.'s are also multinomial. Finally, the joint m.g.f. is derived. In the next section, the bivariate normal distribution is introduced, and it is then shown that what is proposed as a joint p.d.f. is indeed a p.d.f. As a consequence of this proof, it follows that the marginal p.d.f.'s, as well as the conditional ones, are all normal. The section is closed with the derivation of the covariance and the correlation coefficient. In the final section of the chapter, the multivariate normal distribution is introduced without any further discussion, except for the citation of a relevant reference.

9.1 Joint Distribution of k Random Variables and Related Quantities

If instead of two r.v.'s X and Y we have k r.v.'s $X_1, \ldots, X_k$, most of the concepts defined and results obtained in the previous two chapters are carried over to the k-dimensional case in a straightforward way. Thus, the *joint probability distribution function* or just *joint probability distribution* of the r.v.'s $X_1, \ldots, X_k$ is a set function $P_{X_1,\ldots,X_k}$ that assigns values to subsets B of $\Re^k = \Re \times \cdots \times \Re$ (k factors) according to the formula

$$P[(X_1, \ldots, X_k) \in B] = P(\{s \in S; (X_1(s), \ldots, X_k(s)) \in B\}), B \subseteq \Re^k; \quad (9.1)$$

the value assigned is denoted by $P_{X_1,\ldots,X_k}(B)$.

By taking B to be a rectangle of the form $B = (-\infty, x_1] \times \cdots \times (-\infty, x_k]$, relation (9.1) becomes

Introduction to Probability, Second Edition. http://dx.doi.org/10.1016/B978-0-12-800041-0.00009-2

$$
\begin{aligned}
P[(X_1, \ldots, X_k) &\in (-\infty, x_1] \times \cdots \times (-\infty, x_k] \\
&= P(\{s \in S; X_1(s) \leq x_1, \ldots, X_k(s) \leq x_k\}) \\
&= P(X_1 \leq x_1, \ldots, X_k \leq x_k),
\end{aligned}
$$

and it defines a function on $\Re^k$ denoted by $F_{X_1,\ldots,X_k}$ and called the *joint distribution function (joint d.f.)* of the r.v.'s $X_1, \ldots, X_k$.

Remark 1. Comments similar to those made in Remark 1 in Chapter 7 hold here also, as well as a proposition analogous to Proposition 1 in Chapter 7.

Now, let $X_1, \ldots, X_k$ be discrete r.v.'s and let us denote by x_{1,i_1} the values the r.v. X_1 takes on, etc., and by x_{k,i_k} the values the r.v. X_k takes on. There may be either finitely or (countably) infinitely many such values. The respective probabilities are

$$P(X_1 = x_{1,i_1}, \ldots, X_k = x_{k,i_k}),$$

and define the function $f_{X_1,\ldots,X_k}$ as follows:

$$
\begin{aligned}
&f_{X_1,\ldots,X_k}(x_1, \ldots, x_k) \\
&= \begin{cases} P(X_1 = x_{1,i_1}, \ldots, X_k = x_{k,i_k}), & \text{if } x_1 = x_{1,i_1}, \ldots, x_k = x_{k,i_k} \\ 0 & , \text{otherwise.} \end{cases}
\end{aligned} \tag{9.2}
$$

The function $f_{X_1,\ldots,X_k}$ is called the *joint probability density function (joint p.d.f.)* of the r.v.'s $X_1, \ldots, X_k$. Results analogous to those in Proposition 2 in Chapter 7 hold here also.

Next, let $X_1, \ldots, X_k$ be r.v.'s of the continuous type, and suppose there exists a function $f_{X_1,\ldots,X_k}$ such that:

$$
\begin{aligned}
&f_{X_1,\ldots,X_k}(x_1, \ldots, x_k) \geq 0 \text{ for all } x_1, \ldots, x_k \in \Re, \\
&P[(X_1, \ldots, X_k) \in B] = \underbrace{\int \cdots \int}_{B} f_{X_1,\ldots,X_k}(x_1, \ldots, x_k) dx_1 \ldots dx_k, \quad B \subseteq \Re^k, \text{ and} \\
&\underbrace{\int \cdots \int}_{\Re^k} f_{X_1,\ldots,X_k}(x_1, \ldots, x_k) dx_1 \ldots dx_k = 1.
\end{aligned} \tag{9.3}
$$

The function $f_{X_1,\ldots,X_k}$ is called the *joint probability density function (joint p.d.f.)* of the r.v.'s $X_1, \ldots, X_k$. Results analogous to those in Proposition 3 in Chapter 7 hold here also. Furthermore, the comments made in Remark 2 in Chapter 7 hold here as well, properly interpreted.

Marginal d.f.'s and marginal p.d.f.'s can be defined here as well, except that there are many marginal d.f.'s and p.d.f.'s. Thus, if s is a number with $1 \leq s < k$, and if in $F_{X_1,\ldots,X_k}(x_1, \ldots, x_k)$ the last t x's are replaced by ∞ (in the sense they are let to tend to ∞), then what is left is the *marginal joint d.f.* of the r.v.'s $X_1, \ldots, X_s$, $F_{X_1,\ldots,X_s}$, with $s+t = k$. Likewise, if in $f_{X_1,\ldots,X_k}(x_1, \ldots, x_k)$, the last t variables are eliminated through summation (for the discrete case) or integration (for the continuous case), what is left is the *marginal joint p.d.f* of the r.v.'s $X_1, \ldots, X_s$, $f_{X_1,\ldots,X_s}$. Combining joint and marginal joint p.d.f.'s, as in the 2-dimensional case, we obtain a variety of *conditional p.d.f.'s*. Thus, for example,

$$f_{X_{s+1},\ldots,X_k|X_1,\ldots,X_s}(x_{s+1},\ldots,x_k|x_1,\ldots,x_s) = \frac{f_{X_1,\ldots,X_k}(x_1,\ldots,x_k)}{f_{X_1,\ldots,X_s}(x_1,\ldots,x_s)}.$$

Instead of splitting the r.v.'s $X_1, \ldots, X_k$ into two groups consisting of the first s r.v.'s and the last t r.v.'s ($s + t = k$), we may select any s r.v.'s, call them $X_{i_1}, \ldots, X_{i_s}$, for one group; then the other group consists of the r.v.'s $X_{j_1}, \ldots, X_{j_t}$, and the quantities we just defined may also be defined in the present context. Thus, if in $F_{X_1,\ldots,X_k}(x_1,\ldots,x_k)$, t of the x's, $x_{j_1}, \ldots, x_{j_t}$ are replaced by ∞ (in the sense they are let to tend to ∞), then what is left is the *marginal joint d.f.* of the r.v.'s $X_{i_1}, \ldots, X_{i_s}$, $F_{X_{i_1},\ldots,X_{i_s}}$, where $s + t = k$. Likewise, if in $f_{X_1,\ldots,X_k}(x_1,\ldots,x_k)$, $x_{j_1}, \ldots, x_{j_t}$ are eliminated through summation (for the discrete case) or integration (for the continuous case), what is left is the *marginal joint p.d.f.* of the r.v.'s $X_{i_1}, \ldots, X_{i_s}$, $f_{X_{i_1},\ldots,X_{i_s}}$. Combining joint and marginal joint p.d.f.'s, as in the 2-dimensional case, we obtain a variety of *conditional p.d.f.'s*. Thus, for example,

$$f_{X_{j_1},\ldots,X_{j_t}|X_{i_1},\ldots,X_{i_s}}(x_{j_1},\ldots,x_{j_t}|x_{i_1},\ldots,x_{i_s}) = \frac{f_{X_1,\ldots,X_k}(x_1,\ldots,x_k)}{f_{X_{i_1},\ldots,X_{i_s}}(x_{i_1},\ldots,x_{i_s})}. \tag{9.4}$$

Utilizing conditional p.d.f.'s, we can define *conditional expectations* and *conditional variances*, as in the 2-dimensional case (see relations (7.9), (7.10), (7.12), and (7.13) in Chapter 7).

For a (real-valued) function g defined on $\Re^k$, the expectation of the r.v. $g(X_1, \ldots, X_k)$ is defined in a way analogous to that in relation (8.1) in Chapter 8, and the validity of properties (8.3) and (8.6) in Chapter 8 is immediate.

In particular, provided the expectations involved exist:

$$E(c_1X_1 + \cdots + c_kX_k + d) = c_1EX_1 + \cdots + c_kEX_k + d, \tag{9.5}$$

$c_1, \ldots, c_k, d$ constants.

By choosing $g(X_1, \ldots, X_k) = \exp(t_1X_1 + \cdots + t_kX_k)$, $t_1, \ldots, t_k \in \Re$, the resulting expectation (assuming it is finite) is the *joint m.g.f.* of $X_1, \ldots, X_k$; that is,

$$M_{X_1,\ldots,X_k}(t_1,\ldots,t_k) = Ee^{t_1X_1+\cdots+t_kX_k}, \quad (t_1,\ldots,t_k) \in C \subseteq \Re^k C, \tag{9.6}$$

with C containing the origin in $\Re^k$.

The appropriate versions of properties (8.15) and (8.17) in Chapter 8 become here:

$$M_{c_1X_1+d_1,\ldots,c_kX_k+d_k}(t_1,\ldots,t_k) = e^{d_1t_1+\cdots+d_kt_k}M_{X_1,\ldots,X_k}(c_1t_1,\ldots,c_kt_k), \tag{9.7}$$

where $c_1, \ldots, c_k$ and $d_1, \ldots, d_k$ are constants, and:

$$\frac{\partial^{n_1+\cdots+n_k}}{\partial^{n_1}t_1\ldots\partial^{n_k}t_k}M_{X_1,\ldots,X_k}(t_1,\ldots,t_k)|_{t_1=\cdots=t_k=0} = E(X_1^{n_1}\ldots X_k^{n_k}), \tag{9.8}$$

for ≥ 0 integers $n_1, \ldots, n_k$.

The variance of $g(X_1, \ldots, X_k)$ is defined as in Definition 1(ii) in Chapter 8. Finally, the appropriate version of Theorem 3 in Chapter 8 becomes here as follows.

Theorem 1. For k r.v.'s $X_1, \ldots, X_k$ with finite first and second moments, and (positive) standard deviations σ_{X_i}, $i = 1, \ldots, k$:

$$Var(X_1 + \cdots + X_k) = \sum_{i=1}^{k} \sigma_{X_i}^2 + 2 \sum_{1 \leq i < j \leq k} Cov(X_i, X_j)$$

$$= \sum_{i=1}^{k} \sigma_{X_i}^2 + 2 \sum_{1 \leq i < j \leq k} \sigma_{X_i}\sigma_{X_j}\rho(X_i, X_j), \tag{9.9}$$

and

$$Var(X_1 + \cdots + X_k) = \sum_{i=1}^{k} \sigma_{X_i}^2 \tag{9.10}$$

if the X_i's are pairwise uncorrelated; that is, $\rho(X_i, X_j) = 0$ for $i \neq j$.

Exercises.

1.1 If the r.v.'s X_1, X_2, X_3 have the joint p.d.f. $f_{X_1,X_2,X_3}(x_1, x_2, x_3) = c^3 e^{-c(x_1+x_2+x_3)}$, $x_1 > 0, x_2 > 0, x_3 > 0$ $(c > 0)$, determine:

- **(i)** The constant c.
- **(ii)** The marginal p.d.f.'s f_{X_1}, f_{X_2}, and f_{X_3}.
- **(iii)** The conditional joint p.d.f. of X_1 and X_2, given X_3.
- **(iv)** The conditional p.d.f. of X_1, given X_2 and X_3.

1.2 Determine the joint m.g.f. of the r.v.'s X_1, X_2, X_3 with p.d.f. $f_{X_1,X_2,X_3}(x_1, x_2, x_3) = c^3 e^{-c(x_1+x_2+x_3)}$, $x_1 > 0, x_2 > 0, x_3 > 0$ (c any positive constant, see also Exercise 1.1).

1.3 (*Cramér–Wold device*) Show that if we know the joint distribution of the r.v.'s $X_1, \ldots, X_n$, then we can determine the distribution of any linear combination $c_1X_1 + \cdots + c_nX_n$ of $X_1, \ldots, X_n$, where $c_1, \ldots, c_n$ are constants. Conversely, if we know the distribution of all linear combinations just described, then we can determine the joint distribution of $X_1, \ldots, X_n$.

Hint. Use the m.g.f. approach.

1.4 If the r.v.'s $X_1, \ldots, X_m$ and $Y_1, \ldots, Y_n$ have finite second moments, then show that:

$$Cov\left(\sum_{i=1}^{m} X_i, \sum_{j=1}^{n} Y_j\right) = \sum_{i=1}^{m}\sum_{j=1}^{n} Cov(X_i, Y_j).$$

Hint. Use the definition of the covariance and the linearity property of the expectation.

1.5 Refer to Exercise 2.32 in Chapter 6, and compute:

(i) The expected amount to be paid as a refund by selling n pieces of equipment, in terms of C.
(ii) The numerical value in parts (i) for $C = \$2{,}400$ and $n = 100$.

9.2 Multinomial Distribution

The multinomial distribution is a straightforward generalization of the binomial distribution and is based on a *multinomial* experiment, which itself is a straightforward generalization of a binomial experiment. Here, instead of 2, there are k (mutually exclusive) possible outcomes, $O_1, \ldots, O_k$, say, occurring with respective probabilities $p_1, \ldots, p_k$.

Simple examples of multinomial experiments are those of rolling a die (with 6 possible outcomes); selecting (with replacement) r balls from a collection of $n_1 + \cdots + n_k$ balls, so that n_i balls have the number i written on them, $i = 1, \ldots, k$; selecting (with replacement) r objects out of a collection of objects of which n_1 are in good condition, n_2 have minor defects, and n_3 have serious defects; classifying n individuals according to their blood type, etc.

Suppose a multinomial experiment is carried out independently n times and the probabilities $p_1, \ldots, p_k$ remain the same throughout. Denote by X_i the r.v. of the number of times outcome O_i occurs, $i = 1, \ldots, k$. Then the joint p.d.f. of $X_1, \ldots, X_k$ is given by:

$$f_{X_1,\ldots,X_k}(x_1, \ldots, x_k) = \frac{n!}{x_1! \ldots x_k!} p_1^{x_1} \cdots p_k^{x_k}, \tag{9.11}$$

where $x_1, \ldots, x_k$ are ≥ 0 integers with $x_1 + \cdots + x_k = n$, and, of course, $0 < p_i < 1$, $i = 1, \ldots, k$, $p_1 + \cdots + p_k = 1$. The distribution given by (9.11) is the *multinomial* with *parameters n* and $p_1, \ldots, p_k$, and the r.v.'s $X_1, \ldots, X_k$ are said to have (jointly) the *multinomial* distribution with these parameters.

That the right-hand side of (9.11) is the right formula for the joint probabilities $P(X_1 = x_1, \ldots, X_k = x_k)$ ensues as follows: By independence, the probability that O_i occurs n_i times, $i = 1, \ldots, k$, in specified positions, is given by: $p_1^{x_1} \ldots p_k^{x_k}$ regardless of the positions of occurrence of O_i's. The different ways of choosing the n_i positions for the occurrence of O_i, $i = 1, \ldots, k$, is equal to: $\binom{n}{n_1}\binom{n-n_1}{n_2}\cdots\binom{n-n_1-\cdots-n_{k-1}}{n_k}$ (by Theorem 1 in Chapter 2). Writing out each term in factorial form and making the obvious cancellations, we arrive at: $n!/(x_1! \ldots x_k!)$ (see also Exercise 2.1).

For illustrative purposes, let us consider the following example.

Example 1. A fair die is rolled independently 10 times. Find the probability that faces #1 through #6 occur the following respective number of times: 2, 1, 3, 1, 2, and 1.

Discussion. By letting X_i be the r.v. denoting the number of occurrences of face $i, i = 1, \ldots, 6$, we have:

$$f_{X_1,\ldots,X_6}(2,1,3,1,2,1) = \frac{10!}{2!1!3!1!2!1!}(1/6)^{10} = \frac{4{,}725}{1{,}889{,}568} \simeq 0.003.$$

In a multinomial distribution, all marginal p.d.f.'s and all conditional p.d.f.'s are also multinomial. More precisely, we have the following result.

Theorem 2. Let $X_1, \ldots, X_k$ be multinomially distributed with parameters n and $p_1, \ldots, p_k$, and for $1 \leq s < k$, let $X_{i_1}, \ldots, X_{i_s}$ be any r.v.'s and let $Y = n - (X_1 + \cdots + X_s)$ and $q = 1 - (p_1 + \cdots + p_s)$. Then,

(i) The r.v.'s $X_1, \ldots, X_s, Y$ are distributed multinomially with parameters n and $p_1, \ldots, p_k, q$.

(ii) The conditional distribution of $X_{s+1}, \ldots, X_k$, given $X_1 = x_1, \ldots, X_s = x_s$, is multinomial with parameters $n - r$ and $\frac{p_{s+1}}{q}, \ldots, \frac{p_k}{q}$, where $r = x_1 + \cdots + x_s$.

Proof.

(i) For nonnegative integers $x_1, \ldots, x_s$ with $x_1 + \cdots + x_s = r \leq n$, we have:

$$\begin{aligned} f_{X_1,\ldots,X_s}(x_1,\ldots,x_s) &= P(X_1 = x_1, \ldots, X_s = x_s) \\ &= P(X_1 = x_1, \ldots, X_s = x_s, Y = n - r) \\ &= \frac{n!}{x_1! \ldots x_s!(n-r)!} p_1^{x_1} \cdots p_s^{x_s} q^{n-r}. \end{aligned}$$

(ii) For nonnegative integers $x_{s+1}, \ldots, x_k$ with $x_{s+1} + \cdots + x_k = n - r$, we have:

$$\begin{aligned} & f_{X_{s+1},\ldots,X_k|X_1,\ldots,X_s}(x_{s+1},\ldots,x_k|x_1,\ldots,x_s) \\ &= P(X_{s+1} = x_{s+1}, \ldots, X_k = x_k | X_1 = x_1, \ldots, X_s = x_s) \\ &= \frac{P(X_{s+1} = x_{s+1}, \ldots, X_k = x_k, X_1 = x_1, \ldots, X_s = x_s)}{P(X_1 = x_1, \ldots, X_s = x_s)} \\ &= \frac{P(X_1 = x_1, \ldots, X_s = x_s, X_{s+1} = x_{s+1}, \ldots, X_k = x_k)}{P(X_1 = x_1, \ldots, X_s = x_s)} \\ &= \frac{\frac{n!}{x_1! \ldots x_s! x_{s+1}! \ldots x_k!} p_1^{x_1} \cdots p_s^{x_s} p_{s+1}^{x_{s+1}} \cdots p_k^{x_k}}{\frac{n!}{x_1! \ldots x_s!(n-r)!} p_1^{x_1} \cdots p_s^{x_s} q^{n-r}}, \quad \text{(by part (i))} \\ &= \frac{(n-r)!}{x_{s+1}! \ldots x_k!} \left(\frac{p_{s+1}}{q}\right)^{x_{s+1}} \cdots \left(\frac{p_k}{q}\right)^{x_k}, \end{aligned}$$

(since $x_{s+1} + \cdots + x_k = n - r$). ▲

In the theorem above, the s r.v.'s need not be the first s, but any s r.v.'s such as $X_{i_1}, \ldots, X_{i_s}$, say, where $1 \leq i_1 < i_2 < \cdots < i_s \leq k$. Then $Y = n - (X_{i_1} + \cdots + X_{i_s})$ and $q = 1 - (p_{i_1} + \cdots + p_{i_k})$. Consequently, Theorem 2 becomes as follows, and its proof runs along the same lines as that of Theorem 2.

Theorem 2′. Let $X_{i_1}, \ldots, X_{i_s}$ be multinomially distributed with parameters n and $p_1, \ldots, p_k$, and for $1 \leq s < k$, let $Xi_1, \ldots Xi_s$ be any r.v.'s and let $Y = n - (X_{i_1} + \cdots + X_{i_s})$ and $q = 1 - (p_{i_1} + \cdots + p_{i_s})$. Then:

(i) The r.v.'s $X_{i_1}, \ldots, X_{i_s}, Y$ are distributed multinomially with parameters n and $p_{i_1}, \ldots, p_{i_k}, q$.
(ii) The conditional distribution of $X_{j_1}, \ldots, X_{j_t}$, given $X_{i_1} = x_{i_1}, \ldots, X_{i_s} = x_{i_s}$, is multinomial with parameters $n-r$ and $\frac{p_{j_1}}{q}, \ldots, \frac{p_{j_t}}{q}$, where $r = x_{i_1} + \cdots + x_{i_s}$ and $t = k - s$.

The following examples provide an illustration of Theorem 2′.

Example 2. In reference to Example 1, calculate: $P(X_2 = X_4 = X_6 = 2)$ and $P(X_1 = X_3 = 1, X_5 = 2|X_2 = X_4 = X_6 = 2)$.

Discussion. Here $n = 10,\ r = 6,\ p_2 = p_4 = p_6 = \frac{1}{6}$, and $q = 1 - \frac{3}{6} = \frac{1}{2}$. Thus:

$$P(X_2 = X_4 = X_6 = 2) = \frac{10!}{2!2!2!4!}\left(\frac{1}{6}\right)^6\left(\frac{1}{2}\right)^4 = \frac{4{,}725}{186{,}624} \simeq 0.025,$$

and:

$$P(X_1 = X_3 = 1, X_5 = 2|X_2 = X_4 = X_6 = 2) = \frac{4!}{1!1!2!}\left(\frac{1/6}{1/2}\right)^4 = \frac{4}{27} \simeq 0.148.$$

Example 3. In a genetic experiment, two different varieties of a certain species are crossed and a specific characteristic of the offspring can occur only at three levels, A, B, and C, say. According to a proposed model, the probabilities for A, B, and C are $\frac{1}{12}, \frac{3}{12}$, and $\frac{8}{12}$, respectively. Out of 60 offspring, calculate:

(i) The probability that 6, 18, and 36 fall into levels A, B, and C, respectively.
(ii) The (conditional) probability that 6 and 18 fall into levels A and B, respectively, given that 36 have fallen into level C.

Discussion.

(i) Formula (9.11) applies with $n = 60, k = 3, p_1 = \frac{1}{12}, p_2 = \frac{3}{12}, p_3 = \frac{8}{12}$, $x_1 = 6, x_2 = 18, x_3 = 36$ and yields:

$$P(X_1 = 6, X_2 = 18, X_3 = 36) = \frac{60!}{6!18!36!}\left(\frac{1}{12}\right)^6\left(\frac{3}{12}\right)^{18}\left(\frac{8}{12}\right)^{36} \simeq 0.011.$$

(ii) Here Theorem 2′(ii) applies with $s = 1, t = 2, x_{i_1} = x_3 = 36, x_{j_1} = x_1 = 6, x_{j_2} = x_2 = 18, r = 36$, so that $n - r = 60 - 36 = 24, q = 1 - p_3 =$

$1 - \frac{8}{12} = \frac{4}{12}$, and yields:

$$
\begin{aligned}
P(X_1 = 6, X_2 = 18 | X_3 = 36) &= \frac{(n-r)!}{x_1! x_2!} \left(\frac{p_1}{q}\right)^{x_1} \left(\frac{p_2}{q}\right)^{x_2} \\
&= \frac{(24)!}{6!18!} \left(\frac{\frac{1}{12}}{\frac{4}{12}}\right)^6 \left(\frac{\frac{3}{12}}{\frac{4}{12}}\right)^{18} \\
&= \binom{24}{6} \left(\frac{1}{4}\right)^6 \left(\frac{3}{4}\right)^{18} \\
&= 0.1852 \quad \text{(from the binomial tables)}.
\end{aligned}
$$

An application of formula (9.6) gives the *joint m.g.f.* of $X_1, \ldots, X_k$ as follows, where the summation is over all nonnegative integers $x_1, \ldots, x_k$ with $x_1 + \cdots + x_k = n$ (see also #3 in Table 6 in the Appendix):

$$
\begin{aligned}
M_{X_1,\ldots,X_k}(t_1, \ldots, t_k) &= \sum e^{t_1 x_1 + \cdots + t_k x_k} \frac{n!}{x_1! \cdots x_k!} p_1^{x_1} \cdots p_k^{x_k} \\
&= \sum \frac{n!}{x_1! \ldots x_k!} (p_1 e^{t_1})^{x_1} \ldots (p_k e^{t_k})^{x_k} \\
&= (p_1 e^{t_1} + \cdots + p_k e^{t_k})^n; \text{ i.e.,}
\end{aligned}
$$

$$
M_{X_1,\ldots,X_k}(t_1, \ldots, t_k) = (p_1 e^{t_1} + \cdots + p_k e^{t_k})^n, \qquad t_1, \ldots, t_k \in \Re. \quad (9.12')
$$

By means of (9.8) and (9.12′), we can find the $Cov(X_i, X_j)$ and the $\rho(X_i, X_j)$ for any $1 \leq i < j \leq k$. Indeed, $EX_i = np_i$, $EX_j = np_j$, $Var(X_i) = np_i(1 - p_i)$, $Var(X_j) = np_j(1 - p_j)$ and $E(X_i X_j) = n(n-1)\, p_i p_j$. Therefore:

$$
Cov(X_i, X_j) = -np_i p_j \quad \text{and} \quad \rho(X_i, X_j) = -[p_i p_j / ((1 - p_i)(1 - p_j))]^{1/2}
$$

(see Exercise 2.4 for details).

Exercises.

2.1 For any nonnegative integers $n_1, \ldots, n_k$ with $n_1 + \cdots + n_k = n$, show that

$$
\binom{n}{n_1} \binom{n - n_1}{n_2} \cdots \binom{n - n_1 - \cdots - n_{k-1}}{n_k} = \frac{n!}{n_1! n_2! \ldots n_k!}.
$$

Hint. Write out the terms on the left-hand side as factorials, and recall that $0! = 1$.

2.2 In a store selling TV sets, it is known that 25% of the customers will purchase a TV set of brand A, 40% will purchase a TV set of brand B, and 35% will just be browsing. For a lot of 10 customers:

(i) What is the probability that 2 will purchase a TV set of brand A, 3 will purchase a TV set of brand B, and 5 will purchase neither?

(ii) If it is known that 6 customers did not purchase a TV set, what is the (conditional) probability that 1 of the rest will purchase a TV set of brand A and 3 will purchase a TV set of brand B?

Hint. Part (i) is an application of formula (9.11), and part (ii) is an application of Theorem 2(ii).

2.3 Human blood occurs in 4 types termed A, B, AB, and O with respective frequencies $p_A = 0.40$, $p_B = 0.10$, $p_{AB} = 0.05$, and $p_O = 0.45$. If n donors participate in a blood drive, denote by X_A, X_B, X_{AB}, and X_O the numbers of donors with respective blood types A, B, AB, and O. Then X_A, X_B, X_{AB}, and X_O are r.v.'s having multinomial distribution with parameters n and p_A, p_B, p_{AB}, p_O. Write out the appropriate formulas for the following probabilities:

(i) $P(X_A = x_A, X_B = x_B, X_{AB} = x_{AB}, X_O = x_O)$ for x_A, x_B, x_{AB}, and x_O nonnegative integers with $x_A + x_B + x_{AB} + x_O = n$.
(ii) $P(X_A = x_A, X_B = x_B, X_{AB} = x_{AB})$.
(iii) $P(X_A = x_A, X_B = x_B)$.
(iv) $P(X_A = x_A)$.
(v) $P(X_A = x_A, X_B = x_B, X_{AB} = x_{AB} | X_O = x_O)$.
(vi) $P(X_A = x_A, X_B = x_B | X_{AB} = x_{AB}, X_O = x_O)$.
(vii) $P(X_A = x_A | X_B = x_B, X_{AB} = x_{AB}, X_O = x_O)$.
(viii) Give numerical answers to parts (i)–(vii), if $n = 20$, and $x_A = 8$, $x_B = 2$, $x_{AB} = 1$, $x_O = 9$.

Hint. Part (i) is an application of formula (9.11), and parts (ii)–(viii) are applications of Theorem 2.

2.4 In conjunction with multinomial distribution, show that:

$$EX_i = np_i,\ EX_j = np_j,\ Var(X_i) = np_i(1 - p_i),\ Var(X_j) = np_j(1 - p_j),$$
$$Cov(X_i, X_j) = -np_i p_j \quad \text{and} \quad \rho(X_i, X_j) = -\frac{p_i p_j}{[p_i(1 - p_i)p_j(1 - p_j)]^{1/2}}.$$

Hint. Use the m.g.f. given in formula (9.12′).

2.5 Refer to Exercises 2.3 and 2.4, and for $n = 20$, calculate the quantities:

$$EX_A,\ EX_B,\ EX_{AB},\ EX_O;\ Var(X_A),\ Var(X_B),\ Var(X_{AB}),$$
$$Var(X_O);\ Cov(X_A, X_B),\ Cov(X_A, X_{AB}),\ Cov(X_A, X_O);$$
$$\rho(X_A, X_B),\ \rho(X_A, X_{AB}),\ \rho(X_A, X_O).$$

2.6 In a shipment of 32 TV sets, it is known that 20 are brand new, 10 are used but in working condition, and 2 are defective. Six TV sets are selected at random, and let X_1, X_2, X_3 be the r.v.'s denoting, respectively, the numbers, among the 6, of the new, used and defective sets.

(i) Compute the probability $P(X_1 = 4,\ X_2 = 1,\ X_3 = 1)$.
(ii) Compute the same probability as in part (i) by pretending that the r.v.'s X_1, X_2, X_3 have multinomial distribution.
(iii) What portion of the correct probability in part (i) is the probability in part (ii)?

Hint. For part (i), use Theorem 1 and part (ii) of its corollary in Chapter 2.

2.7 In a certain high school, the distribution of students among the four grades is as follows: 20.00% are freshmen, 23.75% are sophomore, 31.25% are junior, and 25.00% are senior. If 20 students are chosen at random from this high school, compute the following probabilities:

(i) 4 students are freshmen, 5 are sophomores, 6 are junior, and 5 are senior.
(ii) 5 students are senior.
(iii) 11 students are either junior or senior.
(iv) At least 11 students are either junior or senior.

Hint. Assume the multinomial model, and use formula (9.11) and Theorem 2.

2.8 In a mouse, a gene occurs in two states, dominant (A) and recessive (a), and such genes occur in pairs, so that there are the following possibilities: AA, Aa, aA (the same as Aa), and aa. If p is the frequency with which state A occurs, then the frequencies of the configurations AA, Aa, and aa are, respectively, $p^2, 2p(1-p)$, and $(1-p)^2$. If n such pairs of genes are observed independently, and if X_1, X_2, X_3 are the numbers of the observed configurations AA, Aa $(= aA)$, aa, respectively, then:

(i) What is the joint distribution of the X_i's?
If $n = 15$ and $p = 0.75$, compute the following quantities:
(ii) $P(X_1 = 8,\ X_2 = 6,\ X_3 = 1)$.
(iii) $P(X_2 = 6)$.
(iv) $P(X_1 = 8,\ X_3 = 1 | X_2 = 6)$.

2.9 A manufactured item is classified as good, defective but usable, and outright defective in the following respective proportions: 62.5%, 31.25%, and 6.25%. Twenty-five such items are selected at random from the production line, and let X_1, X_2, and X_3 be the r.v.'s denoting the numbers of the good, defective but usable, and outright defective items, respectively. Compute the following quantities:

(i) $P(X_1 = 16,\ X_2 = 7,\ X_3 = 2)$.
(ii) $P(X_1 = 15,\ X_2 = 8)$
(iii) $P(X_3 = 2)$
(iv) $EX_i, \sigma(X_i), i = 1, 2, 3$.

Hint. For part (i), use formula (9.11). For parts (ii) and (iii), use Theorem 2. For part (iv), use either Theorem 2(i) suitably, or formula (9.12′).

2.10 From statistical surveys taken in the past, it is known that a certain TV audience watches the 8:00 o'clock news, or a documentary program, or other programs in the following respective proportions: 31.25%, 25%, and 43.75%. A random sample of size 15 is taken from this audience at 8:00 o'clock. Compute the following probabilities:

(i) The numbers observed, among the 15 selected, are, respectively, 5, 4, and 6.
(ii) The number of those watching a documentary program is at least 3.
(iii) The number of those not watching the news is no more than 5.

Hint. For part (i), use formula (9.11), and for parts (ii) and (iii), use Theorem 2(i) suitably.

2.11 Let X be an r.v. of the continuous type with p.d.f. f, and consider the partition of the real line $\Re$ by k intervals as indicated below.
Set $p_i = \int_{I_i} f(x)dx, i = 1, \ldots, k$. Next, take n independent observations on the r.v. X, and let X_i be the number of outcomes lying in the I_i interval, $i = 1, \ldots, k$.

(i) Specify the joint distribution of the r.v.'s X_i, $i = 1, \ldots, k$.
(ii) Let $X \sim N(0, 1)$, and consider the following intervals:

$$I_1 = (-\infty, -3], \quad I_2 = (-3, -2] \quad I_3 = (-2, -1], \quad I_4 = (-1, 0],$$
$$I_5 = (0, 1], \quad I_6 = (1, 2] \quad I_7 = (2, 3], \quad I_8 = (3, \infty].$$

Calculate the probabilities $p_i = P(X \in I_i)$, $i = 1, \ldots, 8$. (see Figure 9.1).

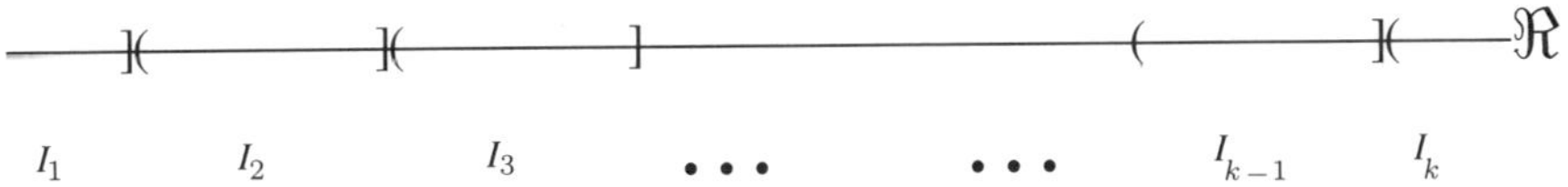

FIGURE 9.1
Partition of the real line $\Re$ into a countable number of intervals.

9.3 Bivariate Normal Distribution

This distribution could have been discussed in Chapter 7, but we chose not to do so in order not to overload that chapter.

The joint distribution of the r.v.'s X and Y is said to be *bivariate normal* with *parameters* μ_1, μ_2 in $\Re$, σ_1, σ_2 positive and $\rho \in [-1, 1]$, if their joint p.d.f. is given by the formula:

$$f_{X,Y}(x, y) = \frac{1}{2\pi\sigma_1\sigma_2\sqrt{1-\rho^2}} e^{-q/2}, \qquad x, y \in \Re, \qquad (9.13')$$

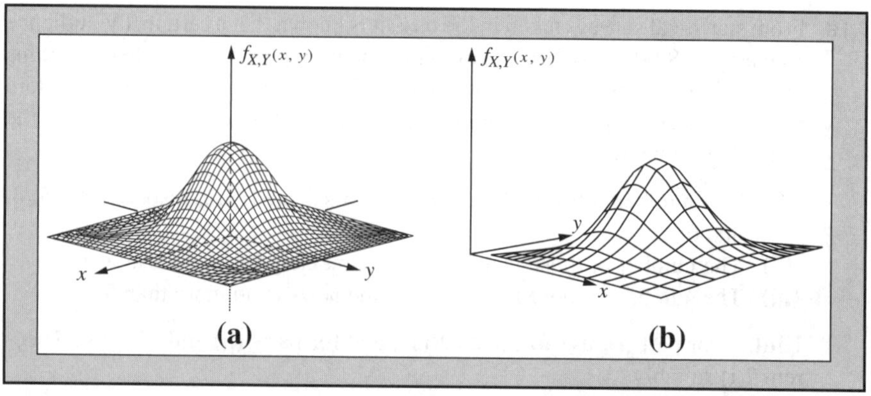

FIGURE 9.2
Graphs of the p.d.f., of bivariate normal distribution: (a) centered at the origin; (b) centered elsewhere in the (x, y)-plane.

where

$$q = \frac{1}{1-\rho^2}\left[\left(\frac{x-\mu_1}{\sigma_1}\right)^2 - 2\rho\left(\frac{x-\mu_1}{\sigma_1}\right)\left(\frac{y-\mu_2}{\sigma_2}\right) + \left(\frac{y-\mu_2}{\sigma_2}\right)^2\right]. \quad (9.14')$$

This distribution is also referred to as *2-dimensional normal*. The shape of $f_{X,Y}$ looks like a bell facing the xy-plane, whose highest point is located at the point $(\mu_1, \mu_2, 1/(2\pi\sigma_1\sigma_2\sqrt{1-\rho^2}))$ (see Figure 9.2).

Many bivariate measurements may be looked upon as having a distribution approximated by a suitable bivariate normal distribution. Heights of fathers, heights of sons; SAT, GPA scores; heights, weights of individuals; voltage, resistance; tire pressure, mileage rate in cars; amount of fertilizer, amount of harvested agricultural commodity; and cholesterol, blood pressure may be such examples.

That $f_{X,Y}$ integrates to 1 and therefore is a p.d.f. is seen by rewriting it in a convenient way. Specifically,

$$\begin{aligned}
&\left(\frac{x-\mu_1}{\sigma_1}\right)^2 - 2\rho\left(\frac{x-\mu_1}{\sigma_1}\right)\left(\frac{y-\mu_2}{\sigma_2}\right) + \left(\frac{y-\mu_2}{\sigma_2}\right)^2 \\
&\quad = \left(\frac{y-\mu_2}{\sigma_2}\right)^2 - 2\left(\rho\frac{x-\mu_1}{\sigma_1}\right)\left(\frac{y-\mu_2}{\sigma_2}\right) + \left(\rho\frac{x-\mu_1}{\sigma_1}\right)^2 \\
&\qquad + (1-\rho^2)\left(\frac{x-\mu_1}{\sigma_1}\right)^2 \\
&\quad = \left[\left(\frac{y-\mu_2}{\sigma_2}\right) - \left(\rho\frac{x-\mu_1}{\sigma_1}\right)\right]^2 + (1-\rho^2)\left(\frac{x-\mu_1}{\sigma_1}\right)^2. \qquad (9.15')
\end{aligned}$$

Furthermore,

$$\begin{aligned}\frac{y-\mu_2}{\sigma_2}-\rho\frac{x-\mu_1}{\sigma_1} &= \frac{y-\mu_2}{\sigma_2}-\frac{1}{\sigma_2}\times\rho\sigma_2\frac{x-\mu_1}{\sigma_1}\\ &= \frac{1}{\sigma_2}\left\{y-\left[\mu_2+\frac{\rho\sigma_2}{\sigma_1}(x-\mu_1)\right]\right\}\\ &= \frac{y-b_x}{\sigma_2},\quad \text{where } b_x=\mu_2+\frac{\rho\sigma_2}{\sigma_1}(x-\mu_1)\end{aligned}$$

(see also Exercise 3.1).

Therefore, the right-hand side of (9.15′) is equal to:

$$\left(\frac{y-b_x}{\sigma_2}\right)^2+(1-\rho^2)\left(\frac{x-\mu_1}{\sigma_1}\right)^2,$$

and hence the exponent becomes:

$$-\frac{(x-\mu_1)^2}{2\sigma_1^2}-\frac{(y-b_x)^2}{2(\sigma_2\sqrt{1-\rho^2})^2}.$$

Then the joint p.d.f. may be rewritten as follows:

$$f_{X,Y}(x,y)=\frac{1}{\sqrt{2\pi}\sigma_1}e^{-\frac{(x-\mu_1)^2}{2\sigma_1^2}}\times\frac{1}{\sqrt{2\pi}(\sigma_2\sqrt{1-\rho^2})}e^{-\frac{(y-b_x)^2}{2(\sigma_2\sqrt{1-\rho^2})^2}}. \tag{9.16′}$$

The first factor on the right-hand side of (9.16′) is the p.d.f. of $N(\mu_1,\sigma_1^2)$, and the second factor is the p.d.f. of $N(b_x,(\sigma_2\sqrt{1-\rho^2})^2)$. Therefore, integration with respect to y produces the marginal $N(\mu_1,\sigma_1^2)$ distribution, which, of course, integrates to 1. So, we have established the following two facts:

$$\int_{-\infty}^{\infty}\int_{-\infty}^{\infty} f_{X,Y}(x,y)dx\,dy=1,\ \text{and}$$
$$X\sim N\big(\mu_1,\sigma_1^2\big),\quad \text{and by symmetry, } Y\sim N\big(\mu_2,\sigma_2^2\big). \tag{9.17′}$$

The results recorded in (9.17′) also reveal the special significance of the parameters μ_1,σ_1^2 and μ_2,σ_2^2. Namely, they are the means and the variances of the (normally distributed) r.v.'s X and Y, respectively. Relations (9.16′) and (9.17′) also provide immediately the conditional p.d.f. $f_{Y|X}$; namely,

$$f_{Y|X}(y|x)=\frac{1}{\sqrt{2\pi}\left(\sigma_2\sqrt{1-\rho^2}\right)}\exp\left[-\frac{(y-b_x)^2}{2\left(\sigma_2\sqrt{1-\rho^2}\right)^2}\right].$$

Thus, in obvious notation:

$$Y|X=x\sim N\left(b_x,\left(\sigma_2\sqrt{1-\rho^2}\right)^2\right),\qquad b_x=\mu_2+\frac{\rho\sigma_2}{\sigma_1}(x-\mu_1), \tag{9.18′}$$

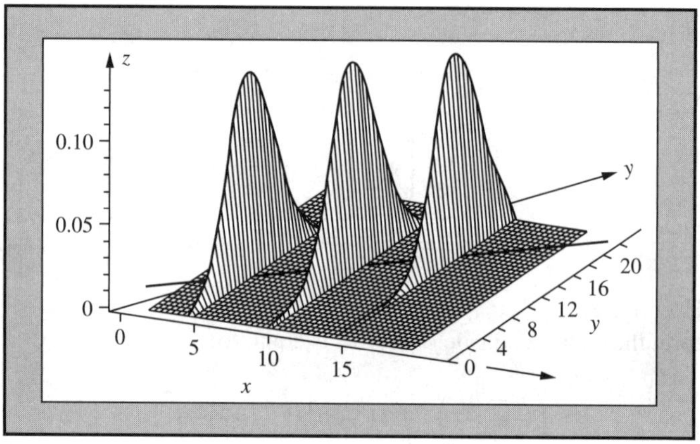

FIGURE 9.3
Conditional probability density functions of the bivariate normal distribution.

and by symmetry:

$$X|Y = y \sim N\left(b_y, \left(\sigma_1\sqrt{1-\rho^2}\right)^2\right), \qquad b_y = \mu_1 + \frac{\rho\sigma_1}{\sigma_2}(y - \mu_2). \qquad (9.19')$$

Gathering together the results obtained above in the form of a theorem, we have then:

Theorem 3.

(i) The function given in (9.13′) is, indeed, a p.d.f. as stated.
(ii) The marginal p.d.f.'s are themselves normal; or the r.v.'s X and Y are normally distributed as specified in (9.17′).
(iii) The conditional p.d.f.'s are also normal as specified in (9.18′) and (9.19′).

In Figure 9.3, the conditional p.d.f. $f_{Y|X}(\cdot|x)$ is depicted for three values of x:$x = 5$, 10, and 15.

Formulas (9.16′), (9.17′), and (9.18′) also allow us to calculate easily the covariance and the correlation coefficient of X and Y. Indeed, by (9.17′):

$$\begin{aligned} E(XY) &= \int_{-\infty}^{\infty}\int_{-\infty}^{\infty} xy\, f_{X,Y}(x,y)dx\,dy = \int_{-\infty}^{\infty} xf_X(x)\left[\int_{-\infty}^{\infty} yf_{Y|X}(y|x)dy\right]dx \\ &= \int_{-\infty}^{\infty} xf_X(x)b_x\,dx = \int_{-\infty}^{\infty} xf_X(x)\left[\mu_2 + \frac{\rho\sigma_2}{\sigma_1}(x-\mu_1)\right]dx \\ &= \mu_2\int_{-\infty}^{\infty} xf_X(x)dx + \frac{\rho\sigma_2}{\sigma_1}\left[\int_{-\infty}^{\infty} x^2 f_X(x)dx - \mu_1\int_{-\infty}^{\infty} xf_X(x)dx\right] \\ &= \mu_1\mu_2 + \frac{\rho\sigma_2}{\sigma_1}(\sigma_1^2 + \mu_1^2 - \mu_1^2) = \mu_1\mu_2 + \rho\sigma_1\sigma_2 \end{aligned}$$

(see also Exercise 3.2). Since we already know that $EX = \mu_1, EY = \mu_2$, and $Var(X) = \sigma_1^2, Var(Y) = \sigma_2^2$, we obtain:

$$Cov(X, Y) = E(XY) - (EX)(EY) = \mu_1\mu_2 + \rho\sigma_1\sigma_2 - \mu_1\mu_2 = \rho\sigma_1\sigma_2,$$

and therefore $\rho(X, Y) = \frac{\rho\sigma_1\sigma_2}{\sigma_1\sigma_2} = \rho$. Thus, we have:

$$Cov(X, Y) = \rho\sigma_1\sigma_2 \quad \text{and} \quad \rho(X, Y) = \rho. \tag{9.20$'$}$$

Relation (9.20′) reveals that the parameter ρ in (9.14′) is actually the correlation coefficient of the r.v.'s X and Y.

Thus, we have the following result.

Theorem 4. In bivariate normal distribution with parameters $\mu_1, \mu_2, \sigma_1, \sigma_2$, and ρ, we have:

$$Cov(X, Y) = \rho\sigma_1\sigma_2 \quad \text{and} \quad \rho(X, Y) = \rho.$$

The following examples provide an illustration for some of the quantities associated with a bivariate normal distribution.

Example 4. If the r.v.'s X_1 and X_2 have bivariate normal distribution with parameters $\mu_1, \mu_2, \sigma_1^2, \sigma_2^2$, and ρ:

(i) Calculate the quantities: $E(c_1X_1 + c_2X_2)$, $Var(c_1X_1 + c_2X_2)$, where c_1, c_2 are constants.

(ii) What the expressions in part (i) become for: $\mu_1 = -1$, $\mu_2 = 3$, $\sigma_1^2 = 4$, $\sigma_2^2 = 9$, and $\rho = \frac{1}{2}$?

Discussion.

(i) $E(c_1X_1 + c_2X_2) = c_1EX_1 + c_2EX_2 = c_1\mu_1 + c_2\mu_2$, e.g., since $X_i \sim N(\mu_i, \sigma_i^2)$, so that $EX_i = \mu_i, i = 1, 2$. Also,

$$\begin{aligned} Var(c_1X_1 + c_2X_2) &= c_1^2\sigma_{X_1}^2 + c_2^2\sigma_{X_2}^2 + 2c_1c_2\sigma_{X_1}\sigma_{X_2}\rho(X_1, X_2) \\ &\qquad \text{(by (9.9) applied for } k = 2) \\ &= c_1^2\sigma_1^2 + c_2^2\sigma_2^2 + 2c_1c_2\sigma_1\sigma_2\rho, \end{aligned}$$

since $X_i \sim N(\mu_i, \sigma_i^2)$, so that $Var(X_i) = \sigma_i^2, i = 1, 2$, and $\rho(X_1, X_2) = \rho$, by (9.20′).

(ii) Here $E(c_1X_1 + c_2X_2) = -c_1 + 3c_2$, and $Var(c_1X_1 + c_2X_2) = 4c_1 + 9c_2 + 2c_1c_2 \times 2 \times 3 \times \frac{1}{2} = 4c_1 + 9c_2 + 6c_1c_2$.

Example 5. Suppose that the heights of fathers and sons are r.v.'s X and Y, respectively, having (approximately) bivariate normal distribution with parameters (expressed in inches) $\mu_1 = 70, \sigma_1 = 2, \mu_2 = 71, \sigma_2 = 2$ and $\rho = 0.90$. If for a given pair (father, son) it is observed that $X = x = 69$, determine:

(i) The conditional distribution of the height of the son.
(ii) The expected height of the son.
(iii) The probability of the height of the son to be more than 72 inches.

Discussion.

(i) According to (9.18′), $Y|X = x \sim N(b_x, (\sigma_2\sqrt{1-\rho^2})^2)$, where:

$$b_x = \mu_2 + \frac{\rho\sigma_2}{\sigma_1}(x - \mu_1) = 71 + 0.90(69 - 70) = 70.1, \quad \text{and}$$
$$\sigma_2\sqrt{1-\rho^2} = 2 \times \sqrt{1 - 0.90^2} = 2 \times \sqrt{0.19} \simeq 0.87.$$

That is, $Y|X = 69$ is distributed as $N(70.1, (0.87)^2)$.

(ii) The (conditional) expectation of Y, given $X = 69$, is equal to $b_{69} = 70.1$.

(iii) The required (conditional) probability is:

$$P(Y > 72 \mid X = 69) = P\left(\frac{Y - b_{69}}{\sigma_2\sqrt{1-\rho^2}} > \frac{72 - 70.1}{0.87}\right) \simeq P(Z > 2.18)$$
$$= 1 - \Phi(2.18) = 1 - 0.985371 = 0.014629.$$

Finally, it can be seen by integration that the joint m.g.f. of X and Y is given by the formula:

$$M_{X,Y}(t_1, t_2) = \exp\left[\mu_1 t_1 + \mu_2 t_2 + \frac{1}{2}\left(\sigma_1^2 t_1^2 + 2\rho\sigma_1\sigma_2 t_1 t_2 + \sigma_2^2 t_2^2\right)\right]; \quad t_1, t_2 \in \Re; \tag{9.21′}$$

we choose not to pursue its justification (which can be found, e.g., in Section 6.5.3 in the book *A Course in Mathematical Statistics*, 2nd edition (1997), Academic Press, by G. G. Roussas). We see, however, easily that

$$\frac{\partial}{\partial t_1} M_{X,Y}(t_1, t_2) = \left(\mu_1 + \sigma_1^2 t_1 + \rho\sigma_1\sigma_2 t_2\right) M_{X,Y}(t_1, t_2),$$

and hence:

$$\frac{\partial^2}{\partial t_1 \partial t_2} M_{X,Y}(t_1, t_2) = \rho\sigma_1\sigma_2 M_{X,Y}(t_1, t_2) + \left(\mu_1 + \sigma_1^2 t_1 + \rho\sigma_1\sigma_2 t_2\right)$$
$$\times \left(\mu_2 + \sigma_2^2 t_2 + \rho\sigma_1\sigma_2 t_1\right) M_{X,Y}(t_1, t_2),$$

which, evaluated at $t_1 = t_2 = 0$, yields: $\rho\sigma_1\sigma_2 + \mu_1\mu_2 = E(XY)$, as we have already seen.

Exercises.

3.1 Elaborate on the expressions in (9.15′), as well as the expressions following (9.15′).

3.2 If the r.v.'s X and Y have bivariate normal distribution with parameters $\mu_1, \mu_2, \sigma_1^2, \sigma_2^2$, and ρ, show that $E(XY) = \mu_1\mu_2 + \rho\sigma_1\sigma_2$.

Hint. Write the joint p.d.f. $f_{X,Y}$ as $f_{Y|X}(y|x)f_X(x)$ and use the fact (see relation (9.18′)) that $E(Y|X = x) = b_x = \mu_2 + \frac{\rho\sigma_2}{\sigma_1}(x - \mu_1)$.

3.3 If the r.v.'s X and Y have bivariate normal distribution , then, by using Exercise 3.2, show that the parameter ρ is, indeed, the correlation coefficient of the r.v.'s X and Y, $\rho = \rho(X, Y)$.

3.4 If the r.v.'s X and Y have joint p.d.f. $f_{X,Y}$, expectations $\mu_1 = EX$, $\mu_2 = EY$ finite, variances $\sigma_1^2 = Var(X)$, $\sigma_2^2 = Var(Y)^2$ finite, and correlation coefficient $\rho = \rho(X, Y)$, and if c_1, c_2 are constants:

(i) Show that $Cov(c_1X, c_2Y) = c_1c_2Cov(X, Y)$.
(ii) Express the expectation $E(c_1X + c_2Y)$ and the variance $Var(c_1X + c_2Y)$ in terms of $c_1, c_2, \mu_1, \mu_2, \sigma_1^2, \sigma_2^2$, and ρ.
(iii) How part (ii) differs from Example 4(i)?

Hint. For the variance part, use Theorem 3 in Chapter 8.

3.5 If the r.v.'s X and Y have bivariate normal distribution , then it is known (see, e.g., relation (11), Section 6.5.3, in the book *A Course in Mathematical Statistics*, 2nd edition (1997), Academic Press, by G. G. Roussas) that the joint m.g.f. of X and Y is given by:

$$M_{X,Y}(t_1, t_2) = \exp\left[\mu_1 t_1 + \mu_2 t_2 + \frac{1}{2}\left(\sigma_1^2 t_1^2 + 2\rho\sigma_1\sigma_2 t_1 t_2 + \sigma_2^2 t_2^2\right)\right], \quad t_1, t_2 \in \Re.$$

Use this m.g.f. to show that:

$$EX = \mu_1, \ EY = \mu_2, \ Var(X) = \sigma_1^2, \ Var(Y) = \sigma_2^2,$$
$$Cov(X, Y) = \rho\sigma_1\sigma_2, \ \text{and} \ \rho(X, Y) = \rho.$$

3.6 Use the joint m.g.f. of the r.v.'s X and Y having a bivariate normal distribution (see Exercise 3.5) to show that:

(i) If X and Y have bivariate normal distribution with parameters $\mu_1, \mu_2, \sigma_1^2, \sigma_2^2$, and ρ, then, for any constants c_1 and c_2, the r.v. $c_1X + c_2Y$ has normal distribution with parameters $c_1\mu_1 + c_2\mu_2$, and $c_1^2\sigma_1^2 + 2c_1c_2\rho\sigma_1\sigma_2 + c_2^2\sigma_2^2$.
(ii) If the r.v. $c_1X + c_2Y$ is normally distributed for any constants c_1 and c_2, then the r.v.'s. X and Y have bivariate normal distribution with parameters $\mu_1 = EX$, $\mu_2 = EY$, $\sigma_1^2 = Var(X)$, $\sigma_2^2 = Var(Y)$, and $\rho = \rho(X, Y)$.

Hint. For part (i), use the m.g.f. given in Exercise 3.5 and regroup the terms appropriately. For part (ii), evaluate the m.g.f. of $t_1X + t_2Y$ for any t_1, t_2 real at 1, and plug in the $E(t_1X + t_2Y)$ and the $Var(t_1X + t_2Y)$.

3.7 Consider the function f defined by:

$$f(x,y)=\begin{cases}\frac{1}{2\pi}e^{-\frac{x^2+y^2}{2}}, & \text{for } (x,y) \text{ outside the square } [-1,1]\times[-1,1]\\ \frac{1}{2\pi}e^{-\frac{x^2+y^2}{2}}+\frac{1}{2\pi e}x^3y^3, & \text{for } (x,y) \text{ in the square } [-1,1]\times[-1,1].\end{cases}$$

(i) Show that f is a non–bivariate normal p.d.f.
(ii) Also, show that both marginals, call them f_1 and f_2, are $N(0,1)$ p.d.f.'s.

Remark: We know that if X, Y have bivariate normal distribution , then the distributions of the r.v.'s X and Y themselves are normal. This exercise shows that the inverse need not be true.

3.8 Let the r.v.'s X and Y have bivariate normal distribution with parameters μ_1, μ_2, σ_1^2, σ_2^2, and ρ, and set $U = X + Y$, $V = X - Y$. Then show that:

(i) The r.v.'s U and V also have bivariate normal distribution with parameters $\mu_1+\mu_2, \mu_1-\mu_2, \tau_1^2=\sigma_1^2+2\rho\sigma_1\sigma_2+\sigma_2^2, \tau_2^2=\sigma_1^2-2\rho\sigma_1\sigma_2+\sigma_2^2$, and $\rho_0=(\sigma_1^2-\sigma_2^2)/\tau_1\tau_2$.
(ii) $U \sim N(\mu_1+\mu_2, \tau_1^2)$, $V \sim N(\mu_1-\mu_2, \tau_2^2)$.
(iii) The r.v.'s U and V are uncorrelated if and only if $\sigma_1^2 = \sigma_2^2$.

Hint. For part (i), start out with the joint m.g.f. of U and V, and express it in terms of the m.g.f. of X and Y. Then use the formula given in Exercise 3.5, and regroup the terms in the exponent suitably to arrive at the desired conclusion. Parts (ii) and (iii) follow from part (i) and known facts about bivariate normal distribution.

3.9 Let the r.v.'s X and Y denote the scores in two tests T_1 and T_2, and suppose that they have bivariate normal distribution with the following parameters: $\mu_1 = 82, \sigma_1 = 10, \mu_2 = 90, \sigma_2 = 9, \rho = 0.75$. Compute the following quantities: $P(Y > 92|X = 84)$, $P(X > Y)$, $P(X + Y > 180)$.

Hint. The first probability is calculated by the fact that we know the conditional distribution of Y, given $X = x$. The last two probabilities are calculated by the fact that the distributions of $X - Y$ and $X + Y$ are given in Exercise 3.8(ii).

3.10 If X and Y have bivariate normal distribution with parameters $\mu_1 = 3.2, \mu_2 = 12, \sigma_1^2 = 1.44, \sigma_2^2 = 16$, and $\rho = 0.7$, determine the following quantities:

(i) $EX, EY, \sigma^2(X), \sigma^2(Y), \rho(X,Y)$, and $Cov(X,Y)$.
(ii) $E(X|Y=10), E(Y|X=3.8), \sigma^2(X|Y=10), \sigma^2(Y|X=3.8)$.
(iii) The distribution of X and the distribution of Y.
(iv) The probabilities $P(0.8 < X < 4.2)$, $P(Y > 14)$.
(v) The conditional distribution of Y, given $X = 3.8$.
(vi) The probabilities $P(X > 3.2|Y = 10)$, $P(Y < 12|X = 3.8)$.

3.11 Let the r.v.'s X and Y denote the heights of pairs (father, son) in a certain age bracket, and suppose that they have bivariate normal distribution with the following parameters (measured in inches as appropriate): $\mu_1 = 67, \sigma_1 = 2, \mu_2 = 68, \sigma_2 = 1, \rho = 0.8$. If it is observed that $X = 69$, compute the following quantities: $E(Y|X = 69)$, conditional s.d. of $Y|X = 69$, $P(Y > 70|X = 69)$, $P(Y > X)$.

3.12 Suppose the r.v.'s X and Y have bivariate normal distribution with parameters $\mu_1, \mu_2, \sigma_1^2, \sigma_2^2, \rho$, and set $U = X + cY$.

(i) Compute the $\sigma^2(U)$ in terms of the parameters involved and c.
(ii) Determine the value of c that minimizes the variance in part (i).
(iii) What is the minimum variance in part (i)?
(iv) How the variance and the minimum variances in parts (ii) and (iii) become when X and Y are independent?

Hint. For part (i), use Theorem 3 in Chapter 8.

3.13 Show that the bivariate normal p.d.f. $f(x, y)$ given by relation (9.13′) is maximized for $x = \mu_1, y = \mu_2$, and the maximum is equal to $1/2\pi\sigma_1\sigma_2\sqrt{1-\rho^2}$.

Hint. Set $\frac{x-\mu_1}{\sigma_1} = u, \frac{y-\mu_2}{\sigma_2} = v$, so that the exponent in the p.d.f. becomes, in terms of u and v : $g(u, v) = u^2 - 2\rho uv + v^2$ (apart from the factor $(1-\rho^2)^{-1}$). Then show that $g(u, v)$ is minimized for $u = v = 0$, which would imply that $f(x, y)$ is maximized for $x = \mu_1, y = \mu_2$.

3.14 Let the r.v.'s X and Y have bivariate normal distribution with parameters $\mu_1, \mu_2, \sigma_1^2, \sigma_2^2, \rho$, and suppose that all parameters are known and the r.v. X is observable. Then we wish to predict Y in terms of a linear expression in X; namely, $\hat{Y} = a + b(X - \mu_1)$, by determining a and b, so that the (average) $E(Y - \hat{Y})^2$ becomes minimum.

(i) Show that the values of a and b that minimize the expression $E(Y - \hat{Y})^2$ are given by: $\hat{a} = \mu_2$ and $\hat{b} = \frac{\rho\sigma_2}{\sigma_1}$.
(ii) The predictor $\hat{Y}$ of Y is given by: $\hat{Y} = \mu_2 + \frac{\rho\sigma_2}{\sigma_1}(x - \mu_1)$. Thus, by setting $X = x$, we have that $\hat{Y}$ at $X = x$ is: $\mu_2 + \frac{\rho\sigma_2}{\sigma_1}(x - \mu_1)$, which is the $E(Y|X = x)$. Hence the $E(Y|X = x)$ is the best predictor of Y (among all predictors of the form $a + b(x - \mu_1)$) in the sense that when x is replaced by X, $E(Y|X)$ minimizes the mean square error $E(Y - \hat{Y})^2$.

Hint. The required determination of a and b can be made through the usual procedure of differentiations. Alternatively, we can write:

$$
\begin{aligned}
E(Y - \hat{Y})^2 &= E[Y - a - b(X - \mu_1)]^2 \\
&= E[(Y - \mu_2) + (\mu_2 - a) - b(X - \mu_1)]^2
\end{aligned}
$$

and proceed in order to find:

$$
\begin{aligned}
E(Y-\hat{Y})^2 &= \sigma_2^2 + (\mu_2 - a)^2 + b^2\sigma_1^2 - 2bCov(X,Y) \\
&= \sigma_2^2 + (\mu_2 - a)^2 + \sigma_1^2[b - \frac{1}{\sigma_1^2}Cov(X,Y)]^2 \\
&\quad - \frac{Cov^2(X,Y)}{\sigma_1^2} \\
&= (\mu_2 - a)^2 + \sigma_1^2[b - \frac{1}{\sigma_1^2}Cov(X,Y)]^2 \\
&\quad + \frac{\sigma_1^2\sigma_2^2 - Cov^2(X,Y)}{\sigma_1^2\sigma_2^2}.
\end{aligned}
$$

Since $Cov^2(X,Y) \le \sigma_1^2\sigma_2^2$, it follows that $E(Y-\hat{Y})^2$ is minimized for $a = \mu_2$ and $b = \frac{1}{\sigma_1^2}Cov(X,Y) = \frac{\rho\sigma_1\sigma_2}{\sigma_1^2} = \frac{\rho\sigma_2}{\sigma_1}$.

3.15 Show that the intersection of the surface represented by the bivariate normal p.d.f. by any plane perpendicular to the z-axis is an ellipse; and it is a circle if and only if $\rho = 0$.

9.4 Multivariate Normal Distribution

This chapter is concluded with a brief reference to the multivariate normal distribution without entering into any details. A relevant reference is given for the interested reader.

The multivariate normal distribution is the generalization of the bivariate normal distribution and can be defined in a number of ways; we choose the one given here. To this end, for $k \ge 2$, let $\boldsymbol{\mu} = (\mu_1, \dots, \mu_k)$ be a vector of constants, and let $\boldsymbol{\Sigma}$ be a $k \times k$ nonsingular matrix, so that the inverse $\boldsymbol{\Sigma}^{-1}$ exists and the determinant $|\boldsymbol{\Sigma}| \ne 0$. Finally, set $\mathbf{X}$ for the vector of r.v.'s $X_1, \dots, X_k$; i.e., $\mathbf{X} = (\mathrm{X}_1, \dots, \mathrm{X}_k)$ and $\mathbf{x} = (\mathrm{x}_1, \dots, \mathrm{x}_k)$ for any point in $\Re^k$. Then, the joint p.d.f. of the X_i's, or the p.d.f. of the random vector $\mathbf{X}$, is said to be *multivariate normal*, or *k-variate normal*, if it is given by the formula:

$$
f_{\mathbf{X}}(\mathbf{x}) = \frac{1}{(2\pi)^{k/2}|\boldsymbol{\Sigma}|^{1/2}} \exp\left[-\frac{1}{2}(\mathbf{x}-\boldsymbol{\mu})\boldsymbol{\Sigma}^{-1}(\mathbf{x}-\boldsymbol{\mu})'\right], \qquad \mathbf{x} \in \Re^k,
$$

where, it is to be recalled that "$'$" stands for transpose.

It can be seen that $EX_i = \mu_i$, $Var(X_i) = \sigma_i^2$ is the (i,i)th element of $\boldsymbol{\Sigma}$, and $Cov(X_i, X_j)$ is the (i,j)th element of $\boldsymbol{\Sigma}$, so that $\boldsymbol{\mu} = (EX_1, \dots, EX_k)$ and $\boldsymbol{\Sigma} = (Cov(X_i, X_j))$, $i, j = 1, \dots, k$. The quantities $\boldsymbol{\mu}$ and $\boldsymbol{\Sigma}$ are called the *parameters* of the distribution. It can also be seen that the joint m.g.f. of the X_i's, or the m.g.f. of the random vector $\mathbf{X}$, is given by:

$$M_{\mathbf{X}}(\mathbf{t}) = \exp\left(\boldsymbol{\mu}\mathbf{t}' + \frac{1}{2}\mathbf{t}\boldsymbol{\Sigma}\mathbf{t}'\right), \qquad \mathbf{t} \in \Re^k.$$

The k-variate normal distribution has properties similar to those of the 2-dimensional normal distribution, and the latter is obtained from the former by taking $\boldsymbol{\mu} = (\mu_1, \mu_2)$ and $\boldsymbol{\Sigma} = \begin{pmatrix} \sigma_1^2 & \rho\sigma_1\sigma_2 \\ \rho\sigma_1\sigma_2 & \sigma_2^2 \end{pmatrix}$, where $\rho = \rho(X_1, X_2)$.

More relevant information can be found, for example, in Chapter 18 of the reference cited in Exercise 3.4.

CHAPTER

Independence of Random Variables and Some Applications

10

This chapter consists of three sections. In the first, we introduce the concept of independence of r.v.'s and establish criteria for proving or disproving independence. Also, its relationship to uncorrelatedness is discussed. In the second section, a random sample of size n is defined, as well as the sample mean and the sample variance; some of their moments are also produced. The main thrust of this section, however, is the discussion of the reproductive property of certain distributions. As a by-product, we also obtain the distribution of the sample mean and of a certain multiple of the sample variance for independent and normally distributed r.v.'s. In the final short section, the distribution of a modified version of the sample variance is derived under normality.

10.1 Independence of Random Variables and Criteria of Independence

In Section 4.2 of Chapter 4, the concept of independence of two events was introduced, and it was suitably motivated and illustrated by means of examples. This concept was then generalized to more than two events. What is done in this section is, essentially, to carry over the concept of independence from events to r.v.'s. To this end, consider first two r.v.'s X_1 and X_2 and the events induced in the sample space $\mathcal{S}$ by each of them separately, as well as by both of them jointly. That is, for subsets B_1, B_2 of $\Re$, let:

$$A_1 = (X_1 \in B_1) = X_1^{-1}(B_1) = \{s \in \mathcal{S};\ X_1(s) \in B_1\}, \tag{10.1}$$

$$A_2 = (X_2 \in B_2) = X_2^{-1}(B_2) = \{s \in \mathcal{S};\ X_2(s) \in B_2\}, \tag{10.2}$$

$$\begin{aligned} A_{12} &= ((X_1, X_2) \in B_1 \times B_2) = (X_1 \in B_1 \,\&\, X_2 \in B_2) = (X_1, X_2)^{-1}(B_1 \times B_2) \\ &= \{s \in \mathcal{S};\ X_1(s) \in B_1 \ \&\ X_2(s) \in B_2\} = A_1 \cap A_2. \end{aligned} \tag{10.3}$$

Then the r.v.'s X_1, X_2 are said to be independent if, for any B_1 and B_2 as before, the corresponding events A_1 and A_2 are independent; that is, $P(A_1 \cap A_2) = P(A_1)P(A_2)$. By (10.1)–(10.3), clearly, this relation is equivalent to:

$$P(X_1 \in B_1, X_2 \in B_2) = P(X_1 \in B_1)P(X_2 \in B_2). \tag{10.4}$$

Introduction to Probability, Second Edition. http://dx.doi.org/10.1016/B978-0-12-800041-0.00010-9

This relation states, in effect, that information regarding one r.v. has no effect on the probability distribution of the other r.v. For example,

$$P(X_1 \in B_1 | X_2 \in B_2) = \frac{P(X_1 \in B_1, X_2 \in B_2)}{P(X_2 \in B_2)}$$
$$= \frac{P(X_1 \in B_1)P(X_2 \in B_2)}{P(X_2 \in B_2)} = P(X_1 \in B_1).$$

Relation (10.4) is taken as the definition of independence of these two r.v.'s, which is then generalized in a straightforward way to k r.v.'s.

Definition 1. Two r.v.'s X_1 and X_2 are said to be *independent* (*statistically* or *stochastically* or in the *probability sense*) if, for any subsets B_1 and B_2 of $\Re$,

$$P(X_1 \in B_1, X_2 \in B_2) = P(X_1 \in B_1)P(X_2 \in B_2).$$

The r.v.'s $X_1, \ldots, X_k$ are said to be *independent* (in the same sense as above) if, for any subsets $B_1, \ldots, B_k$ of $\Re$,

$$P(X_i \in B_i, i = 1, \ldots, k) = \prod_{i=1}^{k} P(X_i \in B_i). \tag{10.5}$$

Nonindependent r.v.'s are said to be *dependent*.

The practical question that now arises is how one checks independence of k given r.v.'s, or lack thereof. This is done by means of the following criterion, referred to as the factorization theorem, because of the form of the expressions involved.

Theorem 1 (Criterion of independence, factorization theorem). For $k \geq 2$, the r.v.'s $X_1, \ldots, X_k$ are independent if and only if any one of the following three relations holds:

(i) $F_{X_1,\ldots,X_k}(x_1, \ldots, x_k) = F_{X_1}(x_1) \cdots F_{X_k}(x_k)$
for all $x_1, \ldots, x_k$ *in* $\Re$. (10.6)

(ii) $f_{X_1,\ldots,X_k}(x_1, \ldots, x_k) = f_{X_1}(x_1) \cdots f_{X_k}(x_k)$
for all $x_1, \ldots, x_k$ *in* $\Re$. (10.7)

(iii) $M_{X_1,\ldots,X_k}(t_1, \ldots, t_k) = M_{X_1}(t_1) \cdots M_{X_k}(t_k)$
for all $t_1, \ldots, t_k$ in a nondegenerate interval containing 0. (10.8)

Before we proceed with the (partial) justification of this theorem, let us refer to Example 1 in Chapter 7 and notice that $f_X(3) = 0.04$, $f_Y(2) = 0.15$, and $f_{X,Y}(3, 2) = 0.02$, so that $f_{X,Y}(3, 2) = 0.02 \neq 0.04 \times 0.15 = 0.006 = f_X(3)f_Y(2)$. Accordingly, the r.v.'s X and Y are *not* independent.

On the other hand, in reference to Example 2 in Chapter 7 (see also Example 6 in the same chapter), we have, for all $x, y > 0$:

$$f_{X,Y}(x, y) = \lambda_1\lambda_2 e^{-\lambda_1 x-\lambda_2 y} = (\lambda_1 e^{-\lambda_1 x})(\lambda_2 e^{-\lambda_2 y}) = f_X(x)f_Y(y),$$

so that $f_{X,Y}(x, y) = f_X(x)f_Y(y)$ for all x and y, and consequently, the r.v.'s X and Y are independent. Three additional examples are discussed below.

Example 1. Examine the r.v.'s X and Y from an independence viewpoint, if their joint p.d.f. is given by: $f_{X,Y}(x, y) = 4xy, 0 < x < 1, 0 < y < 1$ (and 0 otherwise).

Discussion. We will use part (ii) of Theorem 1, for which the marginal p.d.f.'s are needed. To this end, we have:

$$f_X(x) = 4x\int_0^1 y\,dy = 2x, \quad 0 < x < 1;$$

$$f_Y(y) = 4y\int_0^1 x\,dx = 2y, \quad 0 < y < 1.$$

Hence, for all $0 < x < 1$ and $0 < y < 1$, it holds that $2x \times 2y = 4xy$. This relation is also trivially true (both sides are equal to 0) for x and y not satisfying the inequalities $0 < x < 1$ and $0 < y < 1$, so that $f_X(x)f_Y(y) = f_{X,Y}(x, y)$. It follows that X and Y are independent.

Here are two examples where the r.v.'s involved are *not* independent.

Example 2. If the r.v.'s X and Y have joint p.d.f. given by: $f_{X,Y}(x, y) = 2, 0 < x < y < 1$ (and 0 otherwise), check whether these r.v.'s are independent or not.

Discussion. Reasoning as in the previous example, we find:

$$f_X(x) = 2\int_x^1 dy = 2(1 - x), \quad 0 < x < 1;$$

$$f_Y(y) = 2\int_0^y dx = 2y, \quad 0 < y < 1.$$

Then independence of X and Y would require that $4(1 - x)y = 2$ for all $0 < x < y < 1$, which, clearly, need not hold. For example, for $x = \frac{1}{4}, y = \frac{1}{2}, 4(1 - x)y = 4 \times \frac{3}{4} \times \frac{1}{2} = \frac{3}{2} \neq 2$. Thus, the X and Y are not independent.

Example 3. In reference to Exercise 2.21 in Chapter 7, the r.v.'s X and Y have joint p.d.f. $f_{X,Y}(x, y) = 8xy,\ 0 < x \leq y < 1$ (and 0 otherwise), and:

$$f_X(x) = 4x(1 - x^2), \quad 0 < x < 1; \qquad f_Y(y) = 4y^3, \quad 0 < y < 1.$$

Independence of X and Y would require that $4x(1-x^2)\times 4y^3 = 8xy$ or $(1-x^2)y^2 = \frac{1}{2},\ 0 < x \leq y \leq 1$. However, this relation need not be true because, for example, for

$x = \frac{1}{4}$ and $y = \frac{1}{2}$, we have: left-hand side $= \frac{15}{64} \neq \frac{1}{2} =$ right-hand side. So, the r.v.'s X and Y are dependent.

Remark 1. On the basis of Examples 2 and 3, one may surmise the following rule of thumb: If the arguments x and y (for the case of two r.v.'s) do not vary independently of each other, the r.v.'s involved are likely to be dependent.

Partial Proof of Theorem 1. The proof can be only partial, but sufficient for the purposes of this book.

(i) Independence of the r.v.'s $X_1, \ldots, X_k$ means that relation (10.5) is satisfied. In particular, this is true if $B_i = (-\infty, x_i]$, $i = 1, \ldots, k$ which is (10.6). That (10.6) implies (10.5) is a deep probabilistic result dealt with at a much higher level.

(ii) Suppose the r.v.'s are independent, and first assume they are discrete. Then, by taking $B_i = \{x_i\}, i = 1, \ldots, k$ in (10.5), we obtain (10.7). If the r.v.'s are continuous, then consider (10.6) and differentiate both sides with respect to $x_1, \ldots, x_k$, which, once again, leads to (10.7) (for continuity points $x_1, \ldots, x_k$). For the converse, suppose that (10.7) is true; that is, for all $t_1, \ldots, t_k$ in $\Re$,

$$f_{X_1,\ldots,X_k}(t_1, \ldots, t_k) = f_{X_1}(t_1) \cdots f_{X_k}(t_k).$$

Then, if the r.v.'s are discrete, sum over the t_i's from $-\infty$ to x_i, $i = 1, \ldots, k$ to obtain (10.6); if the r.v.'s are continuous, replace the summation operations by integrations in order to obtain (10.6) again. In either case, independence follows.

(iii) Suppose the r.v.'s are independent, and let them be continuous (in the discrete case, integrals are replaced by summation signs). Then $M_{X_1,\ldots,X_k}(t_1, \ldots, t_k) = Ee^{t_1X_1+\cdots+t_kX_k}$

$$\begin{aligned}
&= \int_{-\infty}^{\infty} \cdots \int_{-\infty}^{\infty} e^{t_1x_1+\cdots+t_kx_k} f_{X_1,\ldots,X_k}(x_1, \ldots, x_k)dx_1 \ldots dx_k \\
&= \int_{-\infty}^{\infty} \cdots \int_{-\infty}^{\infty} e^{t_1x_1+\cdots+t_kx_k} f_{X_1}(x_1) \ldots f_{X_k}(x_k)dx_1 \ldots dx_k \text{ (by part (ii))} \\
&= \left[\int_{-\infty}^{\infty} e^{t_1x_1} f_{X_1}(x_1)dx_1\right] \cdots \left[\int_{-\infty}^{\infty} e^{t_kx_k} f_{X_k}(x_k)dx_k\right] \\
&= (Ee^{t_1X_1}) \ldots (Ee^{t_kX_k}) \\
&= M_{X_1}(t_1) \ldots M_{X_k}(t_k).
\end{aligned}$$

The converse is also true, but its proof will not be pursued here (it requires the use of the so-called inversion formula as indicated in the discussion immediately following Theorem 1 of Chapter 5 and Remark 1 of Chapter 8). ▲.

The following result is a consequence of independence, and it is useful in many situations.

Proposition 1. Consider the r.v.'s $X_1, \ldots, X_k$, the functions $g_i : \Re \to \Re, i = 1, \ldots, k$, and suppose all expectations appearing below are finite. Then independence of the r.v.'s $X_1, \ldots, X_k$ implies:

$$E\left[\prod_{i=1}^{k} g_i(X_i)\right] = \prod_{i=1}^{k} Eg_i(X_i). \tag{10.9}$$

Proof. Suppose the r.v.'s are of the continuous type (so that we use integrals; replace them by summations if the r.v.'s are discrete). Then:

$$E\left[\prod_{i=1}^{k} g_i(X_i)\right] = \int_{-\infty}^{\infty} \cdots \int_{-\infty}^{\infty} g_1(x_1)\cdots g_k(x_k) f_{X_1,\ldots,X_k}(x_1,\ldots,x_k)\,dx_1\cdots dx_k$$

$$= \int_{-\infty}^{\infty} \cdots \int_{-\infty}^{\infty} g_1(x_1)\cdots g_k(x_k) f_{X_1}(x_1)\cdots f_{X_k}(x_k)\,dx_1\cdots dx_k$$

(by part (ii) of Theorem 1)

$$= \left[\int_{-\infty}^{\infty} g_1(x_1) f_{X_1}(x_1)\,dx_1\right]\cdots\left[\int_{-\infty}^{\infty} g_k(x_k) f_{X_k}(x_k)\,dx_k\right]$$

$$= Eg_1(X_1)\cdots Eg_k(X_k) = \prod_{i=1}^{k} Eg_i(X_i). \quad \blacktriangle$$

Corollary 1. If the r.v.'s X and Y are independent, then they are uncorrelated. The converse is also true, if the r.v.'s have bivariate normal distribution.

Proof. In (10.9), take $k = 2$, identify X_1 and X_2 with X and Y, respectively, and let $g_1(x) = g_2(x) = x$, $x \in \Re$. Then $E(XY) = (EX)(EY)$, which implies $Cov(X, Y) = 0$ and $\rho(X, Y) = 0$. The converse for bivariate normal distribution follows by means of (9.13), (9.16), and (9.17) in Chapter 9. ▲

Remark 2. That uncorrelated r.v.'s are not, in general, independent may be illustrated by means of examples (see, e.g., Exercise 1.20).

Corollary 2. If the r.v.'s $X_1, \ldots, X_k$ are independent, then:

$$M_{X_1+\cdots+X_k}(t) = M_{X_1}(t)\ldots M_{X_k}(t); \tag{10.10}$$

and, in particular, if they are also identically distributed, then:

$$M_{X_1+\cdots+X_k}(t) = [M_{X_1}(t)]^k. \tag{10.11}$$

Proof. Indeed,

$$\begin{aligned} M_{X_1+\cdots+X_k}(t) &= Ee^{t(X_1+\cdots+X_k)} \\ &= E(e^{tX_1}\ldots e^{tX_k}) \\ &= (Ee^{tX_1})\ldots(Ee^{tX_1}) \quad \text{(by the proposition)} \\ &= M_{X_1}(t)\ldots M_{X_k}(t). \end{aligned}$$

The second assertion follows immediately from the first. ▲

Remark 3. If $X_1, \ldots, X_k$ are independent r.v.'s, then it is intuitively clear that independence should be preserved for suitable functions of the X_i's. For example, if $Y_i = g_i(X_i)$, $i = 1, \ldots, k$, then the r.v.'s $Y_1, \ldots, Y_k$ are also independent. Independence is also preserved if we take different functions of more than one of the X_i's, provided these functions do not include the same X_i's. For instance, if $Y = g(X_{i_1}, \ldots, X_{i_m})$ and $Z = h(X_{j_1}, \ldots, X_{j_n})$, where $1 \leq i_1 < \cdots < i_m \leq k$, $1 \leq j_1 < \cdots < j_n \leq k$ and all $i_1, \ldots, i_m$ are distinct from all $j_1, \ldots, j_n$, then the r.v.'s Y and Z are independent.

One of the stated cases of independence is established below, and other cases are treated likewise (see also Exercise 1.26).

Proposition 2. If the r.v.'s $X_1, \ldots, X_k$ are independent, then so are the r.v.'s $Y_i = g_i(X_i)$, where $g_i : \Re \to \Re$ $i = 1, \ldots, k$.

Proof. By Theorem 1(iii), it suffices to show that

$$M_{Y_1,\ldots,Y_k}(t_1, \ldots, t_k) = M_{Y_1}(t_1) \ldots M_{Y_k}(t_k)$$

for all $t_1, \ldots, t_k$ in a nondegenerate interval containing 0. We have

$$\begin{aligned}
M_{Y_1,\ldots,Y_k}(t_1 \ldots t_k) &= Ee^{t_1 Y_1 + \cdots + t_k Y_k} \\
&= E(e^{t_1 Y_1} \ldots e^{t_k Y_k}) \\
&= (Ee^{t_1 Y_1}) \ldots (Ee^{t_k Y_k}) \\
&\quad \text{(by Proposition 1, where } g_i(X_i) \\
&\quad \text{is replaced by } exp[t_i g_i(X_i)],\ i = 1, \ldots, k) \\
&= M_{Y_1}(t_1) \ldots M_{Y_k}(t_k),
\end{aligned}$$

as was to be seen. ▲

The following result, formulated as a proposition, provides still another interesting application of Theorem 1(i).

Proposition 3. Suppose that certain events occur in time intervals $(t_1, t_2]$ (with $0 \leq t_1 < t_2$) according to the Poisson distribution with parameter $\lambda(t_2 - t_1)$, and that occurrences in nonoverlapping intervals are independent events. Let T_1 be the (waiting) time from time 0 to the occurrence of the 1st event, and for $k = 2, \ldots, n (\geq 2)$, let T_k be the (waiting) time from the $(k-1)$st to the kth event. Then:

(i) The r.v.'s $T_1, \ldots, T_n$ have the negative exponential distribution with parameter λ.
(ii) For $k = 2, \ldots, n$, the r.v.'s T_{k-1} and T_k are independent.
(iii) The conclusion of part (ii) is also true for the r.v.'s $T_1, \ldots, T_n$.

Proof. The proof of part (i) is brief and rigorous, but for parts (ii) and (iii), only an outline is presented, based on some plausibility arguments, unavoidable at this level.

(i) Indeed, for $t > 0$,

$$\begin{aligned}
P(T_1 > t) &= P(\#\text{ of events occurred in } (0, t] = 0) \\
&= e^{-\lambda t} \frac{(\lambda t)^0}{0!} = e^{-\lambda t},
\end{aligned}$$

so that $P(T_1 \leq t) = 1 - e^{\lambda t}$ and $f_{T_1}(t) = \lambda e^{-\lambda t}$ (see also Exercise 2.6 in Chapter 6). For $k \geq 2$ and with t_{k-1} being the time of the occurrence of the $(k-1)$-st event, we have

$$\begin{aligned} P(T_k > t) &= P(\#\text{ of events occurred in } (t_{k-1}, t_{k-1} + t) = 0) \\ &= e^{-\lambda t}, \text{ since } (t_{k-1} + t) - t_{k-1} = t, \text{ and } f_{T_k}(t) = \lambda e^{-\lambda t} \text{ as above.} \end{aligned}$$

(ii) Consider, for example, the r.v.'s T_1 and T_2, and let $t_1, t_2 \geq 0$. Then, with obvious notation,

$$\begin{aligned} P(T_1 > t_1, T_2 > t_2) &= \int_{t_2}^{\infty}\int_{t_1}^{\infty} f_{T_1,T_2}(x_1, x_2)dx_1dx_2 \\ &\quad \times \int_{t_1}^{\infty}\int_{t_2}^{\infty} f_{T_1,T_2}(x_1, x_2)dx_2dx_2 \\ &= \int_{t_1}^{\infty}\left[\int_{t_2}^{\infty} f_{T_2|T_1}(x_2|x_1)dx_2\right] f_{T_1}(x_1)dx_1 \\ &= \int_{t_1}^{\infty} P(T_2 > t_2|T_1 = x_1) f_{T_1}(x_1)dx_1. \end{aligned}$$

Now,

$$P(T_2 > t_2|T_1 = x_1) = P(\#\text{ of events occurred in } (T_1, t_2 + T_1] = 0|T_1 = x_1),$$

and it is intuitively plausible that this is

$$= P(\#\text{ of events occured in } (0, t_2] = 0),$$

since the probability of occurrences of events depends only on the length of the interval, and T_1 depends only on occurrences of events in $(0, T_1]$,

$$= e^{-\lambda t_2}. \text{ Therefore,}$$

$$P(T_1 > t_1, T_2 > t_2) = e^{-\lambda t_2}\int_{t_1}^{\infty} f_{T_1}(x_1)dx_1 = e^{-\lambda t_2}e^{-\lambda t_1}.$$

So,

$$P(T_1 > t_1, T_2 > t_2) = P(T_1 > t_1)P(T_2 > t_2),$$

which (by Exercise 1.13 in Chapter 7) implies that $P(T_1 \leq t_1, T_2 \leq t_2) = P(T_1 \leq t_1)P(T_2 \leq t_2)$, so that the r.v.'s T_1 and T_2 are independent.

(iii) It follows as in part (ii) by using similar plausibility arguments, mathematical induction, and a generalized version of Exercise 1.13 just cited. Details are omitted. ▲

The following result provides still another way of checking independence of k r.v.'s $X_1, \ldots, X_k$. Its significance lies in that in order to check for independence of the r.v.'s $X_1, \ldots, X_k$ all one has to do is to establish a factorization of $f_{X_1,\ldots,X_k}$ as stated in the result. One does not have to ascertain what the factors are. (From the proof of the result, it follows that these factors are multiples of the marginal p.d.f.'s.)

Theorem 1′. The r.v.'s $X_1, \ldots, X_k$ are independent if and only if $f_{X_1,\ldots,X_k}(x_1, \ldots, x_k) = h_1(x_1) \cdots h_k(x_k)$ for all $x_1, \ldots, x_k$ in $\Re$, where h_i is a non-negative function of x_i alone, $i = 1, \ldots, k$.

Proof. Suppose the r.v.'s $X_1, \ldots, X_k$ are independent. Then, by (10.7),

$$f_{X_1,\ldots,X_k}(x_1, \ldots, x_k) = f_{X_1}(x_1) \cdots f_{X_k}(x_1)$$

for all $x_1, \ldots, x_k$ in $\Re$, so that the above factorization holds with $h_i = f_{X_i}$, $i = 1, \ldots, k$. Next, assume that the factorization holds, and suppose that the r.v.'s are continuous. For each fixed $i = 1, \ldots, k$, set

$$c_i = \int_{-\infty}^{\infty} h_i(x_i)\, dx_i,$$

$$\begin{aligned}
\text{so that } c_1 \ldots c_k &= \int_{-\infty}^{\infty} h_1(x_1)\, dx_1 \ldots \int_{-\infty}^{\infty} h_k(x_k)\, dx_k \\
&= \int_{-\infty}^{\infty} \cdots \int_{-\infty}^{\infty} h_1(x_1) \ldots h_k(x_k)\, dx_1 \cdots dx_k \\
&= \int_{-\infty}^{\infty} \cdots \int_{-\infty}^{\infty} f_{X_1,\ldots,X_k}(x_1, \ldots, x_k)\, dx_1 \cdots dx_k \\
&= 1.
\end{aligned}$$

Then, integrating $f_{X_1,\ldots,X_k}(x_1, \ldots, x_k) = h_1(x_1) \ldots h_k(x_k)$ with respect to all x_j's with $j \neq i$, we get

$$\begin{aligned}
f_{X_i}(x_i) &= c_1 \ldots c_{i-1}\, c_{i+1} \ldots c_k h_i(x_i) \\
&= \frac{1}{c_i} h_i(x_i).
\end{aligned}$$

Hence

$$\begin{aligned}
f_{X_1}(x_1) \ldots f_{X_k}(x_k) &= \frac{1}{c_1 \ldots c_k} h_1(x_1) \ldots h_k(x_k) \\
&= h_1(x_1) \ldots h_k(x_k) = f_{X_1,\ldots,X_k}(x_1, \ldots, x_k),
\end{aligned}$$

or $f_{X_1,\ldots,X_k}(x_1, \ldots, x_k) = f_{X_1}(x_1) \ldots f_{X_k}(x_k)$, for all $x_1, \ldots, x_k$ in $\Re$, so that the r.v.'s $X_1, \ldots, X_k$ are independent. The same conclusion holds in case the r.v.'s are discrete by using summations rather than integrations. ▲

Exercises.

1.1 In reference to Exercise 2.3 in Chapter 7, determine whether or not the r.v.'s X and Y are independent. Justify your answer.

1.2 In reference to Exercises 1.1 and 2.1 in Chapter 7, determine whether or not the r.v.'s X and Y are independent.

1.3 The r.v.'s X, Y, and Z have the joint p.d.f. given by: $f_{X,Y,Z}(x, y, z) = \frac{1}{4}$ if $x = 1, y = z = 0$; $x = 0, y = 1, z = 0$; $x = y = 0, z = 1$; $x = y = z = 1$.

(i) Derive the marginal joint p.d.f.'s $f_{X,Y}$, $f_{X,Z}$, $f_{Y,Z}$.
(ii) Derive the marginal p.d.f.'s f_X, f_Y, and f_Z.
(iii) Show that any two of the r.v.'s X, Y, and Z are independent.
(iv) Show that the r.v.'s X, Y, and Z are dependent.

1.4 In reference to Exercise 2.8 in Chapter 7, decide whether or not the r.v.'s X and Y are independent. Justify your answer.

1.5 In reference to Examples 4 and 7 in Chapter 7, investigate whether or not the r.v.'s X and Y are independent and justify your answer.

1.6 Let X and Y be r.v.'s with joint p.d.f. given by:

$$f_{X,Y}(x, y) = \frac{6}{5}(x^2 + y), \qquad 0 \le x \le 1, \quad 0 \le y \le 1.$$

(i) Determine the marginal p.d.f.'s f_X and f_Y.
(ii) Investigate whether or not the r.v.'s X and Y are independent. Justify your answer.

1.7 The r.v.'s X, and Y have joint p.d.f. given by:

$$f_{X,Y}(x, y) = 1, \qquad 0 < x < 1, \quad 0 < y < 1.$$

Then:

(i) Derive the marginal p.d.f.'s f_X, and f_Y.
(ii) Show that X and Y are independent.
(iii) Calculate the probability $P(X + Y < c)$ $(c > 0)$.
(iv) Give the numerical value of the probability in part (iii) for $c = 1/4$.

Hint. For part (iii), you may wish to draw the picture of the set for which $x + y < c$, and compute the probability in terms of c.

1.8 The r.v.'s X, Y, and Z have joint p.d.f. given by:

$$f_{X,Y,Z}(x, y, z) = 8xyz, \qquad 0 < x < 1, \quad 0 < y < 1, \quad 0 < z < 1.$$

(i) Derive the marginal p.d.f.'s f_X, f_Y, and f_Z.
(ii) Show that the r.v.'s X, Y, and Z are independent.
(iii) Calculate the probability $P(X < Y < Z)$.

Hint. In part (iii), can you guess the answer without doing any calculations?

1.9 The r.v.'s X and Y have joint p.d.f. given by:

$$f_{X,Y}(x, y) = c, \quad \text{for } x^2 + y^2 \le 9.$$

(i) Determine the constant c.
(ii) Derive the marginal p.d.f.'s f_X and f_Y.
(iii) Show that the r.v.'s X and Y are dependent.

1.10 The r.v.'s X, Y, and Z have joint p.d.f. given by:

$$f_{X,Y,Z}(x, y, z) = c^3 e^{-c(x+y+z)}, \quad x > 0, \quad y > 0, \quad z > 0.$$

(i) Determine the constant c.
(ii) Derive the marginal joint p.d.f.'s $f_{X,Y}$, $f_{X,Z}$, and $f_{Y,Z}$.
(iii) Derive the marginal p.d.f.'s f_X, f_Y, and f_Z.
(iv) Show that any two of the r.v.'s X, Y, and Z, as well as all three r.v.'s are independent.

1.11 The r.v.'s X and Y have joint p.d.f. given by the following product: $f_{X,Y}(x, y) = g(x)h(y)$, where g and h are nonnegative functions.

(i) Derive the marginal p.d.f.'s f_X and f_Y as functions of g and h, respectively.
(ii) Show that the r.v.'s X and Y are independent.
(iii) If $h = g$, then the r.v.'s are identically distributed.
(iv) From part (iii), conclude that $P(X > Y) = 1/2$, provided the distribution is of the continuous type.

Hint. For part (i), we have $f_X(x) = cg(x)$, $f_Y(y) = \frac{1}{c}h(y)$, where $c = \int_{-\infty}^{\infty} h(y)dy$. Parts (ii) and (iii) follow from part (i). Part (iv) follows either by symmetry or by calculations.

1.12 The life of a certain part in a new automobile is an r.v. X whose p.d.f. is negative exponential with parameter $\lambda = 0.005$ days.

(i) What is the expected life of the part in question?
(ii) If the automobile comes with a spare part whose life is an r.v. Y distributed as X and independent of it, find the p.d.f. of the combined life of the part and its spare.
(iii) What is the probability that $X + Y \geq 500$ days?

1.13 Let the r.v. X be distributed as $U(0, 1)$ and set $Y = -\log X$.

(i) Determine the d.f. of Y and then its p.d.f.
(ii) If the r.v.'s $Y_1, \ldots, Y_n$ are independently distributed as Y, and $Z = Y_1 + \cdots + Y_n$, determine the distribution of the r.v. Z.

Hint. For part (ii), use the m.g.f. approach.

1.14 Let the independent r.v.'s X and Y be distributed as $N(\mu_1, \sigma_1^2)$ and $N(\mu_2, \sigma_2^2)$, respectively, and define the r.v.'s U and V by: $U = aX + b$, $V = cY + d$, where a, b, c, and d are constants.

(i) Use the m.g.f. approach in order to show that:

$$U \sim N(a\mu_1 + b, (a\sigma_1)^2), \quad V \sim N(c\mu_2 + d, (c\sigma_2)^2).$$

(ii) Determine the joint m.g.f. of U and V.
(iii) From parts (i) and (ii), conclude that U and V are independent.

1.15 Let X and Y be independent r.v.'s denoting the lifetimes of two batteries and having negative exponential distribution with parameter λ. Set $T = X + Y$ and:

(i) Determine the d.f. of T by integration, and then the corresponding p.d.f.
(ii) Determine the p.d.f. of T by using the m.g.f. approach.
(iii) For $\lambda = 1/3$, calculate the probability $P(T \leq 6)$.

Hint. For part (i), you may wish to draw the picture of the set for which $x + y < t$ $(t > 0)$.

1.16 Let $X_1, \ldots, X_n$ be independent identically distributed r.v.'s with m.g.f. M, and let $\bar{X} = \frac{1}{n}(X_1 + \cdots + X_n)$. Express the m.g.f. $M_{\bar{X}}$ in terms of M.

1.17 Let X and Y be the r.v.'s denoting the number of sixes when two fair dice are rolled independently 15 times. Then:

(i) Calculate the $E(X + Y)$, $Var(X + Y)$, and the s.d. of $X + Y$.
(ii) Use the Tchebichev inequality to determine a lower bound for the probability: $P(X + Y \leq 10)$.

Hint. For part (ii), use first part (i) to bring it into the form required for the application of the Tchebichev inequality.

1.18 Let p be the proportion of defective computer chips in a very large lot of chips produced over a period of time by a certain manufacturing process. For $i = 1, \ldots, n$, associate with the ith chip the r.v. X_i, where $X_i = 1$ if the i th chip is defective, and $X_i = 0$ otherwise. Then $X_1, \ldots, X_n$ are independent r.v.'s distributed as $B(1, p)$, and let $\bar{X} = \frac{1}{n}(X_1 + \cdots + X_n)$.

(i) Calculate the $E\bar{X}$ and the $Var\,(\bar{X})$ in terms of p and $q = 1 - p$.
(ii) Use the Tchebichev inequality to determine the smallest value of n for which $P(|\bar{X} - p| < 0.1\sqrt{pq}) \geq 0.99$.

1.19 Let the independent r.v.'s $X_1, \ldots, X_n$ be distributed as $P(\lambda)$, and set $\bar{X} = \frac{1}{n}(X_1 + \cdots + X_n)$.

(i) Calculate the $E\bar{X}$ and the $Var\,(\bar{X})$ in terms of λ and n.
(ii) Use the Tchebichev inequality to determine the smallest value of n, in terms of λ and c, for which $P(|\bar{X} - \lambda| < c) \geq 0.95$, for some $c > 0$.
(iii) Give the numerical value of n for $c = \sqrt{\lambda}$ and $c = 0.1\sqrt{\lambda}$.

1.20 The joint distribution of the r.v.'s X and Y is given by:

$y\backslash x$	-1	0	1
-1	α	β	α
0	β	0	β
1	α	β	α

where $\alpha, \beta > 0$ with $\alpha + \beta = 1/4$.

(i) Derive the marginal p.d.f.'s f_X and f_Y.
(ii) Calculate the EX, EY, and $E(XY)$.
(iii) Show that $Cov(X, Y) = 0$.
(iv) Show that the r.v.'s X and Y are dependent.
Remark: Whereas independent r.v.'s are always uncorrelated, this exercise shows that the converse need not be true.

1.21 Refer to Exercise 1.10 here and calculate the following quantities, in terms of c, without any integrations: $E(XY)$, $E(XYZ)$, $Var(X+Y)$, $Var(X+Y+Z)$.

1.22 The i.i.d. r.v.'s $X_1, \ldots, X_n$ have expectation $\mu \in \Re$ and variance $\sigma^2 < \infty$, and set $\bar{X} = \frac{1}{n}(X_1 + \cdots + X_n)$.

(i) Determine the $E\bar{X}$ and the $Var(\bar{X})$ in terms of μ and σ.
(ii) Use the Tchebichev inequality to determine the smallest value of n for which $P(|\bar{X} - \mu| < k\sigma)$ is at least 0.99; take $k = 1, 2, 3$.

1.23 A piece of equipment works on a battery whose lifetime is an r.v. X with expectation μ and s.d. σ. If n such batteries are used successively and independently of each other, denote by $X_1, \ldots, X_n$ their respective lifetimes, so that $\bar{X} = \frac{1}{n}(X_1 + \cdots + X_n)$ is the average lifetime of the batteries. Use the Tchebichev inequality to determine the smallest value of n for which $P(|\bar{X} - \mu| < 0.5\sigma) \geq 0.99$.

1.24 Let $X_1, \ldots, X_n$ be i.i.d. r.v.'s with $EX_1 = \mu \in \Re$ and $Var(X_1) = \sigma^2 < \infty$, and set $\bar{X} = \frac{1}{n}(X_1 + \cdots + X_n)$.

(i) Calculate the $E\bar{X}$ and the $Var\,(\bar{X})$ in terms of μ and σ.
(ii) Use the Tchebichev inequality to determine the smallest value of n, in terms of the positive constant c and α, so that:

$$P(|\bar{X} - \mu| < c\sigma) \geq \alpha \ (0 < \alpha < 1).$$

(iii) What is the numerical value of n in part (ii) if $c = 0.1$ and $\alpha = 0.90$, $\alpha = 0.95$, $\alpha = 0.99$?

Remark: See also Exercise 1.22.

1.25 In reference to Exercise 3.8(iii) in Chapter 9, show that the r.v.'s U and V are independent if and only if $\sigma_1^2 = \sigma_2^2$.

Hint. Use Corollary 1 to Proposition 1.

1.26 Refer to Remark 3 and follow the steps used in the proof of Proposition 3 in order to show that the r.v.'s Y and Z, as defined in Remark 3, are independent.

1.27 The independent r.v.'s $X_1, \ldots, X_n$ have negative exponential distribution with parameter λ (i.e., their common p.d.f. is given by $f(x) = \lambda e^{-\lambda x}$, $x > 0$), and let $U = X_1 + \cdots + X_n$.

(i) Use the m.g.f. approach to determine the distribution of U.
(ii) Determine the EU and $\sigma^2(U)$ in any way you wish.
(iii) If $\lambda = 1/2$, show that $U \sim \chi^2_{2n}$.
(iv) For $\lambda = 1/2$ and $n = 15$, compute the probability $P(U > 20)$.

1.28 Let X and Y be independent r.v.'s distributed as $N(0, 1)$, and set $U = X + Y$.

(i) What is the distribution of U?
(ii) Determine the $Var(U)$, the $Cov(X, U)$, and the $\rho(X, U)$.
(iii) Determine the $M_{X,U}$.
(iv) Use part (iii) to rederive the $Cov(X, U)$.

Hint. For part (iv), use formula (8.17) in Chapter 8.

1.29 Let the r.v.'s X and Y be the lifetimes of two pieces of equipment A and B having negative exponential distribution with respective parameters λ_1 and λ_2; that is, $f_X(x) = \lambda_1 e^{-\lambda_1 x}$, $x > 0$, $f_Y(y) = \lambda_2 e^{-\lambda_2 y}$, $y > 0$; and suppose that X and Y are independent.

(i) Express the probability that the piece of equipment B will outlive the piece of equipment A, in terms of λ_1 and λ_2.
(ii) Compute the probability in part (i) for $\lambda_1 = \lambda$ and $\lambda_2 = 4\lambda$ (for some $\lambda > 0$).

Hint. For part (i), you may wish to draw the picture of the set for which $x < y$.

1.30 The r.v.'s X and Y have joint p.d.f. given by: $f_{X,Y}(x, y) = c$, $-1 \leq x \leq 1$, $-1 \leq y \leq 1$.

(i) Determine the constant c.
(ii) Determine the marginal p.d.f.'s f_X and f_Y.
(iii) Examine whether the r.v.'s X and Y are independent or not.
(iv) Compute the probability $P(X^2 + Y^2 \leq 1)$, both geometrically and by actual calculations.

10.2 The Reproductive Property of Certain Distributions

Independence plays a decisive role in the reproductive property of certain r.v.'s. Specifically, if $X_1, \ldots, X_k$ are r.v.'s having certain distributions, then, if they are also *independent*, it follows that the r.v. $X_1 + \cdots + X_k$ is of the same kind. This is, basically, the content of this section. The tool used to establish this assertion is the m.g.f., and the basic result employed is relation (10.10) resulting from independence of the r.v.'s involved.

First, we derive some general results regarding the sample mean and the sample variance of k r.v.'s, which will be used, in particular, in the normal distribution case discussed below. To this end, n independent and identically distributed (*i.i.d.*) r.v.'s are referred to as forming a *random sample* of size n. Some of their properties are discussed in this section. For any k r.v.'s $X_1, \ldots, X_k$, their *sample mean*, denoted by $\bar{X}_k$ or just $\bar{X}$, is defined by:

$$\bar{X} = \frac{1}{k}\sum_{i=1}^{k} X_i. \tag{10.12}$$

The *sample variance* of the X_i's, denoted by S_k^2 or just S^2, is defined by:

$$S^2 = \frac{1}{k}\sum_{i=1}^{k}(X_i - EX_i)^2,$$

provided the EX_i's are finite. In particular, if $EX_1 = \cdots = EX_k = \mu$, say, then S^2 becomes:

$$S^2 = \frac{1}{k}\sum_{i=1}^{k}(X_i - \mu)^2. \tag{10.13}$$

The r.v.'s defined by (10.12) and (10.13) are most useful when the underlying r.v.'s come from a random sample; that is, they are i.i.d. Below is such a result.

Proposition 4. Let $X_1, \ldots, X_k$ be i.i.d. r.v.'s with (finite) mean μ. Then $E\bar{X} = \mu$. Furthermore, if the X_i's also have (finite) variance σ^2, then $Var\,(\bar{X}) = \frac{\sigma^2}{k}$ and $ES^2 = \sigma^2$.

Proof. The first result follows from relation (9.5) in Chapter 9 by taking $c_1 = \cdots = c_k = 1/k$, and $d = 0$. The second result follows from (9.10) in the same chapter, by way of Corollary 1 to Proposition 2 here, because independence of X_i and X_j, for $i \neq j$, implies $\rho(X_i, X_j) = 0$. For the third result, observe that $E(X_i - \mu)^2 = \sigma^2, \quad i = 1, \ldots, k$, so that

$$E\sum_{i=1}^{k}(X_i - \mu)^2 = \sum_{i=1}^{k} E(X_i - \mu)^2 = n\sigma^2, \quad \text{and } ES^2 = \sigma^2. \quad \blacktriangle$$

The general thrust of the following four results is to the effect that if $X_1, \ldots, X_k$ are independent and have certain distributions, then their sum $X_1 + \cdots + X_k$ has a distribution of the same respective kind. The proof of this statement relies on relation (10.10), as already mentioned.

In the form of motivation for the first result to follow, suppose that one day's productions of three factories, producing the same items (e.g., computer chips) with the same proportion of defectives, are pooled together, and let X be the r.v. denoting the number of defectives. What is the distribution of X? The answer to this question is given by the following result. The theoretical basis of the arguments involved in this result and the subsequent ones is provided by Theorem 1 in Chapter 5.

Theorem 2. Let the r.v.'s $X_1, \ldots, X_k$ be independent and let $X_i \sim B(n_i, p)$ (the same p), $i = 1, \ldots, k$. Then $\sum_{i=1}^k X_i \sim B(\sum_{i=1}^k n_i, p)$.

Proof. By relation (10.10) here, relation (6.2) in Chapter 6, and $t \in \Re$:

$$M_{\sum_{i=1}^k X_i}(t) = \prod_{i=1}^k M_{X_i}(t) = \prod_{i=1}^k (pe^t + q)^{n_i} = (pe^t + q)^{\sum_{i=1}^k n_i},$$

which is the m.g.f. of $B(\sum_{i=1}^k n_i, p)$. Then $\sum_{i=1}^k X_i \sim B(\sum_{i=1}^k n_i, p)$. ▲

Here is an illustrative example.

Example 4. The defective items in two lots of sizes $n_1 = 10$ and $n_2 = 15$ occur independently at the rate of 6.25%. Calculate the probabilities that the total number of defective items: (i) Does not exceed 2; (ii) Is > 5.

Discussion. If X_1 and X_2 are the r.v.'s denoting the numbers of defective items in the two lots, then $X_1 \sim B(10, 0.0625)$, $X_2 \sim B(15, 0.0625)$ and they are independent. Then $X = X_1 + X_2 \sim B(25, 0.0625)$ and therefore: (i) $P(X \leq 2) = 0.7968$ and (ii) $P(X > 5) = 1 - P(X \leq 5) = 0.0038$ (from the binomial tables).

If X_i is the r.v. denoting the number of traffic accidents over a long weekend in the ith county of the state of California, $i = 1, \ldots, k$ then $X = X_1 + \cdots + X_k$ is the total number of traffic accidents in the state of California. But just what is the distribution of X? Here is the answer.

Theorem 3. Let the r.v.'s $X_1, \ldots, X_k$ be independent and let $X_i \sim P(\lambda_i)$, $i = 1, \ldots, k$. Then $\sum_{i=1}^k X_i \sim P(\sum_{i=1}^k \lambda_i)$.

Proof. As above, employ (10.10), and relation (6.9) in Chapter 6 in order to obtain:

$$M_{\sum_{i=1}^k X_i}(t) = \prod_{i=1}^k M_{X_i}(t) = \prod_{i=1}^k \exp(\lambda_i e^t - \lambda_i) = \exp\left[\left(\sum_{i=1}^k \lambda_i\right)e^t - \left(\sum_{i=1}^k \lambda_i\right)\right],$$

which is the m.g.f. of $P(\sum_{i=1}^k \lambda_i)$, so that $\sum_{i=1}^k X_i \sim P\left(\sum_{i=1}^k \lambda_i\right)$. ▲

The theorem is illustrated by the following example.

Example 5. Five radioactive sources independently emit particles at the rate of 0.08 per certain time unit. What is the probability that the total number of particles does not exceed 3 in the time unit considered?

Discussion. In obvious notation, we have here the independent r.v.'s X_i distributed as $P(0.08)$, $i = 1, \ldots, 5$. Then $X = \sum_{i=1}^5 X_i \sim P(0.4)$, and the required probability is: $P(X \leq 3) = 0.999224$ (from thc Poisson tables).

A theorem analogous to Theorems 2 and 3 holds for the normal distribution as well.

Theorem 4. Let the r.v.'s $X_1, \ldots, X_k$ be independent and let $X_i \sim N(\mu_i, \sigma_i^2)$, $i = 1, \ldots, k$. Then $\sum_{i=1}^k X_i \sim N(\sum_{i=1}^k \mu_i, \sum_{i=1}^k \sigma_i^2)$. In particular, if $\mu_1 = \cdots = \mu_k = \mu$ and $\sigma_1 = \cdots = \sigma_k = \sigma$, then $\sum_{i=1}^k X_i \sim N(k\mu, k\sigma^2)$.

Proof. Use relation (10.10), and formula (6.31) in Chapter 6, for $t \in \Re$, in order to obtain:

$$M_{\sum_{i=1}^k X_i}(t) = \prod_{i=1}^k M_{X_i}(t) = \prod_{i=1}^k \exp\left(\mu_i t + \frac{\sigma_i^2}{2}t\right)$$
$$= \exp\left[\left(\sum_{i=1}^k \mu_i\right)t + \frac{\sum_{i=1}^k \sigma_i^2}{2}t\right],$$

which is the m.g.f. of $N(\sum_{i=1}^k \mu_i, \sum_{i=1}^k \sigma_i^2)$, so that $\sum_{i=1}^k X_i \sim N(\sum_{i=1}^k \mu_i, \sum_{i=1}^k \sigma_i^2)$. The special case is immediate. ▲

As an illustration of this result, consider the following example.

Example 6. The rainfall in two locations is measured (in inches over a certain time unit) by two independent and normally distributed r.v.'s X_1 and X_2 as follows: $X_1 \sim N(10, 9)$ and $X_2 \sim N(15, 25)$. What is the probability that the total rainfall: (i) Will exceed 30 inches (which may result in flooding)? (ii) Will be less than 8 inches (which will mean a drought)?

Discussion. If $X = X_1 + X_2$, then $X \sim N(25, 34)$, so that:

$$\text{(i) } P(X > 30) = 1 - P(X \le 30)$$
$$= 1 - P(0 \le X \le 30) = 1 - P\left(-\frac{25}{\sqrt{34}} \le Z \le \tfrac{30-25}{\sqrt{34}}\right)$$
$$= 1 - P\left(Z \le \frac{30-25}{\sqrt{34}}\right) \simeq 1 - \Phi(0.86)$$
$$= 1 - 0.805105 = 0.194895,$$

and

$$\text{(ii) } P(X < 8) = P(0 \le X < 8)$$
$$= P\left(-\frac{25}{\sqrt{34}} \le Z \le \frac{8-25}{\sqrt{34}}\right) = P\left(Z < \frac{8-25}{\sqrt{34}}\right)$$
$$\simeq \Phi(-2.92) = 1 - \Phi(2.92) = 1 - 0.99825 = 0.00175.$$

In Proposition 5 of Chapter 6, it is shown that if $X \sim N(\mu, \sigma^2)$, then $Z = (X - \mu)/\sigma \sim N(0, 1)$. A slightly more general result is stated here as a proposition, and its proof is given in the corollary to Proposition 1 in Chapter 11.

Proposition 5. If $X \sim N(\mu, \sigma^2)$ and $Y = aX + b$, $(a \ne 0)$, then $Y \sim N(a\mu + b, (a\sigma)^2)$. In particular (as already stated), $Z = (X - \mu)/\sigma \sim N(0, 1)$.

On the basis of Proposition 5, Theorem 4 generalizes as follows:

Theorem 4′. Let the r.v.'s $X_1, \ldots, X_k$ be independent, let $X_i \sim N(\mu_i, \sigma_i^2)$, $i = 1, \ldots, k$, and let c_i, $i = 1, \ldots, k$ be constants. Then $\sum_{i=1}^k c_i X_i \sim N(\sum_{i=1}^k c_i \mu_i, \sum_{i=1}^k c_i^2 \sigma_i^2)$.

Proof. By Proposition 5, $c_i X_i \sim N(c_i \mu_i, c_i^2 \sigma_i^2)$, and the r.v.'s $c_i X_i$, $i = 1, \ldots, k$ are independent (by Proposition 3). Then the conclusion follows from Theorem 4. ▲

This theorem has the following corollary.

Corollary. If the r.v.'s $X_1, \ldots, X_k$ are independent and distributed as $N(\mu, \sigma^2)$, then their sample mean $\bar{X} \sim N(\mu, \frac{\sigma^2}{k})$, and $\frac{\sqrt{k}(\bar{X}-\mu)}{\sigma} \sim N(0, 1)$.

Proof. Apply Proposition 5 with $a = 1/k$ and $b = 0$ to get that X_i/k is distributed as $N(\mu/k, \sigma^2/k^2)$. Since the r.v.'s X_i/k, $i = 1, \ldots, k$ are independent (by Proposition 3), Theorem 4′ shows that $\sum_{i=1}^k \frac{X_i}{k} = \frac{1}{k}\sum_{i=1}^k X_i = \bar{X}$ is distributed as $N(\mu, \sigma^2/k)$. Then by Proposition 5 in Chapter 6 as already mentioned (see also Proposition 5 here),

$$\frac{\bar{X} - \mu}{\sigma/\sqrt{k}} = \frac{\sqrt{k}(\bar{X} - \mu)}{\sigma} \sim N(0, 1). \quad ▲$$

Here is an illustrative example of Theorem 4′ and its corollary.

Example 7. Suppose that an academic department in a university offers k undergraduate courses, and assume that the grade in the ith course is an r.v. X_i distributed (approximately) as $N(\mu_i, \sigma_i^2)$. Furthermore, assume that the r.v.'s $X_1, \ldots, X_k$ are independent. What is the distribution of the average $\bar{X} = (X_1 + \cdots + X_k)/k$?

Discussion. Since $\bar{X} = \sum_{i=1}^k \frac{X_i}{k}$, Theorem 4′ applies with $c_i = 1/k$, $i = 1, \ldots, k$ and gives that:

$$\bar{X} \sim N\Big(\frac{\mu_1 + \cdots + \mu_k}{k}, \frac{\sigma_1^2 + \cdots + \sigma_k^2}{k^2}\Big).$$

This is the required distribution of $\bar{X}$. If it so happens that $\mu_1 = \cdots = \mu_k = \mu$, say, and $\sigma_1^2 = \cdots = \sigma_k^2 = \sigma^2$, say, then $\bar{X} \sim N(\mu, \sigma^2/k)$.

The reproducing property we have seen in connection with binomial, Poisson, and normal distributions also holds for chi-square distribution. More precisely, we have:

Theorem 5. Let the r.v.'s $X_1, \ldots, X_k$ be independent, and let $X_i \sim \chi^2_{r_i}$, $i = 1, \ldots, k$. Then $\sum_{i=1}^k X_i \sim \chi^2_{r_1+\cdots+r_k}$.

Proof. Use relation (10.10) here, and formula (6.23) (in Chapter 6), for $t < \frac{1}{2}$, to obtain:

$$M_{\sum_{i=1}^k X_i}(t) = \prod_{i=1}^k M_{X_i}(t) = \prod_{i=1}^k \frac{1}{(1-2t)^{r_i/2}} = \frac{1}{(1-2t)^{(r_1+\cdots+r_k)/2}},$$

which is the m.g.f. of $\chi^2_{r_1+\cdots+r_k}$. ▲

This theorem has a corollary that is stated below. For its justification, as well as for the discussion of an illustrative example, an auxiliary result is needed; it is presented below as a proposition. For an alternative proof of it, see Example 2 in Chapter 11.

Proposition 6. If $Z \sim N(0, 1)$, then $Y = Z^2 \sim \chi_1^2$.

Proof. For $y > 0$, we have:

$$F_Y(y) = P(Y \le y) = P(Z^2 \le y) = P(-\sqrt{y} \le Z \le \sqrt{y})$$
$$= \Phi(\sqrt{y}) - \Phi(-\sqrt{y}) = 2\Phi(\sqrt{y}) - 1,$$

and hence

$$f_Y(y) = \frac{d}{dy}F_Y(y) = 2\frac{d}{dy}\Phi(\sqrt{y}) = 2f_Z(\sqrt{y}) \times \frac{d}{dy}\sqrt{y}$$
$$= 2 \times \frac{1}{2\sqrt{y}} \times \frac{1}{\sqrt{2\pi}}e^{-y/2} = \frac{1}{\sqrt{\pi}2^{1/2}}y^{1/2-1}e^{-y/2}$$
$$= \frac{1}{\Gamma\left(\frac{1}{2}\right)2^{1/2}}y^{1/2-1}e^{-y/2}, \text{ since } \sqrt{\pi} = \Gamma\left(\frac{1}{2}\right)$$

(see relation (6.16) in Chapter 6). However, the last expression on the right-hand side above is the p.d.f. of the χ_1^2 distribution, as was to be seen. ▲

Corollary. Let the r.v.'s $X_1, \ldots, X_k$ be independent and let $X_i \sim N(\mu_i, \sigma_i^2)$, $i = 1, \ldots, k$. Then $\sum_{i=1}^k \left(\frac{X_i-\mu_i}{\sigma_i}\right)^2 \sim \chi_k^2$, and, in particular, if $\mu_1 = \cdots = \mu_k = \mu$ and $\sigma_1^2 = \cdots = \sigma_k^2 = \sigma^2$, then $\frac{kS^2}{\sigma^2} \sim \chi_k^2$, where S^2 is given in (10.13).

Proof. The assumption $X_i \sim N(\mu_i, \sigma_i^2)$ implies that $\frac{X_i-\mu_i}{\sigma_i} \sim N(0, 1)$ by Proposition 5 in Chapter 6 (see also Proposition 5 here). Since independence of X_i, $i = 1, \ldots, k$ implies that of $\left(\frac{X_i-\mu_i}{\sigma_i}\right)^2$, $i = 1, \ldots, k$ (by Proposition 3), the theorem applies, on account of Proposition 6 here, and yields the first assertion. The second assertion follows from the first by taking $\mu_1 = \cdots = \mu_k = \mu$ and $\sigma_1 = \cdots = \sigma_k = \sigma$, and using (10.13) to obtain $\frac{kS^2}{\sigma^2} = \sum_{i=1}^k \left(\frac{X_i-\mu}{\sigma}\right)^2$. ▲

Remark 4. From the fact that $\frac{kS^2}{\sigma^2} \sim \chi_k^2$ and formula (6.23) (in Chapter 6), we have $E\left(\frac{kS^2}{\sigma^2}\right) = k$, $Var\left(\frac{kS^2}{\sigma^2}\right) = 2k$, or $ES^2 = \sigma^2$ and $Var(S^2) = 2\sigma^4/k$.

As an illustration of Theorem 5, consider the following example.

Example 8. Let the r.v. Y_i denote the lifetime of the ith battery in a lot of k identical batteries, and suppose that $Y_1, \ldots, Y_k$ are independently distributed as $N(\mu, \sigma^2)$. Set $Z_i = \frac{Y_i-\mu}{\sigma}$. Then the r.v.'s $Z_1, \ldots, Z_k$ are independently distributed as $N(0, 1)$, and therefore the r.v.'s $X_1, \ldots, X_k$ are independently distributed as χ_1^2, where $X_i = Z_i^2$, $i = 1, \ldots, k$. This is so, by Proposition 6. Then the theorem applies and gives that $X = \sum_{i=1}^k X_i \sim \chi_k^2$. Now, given that X_i represents the (normed) squared deviation of the lifetime of the ith battery from its mean, the sum $\sum_{i=1}^{k} X_i$ represents the totality of such deviations for the k batteries. Expectation, variance, s.d., and probabilities for X can be readily computed by the fact that $X \sim \chi_k^2$.

Remark 5. The last corollary above also provides an explanation of the term "k degrees of freedom" used in conjunction with a chi-square distribution. Namely, k is the number of independent $N(0, 1)$ r.v.'s required, whose sum of squares is distributed as χ_k^2.

Remark 6. This section is concluded with the following comment. Theorems 10.2–10.5 may be misleading in the sense that the sum of independent r.v.'s always has a distribution of the same kind as the summands. That this is definitely not so is illustrated by examples. For instance, if the independent r.v.'s X and Y are $U(0, 1)$, then their sum $X+Y$ is *not* uniform; rather, it is triangular (see Example 3 (continued) in Chapter 11).

Exercises.

2.1 In reference to Exercise 1.17 in this chapter, specify the distribution of the sum $X + Y$, and write out the expression for the exact probability $P(X + Y \leq 10)$.

2.2 If the independent r.v.'s X and Y are distributed as $B(m, p)$ and $B(n, p)$, respectively:

- **(i)** What is the distribution of the r.v. $X + Y$?
- **(ii)** If $m = 8, n = 12$, and $p = 0.25$, what is the numerical value of the probability $P(5 \leq X + Y \leq 15)$?

2.3 The independent r.v.'s $X_1, \ldots, X_n$ are distributed as $B(1, p)$, and let $S_n = X_1 + \cdots + X_n$.

- **(i)** Determine the distribution of the r.v. S_n.
- **(ii)** What is the EX_i and the $Var(X_i)$, $i = 1, \ldots, n$?
- **(iii)** From part (ii) and the definition of S_n, compute the ES_n and $Var(S_n)$.

2.4 Let $X_1, \ldots, X_n$ be i.i.d. r.v.'s with p.d.f. f, and let I be an interval in $\Re$. Let $p = P(X_1 \in I)$.

- **(i)** Express p in terms of the p.d.f. f.
- **(ii)** For k with $1 \leq k \leq n$, express the probability that at least k of $X_1, \ldots, X_n$ take values in the interval I in terms of p and n.
- **(iii)** Simplify the expression in part (ii), if f is the negative exponential p.d.f. with parameter λ and $I = \left(\frac{1}{\lambda}, \infty\right)$.
- **(iv)** Find the numerical value of the probability in part (iii) for $n = 4$ and $k = 2$.

2.5 The breakdown voltage of a randomly chosen diode of a certain type is known to be normally distributed with mean value 40V and s.d. 1.5V.

- **(i)** What is the probability that the breakdown voltage of a single diode is between 39v and 42v?

(ii) If 5 diodes are independently chosen, what is the probability that at least one has a breakdown voltage exceeding 42v?

2.6 Refer to Exercise 1.18 here and set $X = X_1 + \cdots + X_n$.

(i) Justify the statement that $X \sim B(n, p)$.
(ii) Suppose that n is large and p is small (both assumptions quite appropriate in the framework of Exercise 1.18), so that:

$$f(x) = \binom{n}{x} p^x q^{n-x} \simeq e^{-np} \frac{(np)^x}{x!}, \quad x = 0, 1, \ldots$$

If $np = 2$, calculate the approximate values of the probabilities $f(x)$ for $x = 0, 1, 2, 3$, and 4.

Hint. See Exercise 1.25 in Chapter 6.

2.7 The r.v.'s $X_1, \ldots, X_n$ are independent and $X_i \sim P(\lambda_i)$:

(i) What is the distribution of the r.v. $X = X_1 + \cdots + X_n$?
(ii) If $\bar{X} = \frac{1}{n}(X_1 + \cdots + X_n)$, calculate the $E\bar{X}$ and the $Var(\bar{X})$ in terms of $\lambda_1, \ldots, \lambda_n$, and n.
(iii) How do the $E\bar{X}$ and the $Var(\bar{X})$ become when the X_i's in part (i) are distributed as $P(\lambda)$?

2.8 Suppose that the number of no-shows for a scheduled airplane flight is an r.v. X distributed as $P(\lambda)$, and it is known from past experience that on the average, there are 2 no-shows. If there are 5 flights scheduled, compute the probabilities that the total number of no-shows $X = X_1 + \cdots + X_5$ is:

(i) 0.	(v) At most 10.	(ix) 15.
(ii) At most 5.	(vi) 10.	(x) At least 15.
(iii) 5.	(vii) At least 10.	
(iv) At least 5.	(viii) At most 15.	

2.9 The r.v.'s $X_1, \ldots, X_n$ are independent and $X_i \sim P(\lambda_i), i = 1, \ldots, n$. Set $T = \sum_{i=1}^{n} X_i$ and $\lambda = \sum_{i=1}^{n} \lambda_i$, and show that:

(i) The conditional p.d.f. of X_1, given $T = t$, is $B(t, \lambda_1/\lambda)$.
(ii) From part (i), conclude that the conditional p.d.f. of X_i, given $T = t$, is $B(t, \lambda_i/\lambda), i = 2, \ldots, n$ (and, of course, $i = 1$).
(iii) How does the distribution in part (ii) become for $\lambda_1 = \cdots = \lambda_n = c$, say?

2.10 If the independent r.v.'s X and Y are distributed as $N(\mu_1, \sigma_1^2)$ and $N(\mu_2, \sigma_2^2)$, respectively:

(i) Specify the distribution of $X - Y$.
(ii) Calculate the probability $P(X > Y)$ in terms of μ_1, μ_2, σ_1, and σ_2.
(iii) If $\mu_1 = \mu_2$, conclude that $P(X > Y) = 0.5$.

2.11 The $m + n$ r.v.'s $X_1, \ldots, X_m$ and $Y_1, \ldots, Y_n$ are independent and $X_i \sim N(\mu_1, \sigma_1^2)$, $i = 1, \ldots, m$, $Y_j \sim N(\mu_2, \sigma_2^2)$, $j = 1, \ldots, n$. Set $\bar{X} = \frac{1}{m}\sum_{i=1}^m X_i$, $\bar{Y} = \frac{1}{n}\sum_{j=1}^n Y_j$ and:

(i) Calculate the probability $P(\bar{X} > \bar{Y})$ in terms of $m, n, \mu_1, \mu_2, \sigma_1$, and σ_2.
(ii) Give the numerical value of the probability in part (i) when $\mu_1 = \mu_2$, unspecified.

2.12 Let the independent r.v.'s $X_1, \ldots, X_n$ be distributed as $N(\mu, \sigma^2)$ and set

$$X = \sum_{i=1}^n \alpha_i X_i, \; Y = \sum_{j=1}^n \beta_j X_j,$$

where the α_i's and the β_j's are constants. Then:

(i) Determine the p.d.f.'s of the r.v.'s X and Y.
(ii) Show that the joint m.g.f. of X and Y is given by:

$$M_{X,Y}(t_1, t_2) = \exp\left[\mu_1 t_1 + \mu_2 t_2 + \frac{1}{2}\left(\sigma_1^2 t_1^2 + 2\rho\sigma_1\sigma_2 t_1 t_2 + \sigma_2^2 t_2^2\right)\right],$$

where $\mu_1 = \mu\sum_{i=1}^n \alpha_i$, $\mu_2 = \mu\sum_{j=1}^n \beta_j$, $\sigma_1^2 = \sigma^2\sum_{i=1}^n \alpha_i^2$, $\sigma_2^2 = \sigma^2\sum_{j=1}^n \beta_j^2$, $\rho = \frac{\sum_{i=1}^n \alpha_i\beta_i}{(\sum_{i=1}^n \alpha_i^2)^{\frac{1}{2}} \times (\sum_{i=1}^n \beta_i^2)^{\frac{1}{2}}}$.

(iii) From part (ii), conclude that X and Y have bivariate normal distribution with correlation coefficient:

$$\rho(X, Y) = \rho = \frac{\sum_{i=1}^n \alpha_i\beta_i}{(\sum_{i=1}^n \alpha_i^2)^{\frac{1}{2}} \times (\sum_{i=1}^n \beta_i^2)^{\frac{1}{2}}}.$$

(iv) From part (iii), conclude that X and Y are independent if and only if $\sum_{i=1}^n \alpha_i\beta_i = 0$.

Hint. For part (iii), refer to relation (9.22) in Chapter 9.

2.13 Let X and Y be independent r.v.'s distributed as $N(0, \sigma^2)$.

(i) Set $R = \sqrt{X^2 + Y^2}$ and determine the probability: $P(R \leq r)$, for $r > 0$.
(ii) What is the numerical value of $P(R \leq r)$ for $\sigma = 1$ and $r = 1.665$, $r = 2.146$, $r = 2.448$, $r = 2.716$, $r = 3.035$, and $r = 3.255$?

Hint. For part (ii), use the chi-square tables.

2.14 Let $X_1, \ldots, X_n$ be i.i.d. r.v.'s with mean $\mu \in \Re$ and variance $0 < \sigma^2 < \infty$. Use thc Tchebichev inequality:

(i) In determining the smallest value of n, as a function of $c(>0)$ and p, for which $P(|\bar{X}_n - \mu| \leq c\sigma) \geq p$.
(ii) What is the numerical value of n for $p = 0.95$ and $c = 1, 0.5,\ 0.25$?

2.15 Computer chips are manufactured independently by three factories, and let $X_i,\ i = 1, 2, 3$ be the r.v.'s denoting the total numbers of defective items in a day's production by the three factories, respectively. Suppose that the m.g.f. of X_i is given by $M_i(t) = 1/(1-\beta t)^{\alpha_i}, t < \frac{1}{\beta}, i = 1, 2, 3$, and let X be the combined number of defective chips in a day's production.

(i) Express the EX and the $\sigma^2(X)$ in terms of the α_i's and β.
(ii) Compute the numerical values of EX and $\sigma(X)$ for $\beta = 2$ and $\alpha_i = 10^i, i = 1, 2, 3$.

2.16 The blood pressure of an individual taken by an instrument used at home is an r.v. X distributed as $N(\mu, 2\sigma^2)$, whereas the blood pressure of the same individual taken in a doctor's office by a more sophisticated device is an r.v. Y distributed as $N(\mu, \sigma^2)$. If the r.v.'s X and Y are independent, compute the probability that the average $\frac{X+Y}{2}$ lies within 1.5σ from the mean μ.

10.3 Distribution of the Sample Variance under Normality

In the definition of S^2 by (10.13), we often replace μ by the sample mean $\bar{X}$; this is done habitually in statistics, as μ is not really known. Let us denote by $\bar{S}^2$ the resulting quantity; that is,

$$\bar{S}^2 = \frac{1}{k}\sum_{i=1}^{k}(X_i - \bar{X})^2. \tag{10.14}$$

Then it is easy to establish the following identity:

$$\sum_{i=1}^{k}(X_i - \mu)^2 = \sum_{i=1}^{k}(X_i - \bar{X})^2 + k(\bar{X} - \mu)^2, \tag{10.15}$$

or

$$kS^2 = k\bar{S}^2 + [\sqrt{k}(\bar{X} - \mu)]^2. \tag{10.16}$$

Indeed,

$$\sum_{i=1}^{k}(X_i - \mu)^2 = \sum_{i=1}^{k}[(X_i - \bar{X}) + (\bar{X} - \mu)]^2 = \sum_{i=1}^{k}(X_i - \bar{X})^2 + k(\bar{X} - \mu)^2,$$

since $\sum_{i=1}^{k}(X_i - \bar{X})(\bar{X} - \mu) = (\bar{X} - \mu)(k\bar{X} - k\bar{X}) = 0$.

From (10.16), we have, dividing through by σ^2:

$$\frac{kS^2}{\sigma^2} = \frac{k\bar{S}^2}{\sigma^2} + \left[\frac{\sqrt{k}(\bar{X} - \mu)}{\sigma}\right]^2. \tag{10.17}$$

Now $\frac{kS^2}{\sigma^2} \sim \chi_k^2$ and $[\frac{\sqrt{k}(\bar{X}-\mu)}{\sigma}]^2 \sim \chi_1^2$ (by Proposition 6 here) when the r.v.'s $X_1, \ldots, X_k$ are independently distributed as $N(\mu, \sigma^2)$. Therefore, from (10.17), it appears quite feasible that $\frac{k\bar{S}^2}{\sigma^2} \sim \chi_{k-1}^2$. This is, indeed, the case and is the content of the following theorem. This theorem is presently established under an assumption to be justified later on (see Theorem 7 in Chapter 11). The *assumption* is this: If the r.v.'s $X_1, \ldots, X_k$ are independent and distributed as $N(\mu, \sigma^2)$, then the r.v.'s $\bar{X}$ and $\bar{S}^2$ are independent. (The independence of $\bar{X}$ and $\bar{S}^2$ implies then that of $[\frac{\sqrt{k}(\bar{X}-\mu)}{\sigma}]^2$ and $\frac{k\bar{S}^2}{\sigma^2}$, by Proposition 3.)

Theorem 6. Let the r.v.'s $X_1, \ldots, X_k$ be independent and distributed as $N(\mu, \sigma^2)$, and let $\bar{S}^2$ be defined by (10.14). Then $\frac{k\bar{S}^2}{\sigma^2} \sim \chi_{k-1}^2$. Consequently, $E\bar{S}^2 = \frac{k-1}{k}\sigma^2$ and $Var(\bar{S}^2) = \frac{2(k-1)\sigma^4}{k^2}$.

Proof. Consider relation (10.17), take the m.g.f.'s of both sides, use relation (10.10) for two r.v.'s, and the assumption of independence made previously in order to obtain:

$$M_{kS^2/\sigma^2}(t) = M_{k\bar{S}^2/\sigma^2}(t) M_{[\sqrt{k}(\bar{X}-\mu)/\sigma]^2}(t),$$

so that

$$M_{k\bar{S}^2/\sigma^2}(t) = M_{kS^2/\sigma^2}(t) / M_{[\sqrt{k}(\bar{X}-\mu)/\sigma]^2}(t),$$

or

$$M_{k\bar{S}^2/\sigma^2}(t) = \frac{1/(1-2t)^{k/2}}{1/(1-2t)^{1/2}} = \frac{1}{(1-2t)^{(k-1)/2}},$$

which is the m.g.f. of the χ_{k-1}^2 distribution. The second assertion follows immediately from the first and formula (6.23) (in Chapter 6). ▲

Exercises.

3.1 For any r.v.'s $X_1, \ldots, X_n$, set

$$\bar{X} = \frac{1}{n}\sum_{i=1}^{n} X_i \quad \text{and} \quad \bar{S}^2 = \frac{1}{n}\sum_{i=1}^{n}(X_i - \bar{X})^2,$$

and show that:

(i) $n\bar{S}^2 = \sum_{i=1}^{n}(X_i - \bar{X})^2 = \sum_{i=1}^{n} X_i^2 - n\bar{X}^2$.

(ii) If the r.v.'s have common (finite) expectation μ, then:

$$\sum_{i=1}^{n}(X_i - \mu)^2 = \sum_{i=1}^{n}(X_i - \bar{X})^2 + n(\bar{X} - \mu)^2 = n\bar{S}^2 + n(\bar{X} - \mu)^2.$$

(iii) Use part (ii) and refer to Proposition 4 in order to show that:

$$E\left[\frac{1}{n}\sum_{i=1}^{n}(X_i - \mu)^2\right] = E\left[\frac{1}{n-1}\sum_{i=1}^{n}(X_i - \bar{X})^2\right] = \sigma^2.$$

CHAPTER

11 Transformation of Random Variables

This chapter is devoted to transforming a given set of r.v.'s to another set of r.v.'s. The practical need for such transformations will become apparent by means of concrete examples to be cited and/or discussed. The chapter consists of five sections. In the first section, a single r.v. is transformed into another single r.v. In the following section, the number of available r.v.'s is at least two, and they are to be transformed into another set of r.v.'s of the same or smaller number. Two specific applications produce two new distributions, the t-distribution and the F-distribution, which are of great applicability in statistics. A brief account of specific kinds of transformations is given in the subsequent two sections, and the chapter is concluded with a section on order statistics. All r.v.'s we are dealing with in this chapter are of the continuous type, unless otherwise specifically mentioned.

11.1 Transforming a Single Random Variable

Perhaps the best way of introducing the underlying problem here is by means of an example.

Example 1. Suppose that the r.v.'s X and Y represent the temperature in a certain locality measured in degrees Celsius and Fahrenheit, respectively. Then it is known that X and Y are related as follows: $Y = \frac{9}{5}X + 32$.

This simple example illustrates the need for transforming an r.v. X into another r.v. Y, if Celsius degrees are to be transformed into Fahrenheit degrees. In order to complete the discussion:

(i) Let X have d.f. F_X and p.d.f. f_X. Determine the d.f. F_Y and the p.d.f. f_Y of the r.v. Y in terms of F_X and f_X.
(ii) Apply part (i) in the case that $X \sim N(\mu, \sigma^2)$.

Discussion.

(i) We have

$$F_Y(y) = P(Y \leq y) = P\left(\frac{9}{5}X + 32 \leq y\right)$$
$$= P\left(X \leq \frac{5}{9}(y-32)\right) = F_X\left(\frac{5}{9}(y-32)\right),$$

Introduction to Probability, Second Edition. http://dx.doi.org/10.1016/B978-0-12-800041-0.00011-0

so that

$$f_Y(y) = \frac{d}{dy}F_Y(y) = \frac{d}{dy}F_X\left(\frac{5}{9}(y-32)\right) = \frac{5}{9}f_X\left(\frac{5}{9}(y-32)\right).$$

(ii) If $X \sim N(\mu, \sigma^2)$, then

$$F_X(x) = \int_{-\infty}^{x} \frac{1}{\sqrt{2\pi}\sigma} e^{-\frac{(t-\mu)^2}{2\sigma^2}} dt, \quad f_X(x) = \frac{1}{\sqrt{2\pi}\sigma} e^{-\frac{(x-\mu)^2}{2\sigma^2}},$$

so that

$$F_Y(y) = \int_{-\infty}^{\frac{5}{9}(y-32)} \frac{1}{\sqrt{2\pi}\sigma} e^{-\frac{(t-\mu)^2}{2\sigma^2}} dt, \quad \text{and}$$

$$f_Y(y) = \frac{5}{9} \times \frac{1}{\sqrt{2\pi}\sigma} e^{-\frac{\left[\frac{5}{9}(y-32)-\mu\right]^2}{2\sigma^2}}$$

$$= \frac{1}{\sqrt{2\pi}(9\sigma/5)} e^{-\frac{\left[y-\left(\frac{9}{5}\mu+32\right)\right]^2}{2(9\sigma/5)^2}}.$$

It follows that $Y \sim N\left(\frac{9}{5}\mu + 32, (\frac{9\sigma}{5})^2\right)$.

This example is a special case of the following result.

Proposition 1. Let the r.v. X have d.f. F_X and p.d.f. f_X, and let $Y = aX+b$, $(a \neq 0)$. Then:

$$F_Y(y) = \begin{cases} F_X\left(\frac{y-b}{a}\right), & \text{if } a > 0 \\ 1 - F_X\left(\frac{y-b}{a}\right), & \text{if } a < 0, \end{cases}$$

and

$$f_Y(y) = f_X\left(\frac{y-b}{a}\right)\left|\frac{1}{a}\right|.$$

Proof. Indeed,

$$F_Y(y) = P(Y \le y) = P(aX+b \le y) = \begin{cases} P\left(X \le \frac{y-b}{a}\right), & \text{if } a > 0 \\ P\left(X \ge \frac{y-b}{a}\right), & \text{if } a < 0 \end{cases}$$

$$= \begin{cases} F_X\left(\frac{y-b}{a}\right), & \text{if } a > 0 \\ 1 - F_X\left(\frac{y-b}{a}\right), & \text{if } a < 0, \end{cases}$$

and

$$f_Y(y) = \frac{d}{dy}F_Y(y) = \begin{cases} \frac{d}{dy}F_X\left(\frac{y-b}{a}\right), & \text{if } a > 0 \\ \frac{d}{dy}\left[1 - F_X\left(\frac{y-b}{a}\right)\right], & \text{if } a < 0 \end{cases}$$

$$= \begin{cases} f_X\left(\frac{y-b}{a}\right)\frac{1}{a}, & \text{if } a > 0 \\ f_X\left(\frac{y-b}{a}\right)\left(-\frac{1}{a}\right), & \text{if } a < 0 \end{cases}$$

$$= f_X\left(\frac{y-b}{a}\right)\left|\frac{1}{a}\right|, \quad \text{as was to be seen.} \qquad \blacktriangle$$

Corollary. If $X \sim N(\mu, \sigma^2)$, then $Y = aX + b \sim N(a\mu + b, (a\sigma)^2)$.

Proof. With $x = \frac{y-b}{a}$, we get:

$$f_Y(y) = \frac{1}{\sqrt{2\pi}\sigma}\exp\left\{-\frac{[(y-b)/a-\mu]^2}{2\sigma^2}\right\}\frac{1}{|a|}$$
$$= \frac{1}{\sqrt{2\pi}|a\sigma|}\exp\left\{-\frac{[y-(a\mu+b)]^2}{2(a\sigma)^2}\right\}. \quad ▲$$

The transformation $Y = aX + b$ used in Proposition 1 has the property that it is strictly monotone (strictly increasing if $a > 0$, and strictly decreasing if $a < 0$). This observation leads to the following general problem.

Theorem 1. Let X be a r.v. with d.f. F_X and p.d.f. f_X, and let $h : \Re \to \Re$ be a strictly monotone function. Set $Y = h(X)$. Then the d.f. F_Y is given by:

$$F_Y(y) = \begin{cases} F_X[h^{-1}(y)], & \text{if } h \text{ is strictly increasing} \quad (11.1)\\ 1 - F_X[h^{-1}(y)], & \text{if } h \text{ is strictly decreasing.} \quad (11.2)\end{cases}$$

Furthermore, if the function $x = h^{-1}(y)$ is differentiable, then:

$$f_Y(y) = f_X[h^{-1}(y)]\left|\frac{d}{dy}h^{-1}(y)\right|. \tag{11.3}$$

Proof. The function $y = h(x)$ is invertible, and its inverse $x = h^{-1}(y)$ is strictly increasing if h is so, and strictly decreasing if h is so. Therefore,

$$F_Y(y) = P(Y \le y) = P[h(X) \le y]$$
$$= \begin{cases} P[X \le h^{-1}(y)], & \text{if } h \text{ is strictly increasing}\\ P[X \ge h^{-1}(y)], & \text{if } h \text{ is strictly decreasing}\end{cases}$$
$$= \begin{cases} F_X[h^{-1}(y)], & \text{if } h \text{ is strictly increasing} \quad (11.4)\\ 1 - F_X[h^{-1}(y)], & \text{if } h \text{ is strictly decreasing.} \quad (11.5)\end{cases}$$

as stated in relations (11.1) and (11.2).

At this point, recall that $\frac{dx}{dy} = \frac{d}{dy}h^{-1}(y)$ is > 0 if h is strictly increasing, and it is < 0 if h is strictly decreasing. On the basis of this, and by differentiating relations (11.4) and (11.5), we get:

$$f_Y(y) = \frac{d}{dy}F_Y(y) = \frac{d}{dy}F_X[h^{-1}(y)] = f_X[h^{-1}(y)]\frac{d}{dy}h^{-1}(y),$$

if h is strictly increasing, and

$$f_Y(y) = -\frac{d}{dy}F_X[h^{-1}(y)] = -f_X[h^{-1}(y)]\frac{d}{dy}h^{-1}(y)$$
$$= f_X[h^{-1}(y)]\left[-\frac{d}{dy}h^{-1}(y)\right],$$

if h is strictly decreasing.

Since $\frac{d}{dy}h^{-1}(y)$ and $-\frac{d}{dy}h^{-1}(y)$ are both included in the $\left|\frac{d}{dy}h^{-1}(y)\right|$, we get $f_Y(y) = f_X[h^{-1}(y)]\left|\frac{d}{dy}h^{-1}(y)\right|$, as was to be seen according to relation (11.3). ▲

Instead of going through the d.f. (which process requires monotonicity of the transformation $y = h(x)$), under certain conditions, f_Y may be obtained directly from f_X. Such conditions are described in the following theorem.

Theorem 2. Let X be an r.v. with positive and continuous p.d.f. on the set $S \subseteq \Re$, and let $h : S \to T$ (the image of S under h) be a one-to-one transformation, so that the inverse $x = h^{-1}(y)$, $y \in T$, exists. Suppose that, for $y \in T$, the derivative $\frac{d}{dy}h^{-1}(y)$ exists, is continuous, and $\neq 0$. Then the p.d.f. of the r.v. $Y = h(X)$ is given by:

$$f_Y(y) = f_X[h^{-1}(y)]\left|\frac{d}{dy}h^{-1}(y)\right|, \quad y \in T \text{ (and } = 0 \text{ for } y \notin T). \tag{11.6}$$

Proof (rough outline). Let $B = [c, d]$ be an interval in T and suppose B is transformed into the interval $A = [a, b]$ by the inverse transformation $x = h^{-1}(y)$. Then:

$$P(Y \in B) = P[h(X) \in B] = P(X \in A) = \int_A f_X(x)dx.$$

When transforming x into y through the transformation $x = h^{-1}(y)$,

$$\int_A f_X(x)dx = \int_B f_X[h^{-1}(y)]\left|\frac{d}{dy}h^{-1}(y)\right|dy,$$

according to the theory of changing variables in integrals. Thus,

$$P(Y \in B) = \int_B f_X[h^{-1}(y)]\left|\frac{d}{dy}h^{-1}(y)\right|dy,$$

which implies that the integrand is the p.d.f. of Y. ▲

This theorem has already been illustrated by means of Example 1 and Proposition 1.

Remark 1. In the formulation of Theorem 2, it is to be observed that the assumption that $\frac{d}{dy}h^{-1}(y) \neq 0$ on T implies that this derivative will be either always > 0 or always < 0 (because of the assumed continuity of the derivative). But then, $x = h^{-1}(y)$ (and hence $y = h(x)$) is strictly monotone, and hence one-to-one. Thus, the one-to-one *assumption* is superfluous. Although this is true, it will, nevertheless, be retained for the sake of uniformity in the formulation of Theorems 2, 3, 4. In the multi-dimensional case (Theorems 3 and 4) monotonicity becomes meaningless, and the one-to-one assumption is essential.

One of the basic assumptions of Theorem 2 is that the transformation $h : S \to T$ is one-to-one. It may happen, however, that this assumption is violated, but there is a partition of S into two or more sets over which the one-to-one assumption holds. Then Theorem 2 basically still holds, provided it is properly modified. This is done in the following theorem, which is stated without a proof. Its justification follows the same lines of that of Theorem 2, and it is—by and large—a calculus matter.

Theorem 2′. Let X be an r.v. with positive and continuous p.d.f. on the set $S \subseteq \Re$, and suppose that the transformation $h : S \to T$ is *not* one-to-one. Suppose further that when S is partitioned into the pairwise disjoint subsets $S_1, \ldots, S_r$ and h is restricted to S_j and takes values in T_j (the image of S_j under h), then h *is* one-to-one. Denoting by h_j this restriction, we have then: $h_j : S_j \to T_j$ is one-to-one, so that the inverse $x = h_j^{-1}(y)$, $y \in T_j$, exists, $j = 1, \ldots, r$. Finally, we suppose that for any $y \in T_j$, $j = 1, \ldots, r$, the derivatives $\frac{d}{dy}h_j^{-1}(y)$ exist, are continuous, and $\neq 0$. Then the p.d.f. of the r.v. $Y = h(X)$ is determined as follows: Set

$$f_{Y_j}(y) = f_X\left[h_j^{-1}(y)\right]\left|\frac{d}{dy}h_j^{-1}(y)\right|, \quad y \in T_j, \quad j = 1, \ldots, r, \tag{11.7}$$

and for $y \in T$, suppose that y belongs to k of the r T_j's, $1 \leq k \leq r$. Then $f_Y(y)$ is the sum of the corresponding k $f_{Y_j}(y)$'s. Alternatively,

$$f_Y(y) = \sum_{j=1}^{r} \delta_j(y) f_{Y_j}(y), \quad y \in T \quad (\text{and} = 0 \text{ for } y \notin T), \tag{11.8}$$

where $\delta_j(y) = 1$, if $y \in T_j$ and $\delta_j(y) = 0$, if $y \notin T_j$, $j = 1, \ldots, r$.

Remark 2. It is to be noticed that whereas the subsets $S_1, \ldots, S_r$ are pairwise disjoint, their images $T_1, \ldots, T_r$ need not be so. For instance, in Example 2, $S_1 = (-\infty, 0)$, $S_2 = (0, \infty)$ but $T_1 = T_2 = (0, \infty)$.

The theorem is illustrated by the following example.

Example 2. If $Z \sim N(0, 1)$ and $Y = Z^2$, then $Y \sim \chi_1^2$.

Discussion. Here $y = z^2$, $S = \Re$, $T = [0, \infty)$, and the transformation is not one-to-one. However, if we write $S = S_1 \cup S_2$ with $S_1 = (-\infty, 0]$ and $S_2 = (0, \infty)$, then the corresponding images are $T_1 = [0, \infty)$ and $T_2 = (0, \infty)$, and $y_1 : S_1 \to T_1$, $y_2 : S_2 \to T_2$ are one-to-one. Specifically, $z_1 = -\sqrt{y_1}$, $y_1 \geq 0$; $z_2 = \sqrt{y_2}$, $y_2 > 0$. By omitting 0 (which can be done, since $P(Y = 0) = P(Z^2 = 0) = P(Z = 0) = 0$), we have that both y_1 and y_2 are in $(0, \infty)$; denote them by y. Then, essentially, $T_1 = T_2 = (0, \infty)$. Next, $\frac{dz_1}{dy} = -\frac{1}{2\sqrt{y}}$, $\frac{dz_2}{dy} = \frac{1}{2\sqrt{y}}$, so that $|\frac{dz_1}{dy}| = |\frac{dz_2}{dy}| = \frac{1}{2\sqrt{y}}$, $y > 0$. An application of relation (11.7) with $r = 2$ yields:

$$f_{Y_1}(y) = \frac{1}{\sqrt{2\pi}}e^{-\frac{(-\sqrt{y})^2}{2}} \times \left|\frac{1}{-2\sqrt{y}}\right| = \frac{1}{2\sqrt{2\pi}\sqrt{y}}e^{-\frac{y}{2}},$$

$$f_{Y_2}(y) = \frac{1}{\sqrt{2\pi}}e^{-\frac{(\sqrt{y})^2}{2}} \times \frac{1}{2\sqrt{y}} = \frac{1}{2\sqrt{2\pi}\sqrt{y}}e^{-\frac{y}{2}}.$$

Then relation (11.8), for $y > 0$, gives:

$$f_Y(y) = f_{Y_1}(y) + f_{Y_2}(y) = \frac{1}{\sqrt{2\pi}\sqrt{y}}e^{-\frac{y}{2}} = \frac{1}{\sqrt{\pi}2^{1/2}}y^{\frac{1}{2}-1}e^{-\frac{y}{2}} = \frac{1}{\Gamma\left(\frac{1}{2}\right)2^{1/2}}y^{\frac{1}{2}-1}e^{-\frac{y}{2}},$$

since $\sqrt{\pi} = \Gamma\left(\frac{1}{2}\right)$ (by (6.16) (in Chapter 6)). However, the last expression above is the p.d.f. of the χ_1^2 distribution, as was to be seen. ▲

For another illustration of Theorem 2′, the reader is referred to Exercise 1.8.

Remark 3. The result obtained in Example 2 can also be arrived at by means of d.f.'s. For this, see Exercise 1.9 below.

Exercises.

1.1 The (discrete) r.v. X has p.d.f. $f_X(x) = (1-\alpha)\alpha^x$, $x = 0, 1, \ldots$ $(0<\alpha<1)$, and set $Y = X^3$. Determine the p.d.f. f_Y in terms of α.

1.2 Let the r.v.'s X and Y represent the temperature of a certain object in degrees Celsius and Fahrenheit, respectively. Then, it is known that $Y = \frac{9}{5}X + 32$ and $X = \frac{5}{9}Y - \frac{160}{9}$.

- **(i)** If $Y \sim N(\mu, \sigma^2)$, determine the distribution of X.
- **(ii)** If $P(90 \leq Y \leq 95) = 0.95$, then also $P(a \leq X \leq b) = 0.95$, for some $a < b$. Determine the numbers a and b.
- **(iii)** We know that: $P(\mu-\sigma \leq Y \leq \mu+\sigma) \simeq 0.6827 = p_1$, $P(\mu-2\sigma \leq Y \leq \mu+2\sigma) \simeq 0.9545 = p_2$, and $P(\mu-3\sigma \leq Y \leq \mu+3\sigma) \simeq 0.9973 = p_3$. Calculate the intervals $[a_k, b_k]$, $k = 1, 2, 3$, in terms of μ and σ, for which $P(a_k \leq X \leq b_k)$ is, respectively, equal to p_k, $k = 1, 2, 3$.

1.3 Let the r.v. X have p.d.f. f_X positive and continuous on the set $S \subseteq \Re$, and set $U = aX + b$, where a and b are constants and $a > 0$.

- **(i)** Use Theorem 2 in order to derive the p.d.f. f_U.
- **(ii)** If X has negative exponential distribution with parameter λ, show that U has the same kind of distribution with parameter λ/a.
- **(iii)** If $X \sim U(c, d)$, then show that $U \sim U(ac + b, ad + b)$.

1.4 If the r.v. X has negative exponential distribution with parameter λ, set $Y = e^X$ and $Z = \log X$, and determine the p.d.f.'s f_Y and f_Z.

1.5 Let $X \sim U(\alpha, \beta)$ and set $Y = e^X$. Then determine the p.d.f. f_Y. If $\alpha > 0$, set $Z = \log X$ and determine the p.d.f. f_Z.

1.6

- **(i)** If the r.v. X is distributed as $U(0, 1)$ and $Y = -2\log X$, show that Y is distributed as χ_2^2.
- **(ii)** If $X_1, \ldots, X_n$ is a random sample from the $U(0, 1)$ distribution and $Y_i = -2\log X_i$, use part (i) and Theorem 5 in Chapter 10 in order to show that $\sum_{i=1}^n Y_i$ is distributed as χ_{2n}^2.

1.7 If the r.v. X has the p.d.f. $f_X(x) = \frac{1}{\sqrt{2\pi}}x^{-2}e^{-1/2x^2}$, $x \in \Re$, show that the r.v. $Y = \frac{1}{X} \sim N(0, 1)$.

1.8 Suppose that the velocity of a molecule of mass m is an r.v. X with p.d.f. $f_X(x) = \sqrt{\frac{2}{\pi}}x^2 e^{-x^2/2}$, $x > 0$ (the so-called Maxwell distribution). By means of Theorem 2, show that the p.d.f. of the r.v. $Y = \frac{1}{2}mX^2$, which is the kinetic energy of the molecule, is the gamma p.d.f. with $\alpha = \frac{3}{2}$ and $\beta = m$.

Hint. Use relations (6.14) and (6.16) in Chapter 6.

1.9 If the r.v. $X \sim N(0, 1)$, use the d.f. approach in order to show that the r.v. $Y = X^2 \sim \chi_1^2$. That is, derive first the d.f. of the r.v. Y, and then differentiate it in order to obtain the p.d.f.

Hint. Use relation (6.16) in Chapter 6.

1.10 Let the r.v.'s X and Y be as in Exercise 1.8. Then use the d.f. approach in order to determine the p.d.f. of Y. That is, derive first the d.f. of Y, and then differentiate it in order to obtain the p.d.f.

Hint. Use relations (6.14) and (6.16) in Chapter 6.

1.11 Let X be the r.v. denoting the time (in minutes) for an airline reservation desk to respond to a customer's telephone inquiry, and suppose that $X \sim U(t_1, t_2)(0 < t_1 < t_2)$. Then the r.v. $Y = 1/X$ represents the rate at which an inquiry is responded to.

(i) Determine the p.d.f. of Y by means of its d.f.
(ii) Specify the d.f. and the p.d.f. in part (i) for $t_1 = 10, t_2 = 15$.

11.2 Transforming Two or More Random Variables

Often the need arises to transform two or more given r.v.'s to another set of r.v.'s. The following examples illustrate the point.

Example 3. The times of arrival of a bus at two successive bus stops are r.v.'s X_1 and X_2 distributed as $U(\alpha, \beta)$, for two time points $\alpha < \beta$. Calculate the probabilities $P(X_1 + X_2 > x)$ for $2\alpha < x < 2\beta$.

Clearly, this question calls for the determination of the distribution of the r.v. $X_1 + X_2$. (This is done below in Example 3 (continued).)

Or more generally (and more realistically), suppose that a bus makes k stops between its depot and its terminal and that the arrival time at the ith stop is an r.v. $X_i \sim U(\alpha_i, \beta_i), \alpha_i < \beta_i, i = 1, \ldots, k+1$ (where X_{k+1} is the time of arrival at the terminal). Determine the distribution of the duration of the trip $X_1 + \cdots + X_{k+1}$.

Example 4. Consider certain events occurring in every time interval $[t_1, t_2]$ $(0 < t_1 < t_2)$ according to Poisson distribution $P(\lambda(t_2 - t_1))$. Then the waiting times between successive occurrences are independent r.v.'s distributed according to negative exponential distribution with parameter λ. (The fact that the distribution of the waiting times is as described is proved in Exercise 2.6 of Chapter 6; independence of the waiting times is shown in Proposition 1 in Chapter 10). Let X_1 and X_2 be two such times. What is the probability that one would have to wait at least twice as long for the second occurrence than the first? That is, what is the probability $P(X_2 > 2X_1)$?

Here one would have to compute the distribution of the r.v. $X_2 - 2X_1$. (This is done below in Example 4 (continued).)

Below, a brief outline of the theory underpinning the questions posed in the examples is presented. First, consider the case of two r.v.'s X_1 and X_2 having the joint p.d.f. f_{X_1,X_2}. Often the question posed is that of determining the distribution of a function of X_1 and X_2, $h_1(X_1, X_2)$. The general approach is to set $Y_1 = h_1(X_1, X_2)$, and consider another (convenient) transformation $Y_2 = h_2(X_1, X_2)$. Next, determine the joint p.d.f. of Y_1 and Y_2, f_{Y_1,Y_2}, and, finally, compute the (marginal) p.d.f. f_{Y_1}. Conditions under which f_{Y_1,Y_2} is determined by way of f_{X_1,X_2} are given below.

Theorem 3. Consider the r.v.'s X_1 and X_2 with joint p.d.f. f_{X_1,X_2} positive and continuous on the set $S \subseteq \Re^2$, and let h_1, h_2 be two real-valued transformations defined on S; that is, $h_1, h_2 : S \to \Re$, and let T be the image of S under the transformation (h_1, h_2). Suppose that (h_1, h_2) is one-to-one from S onto T. Thus, if we set $y_1 = h_1(x_1, x_2)$ and $y_2 = h_2(x_1, x_2)$, we can solve uniquely for $x_1, x_2 : x_1 = g_1(y_1, y_2), x_2 = g_2(y_1, y_2)$. Suppose further that the partial derivatives $g_{1i}(y_1, y_2) = \frac{\partial}{\partial y_i} g_1(y_1, y_2)$ and $g_{2i}(y_1, y_2) = \frac{\partial}{\partial y_i} g_2(y_1, y_2)$, $i = 1, 2$ exist and are continuous for $(y_1, y_2) \in T$. Finally, suppose that the Jacobian:

$$J = \begin{vmatrix} g_{11}(y_1, y_2) & g_{12}(y_1, y_2) \\ g_{21}(y_1, y_2) & g_{22}(y_1, y_2) \end{vmatrix} \text{is } \neq 0 \text{ on } T.$$

Then the joint p.d.f. of the r.v.'s $Y_1 = h_1(X_1, X_2)$ and $Y_2 = h_2(X_1, X_2)$, f_{Y_1,Y_2}, is given by:

$$f_{Y_1,Y_2}(y_1, y_2) = f_{X_1,X_2}[g_1(y_1, y_2), g_2(y_1, y_2)]|J|, \quad (y_1, y_2) \in T \tag{11.9}$$

(and = 0 for $(y_1, y_2) \notin T$).

The justification of this theorem is entirely analogous to that of Theorem 2 and will be omitted. In any case, its proof is a matter of changing variables in a double integral, which is purely a calculus matter.

In applying Theorem 3, one must be careful in checking that the underlying assumptions hold, and in determining correctly the set T. As an illustration, let us discuss the first part of Example 3.

Example 3. (continued).

Discussion. We have $y_1 = x_1 + x_2$ and let $y_2 = x_2$. Then $x_1 = y_1 - y_2$ and $x_2 = y_2$, so that $\frac{\partial x_1}{\partial y_1} = 1$, $\frac{\partial x_1}{\partial y_2} = -1$, $\frac{\partial x_2}{\partial y_1} = 0$, $\frac{\partial x_2}{\partial y_2} = 1$, and $J = \begin{vmatrix} 1 & -1 \\ 0 & 1 \end{vmatrix} = 1$. For the determination of S and T, see Figures 11.1 and 11.2.

Since $f_{X_1,X_2}(x_1, x_2) = \frac{1}{(\beta-\alpha)^2}$ for $(x_1, x_2) \in S$, we have $f_{Y_1,Y_2}(y_1, y_2) = \frac{1}{(\beta-\alpha)^2}$ for $(y_1, y_2) \in T$; that is, for $2\alpha < y_1 < 2\beta$, $\alpha < y_2 < \beta$, $\alpha < y_1 - y_2 < \beta$ (and = 0 for $(y_1, y_2) \notin T$).

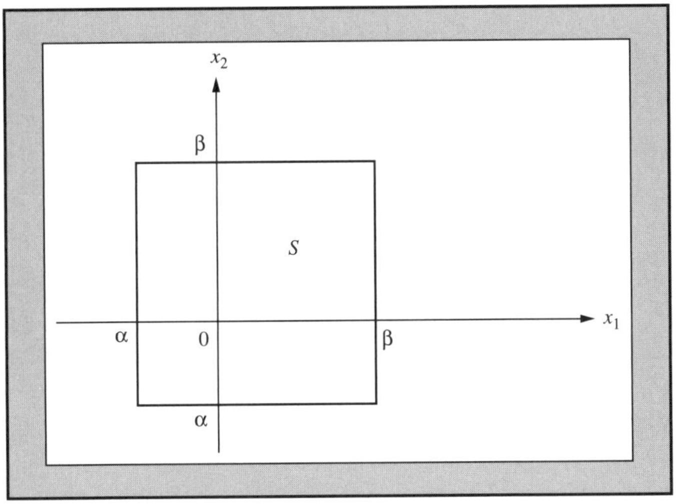

FIGURE 11.1

$S = \{(x_1, x_2) \in \Re^2;\ f_{X_1,X_2}(x_1, x_2) > 0\}$.

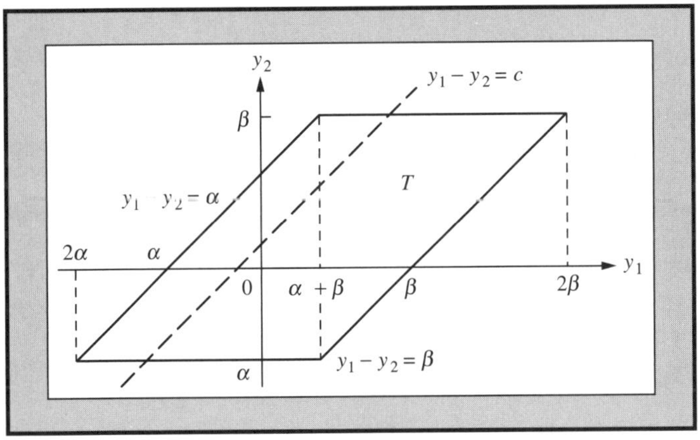

FIGURE 11.2

T = Image of S under the transformation used.

Thus, we get:

$$f_{Y_1,Y_2}(y_1, y_2) = \begin{cases} \frac{1}{(\beta-\alpha)^2}, & 2\alpha < y_1 < 2\beta, \alpha < y_2 < \beta, \alpha < y_1 - y_2 < \beta \\ 0, & \text{otherwise.} \end{cases}$$

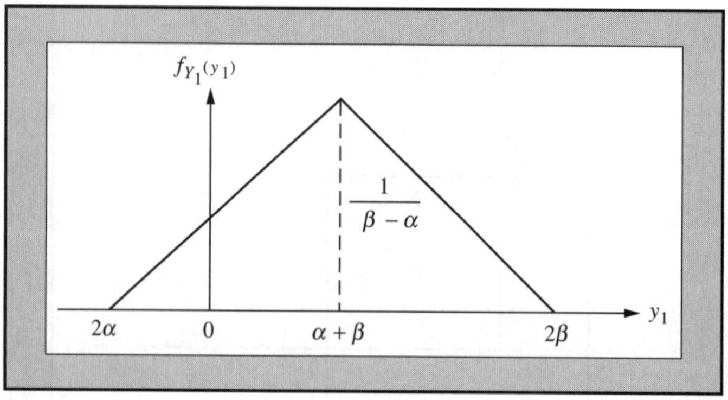

FIGURE 11.3
This density is known as a triangular p.d.f.

Therefore:

$$f_{Y_1}(y_1) = \begin{cases} \frac{1}{(\beta-\alpha)^2}\int_{\alpha}^{y_1-\alpha} dy_2 = \frac{y_1-2\alpha}{(\beta-\alpha)^2}, & \text{for } 2\alpha < y_1 \le \alpha+\beta \\ \frac{1}{(\beta-\alpha)^2}\int_{y_1-\beta}^{\beta} dy_2 = \frac{2\beta-y_1}{(\beta-\alpha)^2}, & \text{for } \alpha+\beta < y_1 \le 2\beta \\ 0, & \text{otherwise.} \end{cases}$$

The graph of f_{Y_1} is given in Figure 11.3.

Example 4. (continued).

Discussion. Here $y_1 = x_2 - 2x_1 = -2x_1 + x_2$ and let $y_2 = x_2$. Then $x_1 = -\frac{1}{2}y_1 + \frac{1}{2}y_2$ and $x_2 = y_2$, so that $J = \begin{vmatrix} -\frac{1}{2} & \frac{1}{2} \\ 0 & 1 \end{vmatrix} = -\frac{1}{2}$ and $|J| = \frac{1}{2}$. Clearly, S is the first quadrant. As for T, we have $y_2 = x_2$, so that $y_2 > 0$. Also, $-\frac{1}{2}y_1 + \frac{1}{2}y_2 = x_1$, so that $-\frac{1}{2}y_1 + \frac{1}{2}y_2 > 0$ or $-y_1 + y_2 > 0$ or $y_2 > y_1$. The conditions $y_2 > 0$ and $y_2 > y_1$ determine T (see Figure 11.4).

Since $f_{X_1,X_2}(x_1,x_2) = \lambda^2 e^{-\lambda(x_1+x_2)}$ $(x_1, x_2 > 0)$, we have

$$f_{Y_1,Y_2}(y_1,y_2) = \frac{\lambda^2}{2} e^{\frac{\lambda}{2}y_1 - \frac{3\lambda}{2}y_2}, \quad (y_1,y_2) \in T$$

(and = 0 otherwise). Therefore $f_{Y_1}(y_1)$ is taken by integrating out y_2. More precisely, for $y_1 < 0$:

$$f_{Y_1}(y_1) = \frac{\lambda^2}{2} e^{\frac{\lambda}{2}y_1} \int_0^{\infty} e^{-\frac{3\lambda}{2}y_2}\, dy_2 = -\frac{\lambda^2}{2} \times \frac{2}{3\lambda} e^{\frac{\lambda}{2}y_1} \times e^{-\frac{3\lambda}{2}y_2}\Big|_0^{\infty}$$
$$= -\frac{\lambda}{3} e^{\frac{\lambda}{2}y_1}(0-1) = \frac{\lambda}{3} e^{\frac{\lambda}{2}y_1},$$

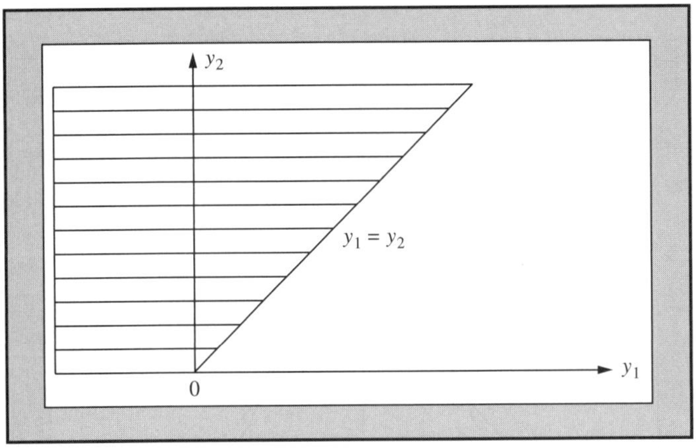

FIGURE 11.4
T is the part of the plane above the y_1-axis and above the main diagonal $y_1 = y_2$.

whereas for $y_1 \geq 0$:

$$f_{Y_1}(y_1) = \frac{\lambda^2}{2} e^{\frac{\lambda}{2} y_1} \int_{y_1}^{\infty} e^{-\frac{3\lambda}{2} y_2}\, dy_2 = -\frac{\lambda}{3} e^{\frac{\lambda}{2} y_1} \times e^{-\frac{3\lambda}{2} y_2} \Big|_{y_1}^{\infty}$$
$$= -\frac{\lambda}{3} e^{\frac{\lambda}{2} y_1} (0 - e^{-\frac{3\lambda}{2} y_1}) = \frac{\lambda}{3} e^{-\lambda y_1}.$$

To summarize:

$$f_{Y_1}(y_1) = \begin{cases} \frac{\lambda}{3} e^{\frac{\lambda}{2} y_1}, & y_1 < 0 \\ \frac{\lambda}{3} e^{-\lambda y_1}, & y_1 \geq 0. \end{cases}$$

Therefore $P(X_2 > 2X_1) = P(X_2 - 2X_1 > 0) = P(Y_1 > 0) = \frac{\lambda}{3} \int_0^{\infty} e^{-\lambda y_1}\, dy_1 = \frac{1}{3} \simeq 0.333$.

Remark 4. To be sure, the preceding probability is also calculated as follows:

$$P(X_2 > 2X_1) = \iint_{(x_2 > 2x_1)} \lambda^2 e^{-\lambda x_1 - \lambda x_2}\, dx_1\, dx_2$$
$$= \int_0^{\infty} \lambda e^{-\lambda x_2} \left(\int_0^{x_2/2} \lambda e^{-\lambda x_1}\, dx_1 \right) dx_2$$
$$= \int_0^{\infty} \lambda e^{-\lambda x_2} (1 - e^{-\frac{\lambda}{2} x_2}) dx_2$$
$$= \int_0^{\infty} \lambda e^{-\lambda x_2}\, dx_2 - \frac{2}{3} \int_0^{\infty} \frac{3\lambda}{2} e^{-\frac{3\lambda}{2} x_2}\, dx_2 = 1 - \frac{2}{3} = \frac{1}{3}.$$

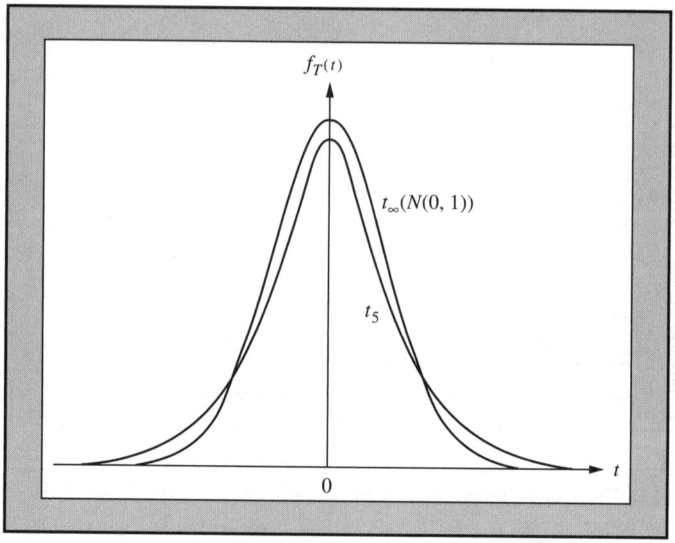

FIGURE 11.5

Two curves of the t probablity density function.

Applications of Theorem 3 lead to two new distributions, which are of great importance in statistics. They are the t-distribution and the F-distribution. The purely probability oriented reader may choose to omit the derivations of their p.d.f.'s.

Definition 1. Let X and Y be two independent r.v.'s distributed as follows: $X \sim N(0, 1)$ and $Y \sim \chi_r^2$, and define the r.v. T by: $T = X/\sqrt{Y/r}$. The r.v. T is said to have the (Student's) *t-distribution* with *r degrees of freedom* (d.f.). The notation used is: $T \sim t_r$. (The term "r degrees of freedom" used is inherited from the χ_r^2 distribution employed.)

The p.d.f. of T, f_T, is given by the formula:

$$f_T(t) = \frac{\Gamma\left[\frac{1}{2}(r+1)\right]}{\sqrt{\pi r}\,\Gamma(r/2)} \times \frac{1}{[1 + (t^2/r)]^{(1/2)(r+1)}}, \quad t \in \Re, \tag{11.10}$$

and its graph (for $r = 5$) is presented in Figure 11.5.

From formula (11.10), it is immediate that f_T is symmetric about 0 and tends to 0 as $t \to \pm\infty$. It can also be seen (see Exercise 2.9) that $f_T(t)$ tends to the p.d.f. of the $N(0, 1)$ distribution as the number r of d.f. tends to ∞. This is depicted in Figure 11.5 by means of the curve denoted by t_∞. Also, it is seen (see Exercise 2.10) that $ET = 0$ for $r \geq 2$, and $Var(T) = \frac{r}{r-2}$ for $r \geq 3$. Finally, the probabilities $P(T \leq t)$, for selected values of t and r, are given by tables (the t-tables). For $r \geq 91$, one may use the tables for standard normal distribution.

DERIVATION OF THE P.D.F. OF T, f_T Regarding the derivation of f_T, we have:

$$f_X(x) = \frac{1}{\sqrt{2\pi}} e^{-(1/2)x^2}, \quad x \in \Re,$$

$$f_Y(y) = \begin{cases} \frac{1}{\Gamma(\frac{1}{2}r)2^{(1/2)r}} y^{(r/2)-1} e^{-y/2}, & y > 0 \\ 0, & y \leq 0. \end{cases}$$

Set $U = Y$ and consider the transformation

$$(h_1, h_2) : \begin{cases} t = \frac{x}{\sqrt{y/r}}; \\ u = y \end{cases} \quad \text{then} \begin{cases} x = \frac{1}{\sqrt{r}} t\sqrt{u} \\ y = u, \end{cases}$$

and

$$J = \begin{vmatrix} \frac{\sqrt{u}}{\sqrt{r}} & \frac{t}{2\sqrt{u}\sqrt{r}} \\ 0 & 1 \end{vmatrix} = \frac{\sqrt{u}}{\sqrt{r}} = |J|.$$

Therefore, for $t \in \Re$, $u > 0$, we get:

$$\begin{aligned} f_{T,U}(t,u) &= \frac{1}{\sqrt{2\pi}} e^{-t^2u/2r} \times \frac{1}{\Gamma(r/2)2^{r/2}} u^{(r/2)-1} e^{-u/2} \times \frac{\sqrt{u}}{\sqrt{r}} \\ &= \frac{1}{\sqrt{2\pi r}\,\Gamma(r/2)2^{r/2}} u^{(1/2)(r+1)-1} \exp\left[-\frac{u}{2}\left(1+\frac{t^2}{r}\right)\right]. \end{aligned}$$

Hence

$$f_T(t) = \int_0^\infty \frac{1}{\sqrt{2\pi r}\,\Gamma(r/2)2^{r/2}} u^{(1/2)(r+1)-1} \exp\left[-\frac{u}{2}\left(1+\frac{t^2}{r}\right)\right] du.$$

Set

$$\frac{u}{2}\left(1+\frac{u^2}{r}\right) = z, \quad \text{so that} \quad u = 2z\left(1+\frac{t^2}{r}\right)^{-1}, \quad du = 2\left(1+\frac{t^2}{r}\right)^{-1} dz,$$

and $z \in [0, \infty)$. Therefore we continue as follows:

$$\begin{aligned} f_T(t) &= \int_0^\infty \frac{1}{\sqrt{2\pi r}\,\Gamma(r/2)2^{r/2}} \left[\frac{2z}{1+(t^2/r)}\right]^{(1/2)(r+1)-1} e^{-z} \frac{2}{1+(t^2/r)} dz \\ &= \frac{1}{\sqrt{2\pi r}\,\Gamma(r/2)2^{r/2}} \frac{2^{(1/2)(r+1)}}{[1+(t^2/r)]^{(1/2)(r+1)}} \int_0^\infty z^{(1/2)(r+1)-1} e^{-z}\, dz \\ &= \frac{1}{\sqrt{\pi r}\,\Gamma(r/2)} \frac{1}{[1+(t^2/r)]^{(1/2)(r+1)}} \Gamma\left[\frac{1}{2}(r+1)\right], \end{aligned}$$

since $\frac{1}{\Gamma[\frac{1}{2}(r+1)]} z^{(1/2)(r+1)-1} e^{-z}$ $(z > 0)$ is the p.d.f. of gamma distribution with parameters $\alpha = \frac{r+1}{2}$ and $\beta = 1$; that is,

$$f_T(t) = \frac{\Gamma[\frac{1}{2}(r+1)]}{\sqrt{\pi r}\,\Gamma(r/2)} \times \frac{1}{[1+(t^2/r)]^{(1/2)(r+1)}}, \quad t \in \Re. \quad \blacktriangle$$

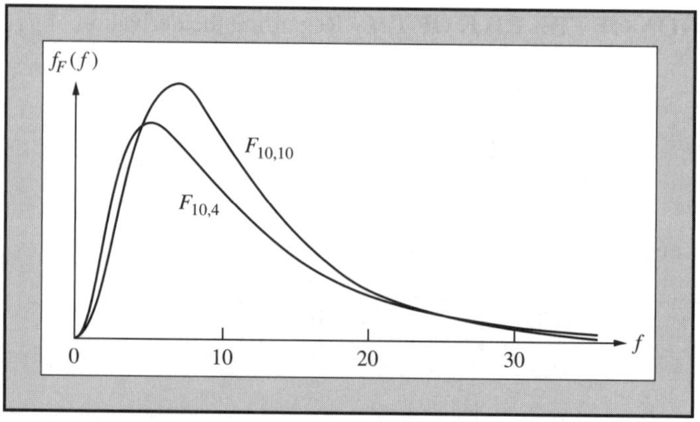

FIGURE 11.6

Two curves of the F probablity density function.

Now, we proceed with the definition of the F-distribution.

Definition 2. Let X and Y be two independent r.v.'s distributed as follows: $X \sim \chi^2_{r_1}$ and $Y \sim \chi^2_{r_2}$, and define the r.v. F by: $F = \frac{X/r_1}{Y/r_2}$. The r.v. F is said to have the *F-distribution* with r_1 and r_2 *degrees of freedom* (d.f.). The notation often used is: $F \sim F_{r_1,r_2}$. (The term "r_1 and r_2 degrees of freedom" used is inherited from the $\chi^2_{r_1}$ and $\chi^2_{r_2}$ distributions employed.)

The p.d.f. of F, f_F, is given by the formula:

$$f_F(f) = \begin{cases} \frac{\Gamma[\frac{1}{2}(r_1+r_2)](r_1/r_2)^{r_1/2}}{\Gamma(\frac{1}{2}r_1)\Gamma(\frac{1}{2}r_2)} \times \frac{f^{(r_1/2)-1}}{[1+(r_1/r_2)f]^{(1/2)(r_1+r_2)}}, & \text{for } f > 0 \\ 0, & \text{for } f \leq 0, \end{cases} \tag{11.11}$$

and its graphs (for $r_1 = 10$, $r_2 = 4$, and $r_1 = r_2 = 10$) are given in Figure 11.6. The probabilities $P(F \leq f)$, for selected values of f and r_1, r_2, are given by tables (the F-tables).

DERIVATION OF THE P.D.F. OF F, f_F The derivation of f_F is based on Theorem 3 and is as follows. For x and $y > 0$, we have:

$$f_X(x) = \frac{1}{\Gamma\left(\frac{1}{2}r_1\right)2^{r_1/2}} x^{(r_1/2)-1} e^{-x/2}, \quad x > 0,$$

$$f_Y(y) = \frac{1}{\Gamma\left(\frac{1}{2}r_2\right)2^{r_2/2}} y^{(r_2/2)-1} e^{-y/2}, \quad y > 0.$$

We set $Z = Y$, and consider the transformation

$$(h_1, h_2) : \begin{cases} f = \frac{x/r_1}{y/r_2} \\ z = y \end{cases} ; \quad \text{then} \begin{cases} x = \frac{r_1}{r_2} f z \\ y = z, \end{cases}$$

and

$$J = \begin{vmatrix} \frac{r_1}{r_2}z & \frac{r_1}{r_2}f \\ 0 & 1 \end{vmatrix} = \frac{r_1}{r_2}z = |J|.$$

For $f, z > 0$, we get:

$$\begin{aligned} f_{F,Z}(f,z) &= \frac{1}{\Gamma\left(\frac{1}{2}r_1\right)\Gamma\left(\frac{1}{2}r_2\right)2^{(1/2)(r_1+r_2)}}\left(\frac{r_1}{r_2}\right)^{(r_1/2)-1} f^{(r_1/2)-1}z^{(r_1/2)-1}z^{(r_2/2)-1} \\ &\quad \times \exp\left(-\frac{r_1}{2r_2}\right)fze^{-z/2}\frac{r_1}{r_2}z \\ &= \frac{(r_1/r_2)^{r_1/2}f^{(r_1/2)-1}}{\Gamma\left(\frac{1}{2}r_1\right)\Gamma\left(\frac{1}{2}r_2\right)2^{(1/2)(r_1+r_2)}}z^{(1/2)(r_1+r_2)-1}\exp\left[-\frac{z}{2}\left(\frac{r_1}{r_2}f+1\right)\right]. \end{aligned}$$

Therefore:

$$\begin{aligned} f_F(f) &= \int_0^\infty f_{F,Z}(f,z)dz \\ &= \frac{(r_1/r_2)^{r_1/2}f^{(r_1/2)-1}}{\Gamma\left(\frac{1}{2}r_1\right)\Gamma\left(\frac{1}{2}r_2\right)2^{(1/2)(r_1+r_2)}}\int_0^\infty z^{(1/2)(r_1+r_2)-1}\exp\left[-\frac{z}{2}\left(\frac{r_1}{r_2}f+1\right)\right]dz. \end{aligned}$$

Set:

$$\begin{aligned} &\frac{z}{2}\left(\frac{r_1}{r_2}f+1\right) = t, \quad \text{so that } z = 2t\left(\frac{r_1}{r_2}f+1\right)^{-1}, \\ &dz = 2\left(\frac{r_1}{r_2}f+1\right)^{-1}dt, \quad t \in [0,\infty). \end{aligned}$$

Thus continuing, we have

$$\begin{aligned} f_F(f) &= \frac{(r_1/r_2)^{r_1/2}f^{(r_1/2)-1}}{\Gamma\left(\frac{1}{2}r_1\right)\Gamma\left(\frac{1}{2}r_2\right)2^{(1/2)(r_1+r_2)}}2^{(1/2)(r_1+r_2)-1}\left(\frac{r_1}{r_2}f+1\right)^{-(1/2)(r_1+r_2)+1} \\ &\quad \times 2\left(\frac{r_1}{r_2}f+1\right)^{-1}\int_0^\infty t^{(1/2)(r_1+r_2)-1}e^{-t}\,dt \\ &= \frac{\Gamma[\frac{1}{2}(r_1+r_2)](r_1/r_2)^{r_1/2}}{\Gamma\left(\frac{1}{2}r_1\right)\Gamma\left(\frac{1}{2}r_2\right)} \times \frac{f^{(r_1/2)-1}}{[1+(r_1/r_2)f]^{(1/2)(r_1+r_2)}}, \end{aligned}$$

since $\frac{1}{\Gamma[\frac{1}{2}(r_1+r_2)]}t^{(1/2)(r_1+r_2)-1}e^{-t}$ $(t > 0)$ is the p.d.f. of gamma distribution with parameters $\alpha = \frac{r_1+r_2}{2}$ and $\beta = 1$. Therefore:

$$f_F(f) = \begin{cases} \frac{\Gamma[\frac{1}{2}(r_1+r_2)](r_1/r_2)^{r_1/2}}{\Gamma(\frac{1}{2}r_1)\Gamma(\frac{1}{2}r_2)} \times \frac{f^{(r_1/2)-1}}{[1+(r_1/r_2)f]^{(1/2)(r_1+r_2)}}, & \text{for } f > 0 \\ 0, & \text{for } f \leq 0. \end{cases} \quad \blacktriangle \tag{11.12}$$

Remark 5.

(i) From the definition of F-distribution, it follows that, if $F \sim F_{r_1,r_2}$, then $\frac{1}{F} \sim F_{r_2,r_1}$.

(ii) If $T \sim t_r$, then $T^2 \sim F_{1,r}$. Indeed, $T = X/\sqrt{Y/r}$, where X and Y are independent, and $X \sim N(0,1)$, $Y \sim \chi_r^2$. But then $T^2 = \frac{X^2}{Y/r} = \frac{X^2/1}{Y/r} \sim F_{1,r}$, since $X^2 \sim \chi_1^2$ and X^2 and Y are independent.

(iii) If $F \sim F_{r_1,r_2}$, then it can be shown (see Exercise 2.11) that

$$EF = \frac{r_2}{r_2 - 2}, \quad \text{for } r_2 \geq 3,$$

$$\text{and } Var(F) = \frac{2r_2^2(r_1 + r_2 - 2)}{r_1(r_2 - 2)^2(r_2 - 4)}, \quad \text{for } r_2 \geq 5.$$

Returning to Theorem 3, it is to be observed that one of the basic assumptions for its validity is that the transformations used are one-to-one. If this assumption is violated, then, under suitable conditions, a version of the theorem still holds true. This was exactly the case in connection with Theorems 2 and 2′. For the sake of completeness, here is a suitable version of Theorem 3.

Theorem 3′. Consider the r.v.'s X_1 and X_2 with joint p.d.f. f_{X_1,X_2} positive and continuous on the set $S \subseteq \Re^2$, let h_1, h_2 be two real-valued transformations defined on S, and let T be the image of S under the transformation (h_1, h_2). Suppose that (h_1, h_2) is *not* one-to-one from S onto T, but there is a partition of S into (pairwise disjoint) subsets $S_1, \ldots, S_r$ such that when (h_1, h_2) is restricted to S_j and takes values on T_j (the image of S_j under (h_1, h_2)), $j = 1, \ldots, r$, then (h_1, h_2) *is* one-to-one. Denoting by (h_{1j}, h_{2j}) this restriction, we have then: $(h_{1j}, h_{2j}) : S_j \to T_j$ *is* one-to-one, $j = 1, \ldots, r$. For $(x_1, x_2) \in S_j$, set $y_1 = h_{1j}(x_1, x_2)$, $y_2 = h_{2j}(x_1, x_2)$, so that $(y_1, y_2) \in T_j$, $j = 1, \ldots, r$. Then we can solve uniquely for $x_1, x_2 : x_1 = g_{j1}(y_1, y_2)$, $x_2 = g_{j2}(y_1, y_2)$, $j = 1, \ldots, r$. Suppose further that the partial derivatives $g_{j11}(y_1, y_2) = \frac{\partial}{\partial y_1} g_{j1}(y_1, y_2)$, $g_{j12}(y_1, y_2) = \frac{\partial}{\partial y_2} g_{j1}(y_1, y_2)$, $g_{j21}(y_1, y_2) = \frac{\partial}{\partial y_1} g_{j2}(y_1, y_2)$, $g_{j22}(y_1, y_2) = \frac{\partial}{\partial y_2} g_{j2}(y_1, y_2)$ exist, are continuous for $(y_1, y_2) \in T_j$, $j = 1, \ldots, r$, and the Jacobian

$$J_j = \begin{vmatrix} g_{j11}(y_1, y_2) & g_{j12}(y_1, y_2) \\ g_{j21}(y_1, y_2) & g_{j22}(y_1, y_2) \end{vmatrix}$$

is $\neq 0$ on T_j for $j = 1, \ldots, r$.

Set:

$$f_{Y_j}(y_1, y_2) = f_{X_1,X_2}[g_{j1}(y_1, y_2), g_{j2}(y_1, y_2)]|J_j|, \quad (y_1, y_2) \in T_j, \;\; j = 1, \ldots, r.$$

Then the joint p.d.f. of the r.v.'s $Y_1 = h_1(X_1, X_2)$, $Y_2 = h_2(X_1, X_2)$, f_{Y_1,Y_2}, is given by:

$$f_{Y_1,Y_2}(y_1, y_2) = \sum_{j=1}^{r} \delta_j(y_1, y_2) f_{Y_j}(y_1, y_2), \quad (y_1, y_2) \in T$$

(and = 0 for $(y_1, y_2) \notin T$), where $\delta_j(y_1, y_2) = 1$, if $(y_1, y_2) \in T_j$ and $\delta_j(y_1, y_2) = 0$, if $(y_1, y_2) \notin T_j$, $j = 1, \dots, r$.

One can formulate versions of Theorems 3, 3′ for $k(>2)$ r.v.'s $X_1, \dots, X_k$. In the following, such versions are formulated for reference purposes.

Theorem 4. Consider the r.v.'s $X_1, \dots, X_k$ with joint p.d.f. $f_{X_1,\dots,X_k}$ positive and continuous on the set $S \subseteq \Re^k$, and let $h_1, \dots, h_k$ be real-valued transformations defined on S; that is, $h_1, \dots, h_k : S \to \Re$, and let T be the image of S under the transformation $(h_1, \dots, h_k)$. Suppose that $(h_1, \dots, h_k)$ is one-to-one from S onto T. Thus, if we set $y_i = h_i(x_1, \dots, x_k), i = 1, \dots, k$, then we can solve uniquely for $x_i, i = 1, \dots, k : x_i = g_i(y_1, \dots, y_k), i = 1, \dots, k$. Suppose further that the partial derivatives $g_{ij}(y_1, \dots, y_k) = \frac{\partial}{\partial y_j} g_i(y_1, \dots, y_k), i, j = 1, \dots, k$ exist and are continuous for $(y_1, \dots, y_k) \in T$. Finally, suppose that the Jacobian

$$J = \begin{vmatrix} g_{11}(y_1, \dots, y_k) & \cdots & g_{1k}(y_1, \dots, y_k) \\ \dots\dots\dots & \cdots & \dots\dots\dots \\ g_{k1}(y_1, \dots, y_k) & \cdots & g_{kk}(y_1, \dots, y_k) \end{vmatrix}$$

is $\neq 0$ on T. Then the joint p.d.f. of the r.v.'s $Y_i = h_i(X_1, \dots, X_k), i = 1, \dots, k$, $f_{Y_1,\dots,Y_k}$, is given by:

$$f_{Y_1,\dots,Y_k}(y_1, \dots, y_k) = f_{X_1,\dots,X_k}[g_1(y_1, \dots, y_k), \dots, g_k(y_1, \dots, y_k)]|J|, \quad (11.13)$$
$$(y_1, \dots, y_k) \in T \quad (\text{and} = 0 \text{ for } (y_1, \dots, y_k) \notin T).$$

This theorem is employed in justifying Theorem 5 in the next section.

A suitable version of the previous result when the transformations $h_1, \dots, h_k$ are not one-to-one is stated below; it will be employed in Theorem 10 in Section 11.5.

Theorem 4′. Let $X_1, \dots, X_k$ be r.v.'s with joint p.d.f. $f_{X_1,\dots,X_k}$ positive and continuous on the set $S \subseteq \Re^k$, and let $h_1, \dots, h_k$ be real-valued transformations defined on S; that is, $h_1, \dots, h_k : S \to \Re$, and let T be the image of S under the transformation $(h_1, \dots, h_k)$. Suppose that $(h_1, \dots, h_k)$ is *not* one-to-one from S onto T but there is a partition of S into (pairwise disjoint) subsets $S_1, \dots, S_r$ such that when $(h_1, \dots, h_k)$ is restricted to S_j and takes values in T_j (the image of S_j under $(h_1, \dots, h_k)$), $j = 1, \dots, r$, then $(h_1, \dots, h_k)$ *is* one-to-one. Denoting by $(h_{1j}, \dots, h_{kj})$ this restriction, we have then: $(h_{1j}, \dots, h_{kj}) : S_j \to T_j$ *is* one-to-one, $j = 1, \dots, r$. For $(x_1, \dots, x_k) \in S_j$, set $y_1 = h_{1j}(x_1, \dots, x_k), \dots, y_k = h_{kj}(x_1, \dots, x_k)$, so that $(y_1, \dots, y_k) \in T_j$, $j = 1, \dots, r$. Then we can solve uniquely for $x_i, i = 1, \dots, k$: $x_i = g_{ji}(y_1, \dots, y_k), i = 1, \dots, k$, for each $j = 1, \dots, r$. Suppose further that the partial derivatives $g_{jil}(y_1, \dots, y_k) = \frac{\partial}{\partial y_l} g_{ji}(y_1, \dots, y_k), i, l = 1, \dots, k, j = 1, \dots, r$ exist, are continuous for $(y_1, \dots, y_k) \in T_j$, $j = 1, \dots, r$, and the Jacobian

$$J_j = \begin{vmatrix} g_{j11}(y_1, \dots, y_k) & \cdots & g_{j1k}(y_1, \dots, y_k) \\ \cdots & \cdots & \cdots \\ g_{jk1}(y_1, \dots, y_k) & \cdots & g_{jkk}(y_1, \dots, y_k) \end{vmatrix}$$

is $\neq 0$ on T_j for $j = 1, \dots, r$.

Set:

$$f_{Y_j}(y_1, \ldots, y_k) = f_{X_1,\ldots,X_k}[g_{j1}(y_1, \ldots, y_k), \ldots, g_{jk}(y_1, \ldots, y_k)]|J_j|,$$
$$(y_1, \ldots, y_k) \in T_j, \quad j = 1, \ldots, r.$$

Then the joint p.d.f. of the r.v.'s $Y_i = h_i(X_1, \ldots, X_k), i = 1, \ldots, k, f_{Y_1,\ldots,Y_k}$, is given by:

$$f_{Y_1,\ldots,Y_k}(y_1, \ldots, y_k) = \sum_{j=1}^{r} \delta_j(y_1, \ldots, y_k) f_{Y_j}(y_1, \ldots, y_k), (y_1, \ldots, y_k) \in T$$
$$(\text{and} = 0 \text{ for } (y_1, \ldots, y_k) \notin T), \tag{11.14}$$

where $\delta_j(y_1, \ldots, y_k) = 1$, if $(y_1, \ldots, y_k) \in T_j$ and $\delta_j(y_1, \ldots, y_k) = 0$, if $(y_1, \ldots, y_k) \notin T_j$, $j = 1, \ldots, r$.

This theorem is employed in justifying Theorem 10 in Section 11.5.

Exercises.

2.1 The r.v.'s X and Y denote the outcomes of one independent throw of two fair dice, and let $Z = X + Y$. Determine the distribution of Z.

2.2 Let the independent r.v.'s X and Y have negative exponential distribution with $\lambda = 1$, and set $U = X + Y$, $V = X/Y$.

- **(i)** Derive the joint p.d.f. $f_{U,V}$.
- **(ii)** Then derive the marginal p.d.f.'s f_U and f_V.
- **(iii)** Show that the r.v.'s U and V are independent.

2.3 Let the independent r.v.'s X and Y have negative exponential distribution with $\lambda = 1$, and set $U = \frac{1}{2}(X + Y)$, $V = \frac{1}{2}(X - Y)$.

- **(i)** Show that the joint p.d.f. of the r.v.'s U and V is given by:

$$f_{U,V}(u, v) = 2e^{-2u}, \quad -u < v < u, \quad u > 0.$$

- **(ii)** Also, show that the marginal p.d.f.'s f_U and f_V are given by:

$$f_U(u) = 4ue^{-2u}, \quad u > 0,$$
$$f_V(v) = \begin{cases} e^{-2v}, & \text{for } v > 0, \\ e^{2v}, & \text{for } v < 0. \end{cases}$$

2.4 Let the independent r.v.'s X and Y have the joint p.d.f. $f_{X,Y}$ positive and continuous on a set S, subset of $\Re^2$, and set $U = aX + b$, $V = cY + d$, where a, b, c, and d are constants with $ac \neq 0$.

(i) Use Theorem 3 in order to show that the joint p.d.f. of U and V is given by:

$$f_{U,V}(u,v) = \frac{1}{|ac|} f_{X,Y}\left(\frac{u-b}{a}, \frac{v-c}{d}\right)$$
$$= \frac{1}{|ac|} f_X\left(\frac{u-b}{c}\right) f_Y\left(\frac{v-c}{d}\right), \quad (u,v) \in T,$$

the image of S under the transformations $u = ax + b$, $v = cy + d$.

(ii) If $X \sim N(\mu_1, \sigma_1^2)$ and $Y \sim N(\mu_2, \sigma_2^2)$, show that $U \sim N(a\mu_1 + b, (a\sigma_1)^2)$, $V \sim N(c\mu_2 + d, (c\sigma_2)^2)$, and that U and V are independent.

2.5 If the independent r.v.'s X and Y are distributed as $N(0, 1)$, set $U = X + Y$, $V = X - Y$, and:

(i) Determine the p.d.f.'s of U and V.
(ii) Show that U and V are independent.
(iii) Compute the probability $P(U < 0, V > 0)$.

2.6 Let X and Y be independent r.v.'s distributed as $N(0, 1)$, and set:

$$U = \frac{1}{\sqrt{2}}(X + Y), \quad V = \frac{1}{\sqrt{2}}(X - Y).$$

(i) Determine the joint p.d.f. of U and V.
(ii) From the joint p.d.f. $f_{U,V}$, infer f_U and f_V without integration.
(iii) Conclude that U and V are also independent.
(iv) How else could you arrive at the p.d.f.'s f_U and f_V?

2.7 Let X and Y be independent r.v.'s distributed as $N(0, \sigma^2)$. Then show that the r.v. $V = X^2 + Y^2$ has negative exponential distribution with parameter $\lambda = 1/2\sigma^2$.

Hint. Use Proposition 6 and Theorem 5 in Chapter 10.

2.8 The independent r.v.'s X and Y have p.d.f. given by: $f_{X,Y}(x, y) = \frac{1}{\pi}$, for $x, y \in \Re$ with $x^2 + y^2 \leq 1$, and let $Z^2 = X^2 + Y^2$. Use polar coordinates to determine the p.d.f. f_{Z^2}.

Hint. Let $Z = +\sqrt{Z^2}$ and set $X = Z \cos \Theta$, $Y = Z \sin \Theta$, where $Z \geq 0$ and $0 < \Theta \leq 2\pi$. First, determine the joint p.d.f. $f_{Z,\Theta}$ and then the marginal p.d.f. f_Z. Finally, by means of f_Z and the transformation $U = Z^2$, determine the p.d.f. $f_U = f_{Z^2}$.

2.9 If the r.v. $X_r \sim t_r$, then the t-tables (at least the ones in this book) do not give probabilities for $r > 90$. For such values, we can use instead the normal tables. The reason for this is that the p.d.f. of X_r converges to the p.d.f. of the $N(0, 1)$

distribution as $r \to \infty$. More precisely,

$$f_{X_r}(t) = \frac{\Gamma\left(\frac{r+1}{2}\right)}{\sqrt{\pi r}\,\Gamma\left(\frac{r}{2}\right)} \times \frac{1}{\left(1+\frac{t^2}{r}\right)^{(r+1)/2}} \underset{r\to\infty}{\longrightarrow} \frac{1}{\sqrt{2\pi}} e^{-t^2/2} \ (t > 0).$$

Hint. In proving this convergence, first observe that

$$\left(1+\frac{t^2}{r}\right)^{(r+1)/2} = \left[\left(1+\frac{t^2}{r}\right)^r\right]^{1/2} \times \left(1+\frac{t^2}{r}\right)^{1/2} \underset{r\to\infty}{\longrightarrow} e^{t^2/2},$$

and then show that

$$\frac{\Gamma\left(\frac{r+1}{2}\right)}{\Gamma\left(\frac{r}{2}\right)} \underset{r\to\infty}{\longrightarrow} \frac{1}{\sqrt{2}},$$

by utilizing the *Stirling formula*. This formula states that:

$$\frac{\Gamma(n)}{\sqrt{2\pi}\, n^{(2n-1)/2} e^{-n}} \to 1 \quad \text{as } n \to \infty.$$

2.10 Let X_r be an r.v. distributed as t with r degrees of freedom: $X_r \sim t_r$ $(r = 1, 2, \ldots)$ whose p.d.f. is given in relation (11.10). Then show that:

(i) EX_r does not exist for $r = 1$.
(ii) $EX_r = 0$ for $r \geq 2$.
(iii) $Var(X_r) = \frac{r}{r-2}$ for $r \geq 3$.

Hint. That EX_r does not exist for $r = 1$ is actually reduced to Exercise 1.16 in Chapter 5. That $EX_r = 0$ for $r \geq 2$ follows by a simple integration. So, all that remains to calculate is EX_r^2. For this purpose, first reduce the original integral to an integral over the interval $(0, \infty)$, by symmetry of the region of integration and the fact that the integrand is an even function. Then, use the transformation $\frac{t^2}{r} = x$, and next the transformation $\frac{1}{1+x} = y$. Except for constants, the integral is then reduced to the form:

$$\int_0^1 y^{\alpha-1}(1-y)^{\beta-1} dy \quad (\alpha > 0, \beta > 0).$$

At this point, use the following fact:

$$\int_0^1 y^{\alpha-1}(1-y)^{\beta-1}\, dy = \frac{\Gamma(\alpha)\Gamma(\beta)}{\Gamma(\alpha+\beta)}.$$

(A proof of this fact may be found, e.g., in Section 3.3.6 (in Chapter 3), of the book *A Course in Mathematical Statistics*, 2nd edition (1997), Academic Press, by G. G. Roussas.) The proof is concluded by using the recursive relation of the gamma function ($\Gamma(\gamma) = (\gamma-1)\Gamma(\gamma-1)$), and the fact that $\Gamma(\frac{1}{2}) = \sqrt{\pi}$.

2.11 Let X_{r_1,r_2} be an r.v. having F-distribution with parameters r_1 and r_2; that is, $X_{r_1,r_2} \sim F_{r_1,r_2}$ with p.d.f. given by (11.12). Then show that:

$$EX_{r_1,r_2} = \frac{r_2}{r_2 - 2}, \quad r_2 \geq 3; \; Var(X_{r_1,r_2}) = \frac{2r_2^2(r_1 + r_2 - 2)}{r_1(r_2 - 2)^2(r_2 - 4)}, \quad r_2 \geq 5.$$

Hint. Start out with the kth moment $EX^k_{r_1,r_2}$, use first the transformation $\frac{r_1}{r_2} f = x$ and second the transformation $\frac{1}{1+x} = y$. Then observe that the integral is expressed in terms of the gamma function, as indicated in the Hint in Exercise 2.10. Thus, the $EX^k_{r_1,r_2}$ is expressed in terms of the gamma function without carrying out any integrations. Specifically, we find:

$$EX^k_{r_1,r_2} = \left(\frac{r_2}{r_1}\right)^k \frac{\Gamma\left(\frac{r_1+2k}{2}\right)\Gamma\left(\frac{r_2-2k}{2}\right)}{\Gamma\left(\frac{r_1}{2}\right)\Gamma\left(\frac{r_2}{2}\right)}, \quad r_2 > 2k.$$

Applying this formula for $k = 1$ (which requires that $r_2 \geq 3$) and $k = 2$ (which requires that $r_2 \geq 5$), and using the recursive property of the gamma function, we determine the required expressions.

11.3 Linear Transformations

In this section, a brief discussion is presented for a specific kind of transformations, linear transformations. The basic concepts and results used here can be found in any textbook on linear algebra.

Definition 3. Suppose the variables $x_1, \ldots, x_k$ are transformed into the variables $y_1, \ldots, y_k$ in the following manner:

$$y_i = \sum_{j=1}^{k} c_{ij}x_j, \quad i = 1, \ldots, k, \tag{11.15}$$

where the c_{ij}'s are real constants. Such a transformation is called a *linear transformation* (all the x_i's enter into the transformation in a linear way, in the first power).

Some terminology and elementary facts from matrix algebra will be used here. Denote by $\mathbf{C}$ the $k \times k$ matrix of the $c_{ij}, i, j = 1, \ldots, k$ constants; that is, $\mathbf{C} = (c_{ij})$, and by $|\mathbf{C}|$ or Δ its determinant. Then it is well known that if $\Delta \neq 0$, one can uniquely solve for x_i in (11.15):

$$x_i = \sum_{j=1}^{k} d_{ij}y_j, \quad i = 1, \ldots, k, \tag{11.16}$$

for suitable constants d_{ij}. Denote by $\mathbf{D}$ the $k \times k$ matrix of the d_{ij}'s and by Δ^* its determinant: $\mathbf{D} = (d_{ij})$, $\Delta^* = |\mathbf{D}|$. Then it is known that $\Delta^* = 1/\Delta$. Among the linear transformations, a specific class is of special importance; it is the class of orthogonal transformations.

A linear transformation is said to be *orthogonal*, if

$$\sum_{j=1}^{k} c_{ij}^2 = 1 \quad \text{and} \quad \sum_{j=1}^{k} c_{ij}c_{i'j} = 0, \quad i, i' = 1, \ldots, k, \ i \neq i', \tag{11.17}$$

or, equivalently (see also Exercise 3.9),

$$\sum_{i=1}^{k} c_{ij}^2 = 1 \quad \text{and} \quad \sum_{i=1}^{k} c_{ij}c_{ij'} = 0, \quad j, j' = 1, \ldots, k, \ j \neq j'. \tag{11.7$'$}$$

Relations (11.17) and (11.7′) simply state that the row (column) vectors of the matrix **C** have norm (length) 1, and any two of them are perpendicular. The matrix **C** itself is also called *orthogonal*. For an orthogonal matrix **C**, it is known that $|\mathbf{C}| = \pm 1$. Also, in the case of an orthogonal matrix **C**, it happens that $d_{ij} = c_{ji}$, $i, j = 1, \ldots, k$; or in matrix notation: $\mathbf{D} = \mathbf{C}'$, where $\mathbf{C}'$ is the *transpose* of **C** (the rows of $\mathbf{C}'$ are the same as the columns of **C**). Thus, in this case:

$$x_i = \sum_{j=1}^{k} c_{ji} y_j, \quad i = 1, \ldots, k. \tag{11.18}$$

Also, under orthogonality, the vectors of the x_i's and of the y_j's have the same norm. To put it differently:

$$\sum_{i=1}^{k} x_i^2 = \sum_{j=1}^{k} y_j^2. \tag{11.19}$$

Example 5. Here are two non-orthogonal matrices:

$$\mathbf{C}_1 = \begin{pmatrix} -1 & 3 & 11 \\ 1 & 5 & -5 \\ 2 & -1 & 8 \end{pmatrix}, \quad \mathbf{C}_2 = \begin{pmatrix} \frac{1}{2} & \frac{1}{2} \\ \frac{1}{2} & -\frac{1}{2} \end{pmatrix}.$$

Also, $|\mathbf{C}_2| = -\frac{1}{2}$. The transformations: $u = \frac{1}{2}x + \frac{1}{2}y$, $v = \frac{1}{2}x - \frac{1}{2}y$, are invertible and give: $x = u + v$, $y = u - v$. The relevant matrix is $\begin{pmatrix} 1 & 1 \\ 1 & -1 \end{pmatrix}$ with determinant $\begin{vmatrix} 1 & 1 \\ 1 & -1 \end{vmatrix} = -2 = 1/|\mathbf{C}_2|$, as it should be. However, $u^2 + v^2 = \frac{1}{2}(x^2 + y^2) \neq x^2 + y^2$.

Example 6. The following matrices are both orthogonal.

$$\mathbf{C}_1 = \begin{pmatrix} -\frac{\sqrt{3}}{3} & \frac{\sqrt{3}}{3} & \frac{\sqrt{3}}{3} \\ \frac{\sqrt{6}}{6} & -\frac{\sqrt{6}}{6} & \frac{2\sqrt{6}}{6} \\ \frac{\sqrt{2}}{2} & \frac{\sqrt{2}}{2} & 0 \end{pmatrix}, \quad \mathbf{C}_2 = \begin{pmatrix} \frac{\sqrt{2}}{2} & \frac{\sqrt{2}}{2} \\ \frac{\sqrt{2}}{2} & -\frac{\sqrt{2}}{2} \end{pmatrix}.$$

Also,

$$\mathbf{C}_1' = \begin{pmatrix} -\frac{\sqrt{3}}{3} & \frac{\sqrt{6}}{6} & \frac{\sqrt{2}}{2} \\ \frac{\sqrt{3}}{3} & -\frac{\sqrt{6}}{6} & \frac{\sqrt{2}}{2} \\ \frac{\sqrt{3}}{3} & \frac{2\sqrt{6}}{6} & 0 \end{pmatrix}, \quad \mathbf{C}_2' = \begin{pmatrix} \frac{\sqrt{2}}{2} & \frac{\sqrt{2}}{2} \\ \frac{\sqrt{2}}{2} & -\frac{\sqrt{2}}{2} \end{pmatrix},$$

and $|\mathbf{C}_1| = |\mathbf{C}_1'| = 1, |\mathbf{C}_2| = |\mathbf{C}_2'| = -1$. The transformations: $u = -\frac{\sqrt{3}}{3}x + \frac{\sqrt{3}}{3}y + \frac{\sqrt{3}}{3}z$, $v = \frac{\sqrt{6}}{6}x - \frac{\sqrt{6}}{6}y + \frac{2\sqrt{6}}{6}z$, $w = \frac{\sqrt{2}}{2}x + \frac{\sqrt{2}}{2}y$ are invertible and give: $x = -\frac{\sqrt{3}}{3}u + \frac{\sqrt{6}}{6}v + \frac{\sqrt{2}}{2}w$, $y = \frac{\sqrt{3}}{3}u - \frac{\sqrt{6}}{6}v + \frac{\sqrt{2}}{2}w$, $z = \frac{\sqrt{3}}{3}u + \frac{2\sqrt{6}}{6}v$. Furthermore, $x^2 + y^2 + z^2 = u^2 + v^2 + w^2$, as it should be. Also, the transformations: $u = \frac{\sqrt{2}}{2}x + \frac{\sqrt{2}}{2}y$, $v = \frac{\sqrt{2}}{2}x - \frac{\sqrt{2}}{2}y$ are invertible and give: $x = \frac{\sqrt{2}}{2}u + \frac{\sqrt{2}}{2}v$, $y = \frac{\sqrt{2}}{2}u - \frac{\sqrt{2}}{2}v$, and $u^2 + v^2 = x^2 + y^2$, as expected.

Applying linear transformations in transforming a set of k r.v.'s into another set of k r.v.'s, we have the following theorem.

Theorem 5. Suppose the r.v.'s $X_1, \ldots, X_k$ are transformed into the r.v.'s $Y_1, \ldots, Y_k$ through a linear transformation with the matrix $\mathbf{C} = (c_{ij})$ and $|\mathbf{C}| = \Delta \neq 0$. Let $S \subseteq \Re^k$ be the set over which the joint p.d.f. of $X_1, \ldots, X_k$, $f_{X_1,\ldots,X_k}$, is positive and continuous, and let T be the image of S under the linear transformation. Then:

(i) The joint p.d.f. of $Y_1, \ldots, Y_k$, $f_{Y_1,\ldots,Y_k}$, is given by:

$$f_{Y_1,\ldots,Y_k}(y_1, \ldots, y_k) = f_{X_1,\ldots,X_k}\left(\sum_{j=1}^{k} d_{1j}y_j, \ldots, \sum_{j=1}^{k} d_{kj}y_j\right)\frac{1}{|\Delta|}, \tag{11.20}$$

for $(y_1, \ldots, y_k) \in T$ (and = 0 otherwise), where the d_{ij}'s are as in (11.16).

(ii) In particular, if $\mathbf{C}$ is *orthogonal*, then:

$$f_{Y_1,\ldots,Y_k}(y_1, \ldots, y_k) = f_{X_1,\ldots,X_k}\left(\sum_{j=1}^{k} c_{j1}y_j, \ldots, \sum_{j=1}^{k} c_{jk}y_j\right), \tag{11.21}$$

for $(y_1, \ldots, y_k) \in T$ (and = 0 otherwise); also,

$$\sum_{j=1}^{k} Y_j^2 = \sum_{i=1}^{k} X_i^2; \tag{11.22}$$

Proof.

(i) Relation (11.20) follows from Theorem 4.

(ii) Relation (11.21) follows from (11.20) and orthogonality of the transformation; and (11.22) is a restatement of (11.19). ▲

Next, we specialize this result to the case that the r.v.'s $X_1, \ldots, X_k$ are normally distributed and independent.

Theorem 6. Let the independent r.v.'s $X_1, \ldots, X_k$ be distributed as follows: $X_i \sim N(\mu_i, \sigma^2)$, $i = 1, \ldots, k$, and suppose they are transformed into the r.v.'s $Y_1, \ldots, Y_k$ by means of an *orthogonal* transformation **C**. Then the r.v.'s $Y_1, \ldots, Y_k$ are also independent and normally distributed as follows:

$$Y_i \sim N\left(\sum_{j=1}^{k} c_{ij}\mu_j, \sigma^2\right), \quad i = 1, \ldots, k; \tag{11.23}$$

also,

$$\sum_{j=1}^{k} Y_j^2 = \sum_{i=1}^{k} X_i^2. \tag{11.24}$$

Proof. From the transformations $Y_i = \sum_{j=1}^{k} c_{ij}X_j$ (and independence), it is immediate that each Y_i is normally distributed with mean $EY_i = \sum_{j=1}^{k} c_{ij}\mu_j$ and variance $Var(Y_i) = \sum_{j=1}^{k} c_{ij}^2\sigma^2 = \sigma^2 \sum_{j=1}^{k} c_{ij}^2 = \sigma^2$. So the only thing to be justified is the assertion of independence. From the normality assumption on the X_i's, we have:

$$f_{X_1,\ldots,X_k}(x_1, \ldots, x_k) = \left(\frac{1}{\sqrt{2\pi}\sigma}\right)^k \exp\left[-\frac{1}{2\sigma^2}\sum_{i=1}^{k}(x_i - \mu_i)^2\right]. \tag{11.25}$$

Then, since **C** is orthogonal, (11.21) applies and gives, by means of (11.25):

$$f_{Y_1,\ldots,Y_k}(y_1, \ldots, y_k) = \left(\frac{1}{\sqrt{2\pi}\sigma}\right)^k \exp\left[-\frac{1}{2\sigma^2}\sum_{i=1}^{k}\left(\sum_{j=1}^{k} c_{ji}y_j - \mu_i\right)^2\right]. \tag{11.26}$$

Thus, the proof is completed by establishing the following algebraic relation:

$$\sum_{i=1}^{k}\left(\sum_{j=1}^{k} c_{ji}y_j - \mu_i\right)^2 = \sum_{i=1}^{k}\left(y_i - \sum_{j=1}^{k} c_{ij}\mu_j\right)^2 \tag{11.27}$$

(see Exercise 3.1). Relation (11.24) is a restatement of relation (11.19). ▲

Example 7. As an application of Theorem 6, refer to Example 6, and first transform the independent and $N(0, 1)$ distributed r.v.'s X, Y, Z into the r.v.'s U, V, W by means of the (orthogonal) transformation $\mathbf{C}_1$.

Discussion. From

$$f_{X,Y,Z}(x, y, z) = \left(\frac{1}{\sqrt{2\pi}}\right)^3 \exp(x^2 + y^2 + z^2),$$

and the fact that $x^2 + y^2 + z^2 = u^2 + v^2 + w^2$, we get:

$$f_{U,V,W}(u, v, w) = \left(\frac{1}{\sqrt{2\pi}}\right)^3 \exp(u^2 + v^2 + w^2).$$

It follows that U, V, W are independent, distributed as $N(0, 1)$, and, of course, $U^2 + V^2 + W^2 = X^2 + Y^2 + Z^2$.

Next, let the independent r.v.'s X and Y be distributed as follows: $X \sim N(\mu_1, \sigma^2)$, $Y \sim N(\mu_2, \sigma^2)$, so that:

$$f_{X,Y}(x, y) = \left(\frac{1}{\sqrt{2\pi}\sigma}\right)^2 \exp\left\{-\frac{1}{2\sigma^2}\left[(x-\mu_1)^2 + (y-\mu_2)^2\right]\right\}.$$

Transform X, Y into the r.v.'s U, V by means of the (orthogonal) transformation $\mathbf{C}_2$ in Example 6, so that:

$$f_{U,V}(u, v) = \left(\frac{1}{\sqrt{2\pi}\sigma}\right)^2 \exp\left\{-\frac{1}{2\sigma^2}\left[\left(\frac{\sqrt{2}}{2}u + \frac{\sqrt{2}}{2}v - \mu_1\right)^2 + \left(\frac{\sqrt{2}}{2}u - \frac{\sqrt{2}}{2}v - \mu_2\right)^2\right]\right\}.$$

However,

$$\left(\frac{\sqrt{2}}{2}u + \frac{\sqrt{2}}{2}v - \mu_1\right)^2 + \left(\frac{\sqrt{2}}{2}u - \frac{\sqrt{2}}{2}v - \mu_2\right)^2 = u^2 + v^2 - \sqrt{2}(u+v)\mu_1 - \sqrt{2}(u-v)\mu_2 + \mu_1^2 + \mu_2^2$$

$$= \left\{u^2 - 2 \times \frac{\sqrt{2}}{2}u(\mu_1+\mu_2) + \left[\frac{\sqrt{2}}{2}(\mu_1+\mu_2)\right]^2\right\} + \left\{v^2 - 2 \times \frac{\sqrt{2}}{2}v(\mu_1-\mu_2) + \left[\frac{\sqrt{2}}{2}(\mu_1-\mu_2)\right]^2\right\}$$

$$= \left[u - \left(\frac{\sqrt{2}}{2}\mu_1 + \frac{\sqrt{2}}{2}\mu_2\right)\right]^2 + \left[v - \left(\frac{\sqrt{2}}{2}\mu_1 - \frac{\sqrt{2}}{2}\mu_2\right)\right]^2,$$

since, $\left[\frac{\sqrt{2}}{2}(\mu_1+\mu_2)\right]^2 + \left[\frac{\sqrt{2}}{2}(\mu_1-\mu_2)\right]^2 = \mu_1^2 + \mu_2^2$.

It follows that $U \sim N\left(\frac{\sqrt{2}}{2}\mu_1 + \frac{\sqrt{2}}{2}\mu_2, \sigma^2\right)$, $V \sim N\left(\frac{\sqrt{2}}{2}\mu_1 - \frac{\sqrt{2}}{2}\mu_2, \sigma^2\right)$, and they are independent, as Theorem 6 dictates.

Finally, suppose the orthogonal matrix $\mathbf{C}$ in Theorem 6 is chosen to be as follows:

$$\mathbf{C} = \begin{pmatrix} 1/\sqrt{k} & 1/\sqrt{k} & \cdots\cdots & \cdots & \cdots\cdots & 1/\sqrt{k} \\ 1/\sqrt{2\times 1} & -1/\sqrt{2\times 1} & 0 & \cdots & \cdots\cdots & 0 \\ 1/\sqrt{3\times 2} & 1/\sqrt{3\times 2} & -2/\sqrt{3\times 2} & 0 & \cdots\cdots & 0 \\ \cdots & \cdots & \cdots\cdots & \cdots & \cdots\cdots & \cdots \\ 1/\sqrt{k(k-1)} & 1/\sqrt{k(k-1)} & \cdots\cdots & \cdots & 1/\sqrt{k(k-1)} & -(k-1)/\sqrt{k(k-1)} \end{pmatrix}.$$

That is, the elements of **C** are given by the expressions:

$$
\begin{aligned}
c_{1j} &= 1/\sqrt{k}, \quad j = 1, \ldots, k, \\
c_{ij} &= 1/\sqrt{i(i-1)}, \quad \text{for } i = 2, \ldots, k \ \text{ and } \ j = 1, \ldots, i-1, \\
&\qquad \text{and } 0 \text{ for } j = i+1, \ldots, k, \\
c_{ii} &= -(i-1)/\sqrt{i(i-1)}, \quad i = 2, \ldots, k.
\end{aligned}
$$

From these expressions, it readily follows that $\sum_{j=1}^{k} c_{ij}^2 = 1$ for all $i = 1, \ldots, k$, and $\sum_{j=1}^{k} c_{ij}c_{i'j} = 0$ for all $i, i' = 1, \ldots, k$, with $i \neq i'$, so that **C** is, indeed, orthogonal (see also Exercise 3.2). By means of **C** and Theorem 6, we may now establish the following result.

Lemma 1. If $Z_1, \ldots, Z_k$ are independent r.v.'s distributed as $N(0, 1)$, then $\bar{Z}$ and $\sum_{i=1}^{k}(Z_i - \bar{Z})^2$ are independent, where $\bar{Z} = k^{-1}\sum_{i=1}^{k} Z_i$.

Proof. Transform $Z_1, \ldots, Z_k$ into the r.v.'s $Y_1, \ldots, Y_k$ by means of **C**; that is,

$$
\begin{aligned}
Y_1 &= \frac{1}{\sqrt{k}}Z_1 + \frac{1}{\sqrt{k}}Z_2 + \cdots + \frac{1}{\sqrt{k}}Z_k \\
Y_2 &= \frac{1}{\sqrt{2\times 1}}Z_1 - \frac{1}{\sqrt{2\times 1}}Z_2 \\
Y_3 &= \frac{1}{\sqrt{3\times 2}}Z_1 + \frac{1}{\sqrt{3\times 2}}Z_2 - \frac{2}{\sqrt{3\times 2}}Z_3 \\
&\vdots \\
Y_k &= \frac{1}{\sqrt{k(k-1)}}Z_1 + \frac{1}{\sqrt{k(k-1)}}Z_2 + \cdots + \frac{1}{\sqrt{k(k-1)}}Z_{k-1} - \frac{k-1}{\sqrt{k(k-1)}}Z_k.
\end{aligned}
$$

Then, by Theorem 6, the r.v.'s $Y_1, \ldots, Y_k$ are independently distributed as $N(0, 1)$, whereas by (11.24):

$$\sum_{j=1}^{k} Y_j^2 = \sum_{i=1}^{k} Z_i^2.$$

However, $Y_1 = \sqrt{k}\bar{Z}$, so that:

$$\sum_{j=2}^{k} Y_j^2 = \sum_{j=1}^{k} Y_j^2 - Y_1^2 = \sum_{i=1}^{k} Z_i^2 - (\sqrt{k}\bar{Z})^2 = \sum_{i=1}^{k} Z_i^2 - k\bar{Z}^2 = \sum_{i=1}^{k}(Z_i - \bar{Z})^2.$$

On the other hand, $\sum_{j=2}^{k} Y_j^2$ and Y_1 are independent; equivalently, $\sum_{i=1}^{k}(Z_i - \bar{Z})^2$ and $k\bar{Z}$ are independent, or:

$$\bar{Z} \ \text{ and } \ \sum_{i=1}^{k}(Z_i - \bar{Z})^2 \ \text{ are independent.} \quad \blacktriangle \qquad (11.28)$$

This last conclusion is now applied as follows.

Theorem 7. Let $X_1, \ldots, X_k$ be independent r.v.'s distributed as $N(\mu, \sigma^2)$. Then the sample mean $\bar{X} = \frac{1}{k}\sum_{i=1}^{k} X_i$ and the sample variance $S^2 = \frac{1}{k}\sum_{i=1}^{k}(X_i - \bar{X})^2$ are independent.

Proof. The assumption that $X_i \sim N(\mu, \sigma^2)$ implies that $\frac{X_i-\mu}{\sigma} \sim N(0, 1)$. By setting $Z_i = (X_i - \mu)/\sigma, i = 1, \ldots, k$, the Z_i's are as in Lemma 1 and therefore (11.28) applies. Since

$$\bar{Z} = \frac{1}{k}\sum_{i=1}^{k}\left(\frac{X_i - \mu}{\sigma}\right) = \frac{1}{\sigma}(\bar{X} - \mu), \quad \text{and}$$

$$\sum_{i=1}^{k}(Z_i - \bar{Z})^2 = \sum_{i=1}^{k}\left(\frac{X_i - \mu}{\sigma} - \frac{\bar{X} - \mu}{\sigma}\right)^2 = \frac{1}{\sigma^2}\sum_{i=1}^{k}(X_i - \bar{X})^2,$$

it follows that $\frac{1}{\sigma}(\bar{X} - \mu)$ and $\frac{1}{\sigma^2}\sum_{i=1}^{k}(X_i - \bar{X})^2$ are independent, and likewise that $\bar{X}$ and $\frac{1}{k}\sum_{i=1}^{k}(X_i - \bar{X})^2$ are independent. ▲

Exercises.

3.1 Establish relation (11.27) in the proof of Theorem 6.

Hint. Expand the left-hand side and the right-hand side in (11.27), use orthogonality, and show that the common value of both sides is:

$$\sum_{j=1}^{k} y_j^2 + \sum_{j=1}^{k} \mu_j^2 - 2\sum_{j=1}^{k}\sum_{i=1}^{k} c_{ji}\mu_i\mu_j.$$

3.2 Show that the matrix with row elements given by:

$$\begin{aligned} c_{1j} &= 1/\sqrt{k}, \quad j = 1, \ldots, k, \\ c_{ij} &= 1/\sqrt{i(i-1)}, \quad i = 2, \ldots, k \quad \text{and} \quad j = 1, \ldots, i-1, \\ &\quad \text{and } 0 \text{ for } j = i+1, \ldots, k, \\ c_{ii} &= -(i-1)/\sqrt{i(i+1)}, \quad i = 2, \ldots, k \end{aligned}$$

is orthogonal.

3.3 Let X_1, X_2, X_3 be independent r.v.'s such that $X_i \sim N(\mu_i, \sigma^2)$, $i = 1, 2, 3$, and set:

$$\begin{aligned} Y_1 &= -\frac{1}{\sqrt{2}}X_1 + \frac{1}{\sqrt{2}}X_2, \\ Y_2 &= -\frac{1}{\sqrt{3}}X_1 - \frac{1}{\sqrt{3}}X_2 + \frac{1}{\sqrt{3}}X_3, \\ Y_3 &= \frac{1}{\sqrt{6}}X_1 + \frac{1}{\sqrt{6}}X_2 + \frac{2}{\sqrt{6}}X_3. \end{aligned}$$

Then:

(i) Show that the r.v.'s Y_1, Y_2, Y_3 are independent normally distributed with variance σ^2 and respective means:

$$EY_1 = \frac{1}{\sqrt{2}}(-\mu_1 + \mu_2), \quad EY_2 = \frac{1}{\sqrt{3}}(-\mu_1 - \mu_2 + \mu_3),$$

$$EY_3 = \frac{1}{\sqrt{6}}(\mu_1 + \mu_2 + 2\mu_3).$$

(ii) If $\mu_1 = \mu_2 = \mu_3 = 0$, then show that $\frac{1}{\sigma^2}(Y_1^2 + Y_2^2 + Y_3^2) \sim \chi_3^2$.

Hint. For part (i), prove that the transformation employed is orthogonal, and then use Theorem 6 to conclude independence of Y_1, Y_2, Y_3. That the means and the variance are as described follows either from Theorem 6 or directly. Part (ii) follows from part (i) and the assumption that $\mu_1 = \mu_2 = \mu_3 = 0$.

3.4 If the r.v.'s X and Y have bivariate normal distribution with parameters $\mu_1, \mu_2, \sigma_1^2, \sigma_2^2$, and ρ, then the r.v.'s $U = \frac{X-\mu_1}{\sigma_1}$, $V = \frac{Y-\mu_2}{\sigma_2}$ have bivariate normal distribution with parameters 0, 0, 1, 1, and ρ; and vice versa.

Hint. For the converse part, you just reverse the process.

3.5 If the r.v.'s X and Y have bivariate normal distribution with parameters 0, 0, 1, 1, and ρ, then the r.v.'s cX and dY have bivariate normal distribution with parameters 0, 0, c^2, d^2, and ρ_0, where $\rho_0 = \rho$ if $cd > 0$, and $\rho_0 = -\rho$ if $cd < 0$; c and d are constants with $cd \neq 0$.

3.6 Let the r.v.'s X and Y have bivariate normal distribution with parameters 0, 0, 1, 1, and ρ, and set: $U = X + Y$, $V = X - Y$. Then show that:

(i) The r.v.'s U and V also have bivariate normal distribution with parameters 0, 0, $2(1+\rho)$, $2(1-\rho)$, and 0.

(ii) From part (i), conclude that the r.v.'s U and V are independent.

(iii) From part (i), also conclude that: $U \sim N(0, 2(1+\rho))$, $V \sim N(0, 2(1-\rho))$.

3.7 Let the r.v.'s X and Y have bivariate normal distribution with parameters $\mu_1, \mu_2, \sigma_1^2, \sigma_2^2$, and ρ, and set:

$$U = \frac{X - \mu_1}{\sigma_1}, \quad V = \frac{Y - \mu_2}{\sigma_2}.$$

Then:

(i) Determine the joint distribution of the r.v.'s U and V. (See also Exercise 3.4.)

(ii) Show that $U + V$ and $U - V$ have bivariate normal distribution with parameters 0, 0, $2(1+\rho)$, $2(1-\rho)$, and 0, and are independent. Also, $U + V \sim N(0, 2(1+\rho))$, $U - V \sim N(0, 2(1-\rho))$.

(iii) For $\sigma_1^2 = \sigma_2^2 = \sigma^2$, say, conclude that the r.v.'s $X + Y$ and $X - Y$ are independent.

Remark: Actually the converse of part (iii) is also true; namely, if X and Y have bivariate normal distribution $N(\mu_1, \mu_2, \sigma_1^2, \sigma_2^2, \rho)$, then independence of $X+Y$ and $X-Y$ implies $\sigma_1^2 = \sigma_2^2$. The justification of this statement is easier by means of m.g.f.'s, and it was actually discussed in Exercise 1.25 of Chapter 10.

3.8 Let the independent r.v.'s $X_1, \ldots, X_n$ be distributed as $N(\mu, \sigma^2)$, and suppose that $\mu = k\sigma\,(k > 0)$. Set:

$$\bar{X} = \frac{1}{n}\sum_{i=1}^{n} X_i, \quad S^2 = \frac{1}{n-1}\sum_{i=1}^{n}(X_i - \bar{X})^2.$$

Then:

(i) Determine an expression for the probability:

$$P(a\mu < \bar{X} < b\mu, \quad 0 < S^2 < c\sigma^2),$$

where a, b, and c are constants, $a < b$ and $c > 0$.

(ii) Give the numerical value of the probability in part (i) if $a = \frac{1}{2}$, $b = \frac{3}{2}$, $c = 1.487$, $k = 1.5$, and $n = 16$.

Hint. Use independence of $\bar{X}$ and S^2 provided by Theorem 7. Also, use the fact that $\frac{(n-1)S^2}{\sigma^2} \sim \chi^2_{n-1}$ by Theorem 6 in Chapter 10 (where S^2 is denoted by $\bar{S}^2$).

3.9 Use matrix notation (and usual properties of matrices) in order to establish equivalence of relations (11.17) and (11.7′).

11.4 The Probability Integral Transform

In this short section, a very special type of transformation is considered, the so-called *probability integral transform*. By means of this transformation, two results are derived. Roughly speaking, these results state that if $X \sim F$ and $Y = F(X)$, then, somewhat surprisingly, Y is always distributed as $U(0, 1)$. Furthermore, for a given d.f. F, there is always an r.v. $X \sim F$; this r.v. is given by $X = F^{-1}(Y)$, where $Y \sim U(0, 1)$ and F^{-1} is the inverse function of F. To facilitate the derivations, F will be assumed to be (strictly) increasing.

Theorem 8. For a continuous and strictly increasing d.f. F, let $X \sim F$ and set $Y = F(X)$. Then $Y \sim U(0, 1)$.

Proof. Since $0 \leq F(X) \leq 1$, it suffices to consider $y \in [0, 1]$. Then

$$\begin{aligned} P(Y \leq y) &= P[F(X) \leq y] = P\{F^{-1}[F(X)] \leq F^{-1}(y)\} \\ &= P[X \leq F^{-1}(y)] = F[F^{-1}(y)] = y, \end{aligned}$$

so that $Y \sim U(0, 1)$. ▲

Theorem 9. Let F be a given continuous and strictly increasing d.f., and let the r.v. $Y \sim U(0, 1)$. Define the r.v. X by: $X = F^{-1}(Y)$. Then $X \sim F$.

Proof. For $x \in \Re$,

$$\begin{aligned}P(X \leq x) &= P[F^{-1}(Y) \leq x] = P\{F[F^{-1}(Y)] \leq F(x)\}\\ &= P[Y \leq F(x)] = F(x),\end{aligned}$$

as was to be seen. ▲

In the form of a verification of Theorems 8 and 9, consider the following simple examples.

Example 8. Let the r.v. X have negative exponential distribution with parameter λ. Then, for $x > 0$, $F(x) = 1 - e^{-\lambda x}$. Let Y be defined by: $Y = 1 - e^{-\lambda X}$. Then Y should be $\sim U(0, 1)$.

Discussion. Indeed, for $0 < y < 1$,

$$\begin{aligned}P(Y \leq y) &= P(1 - e^{-\lambda X} \leq y) = P(e^{-\lambda X} \geq 1 - y) = P[-\lambda X \geq \log(1 - y)]\\ &\qquad \text{(where, as always, log stands for the natural logarithm)}\\ &= P\left[X \leq -\frac{1}{\lambda}\log(1 - y)\right]\\ &= 1 - \exp\left\{(-\lambda)\left[-\frac{1}{\lambda}\log(1 - y)\right]\right\}\\ &= 1 - \exp[\log(1 - y)] = 1 - (1 - y) = y,\end{aligned}$$

as was to be seen.

Example 9. Let F be the d.f. of negative exponential distribution with parameter λ, so that $F(x) = 1 - e^{-\lambda x}$, $x > 0$. Let $y = 1 - e^{-\lambda x}$ and solve for x to obtain $x = -\frac{1}{\lambda}\log(1 - y)$, $0 < y < 1$. Let $Y \sim U(0, 1)$ and define the r.v. X by: $X = -\frac{1}{\lambda}\log(1 - Y)$. Then X should be $\sim F$.

Discussion. Indeed,

$$\begin{aligned}P(X \leq x) &= P\left[-\frac{1}{\lambda}\log(1 - Y) \leq x\right] = P[\log(1 - Y) \geq -\lambda x]\\ &= P(1 - Y \geq e^{-\lambda x}) = P(Y \leq 1 - e^{-\lambda x}) = 1 - e^{-\lambda x},\end{aligned}$$

as was to be seen.

Exercises.

4.1

(i) Let X be an r.v. with continuous and strictly increasing d.f. F, and define the r.v. Y by $Y = F(X)$. Then use Theorem 2 in order to show that $Z = -2\log(1 - Y) \sim \chi^2_2$.

(ii) If $X_1, \ldots, X_n$ is a random sample with d.f. F as described in part (i), and if $Y_i = F(X_i), i = 1, \ldots, n$, then show that the r.v. $U = \sum_{i=1}^{n} Z_i \sim \chi^2_{2n}$, where $Z_i = -2\log(1 - Y_i), i = 1, \ldots, n$.

Hint. For part (i), use Theorem 8, according to which $Y \sim U(0, 1)$. For part (ii), use part (i) and Theorem 5 in Chapter 10.

11.5 Order Statistics

In this section, an unconventional kind of transformation is considered, which, when applied to r.v.'s, leads to the so-called *order statistics*. For the definition of the transformation, consider n distinct numbers $x_1, \ldots, x_n$ and order them in ascending order. Denote by $x_{(1)}$ the smallest number: $x_{(1)} =$ smallest of $x_1, \ldots, x_n$; by $x_{(2)}$ the second smallest, and so on until $x_{(n)}$ is the nth smallest or, equivalently, the largest of the x_i's. In a summary form, we write: $x_{(j)} =$ the jth smallest of the numbers $x_1, \ldots, x_n$, where $j = 1, \ldots, n$. Then, clearly, $x_{(1)} < x_{(2)} < \cdots < x_{(n)}$. For simplicity, set $y_j = x_{(j)}, j = 1, \ldots, n$, so that again $y_1 < y_2 < \cdots < y_n$. The transformation under consideration is the one that transforms the x_i's into the y_j's in the way just described.

This transformation now applies to n r.v.'s as follows.

Let $X_1, X_2, \ldots, X_n$ be i.i.d. r.v.'s with d.f. F. The *jth order statistic* of $X_1, X_2, \ldots, X_n$ is denoted by $X_{(j)}$, or Y_j for easier writing, and is defined as follows:

$$Y_j = j\text{th smallest of the } X_1, X_2, \ldots, X_n, \quad j = 1, \ldots, n;$$

(i.e., for each $s \in \mathcal{S}$, look at $X_1(s), X_2(s), \ldots, X_n(s)$, and then $Y_j(s)$ is defined to be the jth smallest among the numbers $X_1(s), X_2(s), \ldots, X_n(s), j = 1, 2, \ldots, n$). It follows that $Y_1 \leq Y_2 \leq \cdots \leq Y_n$, and, in general, the Y_j's are not independent.

We assume now that the X_i's are of the continuous type with p.d.f. f such that $f(x) > 0, (-\infty \leq) a < x < b (\leq \infty)$ and zero otherwise. One of the problems we are concerned with is that of finding the joint p.d.f. of the Y_j's. By means of Theorem 4$'$, it will be established that:

Theorem 10. If $X_1, \ldots, X_n$ are i.i.d. r.v.'s with p.d.f. f, which is positive and continuous for $(-\infty \leq) a < x < b (\leq \infty)$ and 0 otherwise, then the joint p.d.f. of the order statistics $Y_1, \ldots, Y_n$ is given by:

$$g(y_1, \ldots, y_n) = \begin{cases} n! f(y_1) \cdots f(y_n), & a < y_1 < y_2 < \cdots < y_n < b \\ 0, & \text{otherwise.} \end{cases} \tag{11.29}$$

Proof. The proof is carried out explicitly for $n = 2$ and $n = 3$, but it is easily seen, with the proper change in notation, to be valid in the general case as well. First, consider the case $n = 2$. Since

$$P(X_1 = X_2) = \iint_{(x_1 = x_2)} f(x_1) f(x_2) dx_1 dx_2 = \int_a^b \int_{x_2}^{x_2} f(x_1) f(x_2) dx_1\, dx_2 = 0,$$

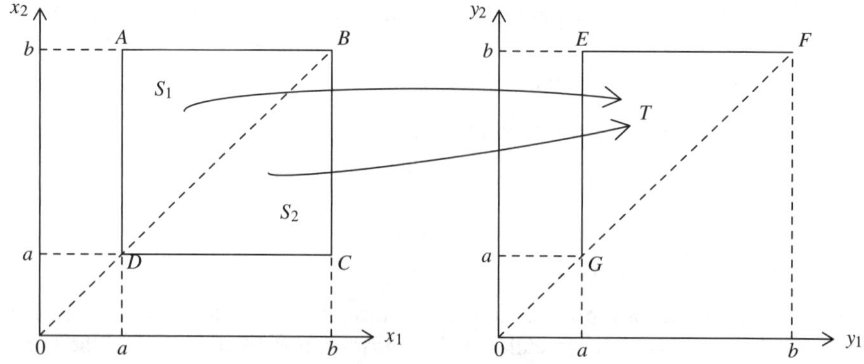

FIGURE 11.7

The set ABCD of positivity of the joint p.d.f. of X_1, X_2 is partitioned into the disjoined sets S_1 and S_2. Both sets S_1 ans S_2 are mapped onto T under the transformation of ordering X_1 and X_2.

we may assume that the joint p.d.f., $f(\cdot,\cdot)$, of X_1, X_2 is 0 for $x_1 = x_2$. Thus, $f(\cdot,\cdot)$ is positive on the rectangle $ABCD$ except for its diagonal DB, call this set S.

Write $S = S_1 \cup S_2$ as in the Figure 11.7. Points (x_1, x_2) in S_1 are mapped into the region T consisting of the triangle EFG (except for the side GF) depicted in Figure 11.7. This is so, because $x_1 < x_2$, so that $y_1 = x_1, y_2 = x_2$. For points (x_1, x_2) in S_2, we have $x_1 > x_2$, so that $y_1 = x_2, y_2 = x_1$, and the point (x_1, x_2) is also mapped into T as indicated in the figures. On $S_1 : y_1 = x_1, y_2 = x_2$, so that $x_1 = y_1, x_2 = y_2$, and $J = \begin{vmatrix} 1 & 0 \\ 0 & 1 \end{vmatrix} = 1$. Since $f(x_1, x_2) = f(x_1)f(x_2)$, it follows that $f_{Y_1,Y_2}(y_1, y_2) = f(y_1)f(y_2)$. On $S_2 : y_1 = x_2, y_2 = x_1$, so that $x_1 = y_2, x_2 = y_1$, and $J = \begin{vmatrix} 0 & 1 \\ 1 & 0 \end{vmatrix} = -1$. It follows that $f_{Y_1,Y_2}(y_1, y_2) = f(y_2)f(y_1)$. Therefore, by Theorem 4′, $f_{Y_1,Y_2}(y_1, y_2) = 2f(y_1)f(y_2) = 2!f(y_1)f(y_2)$, or $g(y_1, y_2) = 2!f(y_1)f(y_2), a < y_1 < y_2 < b$.

For $n = 3$, again since for $i \neq j$,

$$P(X_i = X_j) = \iint_{(x_i = x_j)} f(x_i)f(x_j)dx_i\,dx_j = \int_a^b \int_{x_j}^{x_j} f(x_i)f(x_j)dx_i\,dx_j = 0,$$

and therefore $P(X_i = X_j = X_k) = 0$ for $i \neq j \neq k$, we may assume that the joint p.d.f., $f(\cdot,\cdot,\cdot)$, of X_1, X_2, X_3 is zero, if at least two of the arguments x_1, x_2, x_3 are equal. Thus, we have:

$$f(x_1, x_2, x_3) = \begin{cases} f(x_1)f(x_2)f(x_3), & a < x_1 \neq x_2 \neq x_3 < b \\ 0, & \text{otherwise.} \end{cases}$$

Therefore $f(x_1, x_2, x_3)$ is positive on the set S, where:

$$S = \left\{(x_1, x_2, x_3) \in \Re^3; \quad a < x_i < b, \quad i = 1, 2, 3, \quad x_1, x_2, x_3 \text{ all different}\right\}.$$

Let $S_{ijk} \subset S$ be defined by:

$$S_{ijk} = \{(x_1, x_2, x_3); \quad a < x_i < x_j < x_k < b\}, \quad i, j, k = 1, 2, 3, \quad i \neq j \neq k.$$

Then we have that these six sets are pairwise disjoint and (essentially):

$$S = S_{123} \cup S_{132} \cup S_{213} \cup S_{231} \cup S_{312} \cup S_{321}.$$

Now on each one of the S_{ijk}'s there exists a one-to-one transformation from the x_i's to the y_i's defined as follows:

$$\begin{aligned}
S_{123} &: y_1 = x_1, \ y_2 = x_2, \ y_3 = x_3 \\
S_{132} &: y_1 = x_1, \ y_2 = x_3, \ y_3 = x_2 \\
S_{213} &: y_1 = x_2, \ y_2 = x_1, \ y_3 = x_3 \\
S_{231} &: y_1 = x_2, \ y_2 = x_3, \ y_3 = x_1 \\
S_{312} &: y_1 = x_3, \ y_2 = x_1, \ y_3 = x_2 \\
S_{321} &: y_1 = x_3, \ y_2 = x_2, \ y_3 = x_1.
\end{aligned}$$

Solving for the x_i's, we have then:

$$\begin{aligned}
S_{123} &: x_1 = y_1, \ x_2 = y_2, \ x_3 = y_3 \\
S_{132} &: x_1 = y_1, \ x_2 = y_3, \ x_3 = y_2 \\
S_{213} &: x_1 = y_2, \ x_2 = y_1, \ x_3 = y_3 \\
S_{231} &: x_1 = y_3, \ x_2 = y_1, \ x_3 = y_2 \\
S_{312} &: x_1 = y_2, \ x_2 = y_3, \ x_3 = y_1 \\
S_{321} &: x_1 = y_3, \ x_2 = y_2, \ x_3 = y_1.
\end{aligned}$$

The Jacobians are thus given by:

$$S_{123} : J_{123} = \begin{vmatrix} 1 & 0 & 0 \\ 0 & 1 & 0 \\ 0 & 0 & 1 \end{vmatrix} = 1, \qquad S_{231} : J_{231} = \begin{vmatrix} 0 & 0 & 1 \\ 1 & 0 & 0 \\ 0 & 1 & 0 \end{vmatrix} = 1,$$

$$S_{132} : J_{132} = \begin{vmatrix} 1 & 0 & 0 \\ 0 & 0 & 1 \\ 0 & 1 & 0 \end{vmatrix} = -1, \qquad S_{312} : J_{312} = \begin{vmatrix} 0 & 1 & 0 \\ 0 & 0 & 1 \\ 1 & 0 & 0 \end{vmatrix} = 1,$$

$$S_{213} : J_{213} = \begin{vmatrix} 0 & 1 & 0 \\ 1 & 0 & 0 \\ 0 & 0 & 1 \end{vmatrix} = -1, \qquad S_{321} : J_{321} = \begin{vmatrix} 0 & 0 & 1 \\ 0 & 1 & 0 \\ 1 & 0 & 0 \end{vmatrix} = -1.$$

Hence $|J_{123}| = \cdots = |J_{321}| = 1$, and Theorem 4′ gives

$$g(y_1, y_2, y_3) = \begin{cases} f(y_1)f(y_2)f(y_3) + f(y_1)f(y_3)f(y_2) + f(y_2)f(y_1)f(y_3) \\ +f(y_3)f(y_1)f(y_2) + f(y_2)f(y_3)f(y_1) + f(y_3)f(y_2)f(y_1), \\ \qquad\qquad\qquad\qquad a < y_1 < y_2 < y_3 < b \\ 0, \qquad\qquad\qquad\qquad \text{otherwise:} \end{cases}$$

That is,

$$g(y_1, y_2, y_3) = \begin{cases} 3!f(y_1)f(y_2)f(y_3), & a < y_1 < y_2 < y_3 < b \\ 0, & \text{otherwise:} \end{cases} \quad \blacktriangle$$

Note that the proof in the general case is exactly the same. One has $n!$ regions forming S, one for each permutation of the integers 1 through n. From the definition of a determinant and the fact that each row and column contains exactly one 1 and the rest all 0, it follows that the $n!$ Jacobians are either 1 or -1 and the remaining part of the proof is identical to the one just given except that one adds up $n!$ like terms instead of 3!.

The theorem is illustrated by the following two examples.

Example 10. Let $X_1, \ldots, X_n$ be i.i.d. r.v.'s distributed as $N(\mu, \sigma^2)$. Then the joint p.d.f. of the order statistics $Y_1, \ldots, Y_n$ is given by

$$g(y_1, \ldots, y_n) = n!\left(\frac{1}{\sqrt{2\pi}\sigma}\right)^n \exp\left[-\frac{1}{2\sigma^2}\sum_{j=1}^{n}(y_j - \mu)^2\right],$$

if $-\infty < y_1 < \cdots < y_n < \infty$, and zero otherwise.

Example 11. Let $X_1, \ldots, X_n$ be i.i.d. r.v.'s distributed as $U(\alpha, \beta)$. Then the joint p.d.f. of the order statistics $Y_1, \ldots, Y_n$ is given by

$$g(y_1, \ldots, y_n) = \frac{n!}{(\beta - \alpha)^n},$$

if $\alpha < y_1 < \cdots < y_n < \beta$, and zero otherwise.

From the joint p.d.f. in (11.29), it is relatively easy to derive the p.d.f. of Y_j for any j, as well as the joint p.d.f. of Y_i and Y_j for any $1 \leq i < j \leq n$. We restrict ourselves to the derivation of the distributions of Y_1 and Y_n alone.

Theorem 11. Let $X_1, \ldots, X_n$ be i.i.d. r.v.'s with d.f. F and p.d.f. f, which is positive and continuous for $(-\infty \leq) a < x < b (\leq \infty)$ and zero otherwise, and let $Y_1, \ldots, Y_n$ be the order statistics. Then the p.d.f.'s g_1 and g_n of Y_1 and Y_n, respectively, are given by:

$$g_1(y_1) = \begin{cases} n[1 - F(y_1)]^{n-1} f(y_1), & a < y_1 < b \\ 0, & \text{otherwise,} \end{cases} \tag{11.30}$$

and

$$g_n(y_n) = \begin{cases} n[F(y_n)]^{n-1} f(y_n), & a < y_n < b \\ 0, & \text{otherwise.} \end{cases} \tag{11.31}$$

Proof. First, derive the d.f.'s involved and then differentiate them to obtain the respective p.d.f.'s. To this end,

$$\begin{aligned} G_n(y_n) &= P(Y_n \leq y_n) = P[\max(X_1, \ldots, X_n) \leq y_n] \\ &= P(\text{all } X_1, \ldots, X_n \leq y_n) = P(X_1 \leq y_n, \ldots, X_n \leq y_n) \\ &= P(X_1 \leq y_n) \cdots P(X_n \leq y_n) \quad \text{(by the independence of the } X_i's) \\ &= [F(y_n)]^n. \end{aligned}$$

That is, $G_n(y_n) = [F(y_n)]^n$, so that:

$$g_n(y_n) = \frac{d}{dy_n} G_n(y_n) = n[F(y_n)]^{n-1} \frac{d}{dy_n} F(y_n) = n[F(y_n)]^{n-1} f(y_n).$$

Likewise,

$$\begin{aligned} 1 - G_1(y_1) &= P(Y_1 > y_1) = P[\min(X_1, \ldots, X_n) > y_1] \\ &= P(\text{all } X_1, \ldots, X_n > y_1) = P(X_1 > y_1, \ldots, X_n > y_1) \\ &= P(X_1 > y_1) \cdots P(X_n > y_1) \quad (\text{by the independence of the } X_i'\text{s}) \\ &= [1 - P(X_1 \le y_1)] \cdots [1 - P(X_1 \le y_1)] = [1 - F(y_1)]^n. \end{aligned}$$

That is, $1 - G_1(y_1) = [1 - F(y_1)]^n$, so that

$$\begin{aligned} -g_1(y_1) &= \frac{d}{dy_1}[1 - G_1(y_1)] = n[1 - F(y_1)]^{n-1} \frac{d}{dy_1}[1 - F(y_1)] \\ &= n[1 - F(y_1)]^{n-1}[-f(y_1)] = -n[1 - F(y_1)]^{n-1} f(y_1), \end{aligned}$$

and hence

$$g_1(y_1) = n[1 - F(y_1)]^{n-1} f(y_1). \quad \blacktriangle$$

As an illustration of the theorem, consider the following example.

Example 12. Let the independent r.v.'s $X_1, \ldots, X_n$ be distributed as $U(0, 1)$. Then, for $0 < y_1, y_n < 1$:

$$g_1(y_1) = n(1 - y_1)^{n-1} \quad \text{and} \quad g_n(y_n) = n y_n^{n-1}.$$

Discussion. Here, for $0 < x < 1$, $f(x) = 1$ and $F(x) = x$. Therefore relations (11.30) and (11.31) give, for $0 < y_1, y_n < 1$:

$$g_1(y_1) = n(1 - y_1)^{n-1} \times 1 = n(1 - y_1)^{n-1} \quad \text{and} \quad g_n(y_n) = n y_n^{n-1},$$

as asserted.

As a further illustration of the theorem, consider the following example, which is of interest in its own right.

Example 13. If $X_1, \ldots, X_n$ are independent r.v.'s having negative exponential distribution with parameter λ, then Y_1 has also negative exponential distribution with parameter $n\lambda$.

Discussion. Here $f(x) = \lambda e^{-\lambda x}$ and $F(x) = 1 - e^{-\lambda x}$ for $x > 0$. Then, for $y_1 > 0$, formula (11.30) yields:

$$g_1(y_1) = n(e^{-\lambda y_1})^{n-1} \times \lambda e^{-\lambda y_1} = (n\lambda)e^{-(n-1)y_1} e^{-\lambda y_1} = (n\lambda)e^{-(n\lambda)y_1},$$

as was to be seen.

Example 14.

(i) In a complex system, n identical components are connected serially, so that the system works if and only if all n components function. If the lifetime of said components is described by an r.v. X with d.f. F and positive and continuous p.d.f. f for $(-\infty \le) a < x < b (\le \infty)$, write out the expression for the probability that the system functions for at least t time units.

(ii) Do the same as in part (i), if the components are connected in parallel, so that the system functions if and only if at least one of the components works.

(iii) Simplify the expressions in parts (i) and (ii), if f is negative exponential with parameter λ.

Discussion.

(i) Clearly, P(system works for at least t time units)

$$
\begin{aligned}
&= P(X_1 \ge t, \ldots, X_n \ge t) && \text{(where } X_i \text{ is the lifetime of the } i\text{th component)}\\
&= P(Y_1 \ge t) && \text{(where } Y_1 \text{ is the smallest-order statistic)}\\
&= \textstyle\int_t^\infty g_1(y)\,dy && \text{(where } g_1 \text{ is the p.d.f. of } Y_1)\\
&= \textstyle\int_t^\infty n[1-F(y)]^{n-1} f(y)\,dy && \text{(by (11.30))}
\end{aligned}
$$

(ii) Here
P(system works for at least t time units)

$$
\begin{aligned}
&= P(\text{at least one of } X_1, \ldots, X_n \ge t)\\
&= P(Y_n \ge t) && \text{(where } Y_n \text{ is the largest-order statistic)}\\
&= \textstyle\int_t^\infty g_n(y)\,dy && \text{(where} g_n \text{ is the p.d.f. of } Y_n)\\
&= \textstyle\int_t^\infty n[F(y)]^{n-1} f(y)\,dy && \text{(by (11.31)).}
\end{aligned}
$$

(iii) Here $F(y) = 1 - e^{-\lambda y}$ and $f(y) = \lambda e^{-\lambda y}$ $(y > 0)$ from Example 13. Also, from the same example, the p.d.f. of Y_1 is $g_1(y) = (n\lambda)e^{-(n\lambda)y}$, so that (11.30) gives:

$$
\begin{aligned}
P(Y_1 \ge t) &= \int_t^\infty (n\lambda)e^{-(n\lambda)y}\,dy\\
&= -\int_t^\infty de^{-(n\lambda)y}\,dy\\
&= -e^{-(n\lambda)y}\Big|_t^\infty = e^{-n\lambda t},
\end{aligned}
$$

and, by (11.31),

$$
P(Y_n \ge t) = \int_t^\infty n(1-e^{-\lambda y})^{n-1}\lambda e^{-\lambda y}\,dy.
$$

However,

$$
\begin{aligned}
&\int_t^\infty n(1-e^{-\lambda y})^{n-1}\lambda e^{-\lambda y}\,dy\\
&= -n\int_t^\infty (1-e^{-\lambda y})^{n-1} de^{-\lambda y}\\
&= n\int_t^\infty (1-e^{-\lambda y})^{n-1} d(1-e^{-\lambda y})\\
&= \int_t^\infty d(1-e^{-\lambda y})^n\\
&= (1-e^{-\lambda y})^n\Big|_t^\infty = 1-(1-e^{-\lambda t})^n.
\end{aligned}
$$

For example for $n = 2$, $P(Y_2 \geq t) = 2e^{-\lambda t} - e^{-2\lambda t}$.

Exercises.

5.1 Let $X_1, \ldots, X_n$ be independent r.v.'s with p.d.f. $f(x) = cx^{-(c+1)}$, $x > 1 (c > 0)$, and set $U = Y_1 = \min(X_1, \ldots, X_n)$, $V = Y_n = \max(X_1, \ldots, X_n)$.

(i) Determine the d.f. F corresponding to the p.d.f. f.
(ii) Use Theorem 11 to determine the p.d.f.'s f_U and f_V in terms of c.

5.2 Refer to Example 12 and calculate the expectations EY_1 and EY_n, and determine the lim EY_1 and lim EY_n as $n \to \infty$.

5.3 Let Y_1 and Y_n be the smallest- and the largest-order statistics based on a random sample $X_1, \ldots, X_n$ from the $U(\alpha, \beta)$ $(\alpha < \beta)$ distribution.

(i) For $n = 3$ and $n = 4$, show that the joint p.d.f. of Y_1 and Y_n is given, respectively, by:

$$
g_{13}(y_1, y_3) = \frac{3 \times 2}{(\beta-\alpha)^2}(y_3 - y_1), \quad \alpha < y_1 < y_3 < \beta,
$$

$$
g_{14}(y_1, y_4) = \frac{4 \times 3}{(\beta-\alpha)^3}(y_4 - y_1)^2, \quad \alpha < y_1 < y_4 < \beta.
$$

(ii) Generalize the preceding results and show that:

$$
g_{1n}(y_1, y_n) = \frac{n(n-1)}{(\beta-\alpha)^n}(y_n - y_1)^{n-2}, \quad \alpha < y_1 < y_n < \beta.
$$

Hint. For part (ii), all one has to do is to calculate the integrals:

$$
\int_{y_1}^{y_n}\int_{y_1}^{y_{n-1}} \cdots \int_{y_1}^{y_4}\int_{y_1}^{y_3} dy_2\, dy_3 \cdots dy_{n-2}\, dy_{n-1},
$$

which is done one at a time; also, observe the pattern emerging.

5.4 Let Y_1 and Y_n be the smallest- and the largest-order statistics based on a random sample $X_1, \ldots, X_n$ from the $U(0, 1)$ distribution. Then show that:

$$Cov(Y_1, Y_n) = \frac{1}{(n+1)^2(n+2)}.$$

Hint. Use the joint p.d.f. taken from Exercise 5.3(ii) for $\alpha = 0$ and $\beta = 1$.

5.5 If Y_1 and Y_n are the smallest- and the largest-order statistics based on a random sample $X_1, \ldots, X_n$ from the $U(0, 1)$ distribution:

(i) Show that the p.d.f. of the *sample range* $R = Y_n - Y_1$ is given by:

$$f_R(r) = n(n-1)r^{n-2}(1-r), \quad 0 < r < 1.$$

(ii) Also, calculate the expectation ER and determine its limit as $n \to \infty$.

Hint. Use Exercise 5.3(ii) with $\alpha = 0, \beta = 1$.

5.6 Refer to Example 13 and set $Z = nY_1$. Then show that Z is distributed as the X_i's.

5.7 The lifetimes of two batteries are independent r.v.'s X and Y with negative exponential distribution with parameter λ. Suppose that the two batteries are connected serially, so that the system works if and only if both work.

(i) Use Example 14 (with $n = 2$) to calculate the probability that the system works beyond time $t > 0$.
(ii) What is the expected lifetime of the system?
(iii) How do parts (i) and (ii) become for $\lambda = 1/3$?

5.8 Let Y_1 and Y_n be the smallest- and the largest-order statistics based on a random sample $X_1, \ldots, X_n$ from negative exponential distribution with parameter λ. Then, by Example 13, $g_1(y_1) = (n\lambda)e^{-(n\lambda)y_1}$, $y_1 > 0$.

(i) Use relation (11.31) (with $a = 0$ and $b = \infty$) to determine the p.d.f. g_n of the r.v. Y_n.
(ii) Calculate the EY_n for $n = 2$ and $n = 3$.

5.9

(i) Refer to Exercise 5.8(i) and show that:

$$EY_n = \frac{n}{\lambda}\sum_{r=0}^{n-1}(-1)^{n-r-1}\frac{\binom{n-1}{r}}{(n-r)^2}.$$

(ii) Apply part (i) for $n = 2$ and $n = 3$ to recover the values found in Exercise 5.8 (ii).

Hint. Consider the binomial expansion: $(a+b)^k = \sum_{r=0}^{k}\binom{k}{r}a^r b^{k-r}$ and apply it to: $(1-e^{-\lambda y})^{n-1}$ for $a = 1$, $b = -e^{-\lambda y}$, and $k = n-1$. Then carry out the multiplications indicated and integrate term by term.

5.10 Let $X_1, \ldots, X_n$ be a random sample of size n of the continuous type with d.f. F and p.d.f. f, positive and continuous in $-\infty \le a < x < b \le \infty$, and let Y_1 and Y_n be the smallest- and the largest-order statistics of the X_i's. Use relation (11.29) to show that the joint p.d.f. g_{1n} of the r.v.'s Y_1 and Y_n is given by the expression:

$$g_{1n}(y_1, y_n) = n(n-1)[F(y_n) - F(y_1)]^{n-2} f(y_1) f(y_n), \quad a < y_1 < y_n < b.$$

Hint. The p.d.f. g_{1n} is obtained by integrating $g(y_1, \ldots, y_n)$ in (11.29) with respect to $y_{n-1}, y_{n-2}, \ldots, y_2$ as indicated below:

$$g_{1n}(y_1, y_n) = n! f(y_1) f(y_n) \int_{y_1}^{y_n} \cdots \int_{y_{n-3}}^{y_n} \int_{y_{n-2}}^{y_n} f(y_{n-1}) f(y_{n-2}) \\ \times \cdots f(y_2) dy_{n-1} dy_{n-2} \cdots dy_2.$$

However,

$$\int_{y_{n-2}}^{y_n} f(y_{n-1}) dy_{n-1} = F(y_n) - F(y_{n-2}) = \frac{[F(y_n) - F(y_{n-2})]^1}{1!},$$

$$\int_{y_{n-3}}^{y_n} \frac{[F(y_n) - F(y_{n-2})]^1}{1!} f(y_{n-2})\, dy_{n-2}$$

$$= -\int_{y_{n-3}}^{y_n} \frac{[F(y_n) - F(y_{n-2})]^1}{1!} d[F(y_n)$$

$$-F(y_{n-2})] = -\frac{[F(y_n) - F(y_{n-2})]^2}{2!}\Bigg|_{y_{n-3}}^{y_n} = \frac{[F(y_n) - F(y_{n-3})]^2}{2!},$$

and continuing on like this, we finally get:

$$\int_{y_1}^{y_n} \frac{[F(y_n) - F(y_2)]^{n-3}}{(n-3)!} f(y_2)\, dy_2$$

$$= -\int_{y_1}^{y_n} \frac{[F(y_n) - F(y_2)]^{n-3}}{(n-3)!} d[F(y_n) - F(y_2)]$$

$$= -\frac{[F(y_n) - F(y_2)]^{n-2}}{(n-2)!}\Bigg|_{y_1}^{y_n} = \frac{[F(y_n) - F(y_1)]^{n-2}}{(n-2)!}.$$

Since $\frac{n!}{(n-2)!} = n(n-1)$, the result follows.

5.11 In reference to Exercise 5.10, determine the joint p.d.f. g_{1n} if the r.v.'s $X_1, \ldots, X_n$ are distributed as $U(0, 1)$.

CHAPTER

12 Two Modes of Convergence, the Weak Law of Large Numbers, the Central Limit Theorem, and Further Results

This chapter introduces two modes of convergence for sequences of r.v.'s—convergence in distribution and convergence in probability—and then investigates their relationship.

A suitable application of these convergences leads to the most important results in this chapter, which are the Weak Law of Large Numbers and the Central Limit Theorem. These results are illustrated by concrete examples, including numerical examples in the case of the Central Limit Theorem.

In the final section of the chapter, it is shown that convergence in probability is preserved under continuity. This is also the case for convergence in distribution, but there will be no elaboration here. These statements are illustrated by two general results and a specific application.

The proofs of some of the theorems stated are given in considerable detail; in some cases, only a rough outline is presented, whereas in other cases, we restrict ourselves to the statements of the theorems alone.

12.1 Convergence in Distribution and in Probability

In all that follows, $X_1, \ldots, X_n$ are i.i.d. r.v.'s, which may be either discrete or continuous. In applications, these r.v.'s represent n independent observations on an r.v. X, associated with an underlying phenomenon that is of importance to us. In a probabilistic/statistical environment, our interest lies in knowing the distribution of X, whether it is represented by the probabilities $P(X \in B)$, $B \subseteq \Re$, or the d.f. F of the X_i's, or their p.d.f. f. In practice, this distribution is unknown to us. Something then that would be desirable would be to approximate in some sense, the unknown distribution, by a known distribution. In this section, the foundation is laid for such an approximation.

Definition 1. Let $Y_1, \ldots, Y_n$ be r.v.'s with respective d.f.'s. $F_1, \ldots, F_n$. The r.v.'s may be either discrete or continuous and need be neither independent nor identically distributed. Also, let Y be an r.v. with d.f. G. We say that the sequence of r.v.'s

Introduction to Probability, Second Edition. http://dx.doi.org/10.1016/B978-0-12-800041-0.00012-2

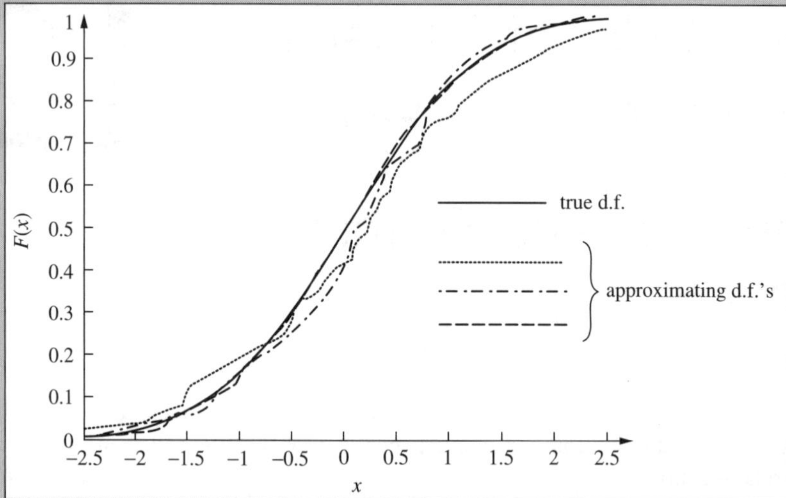

FIGURE 12.1

The d.f. represented by the solid curve is approximated by the d.f.'s represented by the ·······, · − · − · − ·, and − − − − − curves.

$\{Y_n\}$, $n \geq 1$, *converges in distribution* to the r.v. Y as $n \to \infty$ and write $Y_n \underset{n\to\infty}{\overset{d}{\longrightarrow}} Y$, if $F_n(x) \underset{n\to\infty}{\longrightarrow} G(x)$ for all continuity points x of G. (See also Figure 12.1.)

The following example illustrates the definition.

Example 1. For $n \geq 1$, let the d.f.'s F_n and the d.f. G be given by:

$$F_n(x) = \begin{cases} 0, \text{ if } x < 1 - \frac{1}{n} \\ \frac{1}{2}, \text{ if } 1 - \frac{1}{n} \leq x < 1 + \frac{1}{n} \\ 1, \text{ if } x \geq 1 + \frac{1}{n} \end{cases}, \qquad G(x) = \begin{cases} 0, \text{ if } x < 1 \\ 1, \text{ if } x \geq 1, \end{cases}$$

and discuss whether or not $F_n(x)$ converges to $G(x)$ as $n \to \infty$. (See also Figure 12.2.)

Discussion. The d.f. G is continuous everywhere except for the point $x = 1$. For $x < 1$, let $n_0 > 1/(1 - x)$. Then $x < 1 - \frac{1}{n_0}$ and $x < 1 - \frac{1}{n}$ for all $n \geq n_0$. Thus, $F_n(x) = 0, n \geq n_0$. For $x > 1$, let $n_0 \geq 1/(x - 1)$. Then $x \geq 1 + \frac{1}{n_0}$ and $x \geq 1 + \frac{1}{n}$ for all $n \geq n_0$, so that $F_n(x) = 1, n \geq n_0$. Thus, for $x \neq 1$, $F_n(x) \to G(x)$, so, if Y_n and Y are r.v.'s such that $Y_n \sim F_n$ and $Y \sim G$, then $Y_n \underset{n\to\infty}{\overset{d}{\longrightarrow}} Y$.

Remark 1. The example also illustrates the point that if x is a discontinuity point of G, then $F_n(x)$ need not converge to $G(x)$. In Example 1, $F_n(1) = \frac{1}{2}$ for all n, and $G(1) = 1$.

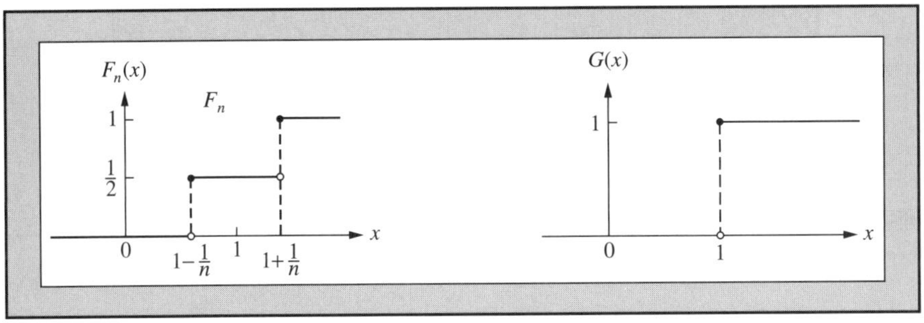

FIGURE 12.2
The d.f. G is approximated by the d.f.'s F_n at all points $x \neq 1$.

The idea, of course, behind Definition 1 is the approximation of the (presumably unknown) probability $P(Y \leq x) = G(x)$ by the (presumably known) probabilities $P(Y_n \leq x) = F_n(x)$, for large enough n. Convergence in distribution also allows the approximation of probabilities of the form $P(x < Y \leq y)$ by the probabilities $P(x < Y_n \leq y)$, for x and y continuity points of G with $x < y$. This is so because:

$$P(x < Y_n \leq y) = P(Y_n \leq y) - P(Y_n \leq x) = F_n(y) - F_n(x)$$
$$\underset{n\to\infty}{\longrightarrow} G(y) - G(x) = P(x < Y \leq y).$$

Whereas convergence in distribution allows the comparison of certain probabilities, calculated in terms of the *individual* r.v.'s Y_n and Y, it does not provide evaluation of probabilities calculated on the *joint* behavior of Y_n and Y. This is taken care of to a satisfactory extent by the following mode of convergence.

Definition 2. The sequence of r.v.'s $\{Y_n\}$, $n \geq 1$, *converges in probability* to the r.v. Y as $n \to \infty$, if, for every $\varepsilon > 0$, $P(|Y_n - Y| > \varepsilon) \underset{n\to\infty}{\longrightarrow} 0$; equivalently, $P(|Y_n - Y| \leq \varepsilon) \underset{n\to\infty}{\longrightarrow} 1$. The notation used is: $Y_n \overset{P}{\underset{n\to\infty}{\longrightarrow}} Y$.

Thus, if the event $A_n(\varepsilon)$ is defined by: $A_n(\varepsilon) = \{s \in \mathcal{S};\ Y_n(s) - \varepsilon \leq Y(s) \leq Y_n(s) + \varepsilon\}$ (i.e., the event for which the r.v. Y is within ε from the r.v. Y_n, see also Figure 12.3), then $P(A_n)(\varepsilon) \underset{n\to\infty}{\longrightarrow} 1$ for every $\varepsilon > 0$. Equivalently, $P(A_n^c(\varepsilon)) = P(\{s \in \mathcal{S};\ Y(s) < Y_n(s) - \varepsilon \text{ or } Y(s) > Y_n(s) + \varepsilon\}) \underset{n\to\infty}{\longrightarrow} 0$.

The probability that Y lies within a small neighborhood around Y_n, such as $[Y_n - \varepsilon, Y_n + \varepsilon]$, is as close to 1 as one pleases, provided n is sufficiently large.

It is rather clear that convergence in probability is stronger than convergence in distribution. That this is, indeed, the case is illustrated by the following example, where we have convergence in distribution but not in probability (see Figures 12.3).

Example 2. Let $\mathcal{S} = \{1, 2, 3, 4\}$, and on the subsets of $\mathcal{S}$, let P be the discrete uniform probability function. Define the following r.v.'s:

$$X_n(1) = X_n(2) = 1, \quad X_n(3) = X_n(4) = 0, \qquad n = 1, 2, \ldots,$$

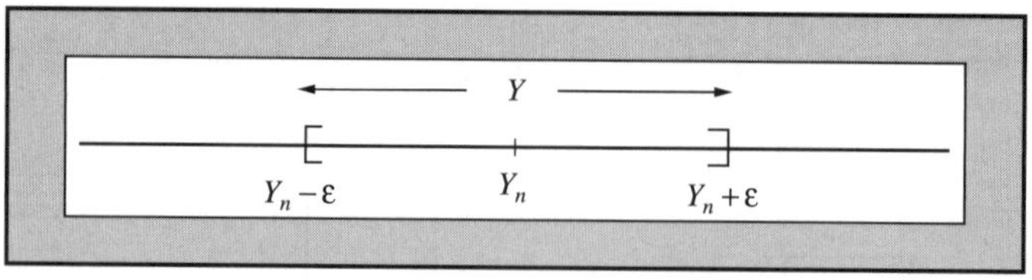

FIGURE 12.3
$(|Y_n - Y| \leq \varepsilon) = (Y_n - \varepsilon \leq Y \leq Y_n + \varepsilon)$.

and

$$X(1) = X(2) = 0, \qquad X(3) = X(4) = 1.$$

Discussion. Then:

$$|X_n(s) - X(s)| = 1 \quad \text{for all } s \in \mathcal{S}.$$

Hence X_n does *not* converge in probability to X, as $n \to \infty$. Now,

$$F_n(x) = \begin{cases} 0, & x < 0 \\ \frac{1}{2}, & 0 \leq x < 1 \\ 1, & x \geq 1 \end{cases}, \qquad G(x) = \begin{cases} 0, & x < 0 \\ \frac{1}{2}, & 0 \leq x < 1 \\ 1, & x \geq 1, \end{cases}$$

so that $F_n(x) = G(x)$ for all $x \in \Re$. Thus, trivially, $F_n(x) \underset{n\to\infty}{\longrightarrow} G(x)$ for all continuity points of G; that is, $X_n \overset{d}{\underset{n\to\infty}{\longrightarrow}} X$, but X_n does not converge in probability to X.

The precise relationship between convergence in distribution and convergence in probability is stated in the following theorem.

Theorem 1. Let $\{Y_n\}, n \geq 1$, be a sequence of r.v.'s and let Y be an r.v. Then $Y_n \overset{P}{\underset{n\to\infty}{\longrightarrow}} Y$ always implies $Y_n \overset{d}{\underset{n\to\infty}{\longrightarrow}} Y$. The converse is not true in general (as illustrated by Example 2). However, it is true if $P(Y = c) = 1$, where c is a constant. That is, $Y_n \overset{d}{\underset{n\to\infty}{\longrightarrow}} c$ implies $Y_n \overset{P}{\underset{n\to\infty}{\longrightarrow}} c$, so that $Y_n \overset{P}{\underset{n\to\infty}{\longrightarrow}} c$ if and only if $Y_n \overset{d}{\underset{n\to\infty}{\longrightarrow}} c$.

Proof (outline). That $Y_n \overset{P}{\underset{n\to\infty}{\longrightarrow}} Y$ implies $Y_n \overset{d}{\underset{n\to\infty}{\longrightarrow}} Y$ is established by employing the concepts of lim inf (limit inferior) and lim sup (limit superior) of a sequence of numbers, and we choose not to pursue it. For the proof of the fact that $Y_n \overset{d}{\underset{n\to\infty}{\longrightarrow}} c$ implies $Y_n \overset{P}{\underset{n\to\infty}{\longrightarrow}} c$, observe that $F(x) = 0$, for $x < c$ and $F(x) = 1$ for $x \geq c$, where F is the d.f. of c so that $c - \varepsilon$ and $c + \varepsilon$ are continuity points of F for all $\varepsilon > 0$. But $P(|Y_n - c| \leq \varepsilon) = P(c - \varepsilon \leq Y_n \leq c + \varepsilon) = P(Y_n \leq c + \varepsilon) - P(Y_n < c - \varepsilon) = F_n(c + \varepsilon) - P(Y_n < c - \varepsilon)$. However, $F_n(c + \varepsilon) \underset{n\to\infty}{\longrightarrow} 1$ and $P(Y_n < c - \varepsilon) \leq P(Y_n \leq c - \varepsilon) = F_n(c - \varepsilon) \underset{n\to\infty}{\longrightarrow} 0$, so that $P(Y_n < c - \varepsilon) \underset{n\to\infty}{\longrightarrow} 0$. Thus, $P(|Y_n - c| \leq \varepsilon) \underset{n\to\infty}{\longrightarrow} 1$ or $Y_n \overset{P}{\underset{n\to\infty}{\longrightarrow}} c$. ▲

According to Definition 1, in order to establish that $Y_n \xrightarrow[n\to\infty]{d} Y$, all one has to do is to prove the (pointwise) convergence $F_n(x) \underset{n\to\infty}{\longrightarrow} F(x)$ for every continuity point x of F. As is often the case, however, definitions do not lend themselves to checking the concepts defined. This also holds here. Accordingly, convergence in distribution is delegated to convergence of m.g.f.'s, which, in general, is a much easier task to perform. That this can be done is based on the following deep probabilistic result. Its justification is omitted entirely.

Theorem 2 (Continuity Theorem). For $n = 1, 2, \ldots$, let Y_n and Y be r.v.'s with respective d.f.'s F_n and F, and respective m.g.f.'s M_n and M (which are assumed to be finite at least in an interval $(-c, c)$, some $c > 0$). Then:

(i) If $F_n(x) \underset{n\to\infty}{\longrightarrow} F(x)$ for all continuity points x of F, it follows that $M_n(t) \underset{n\to\infty}{\longrightarrow} M(t)$ for all $t \in (-c, c)$.

(ii) If $M_n(t) \underset{n\to\infty}{\longrightarrow} M(t)$, $t \in (-c, c)$, then $F_n(x) \underset{n\to\infty}{\longrightarrow} F(x)$ for all continuity points x of F.

Thus, according to this result, $Y_n \xrightarrow[n\to\infty]{d} Y$ or, equivalently, $F_n(x) \underset{n\to\infty}{\longrightarrow} F(x)$ for all continuity points x of F, if and only if $M_n(t) \underset{n\to\infty}{\longrightarrow} M(t)$, $t \in (-c, c)$, some $c > 0$. The fact that convergence of m.g.f.'s implies convergence of the respective d.f.'s is the most useful part from a practical viewpoint.

Exercises.

1.1 For $n = 1, 2, \ldots$, let X_n be an r.v. with d.f. F_n defined by: $F_n(x) = 0$ for $x < n$, and $F_n(x) = 1$ for $x \geq n$. Then show that $F_n(x) \underset{n\to\infty}{\longrightarrow} F(x)$, which is identically 0 in $\Re$ and hence it is *not* a d.f. of an r.v.

1.2 Let $\{X_n\}$, $n \geq 1$, be r.v.'s with X_n taking the values 1 and 0 with respective probabilities p_n and $1 - p_n$; that is, $P(X_n = 1) = p_n$ and $P(X_n = 0) = 1 - p_n$. Then show that $X_n \xrightarrow[n\to\infty]{P} 0$, if and only if $p_n \underset{n\to\infty}{\longrightarrow} 0$.

Hint. Just elaborate on Definition 2.

1.3 For $n = 1, 2, \ldots$, let X_n be an r.v. distributed as $B(n, p_n)$ and suppose that $np_n \underset{n\to\infty}{\longrightarrow} \lambda \in (0, \infty)$. Then show that $X_n \xrightarrow[n\to\infty]{d} X$, where X is an r.v. distributed as $P(\lambda)$, by showing that $M_{X_n}(t) \underset{n\to\infty}{\longrightarrow} M_X(t)$, $t \in \Re$.

Remark: This is an application of Theorem 2(ii).

1.4 Let $Y_{1,n}$ and $Y_{n,n}$ be the smallest and the largest order statistics, respectively, based on the random sample $X_1, \ldots, X_n$ from the $U(0, 1)$ distribution. Then show that:
(i) $Y_{1,n} \xrightarrow[n\to\infty]{P} 0$; (ii) $Y_{n,n} \xrightarrow[n\to\infty]{P} 1$.

Hint. For $\varepsilon > 0$, calculate the probabilities: $P(|Y_{1,n}| > \varepsilon)$ and $P(|Y_{n,n} - 1| \leq \varepsilon)$ and show that they tend to 0 and 1, respectively, as $n \to \infty$. Use the p.d.f.'s of $Y_{1,n}$ and $Y_{n,n}$ determined in Example 12 of Chapter 11.

1.5 In refer to Exercise 1.4, set: $U_n = nY_{1,n}$, $V_n = n(1 - Y_{n,n})$ and let U and V be r.v.'s having negative exponential distribution with parameter $\lambda = 1$. Then:

(i) Derive the p.d.f.'s of the r.v.'s U_n and V_n.

(ii) Derive the d.f.'s of the r.v.'s U_n and V_n, and show that $U_n \underset{n\to\infty}{\overset{d}{\longrightarrow}} U$ by showing that:

$$F_{U_n}(u) \underset{n\to\infty}{\longrightarrow} F_U(u), \quad u \in \Re.$$

Likewise for V_n.

Hint. For part (ii), refer to #6 in Table 6 in the Appendix.

1.6 We say that a sequence $\{X_n\}$, $n \geq 1$, of r.v.'s converges to an r.v. X in *quadratic mean* and write:

$$X_n \underset{n\to\infty}{\overset{q.m.}{\longrightarrow}} X \quad \text{or} \quad X_n \underset{n\to\infty}{\overset{(2)}{\longrightarrow}} X, \quad \text{if } E(X_n - X)^2 \underset{n\to\infty}{\longrightarrow} 0.$$

Now, if $X_1, \ldots, X_n$ are i.i.d. r.v.'s with (finite) expectation μ and (finite) variance σ^2, show that the sample mean $\bar{X}_n \underset{n\to\infty}{\overset{q.m.}{\longrightarrow}} \mu$.

1.7 In the first part of the proof of Theorem 1 of Chapter 8 (see also its proof, part (i)), the following version of the *Cauchy–Schwarz* inequality was established: For any two r.v.'s X and Y with $EX = EY = 0$ and $Var(X) = Var(Y) = 1$, it holds: $|E(XY)| \leq 1$. (This is actually only part of said inequality.) Another more general version of this inequality is the following: For any two r.v.'s X and Y with finite expectations and variances, it holds: $|E(XY)| \leq E|XY| \leq E^{1/2}|X|^2 \times E^{1/2}|Y|^2$.

(i) Prove the inequality in this general setting.

(ii) For any r.v. X, show that $|EX| \leq E|X| \leq E^{1/2}|X|^2$.

Hint. For part (i), use the obvious result $(x \pm y)^2 = x^2 + y^2 \pm 2xy \geq 0$ in order to conclude that $-\frac{1}{2}(x^2 + y^2) \leq xy \leq \frac{1}{2}(x^2 + y^2)$ and hence $|xy| \leq \frac{1}{2}(x^2 + y^2)$. Next, replace x by $X/E^{1/2}|X|^2$, and y by $Y/E^{1/2}|Y|^2$ (assuming, of course, that $E|X|^2 > 0$, $E|Y|^2 > 0$, because otherwise the inequality is, trivially, true), and take the expectations of both sides to arrive at the desirable result.

1.8 Let $\{X_n\}$ and $\{Y_n\}$, $n \geq 1$, be two sequences of r.v.'s such that: $X_n \underset{n\to\infty}{\overset{q.m.}{\longrightarrow}} X$, some r.v., and $X_n - Y_n \underset{n\to\infty}{\overset{q.m.}{\longrightarrow}} 0$. Then show that $Y_n \underset{n\to\infty}{\overset{q.m.}{\longrightarrow}} X$.

Hint. Start out with the $E(Y_n - X)^2$, add X_n in $Y_n - X$ and subtract it off, square out the expression, apply the assumptions made, and then use appropriately the Cauchy–Schwarz inequality discussed in Exercise 1.7.

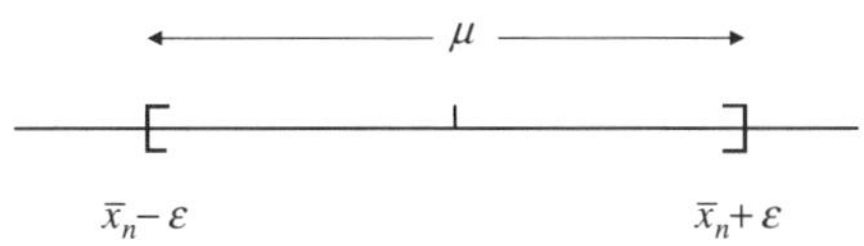

FIGURE 12.4

Parameter μ lies in the interval $[\bar{X}_n - \varepsilon, \bar{X}_n + \varepsilon]$ with high probability for all sufficiently large n.

12.2 The Weak Law of Large Numbers and the Central Limit Theorem

As a first application of the concept of convergence in probability (distribution), we have the so-called *Weak Law of Large Numbers* (WLLN). This result is stated and proved, an interpretation is provided, and then a number of specific applications are presented.

Theorem 3 (Weak Law of Large Numbers, WLLN). Let $X_1, X_2, \ldots$ be i.i.d. r.v.'s with (common) finite expectation μ, and let $\bar{X}_n$ be the sample mean of $X_1, \ldots, X_n$. Then $\bar{X}_n \underset{n\to\infty}{\overset{d}{\longrightarrow}} \mu$, or (on account of Theorem 1) $\bar{X}_n \underset{n\to\infty}{\overset{P}{\longrightarrow}} \mu$.

Thus, the probability that μ lies within a small neighborhood around $\bar{X}_n$, such as $[\bar{X}_n - \varepsilon, \bar{X}_n + \varepsilon]$, is as close to 1 as one pleases, provided n is sufficiently large. (See also Figures 12.4.)

Proof. The proof is a one-line proof, if it happens that the X_i's also have a (common) finite variance σ^2 (which they are not required to have for the validity of the theorem). Since $E\bar{X}_n = \mu$ and $Var(\bar{X}_n) = \frac{\sigma^2}{n}$, the Tchebichev inequality gives, for every $\varepsilon > 0$, $P(|\bar{X}_n - \mu| > \varepsilon) \leq \frac{1}{\varepsilon^2} \times \frac{\sigma^2}{n} \underset{n\to\infty}{\longrightarrow} 0$, so that $\bar{X}_n \underset{n\to\infty}{\overset{P}{\longrightarrow}} \mu$.

Without reference to the variance, one would have to show that $M_{\bar{X}_n}(t) \underset{n\to\infty}{\longrightarrow} M_\mu(t)$ (for $t \in (-c, c)$, some $c > 0$). Let M stand for the (common) m.g.f. of the X_i's. Then use familiar properties of the m.g.f. and independence of the X_i's in order to obtain:

$$M_{\bar{X}_n}(t) = M_{\sum_{i=1}^n X_i}\left(\frac{t}{n}\right) = \prod_{i=1}^n M_{X_i}\left(\frac{t}{n}\right) = \left[M\left(\frac{t}{n}\right)\right]^n.$$

Consider the function $M(z)$, and expand it around $z = 0$ according to Taylor's formula up to terms of first order to get:

$$\begin{aligned} M(z) &= M(0) + \frac{z}{1!}\frac{d}{dz}M(z)|_{z=0} + R(z) \quad \left(\frac{1}{z}R(z) \to 0 \text{ as } z \to 0\right) \\ &= 1 + z\mu + R(z), \end{aligned}$$

since $M(0) = 1$ and $\frac{d}{dz}M(z)|_{z=0} = EX_1 = \mu$. Replacing z by t/n, for fixed t, the last formula becomes:

$$M\left(\frac{t}{n}\right) = 1 + \frac{t}{n}\mu + R\left(\frac{t}{n}\right), \quad \text{where } nR\left(\frac{t}{n}\right) \to 0 \text{ as } n \to \infty.$$

Therefore

$$M_{\bar{X}_n}(t) = \left[1 + \frac{\mu t + nR(\frac{t}{n})}{n}\right]^n,$$

and this converges to $e^{\mu t}$, as $n \to \infty$, by Remark 2 below. Since $e^{\mu t}$ is the m.g.f. of (the degenerate r.v.) μ, we have shown that $M_{\bar{X}_n}(t) \underset{n\to\infty}{\longrightarrow} M_\mu(t)$, as was to be seen. ▲

Remark 2. For every $z \in \Re$, one way of defining the exponential function e^z is:

$$e^z = \lim_{n\to\infty}\left(1 + \frac{z}{n}\right)^n.$$

It is a consequence of this result that, as $n \to \infty$, also $(1 + \frac{z_n}{n})^n \to e^z$ whenever $z_n \to z$. See also #6 in Table 6 in the Appendix.

The interpretation and most common use of the WLLN is that if μ is an unknown entity, which is typically the case in statistics, then μ may be approximated (in the sense of distribution or probability) by the known entity $\bar{X}_n$, for sufficiently large n.

12.2.1 Applications of the WLLN

1. If the independent X_i's are distributed as $B(1, p)$, then $EX_i = p$ and therefore $\bar{X}_n \underset{n\to\infty}{\overset{P}{\longrightarrow}} p$.
2. If the independent X_i's are distributed as $P(\lambda)$, then $EX_i = \lambda$ and therefore $\bar{X}_n \underset{n\to\infty}{\overset{P}{\longrightarrow}} \lambda$.
3. If the independent X_i's are distributed as $N(\mu, \sigma^2)$, then $EX_i = \mu$ and therefore $\bar{X}_n \underset{n\to\infty}{\overset{P}{\longrightarrow}} \mu$.
4. If the independent X_i's are distributed as negative exponential with parameter λ, $f(x) = \lambda e^{-\lambda x}, x > 0$, then $EX_i = 1/\lambda$ and therefore $\bar{X}_n \underset{n\to\infty}{\overset{P}{\longrightarrow}} 1/\lambda$.
   A somewhat more involved application is that of the approximation of an entire d.f. by the so-called empirical d.f. To this effect:
5. Let $X_1, X_2, \ldots, X_n$ be i.i.d. r.v.'s with d.f. F, and define the *empirical d.f.* F_n as follows. For each $x \in \Re$ and each $s \in \mathcal{S}$,

$$F_n(x, s) = \frac{1}{n}[\text{number of } X_1(s), \ldots, X_n(s) \le x].$$

From this definition, it follows immediately that for each fixed $x \in \Re$, $F_n(x, s)$ is an r.v. as a function of s, and for each fixed $s \in \mathcal{S}$, $F_n(x, s)$ is a d.f. as a function of x.

Actually, if we set $Y_i(x, s) = 1$ when $X_i(s) \leq x$, and $Y_i(x, s) = 0$ when $X_i(s) > x$, then $Y_i(x, \cdot), \ldots, Y_i(x, \cdot)$ are r.v.'s that are independent and distributed as $B(1, F(x))$, since $P[Y_i(x, \cdot) = 1] = P(X_i \leq x) = F(x)$. Also, $EY_i(x, \cdot) = F(x)$. Then $F_n(x, s)$ may be rewritten as:

$$F_n(x, s) = \frac{1}{n} \sum_{i=1}^{n} Y_i(x, s), \quad \text{the sample mean of } Y_1(x, s), \ldots, Y_n(x, s).$$

By omitting the sample point s, as is usually the case, we write $F_n(x)$ and $Y_i(x), i = 1, \ldots, n$ rather than $F_n(x, s)$ and $Y_i(x, s), i = 1, \ldots, n$, respectively. Then $F_n(x) \xrightarrow[n\to\infty]{P} F(x)$ for each $x \in \Re$. Thus, for every $x \in \Re$, the value of $F(x)$ of the (potentially unknown) d.f. F is approximated by the (known) values $F_n(x)$ of the r.v.'s $F_n(x)$.

Remark 3. Actually, it can be shown that the convergence $F_n(x) \xrightarrow[n\to\infty]{P} F(x)$ is *uniform* in $x \in \Re$. This implies that for every $\varepsilon > 0$, there is a positive integer $N(\varepsilon)$ *independent* of $x \in \Re$, such that $F_n(x) - \varepsilon < F(x) < F_n(x) + \varepsilon$ with probability as close to 1 as one pleases *simultaneously* for all $x \in \Re$, provided $n > N(\varepsilon)$.

As another application of the concept of convergence in distribution, we obtain, perhaps, the most celebrated theorem of probability theory; it is the so-called *Central Limit Theorem* (CLT), which is stated and proved below. Comments on the significance of the CLT follow, and the section is concluded with applications and numerical examples. Some preliminary work will facilitate the formulation of the theorem. To this end, let $X_1, X_2, \ldots$ be i.i.d. r.v.'s with finite expectation μ and finite and positive variance σ^2, let $\bar{X}_n$ be the sample mean of $X_1, \ldots, X_n$, and denote by S_n the partial sum $\sum_{i=1}^{n} X_i$; that is,

$$S_n = \sum_{i=1}^{n} X_i. \tag{12.1}$$

Then:

$$E\bar{X}_n = \mu, \; Var(\bar{X}_n) = \sigma^2/n, \; ES_n = n\mu, \text{ and } Var(S_n) = n\sigma^2. \tag{12.2}$$

(Although the notation S_n has been used before (relation (10.13) (in Chapter 10)) to denote the sample standard deviation of $X_1, \ldots, X_n$, there should be no confusion; from the context, it should be clear what S_n stands for.)

From (12.1) and (12.2), it follows immediately that:

$$\frac{S_n - ES_n}{\sqrt{Var(S_n)}} = \frac{S_n - n\mu}{\sigma\sqrt{n}} = \frac{\bar{X}_n - \mu}{\sigma/\sqrt{n}} \left(= \frac{\bar{X}_n - E\bar{X}_n}{\sqrt{Var(\bar{X}_n)}} \right) = \frac{\sqrt{n}(\bar{X}_n - \mu)}{\sigma}. \tag{12.3}$$

Then the CLT is stated as follows.

Theorem 4 (Central Limit Theorem, CLT). Let $X_1, X_2, \ldots$ be i.i.d. r.v.'s with finite expectation μ and finite and positive variance σ^2, let $\bar{X}_n$ be the sample mean of $X_1, \ldots, X_n$, and let $S_n = \sum_{i=1}^{n} X_i$. Then:

$$\frac{S_n - n\mu}{\sigma\sqrt{n}} = \frac{\sqrt{n}(\bar{X}_n - \mu)}{\sigma} \xrightarrow[n\to\infty]{d} Z \sim N(0, 1), \tag{12.4}$$

or

$$P\left(\frac{S_n - n\mu}{\sigma\sqrt{n}} \le z\right) = P\left[\frac{\sqrt{n}(\bar{X}_n - \mu)}{\sigma} \le z\right] \underset{n\to\infty}{\longrightarrow} \Phi(z) = \int_{-\infty}^{z} \frac{1}{\sqrt{2\pi}} e^{-\frac{x^2}{2}} dx, \quad z \in \Re. \tag{12.5}$$

(Also, see Remark 4(ii).)

Remark 4.

(i) An interpretation of (12.4) and (12.5) is that, for sufficiently large n:

$$P\left[\frac{\sqrt{n}(\bar{X}_n - \mu)}{\sigma} \le z\right] = P\left(\frac{S_n - n\mu}{\sigma\sqrt{n}} \le z\right) \simeq \Phi(z), \quad z \in \Re. \tag{12.6}$$

Often this approximation is also denoted (rather loosely) as follows:

$$\frac{\sqrt{n}(\bar{X}_n - \mu)}{\sigma} \simeq N(0, 1), \quad \text{or} \quad \bar{X}_n \simeq N\left(\mu, \frac{\sigma^2}{n}\right), \quad \text{or} \quad S_n \simeq N(n\mu, n\sigma^2). \tag{12.7}$$

(ii) Actually, it can be shown that the convergence in (12.5) is *uniform* in $z \in \Re$. That is to say, if we set

$$F_n(z) = P\left[\frac{\sqrt{n}(\bar{X}_n - \mu)}{\sigma} \le z\right] = P\left(\frac{S_n - n\mu}{\sigma\sqrt{n}} \le z\right), \tag{12.8}$$

then

$$F_n(z) \underset{n\to\infty}{\longrightarrow} \Phi(z) \quad \textit{uniformly} \text{ in } z \in \Re. \tag{12.9}$$

To be more precise, for every $\varepsilon > 0$, there exists a positive integer $N(\varepsilon)$ *independent* of $z \in \Re$, such that

$$|F_n(z) - \Phi(z)| < \varepsilon \quad \text{for } n \ge N(\varepsilon) \text{ and } \textit{all } z \in \Re. \tag{12.10}$$

Its justification is provided by Lemma 1, Chapter 8, in the book *A Course in Mathematical Statistics*, 2nd edition (1997), Academic Press, by G. G. Roussas.

(iii) The approximation of the probability $F_n(z)$ by $\Phi(z)$, provided by the CLT, is also referred to as *normal approximation* for obvious reasons.

(iv) On account of (12.6), the CLT also allows for the approximation of probabilities of the form $P(a < S_n \leq b)$ for any $a < b$. Indeed,

$$\begin{aligned} P(a < S_n \leq b) &= P(S_n \leq b) - P(S_n \leq a) \\ &= P\left(\frac{S_n - n\mu}{\sigma\sqrt{n}} \leq \frac{b - n\mu}{\sigma\sqrt{n}}\right) - P\left(\frac{S_n - n\mu}{\sigma\sqrt{n}} \leq \frac{a - n\mu}{\sigma\sqrt{n}}\right) \\ &= P\left(\frac{S_n - n\mu}{\sigma\sqrt{n}} \leq b_n^*\right) - P\left(\frac{S_n - n\mu}{\sigma\sqrt{n}} \leq a_n^*\right), \end{aligned}$$

where

$$a_n^* = \frac{a - n\mu}{\sigma\sqrt{n}} \quad \text{and} \quad b_n^* = \frac{b - n\mu}{\sigma\sqrt{n}}. \tag{12.11}$$

By (12.6), and Remark 4(ii),

$$P\left(\frac{S_n - n\mu}{\sigma\sqrt{n}} \leq b_n^*\right) \simeq \Phi(b_n^*) \quad \text{and} \quad P\left(\frac{S_n - n\mu}{\sigma\sqrt{n}} \leq a_n^*\right) \simeq \Phi(a_n^*),$$

so that

$$P(a < S_n \leq b) \simeq \Phi(b_n^*) - \Phi(a_n^*). \tag{12.12}$$

Formulas (12.12) and (12.11) are always to be used whenever a discrete distribution is approximated by (the standard) normal distribution (CLT) without using continuity correction (see Section 12.2.3).
The uniformity referred to in Remark 4(ii) is what actually validates many of the applications of the CLT. This is the case, for instance, in Remark 4(iv).

(v) So, the convergence in (12.5) is a special case of the convergence depicted in Figure 12.1. In (12.5), the limiting d.f. is Φ and F_n is the d.f. of $\frac{S_n - n\mu}{\sigma\sqrt{n}}$ or $\frac{\sqrt{n}(\bar{X}_n - \mu)}{\sigma}$. This convergence holds for all $x \in \Re$ since Φ is a continuous function in $\Re$.

Here are some illustrative examples of the CLT.

Example 3. From a large collection of bolts which is known to contain 3% defectives, 1,000 are chosen at random. If X is the number of the defective bolts among those chosen, what is the (approximate) probability that X does not exceed 5% of 1,000?

Discussion. With the selection of the ith bolt, associate the r.v. X_i to take the value 1, if the bolt is defective, and 0 otherwise. Then it may be assumed that the r.v.'s $X_i, i = 1, \ldots, 1{,}000$ are independently-distributed as $B(1, 0.03)$. Furthermore, it is clear that $X = \sum_{i=1}^{1{,}000} X_i$. Since 5% of 1,000 is 50, the required probability is: $P(X \leq 50)$. Since $EX_i = 0.03$, $Var(X_i) = 0.03 \times 0.97 = 0.0291$, the CLT gives, by means of (12.11) and (12.12) (with $X = S_{1{,}000}$):

$$\begin{aligned} P(X \leq 50) &= P(0 \leq X \leq 50) = P(-0.5 < X \leq 50) \\ &= P(X \leq 50) - P(X \leq -0.5) \simeq \Phi(b_n^*) - \Phi(a_n^*), \end{aligned}$$

$$\text{where}\quad a_n^* = \frac{-0.5 - 1{,}000 \times 0.03}{\sqrt{1{,}000 \times 0.03 \times 0.97}} = -\frac{30.5}{\sqrt{29.1}} \simeq -\frac{30.5}{5.394} \simeq -5.65,$$

$$b_n^* = \frac{50 - 1{,}000 \times 0.03}{\sqrt{1{,}000 \times 0.03 \times 0.97}} = \frac{20}{\sqrt{29.1}} \simeq \frac{20}{5.394} \simeq 3.71,$$

so that

$$P(X \leq 50) \simeq \Phi(3.71) - \Phi(-5.65) = \Phi(3.71) = 0.999896.$$

Example 4. A certain manufacturing process produces vacuum tubes whose lifetimes in hours are independent r.v.'s with negative exponential distribution with mean 1,500 hours. What is the probability that the total lifetime of 50 tubes will exceed 80,000 hours?

Discussion. If X_i is the r.v. denoting the lifetime of the ith vacuum tube, then X_i, $i = 1, \ldots, 50$ are independently negative exponentially-distributed with $EX_i = \frac{1}{\lambda} = 1{,}500$ and $Var(X_i) = \frac{1}{\lambda^2} = 1{,}500^2$. Since $nEX_i = 50 \times 1{,}500 = 75{,}000$, $\sigma\sqrt{n} = 1{,}500\sqrt{50}$, if we set $S_{50} = \sum_{i=1}^{50} X_i$, then the required probability is:

$$\begin{aligned}
P(S_{50} > 80{,}000) &= 1 - P(S_{50} \leq 80{,}000) \\
&= 1 - P(0 \leq S_{50} \leq 80{,}000) = 1 - P(-0.5 < S_{50} \leq 80{,}000) \\
&= 1 - P(S_{50} \leq 80{,}000) + P(S_{50} \leq -0.5) \\
&\simeq 1 - \Phi\left(\frac{80{,}000 - 75{,}000}{1{,}500\sqrt{50}}\right) + \Phi(-7.07) \\
&\simeq 1 - \Phi\left(\frac{80{,}000 - 75{,}000}{1{,}500\sqrt{50}}\right) \\
&= 1 - \Phi\left(\frac{\sqrt{50}}{15}\right) \simeq 1 - \Phi(0.47) \\
&= 1 - 0.680822 = 0.319178 \simeq 0.319.
\end{aligned}$$

Example 5. Someone is organizing a get-together party and sends out invitations to 60 people. The invitation states that the invitee may bring along another person if he/she so wishes. For the purpose of efficient planning, the party organizer would wish to know:

(i) What is the expected number of guests and the s.d. around this number?
(ii) What is the probability that the number of guests is between 75 and 85, inclusively?

Discussion.

(i) With the ith invitation, associate an r.v. X_i, denoting the number of guests to come in response to that invitation, defined as follows:

$$X_i = \begin{cases} 0, & 3/15 \\ 1, & 4/15, \\ 2, & 8/15 \end{cases} \quad i = 1, \ldots, 60.$$

Then the number of guests is $S_{60} = \sum_{i=1}^{60} X_i$. Here

$$\mu = EX_i = \frac{(1 \times 4) + (2 \times 8)}{15} = \frac{20}{15},$$
$$EX_i^2 = \frac{(1 \times 4) + (4 \times 8)}{15} = \frac{36}{15},$$

so that

$$\sigma^2 = Var(X_i) = \frac{36}{15} - \left(\frac{20}{15}\right)^2 = \frac{140}{225} \quad \text{and} \quad \sigma = \frac{2\sqrt{35}}{15}.$$

Therefore

$$ES_{60} = 60 \times \frac{20}{15} = 80.$$

If we also assume that the r.v.'s $X_1, \ldots, X_{60}$ are independent, then:

$$Var(S_{60}) = 60 \times \frac{140}{225} \quad \text{and} \quad \sqrt{Var(S_{60})} = \frac{4\sqrt{21}}{3} \simeq 6.11.$$

So, the expected number of guests is 80, and the variation around it is about 6 individuals.

(ii) By means of (12.11) and (12.12), we have:

$$P(75 \leq S_{60} \leq 85) = P(74 < S_{60} \leq 85) \simeq \Phi(b_{60}^*) - \Phi(a_{60}^*),$$

where

$$a_{60}^* = \frac{74 - 80}{\frac{4\sqrt{21}}{3}} = \frac{-3\sqrt{21}}{14} \simeq -0.98,$$
$$b_{60}^* = \frac{85 - 80}{\frac{4\sqrt{21}}{3}} = \frac{5\sqrt{21}}{28} \simeq 0.82.$$

Therefore

$$P(75 \leq S_{60} \leq 85) \simeq \Phi(0.82) - \Phi(-0.98) = \Phi(0.82) + \Phi(0.98) - 1$$
$$= 0.793892 + 0.836457 - 1 = 0.630349.$$

The proof of the CLT is based on the same ideas as those used in the proof of the WLLN and goes as follows.

Proof of Theorem 4. Set $Z_i = \frac{X_i - \mu}{\sigma}$, so that $Z_1, \ldots, Z_n$ are i.i.d. r.v.'s with $EZ_i = 0$ and $Var(Z_i) = 1$. Also,

$$\frac{1}{\sqrt{n}} \sum_{i=1}^{n} Z_i = \frac{1}{\sigma\sqrt{n}}(S_n - n\mu) = \frac{\sqrt{n}(\bar{X}_n - \mu)}{\sigma}. \tag{12.13}$$

With F_n defined by (12.8), we wish to show that (12.9) holds (except for the uniformity assertion, with which we will not concern ourselves). By Theorem 2, it suffices to show that, for all t,

$$M_{\sqrt{n}(\bar{X}_n-\mu)/\sigma}(t) \underset{n\to\infty}{\longrightarrow} M_Z(t) = e^{t^2/2}. \tag{12.14}$$

By means of (12.13), and with M standing for the (common) m.g.f. of the Z_i's, we have:

$$\begin{aligned} M_{\sqrt{n}(\bar{X}_n-\mu)/\sigma}(t) &= M_{\frac{1}{\sqrt{n}}\sum_{i=1}^n Z_i}(t) = M_{\sum_{i=1}^n Z_i}\left(\frac{t}{\sqrt{n}}\right) \\ &= \prod_{i=1}^n M_{Z_i}\left(\frac{t}{\sqrt{n}}\right) = \left[M\left(\frac{t}{\sqrt{n}}\right)\right]^n. \end{aligned} \tag{12.15}$$

Expand the function $M(z)$ around $z = 0$ according to Taylor's formula up to terms of second order to get:

$$\begin{aligned} M(z) &= M(0) + \frac{z}{1!}\frac{d}{dz}M(z)|_{z=0} + \frac{z^2}{2!}\frac{d^2}{dz^2}M(z)|_{z=0} + R(z) \\ &= 1 + zEZ_1 + \frac{z^2}{2}EZ_1^2 + R(z) \\ &= 1 + \frac{z^2}{2} + R(z), \quad \text{where } \frac{1}{z^2}R(z) \to 0 \quad \text{as } z \to 0. \end{aligned}$$

In this last formula, replace z by $t/\sqrt{n}$, for fixed t, in order to obtain:

$$M\left(\frac{t}{\sqrt{n}}\right) = 1 + \frac{t^2}{2n} + R\left(\frac{t}{\sqrt{n}}\right), \quad nR\left(\frac{t}{\sqrt{n}}\right) \to 0 \quad \text{as } n \to \infty.$$

Therefore (12.15) becomes:

$$M_{\sqrt{n}(\bar{X}_n-\mu)/\sigma}(t) = \left[1 + \frac{t^2}{2n} + R\left(\frac{t}{\sqrt{n}}\right)\right]^n = \left\{1 + \frac{\frac{t^2}{2}\left[1 + \frac{2n}{t^2}R\left(\frac{t}{\sqrt{n}}\right)\right]}{n}\right\}^n,$$

and this converges to $e^{t^2/2}$, as $n \to \infty$, by Remark 2. This completes the proof of the theorem. ▲

12.2.2 Applications of the CLT

In all of the following applications, it will be assumed that n is sufficiently large, so that the CLT will apply.

Let the independent X_i's be distributed as $B(1, p)$, set $S_n = \sum_{i=1}^n X_i$, and let a, b be integers such that $0 \le a < b \le n$. By an application of the CLT, we wish to find an approximate value to the probability $P(a < S_n \le b)$.

If p denotes the proportion of defective items in a large lot of certain items, then S_n is the number of actually defective items among the n sampled. Therefore approximation of the probability $P(a < S \leq b)$ is meaningful when the binomial tables are not usable (either because of p or because of n or, perhaps, because of both).

Here $EX_i = p, Var(X_i) = pq$ $(q = 1 - p)$, and therefore by (12.12) and (12.11):

$$P(a < S_n \leq b) \simeq \Phi(b_n^*) - \Phi(a_n^*), \quad \text{where} \quad a_n^* = \frac{a - np}{\sqrt{npq}}, \quad b_n^* = \frac{b - np}{\sqrt{npq}}. \tag{12.16}$$

Remark 5. If the required probability is of any one of the forms $P(a \leq S_n \leq b)$ or $P(a \leq S_n < b)$ or $P(a < S_n < b)$, then formula (12.16) applies again, provided the necessary adjustments are first made; namely, $P(a \leq S_n \leq b) = P(a - 1 < S_n \leq b)$, $P(a \leq S_n < b) = P(a - 1 < S_n \leq b - 1)$, $P(a < S_n < b) = P(a < S_n \leq b - 1)$. However, if the underlying distribution is continuous, then $P(a < S_n \leq b) = P(a \leq S_n \leq b) = P(a \leq S_n < b) = P(a < S_n < b)$, and no adjustments are required for the approximation in (12.16) to hold. As a rule of thumb, for the approximation to be valid, both np and $n(1 - p)$ must be ≥ 5.

Example 6 (Numerical). For $n = 100$ and $p = \frac{1}{2}$ or $p = \frac{5}{16}$, find the probability $P(45 \leq S_n \leq 55)$.

Discussion.

(i) For $p = \frac{1}{2}$, it is seen (from tables) that the exact value is equal to: 0.7288. For the normal approximation, we have: $P(45 \leq S_n \leq 55) = P(44 < S_n \leq 55)$ and, by (12.16):

$$a^* = \frac{44 - 100 \times \frac{1}{2}}{\sqrt{100 \times \frac{1}{2} \times \frac{1}{2}}} = -\frac{6}{5} = -1.2, \quad b^* = \frac{55 - 100 \times \frac{1}{2}}{\sqrt{100 \times \frac{1}{2} \times \frac{1}{2}}} = \frac{5}{5} = 1.$$

Therefore $\Phi(b^*) - \Phi(a^*) = \Phi(1) - \Phi(-1.2) = \Phi(1) + \Phi(1.2) - 1 = 0.841345 + 0.884930 - 1 = 0.7263$. So:

Exact value : 0.7288, *Approximate value* : 0.7263,

and the exact probability is underestimated by 0.0025, or the approximating probability is about 99.66% of the exact probability.

(ii) For $p = \frac{5}{16}$, the exact probability is almost 0; 0.0000. For the approximate probability, we find $a^* = 2.75$ and $b^* = 4.15$, so that $\Phi(b^*) - \Phi(a^*) = 0.0030$. Thus:

Exact value : 0.0000, *Approximate value* : 0.0030,

and the exact probability is overestimated by 0.0030.

If the underlying distribution is $P(\lambda)$, then $ES_n = Var(S_n) = n\lambda$ and formulas (12.11) and (12.12) become:

$$P(a < S_n \leq b) \simeq \Phi(b_n^*) - \Phi(a_n^*), \quad a_n^* = \frac{a - n\lambda}{\sqrt{n\lambda}}, \quad b_n^* = \frac{b - n\lambda}{\sqrt{n\lambda}}.$$

The comments made in Remark 5 apply here also.

Example 7 (Numerical). In the Poisson distribution $P(\lambda)$, let n and λ be so that $n\lambda = 16$ and find the probability $P(12 \leq S_n \leq 21)(= P(11 < S_n \leq 21))$.

Discussion. The exact value (found from tables) is: 0.7838. For the normal approximation, we have:

$$a^* = \frac{11 - 16}{\sqrt{16}} = -\frac{5}{4} = -1.25, \quad b^* = \frac{21 - 16}{\sqrt{16}} = \frac{5}{4} = 1.25,$$

so that $\Phi(b^*) - \Phi(a^*) = \Phi(1.25) - \Phi(-1.25) = 2\Phi(1.25) - 1 = 2 \times 0.894350 - 1 = 0.7887$. So:

$$\textit{Exact value}: \ 0.7838, \quad \textit{Approximate value}: \ 0.7887,$$

and the exact probability is overestimated by 0.0049, or the approximating probability is about 100.63% of the exact probability (see also Example 7 (continued)).

12.2.3 The Continuity Correction

When a discrete distribution is approximated by normal distribution, the error committed is easy to see in a geometric picture. This is done, for instance, in Figure 12.5, where the p.d.f. of the $B(10, 0.2)$ distribution is approximated by the p.d.f. of the $N(10 \times 0.2, 10 \times 0.2 \times 0.8) = N(2, 1.6)$ distribution (see relation (12.7)). From the same figure, it is also clear how the approximation may be improved.

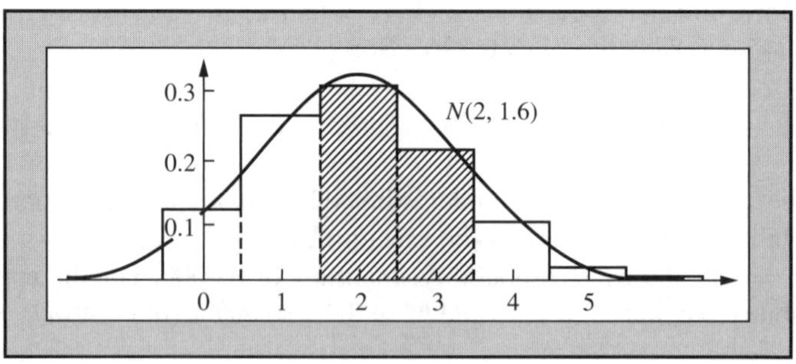

FIGURE 12.5

Exact and approximate values for the probability $P(a \leq S_n \leq b) = P(a - 1 < S_n \leq b) = P(1 < S_n \leq 3)$.

Now

$$P(1 < S_n \leq 3) = P(2 \leq S_n \leq 3) = f_n(2) + f_n(3)$$
$$= \text{shaded area},$$

while the approximation without correction is the area bounded by the normal curve, the horizontal axis, and the abscissas 1 and 3. Clearly, the correction, given by the area bounded by the normal curve, the horizontal axis, and the abscissas 1.5 and 3.5, is closer to the exact area.

Thus, under the conditions of the CLT, and for discrete r.v.'s, $P(a < S_n \leq b) \simeq \Phi(b^*) - \Phi(a^*)$, where $a^* = \frac{a-n\mu}{\sigma\sqrt{n}}$ and $b^* = \frac{b-n\mu}{\sigma\sqrt{n}}$ *without* continuity correction, and $P(a < S_n \leq b) \simeq \Phi(b') - \Phi(a')$, where $a' = \frac{a+0.5-n\mu}{\sigma\sqrt{n}}$ and $b' = \frac{b+0.5-n\mu}{\sigma\sqrt{n}}$ *with* continuity correction.

Summarizing the procedure of using the CLT (or normal approximation) in computing (approximate) probabilities, we have:

For integer-valued r.v.'s and probabilities of the form $P(a \leq S_n \leq b)$, we first rewrite the expression as follows:

$$P(a \leq S_n \leq b) = P(a - 1 < S_n \leq b),$$

and then apply the preceding approximations in order to obtain:

$$P(a \leq S_n \leq b) \simeq \Phi(b^*) - \Phi(a^*), \text{ where}$$

$a^* = \frac{a-1-n\mu}{\sigma\sqrt{n}}$ and $b^* = \frac{b-n\mu}{\sigma\sqrt{n}}$ *without* continuity correction, and $P(a \leq S_n \leq b) \simeq \Phi(b') - \Phi(a')$, where $a' = \frac{a-0.5-n\mu}{\sigma\sqrt{n}}$ and $b' = \frac{b+0.5-n\mu}{\sigma\sqrt{n}}$ *with* continuity correction.

We work in a similar way for the intervals $[a, b)$ and (a, b).

The improvement brought about by the continuity correction is demonstrated by the following numerical examples.

Example 6. (continued).

Discussion.

(i) For $p = \frac{1}{2}$, we get:

$$a' = \frac{44 + 0.5 - 100 \times \frac{1}{2}}{\sqrt{100 \times \frac{1}{2} \times \frac{1}{2}}} = -\frac{5.5}{5} = -1.1,$$

$$b' = \frac{55 + 0.5 - 100 \times \frac{1}{2}}{\sqrt{100 \times \frac{1}{2} \times \frac{1}{2}}} = \frac{5.5}{5} = 1.1,$$

so that:

$$\Phi(b') - \Phi(a') = \Phi(1.1) - \Phi(-1.1) = 2\Phi(1.1) - 1$$
$$= 2 \times 0.864334 - 1 = 0.7286.$$

Thus, we have:

$$\textit{Exact value}: \ 0.7288,$$
$$\textit{Approximate value with continuity correction}: \ 0.7286,$$

and the approximation underestimates the probability by only 0.0002, or the approximating probability (with continuity correction) is about 99.97% of the exact probability.

(ii) For $p = \frac{5}{16}$, we have $a' = 2.86$, $b' = 5.23$ and $\Phi(b') - \Phi(a') = 0.0021$. Then:

$$\textit{Exact value}: \ 0.0000,$$
$$\textit{Approximate value with continuity correction}: \ 0.0021,$$

and the probability is overestimated by only 0.0021.

Example 7. (continued).

Discussion. Here:

$$a' = \frac{11 + 0.5 - 16}{\sqrt{16}} = -\frac{4.5}{4} = -1.125, \quad b' = \frac{21 + 0.5 - 16}{\sqrt{16}} = \frac{5.5}{4} = 1.375,$$

so that:

$$\begin{aligned}\Phi(b') - \Phi(a') &= \Phi(1.375) - \Phi(-1.125)\\ &= \Phi(1.375) + \Phi(1.125) - 1 = 0.7851.\end{aligned}$$

Thus:

$$\textit{Exact value}: \ 0.7838,$$
$$\textit{Approximate value with continuity correction}: \ 0.7851,$$

and the approximation overestimates the probability by only 0.0013, or the approximating probability (with continuity correction) is about 100.17% of the exact probability.

Exercises.

2.1 Let $X_1, \ldots, X_n$ be i.i.d. r.v.'s, and for a positive integer k, suppose that EX_1^k is finite. Form the kth *sample mean* $\bar{X}_n^{(k)}$ defined by:

$$\bar{X}_n^{(k)} = \frac{1}{n}\sum_{i=1}^{n} X_i^k.$$

Then show that:

$$\bar{X}_n^{(k)} \xrightarrow[n\to\infty]{P} EX_1^k.$$

2.2 Let X be an r.v. with p.d.f. $f_X(x) = c\alpha^x, x = 0, 1, \ldots (0 < \alpha < 1)$. Then $c = 1 - \alpha$ by Exercise 3.8 in Chapter 3.

(i) Show that the m.g.f. of X is: $M_X(t) = \frac{1-\alpha}{1-\alpha e^t}$, $t < -\log\alpha$, where as always, log is the natural logarithm.
(ii) Use the m.g.f. to show that $EX = \frac{\alpha}{1-\alpha}$.
(iii) If $X_1, \ldots, X_n$ is a random sample from f_X, show that the WLLN holds by showing that:

$$M_{\bar{X}_n}(t) \underset{n\to\infty}{\longrightarrow} e^{\alpha t/(1-\alpha)} = M_{EX}(t), \quad t < -\log\alpha.$$

Hint. Expand e^t around 0 up to second term, according to Taylor's formula, $e^t = 1 + t + R(t)$, where $\frac{1}{t}R(t) \underset{n\to\infty}{\longrightarrow} 0$, replace t by $\frac{t}{n}$, and use the fact that $(1 + \frac{x_n}{n})^n \to e^x$, if $x_n \to x$ as $n \to \infty$. See also #6 in Table 6 in the Appendix.

2.3 Let the r.v. X be distributed as $B(150, 0.6)$. Then:

(i) Write down the formula for the exact probability $P(X \leq 80)$.
(ii) Use the CLT in order to find an approximate value for the above probability.

Hint. Write $P(X \leq 80) = P(-0.5 < X \leq 80)$, and do not employ the continuity correction.

2.4 A binomial experiment with probability p of a success is repeated independently 1,000 times, and let X be the r.v. denoting the number of successes. For $p = \frac{1}{2}$ and $p = \frac{1}{4}$, find:

(i) The exact probability $P(1{,}000p - 50 \leq X \leq 1{,}000p + 50)$.
(ii) Use the CLT to find an approximate value for this probability.

Hint. For part (i), just write down the right formula. For part (ii), first bring it under the form $P(a < X \leq b)$, and compute the approximate probability, without continuity correction.

2.5 Let $X_1, \ldots, X_{100}$ be independent r.v.'s distributed as $B(1, p)$. Then:

(i) Write out the expression for the exact probability $P(\sum_{i=1}^{100} X_i = 50)$.
(ii) Use of CLT in order to find an approximate value for this probability.
(iii) What is the numerical value of the probability in part (ii) for $p = 0.5$?

Hint. For part (ii), first observe that $P(X = 50) = P(49 < X \leq 50)$, and then apply the CLT, both without and with continuity correction.

2.6 Fifty balanced dice are tossed once, and let X be the r.v. denoting the sum of the upturned spots. Use the CLT to find an approximate value of the probability $P(150 \leq X \leq 200)$.

Hint. With the ith die, associate the r.v. X_i, which takes on the values 1 through 6, each with probability 1/6. These r.v.'s may be assumed to be independent and $X = \sum_{i=1}^{50} X_i$. Next, write $P(150 \le X \le 200) = P(149 < X \le 200)$ and use approximation without and with continuity correction.

2.7 One thousand cards are drawn (with replacement) from a standard deck of 52 playing cards, and let X be the r.v. denoting the total number of aces drawn. Use the CLT to find an approximate value of the probability $P(65 \le X \le 90)$.

Hint. Write $P(65 \le X \le 90) = P(64 < X \le 90)$ and use approximation, both without and with continuity correction.

2.8 From a large collection of bolts that is known to contain 3% defectives, 1,000 are chosen at random, and let X be the r.v. denoting the number of defective bolts among those chosen. Use the CLT to find an approximate value of the probability that X does not exceed 5% of 1,000.

Hint. With the ith bolt drawn, associate the r.v. X_i, which takes on the value 1 if the bolt drawn is defective, and 0 otherwise. Since the collection of bolts is large, we may assume that after each drawing, the proportion of the remaining defective bolts remains (approximately) the same. This implies that the independent r.v.'s $X_1, \ldots, X_{1,000}$ are distributed as $B(1, 0.03)$ and that $X = \sum_{i=1}^{1,000} X_i \sim B(1{,}000, 0.3)$. Next, write $P(X \le 50) = P(-0.5 < X \le 50)$ and use approximation, both without and with continuity correction.

2.9 A manufacturing process produces defective items at the constant (but unknown to us) proportion p. Suppose that n items are sampled independently, and let X be the r.v. denoting the number of defective items among the n, so that $X \sim B(n, p)$. Determine the smallest value of the sample size n, so that

$$P\left(\left|\frac{X}{n} - p\right| < 0.05\sqrt{pq}\right) \ge 0.95 \quad (q = 1 - p):$$

(i) By utilizing the CLT (without continuity correction).
(ii) By using the Tchebichev inequality.
(iii) Compare the answers in parts (i) and (ii).

2.10 Suppose that 53% of the voters favor a certain legislative proposal. How many voters must be sampled so that the observed relative frequency of those favoring the proposal will not differ from the assumed frequency by more than 2% with probability 0.99?

Hint. With the ith voter sampled, associate the r.v. X_i, which takes on the value 1 if the voter favors the proposal, and 0 otherwise. Then it may be assumed that the r.v.'s $X_1, \ldots, X_n$ are independent and their common distribution is $B(1, 0.53)$. Furthermore, the number of voters favoring the proposal is $X = \sum_{i=1}^{n} X_i$. Use the CLT (without continuity correction) in order to find an approximate value to the required probability.

2.11 In playing a game, you win or lose \$1 with probability 0.5, and you play the game independently 1,000 times. Use the CLT to find an approximate value of the probability that your fortune (i.e., the total amount you won or lost) is at least \$10.

Hint. With the ith game, associate the r.v. X_i, which takes on the value 1 if \$1 is won, and -1 if \$1 is lost. Then the r.v.'s $X_1, \ldots, X_{1,000}$ are independent, and the fortune X is given by $\sum_{i=1}^{1,000} X_i$. Write $P(X \geq 10) = P(10 \leq X \leq 1,000)$ and do not use continuity correction.

2.12 It is known that the number of misprints in a page of a certain publication is an r.v. X having Poisson distribution with parameter λ. If $X_1, \ldots, X_n$ are the misprints counted in n pages, use the CLT to determine the (approximate) probability that the total number of misprints is:

(i) Not more than $n\lambda$.
(ii) Not less than $n\lambda$.
(iii) Not less than $n\lambda/2$, but not more than $3n\lambda/4$.
(iv) Give the numerical values in parts (i)–(iii) for $n\lambda = 100$ (which may be interpreted, for example, as one misprint per 4 pages ($\lambda = 0.25$) in a book of 400 pages).

Hint. In all cases, first bring it under the form $P(a < S_n \leq b)$, and then use approximation, without continuity correction. In part (ii), consider two cases: $n\lambda$ is *not* an integer and $n\lambda$ *is* an integer. Do the same in part (iii) for $n\lambda/2$.

2.13 Let the r.v. X be distributed as $P(100)$. Then:

(i) Write down the formula for the exact probability $P(X \leq 116)$.
(ii) Use the CLT appropriately to find an approximate value for the above probability. (Do not use the continuity correction.)

Hint. Select n large and λ small, so that $n\lambda = 100$ and look at X as the sum $\sum_{i=1}^{n} X_i$ of n independent r.v.'s $X_1, \ldots, X_n$ distributed as $P(\lambda)$.

2.14 A certain manufacturing process produces vacuum tubes whose lifetimes in hours are independently distributed r.v.'s with negative exponential distribution with mean 1,500 hours. Use the CLT to find an approximate value for the probability that the total life of 50 tubes will exceed 80,000 hours.

2.15 The lifespan of an electronic component in a (complicated) system is an r.v. X having negative exponential distribution with parameter λ.

(i) What is the probability that said lifespan will be at least t time units?
(ii) If the independent r.v.'s $X_1, \ldots, X_n$ represent the lifespans of n spare items such as the one described above, then $Y = \sum_{i=1}^{n} X_i$ is the combined lifespan of these n items. Use the CLT to find an approximate value of the probability $P(t_1 \leq Y \leq t_2)$, where $0 < t_1 < t_2$ are given time units.

(iii) Compute the answer (in terms of λ) in part (i), if $t = -\log(0.9)/\lambda$ (where as always, log is the natural logarithm).

(iv) Do the same for part (ii), if $\lambda = 1/10$, $n = 36$, $t_1 = 300$, and $t_2 = 420$.

2.16 Let the independent r.v.'s $X_1, \ldots, X_n$ be distributed as $U(0, 1)$.

(i) Use the CLT to find an approximate value for the probability $P(a \leq \bar{X} \leq b)$ in terms of a and b $(a < b)$.

(ii) What is the numerical value of this probability for $n = 12$, $a = 7/16$, and $b = 9/16$?

2.17 If the independent r.v.'s $X_1, \ldots, X_{12}$ are distributed as $U(0, \theta)$ $(\theta > 0)$, use the CLT to show that the probability $P(\frac{\theta}{4} < \bar{X} < \frac{3\theta}{4})$ is approximately equal to 0.9973.

2.18 A manufacturing process produces 1/2-inch ball bearings, which are assumed to be satisfactory if their diameter lies in the interval 0.5 ± 0.0006 and defective otherwise. Let X_i, $i = 1, \ldots, n$ be the diameters of n ball bearings. If $EX_i = \mu = 0.5$ inch and $\sigma(X_i) = \sigma = 0.0005$ inch, use the CLT to determine the smallest value of n for which $P(|\bar{X} - \mu| \leq 0.0001) = 0.99$, where $\bar{X}$ is the sample mean of the X_i's.

2.19 The i.i.d. r.v.'s $X_1, \ldots, X_{100}$ have (finite) mean μ and variance $\sigma^2 = 4$. Use the CLT (without continuity correction) to determine the value of the constant c for which $P(|\bar{X} - \mu| \leq c) = 0.90$, where $\bar{X}$ represents the sample mean of the X_i's.

2.20 Let $X_1, \ldots, X_n$ be i.i.d. r.v.'s with (finite) expectation μ and (finite and positive) variance σ^2, and let $\bar{X}_n$ be the sample mean of the X_i's. Determine the smallest value of the sample size n, in terms of k and p, for which $P(|\bar{X}_n - \mu| < k\sigma) \geq p$, where $p \in (0, 1)$, $k > 0$. Do so by using:

(i) the CLT (without continuity correction) and

(ii) the Tchebichev inequality; then

(iii) find the numerical values of n in parts (i) and (ii) for $p = 0.90, 0.95, 0.99$ and $k = 0.50, 0.25, 0.10$ for each value of p.

2.21 A certain manufacturing process produces light bulbs whose lifespan (in hours) is an r.v. X that has $EX = 2{,}000$ and $\sigma(X) = 200$, but it is not necessarily normally distributed. Also, consider another manufacturing process producing light bulbs whose mean lifespan is claimed to be 10% higher than the mean lifespan of the bulbs produced by the existing process; it is assumed that the s.d. remains the same for the new process. How many bulbs manufactured by the new process must be examined to establish the claim of their superiority (should that be the case) with probability 0.95?

Hint. Let Y be the r.v. denoting the lifespan of a light bulb manufactured by the new process. We do not necessarily assume that Y is normally distributed. If the claim made is correct, then $EY = 2{,}000 + 10\% \times 2{,}000 = 2{,}200$, whereas $\sigma(Y) = 200$. A random sample from Y produces the sample mean $\bar{Y}_n$

for which $E\bar{Y}_n = 2{,}200$ (under the claim) and $Var(\bar{Y}_n) = 200^2/n$, and we must determine n, so that $P(\bar{Y}_n > 2{,}000) = 0.95$. If the new process were the same as the old one, then, for all sufficiently large n, $P(\bar{Y}_n > 2{,}000) \simeq 0.50$. So, if $P(\bar{Y}_n > 2{,}000) = 0.95$, the claim made would draw support. Use the CLT without continuity correction.

2.22

(i) Consider the i.i.d. r.v.'s $X_1, \ldots, X_n$ and $Y_1, \ldots, Y_n$ with expectation μ and variance σ^2, both finite, and let $\bar{X}_n$ and $\bar{Y}_n$ be the respective sample means. Use the CLT (without continuity correction) to determine the sample size n, so that $P(|\bar{X}_n - \bar{Y}_n| \leq 0.25\sigma) = 0.95$.

(ii) Let the random samples $X_1, \ldots, X_n$ and $Y_1, \ldots, Y_n$ be as in part (i), but we do not assume that they are coming from the same distribution. We do assume, however, that they have the same mean and the same variance σ^2, both finite. Then determine n as required above by using the Tchebichev inequality.

Hint. Set $Z_i = X_i - Y_i$ and then work as in Exercise 2.20(ii) with the i.i.d. r.v.'s $Z_1, \ldots, Z_n$. Finally, revert to the X_i's and the Y_i's.

2.23 Let X_i, $i = 1, \ldots, n$, Y_i, $i = 1, \ldots, n$ be independent r.v.'s such that the X_i's are identically distributed with $EX_i = \mu_1$, $Var(X_i) = \sigma^2$, both finite, and the Y_i's are identically distributed with $EY_i = \mu_2$ and $Var(Y_i) = \sigma^2$, both finite. If $\bar{X}_n$ and $\bar{Y}_n$ are the respective sample means of the X_i's and the Y_i's, then:

(i) Show that $E(\bar{X}_n - \bar{Y}_n) = \mu_1 - \mu_2$, $Var(\bar{X}_n - \bar{Y}_n) = \frac{2\sigma^2}{n}$.

(ii) Use the CLT to show that $\frac{\sqrt{n}[(\bar{X}_n - \bar{Y}_n) - (\mu_1 - \mu_2)]}{\sigma\sqrt{2}}$ is asymptotically distributed as $N(0, 1)$.

Hint. Set $Z_i = X_i - Y_i$ and work with the i.i.d. r.v.'s $Z_1, \ldots, Z_n$; then revert to the X_i's and the Y_i's.

2.24 Within a certain period of time, let n be the number of health claims of an insurance company, and suppose that the sizes of the claims are independent r.v.'s $X_1, \ldots, X_n$ having negative exponential distribution with parameter λ; that is, $f(x) = \lambda e^{-\lambda x}$, $x > 0$. Let P be the premium charged for each policy, and set $S_n = X_1 + \cdots + X_n$. If the total amount of claims is not to exceed the total premium for the n policies sold with probability p:

(i) Express the premium P in terms of n, λ, and p.

(ii) What is the value of P in part (i) for $n = 10{,}000$, $\lambda = 1/1{,}000$ and $p = 0.99$?

Hint. Employ the CLT.

2.25 The lifetime of a light bulb is an r.v. X having negative exponential distribution with parameter $\lambda = 0.2$ hours (i.e., the p.d.f. of X is given by $f(x) = \lambda e^{-\lambda x}$, $x > 0$ $(\lambda = 0.2)$). If $X_1, \ldots, X_n$ are the independent lifetimes of n such light bulbs:

(i) Determine the smallest value of n (in terms of the constant $c > 0$ and p), so that $P(|\bar{X}_n - EX_1| \leq c) \geq p$.

(ii) What is the numerical value of n for $c = 1$ and $p = 0.950$?

Hint. Use the CLT.

2.26 Certain measurements are rounded up to the nearest integer, and let X be the r.v. denoting the difference between an actual measurement and its rounded-up value. It is assumed that $X \sim U(-0.5, 0.5)$. For a random sample of size $n = 100$, compute the probability that the sample mean and the true mean do not differ in absolute value by more than 0.1.

Hint. Use the CLT.

12.3 Further Limit Theorems

Convergence in probability enjoys some of the familiar properties of the usual pointwise convergence. One such property is stated below in the form of a theorem whose proof is omitted.

Theorem 5.

(i) For $n \geq 1$, let X_n and X be r.v.'s such that $X_n \xrightarrow[n\to\infty]{P} X$, and let g be a continuous real-valued function; that is, $g : \Re \to \Re$ continuous. Then the r.v.'s $g(X_n)$, $n \geq 1$, also converge in probability to $g(X)$; that is, $g(X_n) \xrightarrow[n\to\infty]{P} g(X)$.
More generally:

(ii) For $n \geq 1$, let X_n, Y_n, X, and Y be r.v.'s such that $X_n \xrightarrow[n\to\infty]{P} X$, $Y_n \xrightarrow[n\to\infty]{P} Y$, and let g be a continuous real-valued function; that is, $g : \Re^2 \to \Re$ continuous. Then the r.v.'s. $g(X_n, Y_n)$, $n \geq 1$, also converge in probability to $g(X, Y)$; that is, $g(X_n, Y_n) \xrightarrow[n\to\infty]{P} g(X, Y)$.

(This part also generalizes in an obvious manner to k sequences $\{X_n^{(i)}\}$, $n \geq 1$, $i = 1, \ldots, k$.)

To this theorem, there is the following important corollary.

Corollary. If $X_n \xrightarrow[n\to\infty]{P} X$ and $Y_n \xrightarrow[n\to\infty]{P} Y$, then:

(i) $aX_n + bY_n \xrightarrow[n\to\infty]{P} aX + bY$, where a and b are constants; and, in particular, $X_n + Y_n \xrightarrow[n\to\infty]{P} X + Y$.

(ii) $X_nY_n \xrightarrow[n\to\infty]{P} XY$.

(iii) $\frac{X_n}{Y_n} \xrightarrow[n\to\infty]{P} \frac{X}{Y}$, provided $P(Y_n \neq 0) = P(Y \neq 0) = 1$.

Proof. Although the proof of the theorem was omitted, the corollary can be proved. Indeed, all one has to do is to take: $g : \Re^2 \to \Re$ as follows, respectively, for parts (i)–(iii), and observe that it is continuous: $g(x, y) = ax + by$ (and, in particular, $g(x, y) = x + y$); $g(x, y) = xy$; $g(x, y) = x/y$, $y \neq 0$. ▲

Actually, a special case of the preceding corollary also holds for convergence in distribution. Specifically, we have

Theorem 6 (Slutsky). Let $X_n \xrightarrow[n\to\infty]{d} X$ and let $Y_n \xrightarrow[n\to\infty]{d} c$, a constant c rather than a (proper) r.v. Y. Then:

(i) $X_n + Y_n \xrightarrow[n\to\infty]{d} X + c$;

(ii) $X_nY_n \xrightarrow[n\to\infty]{d} cX$;

(iii) $\frac{X_n}{Y_n} \xrightarrow[n\to\infty]{d} \frac{X}{c}$, provided $P(Y_n \neq 0) = 1$ and $c \neq 0$.

In terms of d.f.'s, these convergences are written as follows, always as $n \to \infty$ and for all $z \in \Re$ for which: $z - c$ is a continuity point of F_X for part (i); z/c is a continuity point of F_X for part (ii); cz is a continuity point of F_X for part (iii):

$$P(X_n + Y_n \leq z) \to P(X + c \leq z) = P(X \leq z - c), \quad \text{or}$$

$$F_{X_n+Y_n}(z) \to F_X(z - c);$$

$$P(X_nY_n \leq z) \to P(cX \leq z) = \begin{cases} P\left(X \leq \frac{z}{c}\right), & c > 0 \\ P\left(X \geq \frac{z}{c}\right), & c < 0 \end{cases}, \quad \text{or}$$

$$F_{X_nY_n}(z) \to \begin{cases} F_X\left(\frac{z}{c}\right), & c > 0 \\ 1 - P\left(X < \frac{z}{c}\right)\left(= 1 - P\left(X \leq \frac{z}{c}\right) = 1 - F_X\left(\frac{z}{c}\right)\right), & c < 0; \end{cases}$$

$$P\left(\frac{X_n}{Y_n} \leq z\right) \to P\left(\frac{X}{c} \leq z\right) = \begin{cases} P(X \leq cz), & c > 0 \\ P(X \geq cz), & c < 0 \end{cases}, \quad \text{or}$$

$$F_{\frac{X_n}{Y_n}}(z) \to \begin{cases} F_X(cz), & c > 0 \\ 1 - P(X < cz)(= 1 - P(X \leq cz) = 1 - F_X(cz)), & c < 0. \end{cases}$$

The proof of this theorem, although conceptually not complicated, is nevertheless long and is omitted. Recall, however, that $Y_n \xrightarrow[n\to\infty]{d} c$ if and only if $Y_n \xrightarrow[n\to\infty]{P} c$, and this is the way the convergence of Y_n is often stated in the theorem.

As a simple concrete application of Theorem 6, consider the following example.

Example 8. For $n \to \infty$, suppose that $X_n \xrightarrow{d} X \sim N(\mu, \sigma^2)$, and let c_n, c, d_n, and d be constants such that $c_n \to c$ and $d_n \to d$. Then $c_nX_n + d_n \xrightarrow{d} Y \sim N(c\mu + d, c^2\sigma^2)$.

Discussion. As $n \to \infty$, trivially, $c_n \xrightarrow{d} c$ and $d_n \xrightarrow{d} d$, so that, by Theorem 6(ii), $c_n X_n \xrightarrow{d} cX$, and by Theorem 6(i), $c_n X_n + d_n \xrightarrow{d} cX + d$. However, $X \sim N(\mu, \sigma^2)$ implies that $cX + d \sim N(c\mu + d, c^2\sigma^2)$. Thus, $c_n X_n + d_n \xrightarrow{d} cX + d = Y \sim N(c\mu + d, c^2\sigma^2)$.

The following result is an application of Theorems 5 and 6 and is of much use in statistical inference. For its formulation, let $X_1, \ldots, X_n$ be i.i.d. r.v.'s with finite mean μ and finite and positive variance σ^2, and let $\bar{X}_n$ and S_n^2 be the sample mean and the "adjusted" (in the sense that μ is replaced by $\bar{X}_n$) sample variance (which we have denoted by $\bar{S}_n^2$ in relation (10.14) of Chapter 10); that is, $\bar{X}_n = \frac{1}{n}\sum_{i=1}^n X_i$, $S_n^2 = \frac{1}{n}\sum_{i=1}^n (X_i - \bar{X}_n)^2$. Then:

Theorem 7. Under the assumptions just made and the notation introduced, it holds: (i) $S_n^2 \xrightarrow[n\to\infty]{P} \sigma^2$; (ii) $\frac{\sqrt{n}(\bar{X}_n - \mu)}{S_n} \xrightarrow[n\to\infty]{d} Z \sim N(0, 1)$.

Proof.

(i) Recall that $\sum_{i=1}^n (X_i - \bar{X}_n)^2 = \sum_{i=1}^n X_i^2 - n\bar{X}_n^2$, so that $S_n^2 = \frac{1}{n}\sum_{i=1}^n X_i^2 - \bar{X}_n^2$. Since $EX_i^2 = Var(X_i) + (EX_i)^2 = \sigma^2 + \mu^2$, the WLLN applies to the i.i.d. r.v.'s $X_1^2, \ldots, X_n^2$ and gives: $\frac{1}{n}\sum_{i=1}^n X_i^2 \xrightarrow[n\to\infty]{P} \sigma^2 + \mu^2$. Also, $\bar{X}_n \xrightarrow[n\to\infty]{P} \mu$, by the WLLN again, and then $\bar{X}_n^2 \xrightarrow[n\to\infty]{P} \mu^2$ by Theorem 5(i). Then, by Theorem 5(ii),

$$\frac{1}{n}\sum_{i=1}^n X_i^2 - \bar{X}_n^2 \xrightarrow[n\to\infty]{P} (\sigma^2 + \mu^2) - \mu^2 = \sigma^2,$$

which is what part (i) asserts.

(ii) Part (i) and Theorem 5(i) imply that $S_n \xrightarrow[n\to\infty]{P} \sigma$, or $\frac{S_n}{\sigma} \xrightarrow[n\to\infty]{P} 1$. By Theorem 4, $\frac{\sqrt{n}(\bar{X}_n - \mu)}{\sigma} \xrightarrow[n\to\infty]{d} Z \sim N(0, 1)$. Then Theorem 6(iii) applies and gives:

$$\frac{\sqrt{n}(\bar{X}_n - \mu)/\sigma}{S_n/\sigma} = \frac{\sqrt{n}(\bar{X}_n - \mu)}{S_n} \xrightarrow[n\to\infty]{d} Z \sim N(0, 1). \quad \blacktriangle$$

Remark 6. Part (ii) of the theorem states, in effect, that for sufficiently large n, σ may be replaced in the CLT by the adjusted sample standard deviation S_n and the resulting expression still has a distribution that is close to the $N(0, 1)$ distribution.

The WLLN states that $\bar{X}_n \xrightarrow[n\to\infty]{d} \mu$, which, for a real-valued continuous function g, implies that:

$$g(\bar{X}_n) \xrightarrow[n\to\infty]{P} g(\mu).$$

On the other hand, the CLT states that:

$$\frac{\sqrt{n}(\bar{X}_n - \mu)}{\sigma} \xrightarrow[n\to\infty]{d} N(0, 1) \quad \text{or} \quad \sqrt{n}(\bar{X}_n - \mu) \xrightarrow[n\to\infty]{d} N(0, \sigma^2). \tag{12.17}$$

Then the question is what happens to the distribution of $g(\bar{X}_n)$. In other words, is there a result analogous to (12.17) when the distribution of $g(\bar{X}_n)$ is involved? The question is answered by the following result.

Theorem 8. Let $X_1, \ldots, X_n$ be i.i.d. r.v.'s with finite mean μ and variance $\sigma^2 \in (0, \infty)$, and let $g : \Re \to \Re$ be differentiable with derivative g' continuous at μ. Then:

$$\sqrt{n}[g(\bar{X}_n) - g(\mu)] \xrightarrow[n\to\infty]{d} N(0, [\sigma g'(\mu)]^2). \tag{12.18}$$

The proof of this result involves the employment of some of the theorems established in this chapter, including the CLT, along with a Taylor expansion. The proof itself will not be presented, and this section will be concluded with an application to this theorem. The method of establishing asymptotic normality for $g(\bar{X}_n)$ is often referred to as the *delta method*, and it also applies in cases more general than the one described here.

Application to this Theorem. Let the independent r.v.'s $X_1, \ldots, X_n$ be distributed as $B(1, p)$. Then

$$\sqrt{n}[\bar{X}_n(1 - \bar{X}_n) - pq] \xrightarrow[n\to\infty]{d} N(0, pq(1 - 2p)^2) \quad (q = 1 - p). \tag{12.19}$$

Proof. Here $\mu = p, \sigma^2 = pq$, and $g(x) = x(1 - x)$, so that $g'(x) = 1 - 2x$. Since $g(\bar{X}_n) = \bar{X}_n(1 - \bar{X}_n)$, $g(\mu) = p(1 - p) = pq$, and $g'(\mu) = 1 - 2p$, the convergence in (12.18) becomes as stated in (12.19). ▲

Exercises.

3.1 Let $X_1, \ldots, X_n$ be i.i.d. r.v.'s with finite $EX_i = \mu$, and $Var(X_i) = \sigma^2 \in (0, \infty)$ so that the CLT holds; that is,

$$\frac{\sqrt{n}(\bar{X}_n - \mu)}{\sigma} \xrightarrow[n\to\infty]{d} Z \sim N(0, 1), \quad \text{where } \bar{X}_n = \frac{1}{n}\sum_{i=1}^{n} X_i.$$

Then use Theorem 6 in order to show that the WLLN also holds.

3.2 Let $X_1, \ldots, X_n$ be i.i.d. r.v.'s with finite $EX_i = \mu$, and finite $Var(X_i) = \sigma^2$. Then use the identity (see also Exercise 3.1(i) in Chapter 10)

$$\sum_{i=1}^{n}(X_i - \bar{X}_n)^2 = \sum_{i=1}^{n} X_i^2 - n\bar{X}_n^2,$$

the WLLN, and Theorems 5, 6, and 1 in order to show that:

$$\frac{1}{n-1}\sum_{i=1}^{n}(X_i - \bar{X}_n)^2 \xrightarrow[n\to\infty]{P} \sigma^2.$$

CHAPTER

An Overview of Statistical Inference

13

A review of the previous chapters reveals that the main objectives throughout have been those of calculating probabilities or certain summary characteristics of a distribution, such as mean, variance, median, and mode. However, for these calculations to result in numerical answers, it is a prerequisite that the underlying distribution be completely known. Typically, this is rarely, if ever, the case. The reason for this is that the parameters that appear, for example, in the functional form of the p.d.f. of a distribution are simply unknown to us. The only thing known about them is that they lie in specified sets of possible values for these parameters, the *parameter space*.

It is at this point where statistical inference enters the picture. Roughly speaking, the aim of statistical inference is to make certain determinations with regard to the unknown constants (*parameters*) figuring in the underlying distribution. This is to be done on the basis of data, represented by the observed values of a random sample drawn from said distribution. Actually, this is the so-called *parametric statistical inference*, as opposed to the *nonparametric statistical inference*. The former is applicable to distributions, which are completely determined by the knowledge of a finite number of parameters. The latter applies to distributions not determined by any finite number of parameters.

The remaining part of this chapter is concerned with a brief overview of statistical inference, and mostly with parametric statistical inference. Within the framework of parametric statistical inference, there are three main objectives, depending on what kind of determinations we wish to make with regard to the parameters. If the objective is to arrive at a number, by means of the available data, as the value of an unknown parameter, then we are talking about *point estimation*. If, on the other hand, we are satisfied with the statement that an unknown parameter lies within a known random interval (i.e., an interval with r.v.'s as its end-points) with high prescribed probability, then we are dealing with *interval estimation* or *confidence intervals*. Finally, if the objective is to decide that an unknown parameter lies in a specified subset of the parameter space, then we are in the area of *testing hypotheses*.

These three subjects—point estimation, interval estimation, and testing hypotheses—are briefly discussed in the following three sections. In the subsequent three sections, it is pointed out what the statistical inference issues are in specific models—a *regression model* and two *analysis of variance models*. The final section touches upon some aspects of *nonparametric statistical inference*.

Introduction to Probability, Second Edition. http://dx.doi.org/10.1016/B978-0-12-800041-0.00013-4

13.1 The Basics of Point Estimation

The problem here, briefly stated, is as follows. Let X be an r.v. with a p.d.f. f of known functional form, which, however, involves a parameter. This is the case, for instance, in binomial distribution $B(1, p)$, Poisson distribution $P(\lambda)$, negative exponential $f(x) = \lambda e^{-\lambda x}$, $x > 0$ distribution, uniform distribution $U(0, \alpha)$, and normal distribution $N(\mu, \sigma^2)$ with one of the quantities μ and σ^2 known. The *parameter* is usually denoted by θ, and the set of its possible values is denoted by Ω and is called the *parameter space*. In order to emphasize the fact that the p.d.f. depends on θ, we write $f(\cdot; \theta)$. Thus, in the distributions mentioned above, we have for the respective p.d.f.'s and the parameter spaces:

$$f(x; \theta) = \theta^x (1-\theta)^{1-x}, \quad x = 0, 1, \quad \theta \in \Omega = (0, 1).$$

The situations described in Examples 5, 6, 8, and 9 of Chapter 1 may be described by a binomial distribution.

$$f(x; \theta) = \frac{e^{-\theta}\theta^x}{x!}, \quad x = 0, 1, \ldots, \quad \theta \in \Omega = (0, \infty).$$

A Poisson distribution can be used appropriately in the case described in Example 11 of Chapter 1.

$$f(x; \theta) = \theta e^{-\theta x}, \quad x > 0, \quad \theta \in \Omega = (0, \infty).$$

For an application of a negative exponential distribution, see, Example 15 in Chapter 1.

$$f(x; \theta) = \begin{cases} \frac{1}{\theta}, & 0 < x < \theta \\ 0, & \text{otherwise}, \end{cases} \qquad \theta \in \Omega = (0, \infty).$$

$$f(x; \theta) = \frac{1}{\sqrt{2\pi}\sigma} e^{-\frac{(x-\theta)^2}{2\sigma^2}}, \quad x \in \Re, \quad \theta \in \Omega = \Re, \quad \sigma^2 \text{ known},$$

and

$$f(x; \theta) = \frac{1}{\sqrt{2\pi\theta}} e^{-\frac{(x-\mu)^2}{2\theta}}, \quad x \in \Re, \quad \theta \in \Omega = (0, \infty), \quad \mu \text{ known}.$$

Normal distributions are suitable for modeling the situations described in Example 14 of Chapter 1.

Our objective is to draw a random sample of size n, $X_1, \ldots, X_n$, from the underlying distribution, and on the basis of it to construct a *point estimate* (*or estimator*) for θ; that is, a statistic $\hat{\theta} = \hat{\theta}(X_1, \ldots, X_n)$, which is used for estimating θ, where a *statistic* is a known function of the random sample $X_1, \ldots, X_n$. If $x_1, \ldots, x_n$ are the actually observed values of the r.v.'s $X_1, \ldots, X_n$, respectively, then the observed value of our estimate has the numerical value $\hat{\theta}(x_1, \ldots, x_n)$. The observed values $x_1, \ldots, x_n$ are also referred to as *data*. Then, on the basis of the available data, it is declared that the value of θ is $\hat{\theta}(x_1, \ldots, x_n)$ from among all possible points in Ω. A point estimate is often referred to just as an estimate, and the notation $\hat{\theta}$ is used indiscriminately, both for the estimate $\hat{\theta}(X_1, \ldots, X_n)$ (which is an r.v.) and for its observed value $\hat{\theta}(x_1, \ldots, x_n)$ (which is just a number).

The only obvious restriction on $\hat{\theta}(x_1, \ldots, x_n)$ is that it lies in Ω for all possible values of $X_1, \ldots, X_n$. Apart from it, there are any number of estimates one may construct—thus, the need to assume certain principles and/or invent methods for constructing $\hat{\theta}$. Perhaps the most widely accepted principle is the so-called *maximum likelihood* (ML). This principle dictates that we form the joint p.d.f. of the x_i's, for the observed values of the X_i's, look at this joint p.d.f. as a function of θ (the *likelihood function*), and maximize the likelihood function with respect to θ. The maximizing point (assuming it exists and is unique) is a function of $x_1, \ldots, x_n$, and is what we call the *Maximum Likelihood Estimate* (MLE) of θ. The notation used for the likelihood function is $L(\theta \mid x_1, \ldots, x_n)$. Then, we have that:

$$L(\theta|x_1, \ldots, x_n) = f(x_1; \theta) \cdots f(x_n; \theta), \quad \theta \in \Omega.$$

Another principle often used in constructing an estimate for θ is that of *unbiasedness*. In this context, an estimate is usually denoted by $U = U(X_1, \ldots, X_n)$. Then the principle of unbiasedness dictates that U should be constructed so as to be *unbiased*; that is, its expectation (mean value) should always be θ, no matter what the value of θ in Ω. More formally, $E_\theta U = \theta$ for all $\theta \in \Omega$. (In the expectation sign E, the parameter θ was inserted to indicate that this expectation does depend on θ, since it is calculated by using the p.d.f. $f(\cdot; \theta)$.) Now, it is intuitively clear that in comparing two unbiased estimates, one would pick the one with the smaller variance, since it would be more closely concentrated around its mean θ. Envision the case that within the class of all unbiased estimates, there exists one that has the smallest variance (and that is true for all $\theta \in \Omega$). Such an estimate is called a *Uniformly Minimum Variance Unbiased* (UMVU) estimate and is, clearly, a desirable estimate.

The principle (or rather the method) based on sample moments is another way of constructing estimates. The *method of moments*, in the simplest case, dictates to form the sample mean $\overline{X}$ and equate it with the (theoretical) mean $E_\theta X$. Then solve for θ (assuming it can be done, and, indeed, uniquely) in order to arrive at a *moment estimate* of θ.

A much more sophisticated method of constructing estimates of θ is the so-called *decision-theoretic* method. This method calls for the introduction of a host of concepts, terminology, and notation, and it will not be pursued any further here.

Finally, another relatively popular method (in particular, in the context of certain models) is *Least Squares (LS)*, based on the *Least Squares* Principle. The LS method leads to the construction of an estimate for θ, the *Least Squares Estimate* (LSE) of θ, through a minimization (with respect to θ) of the sum of certain squares. This sum of squares represents squared deviations between what we actually observe after experimentation is completed and what we would expect to have on the basis of an assumed model.

In all of the preceding discussion, it was assumed that the underlying p.d.f. depended on a single parameter, which was denoted by θ. It may very well be the case that there are two or more parameters involved. This may happen, for instance, in the uniform distribution $U(\alpha, \beta)$, $-\infty < \alpha < \beta < \infty$, where both α and β are unknown; the normal distribution, $N(\mu, \sigma^2)$, where both μ and σ^2 are unknown; and it also happens in the multinomial distribution, where the number of parameters is k, $p_1, \ldots, p_k$ (or more precisely, $k-1$, since the kth parameter, e.g., $p_k = 1 - p_1 - \cdots - p_{k-1}$). For instance, Example 16 in Chapter 1, and Examples 1 and 3 in Chapter 9 refer to situations where a multinomial distribution is appropriate. In such multiparameter cases, one simply applies to each parameter separately what was said above for a single parameter. The alternative option, to use the vector notation for the parameters involved, does simplify things in a certain way, but also introduces some complications in other ways.

13.2 The Basics of Interval Estimation

Suppose we are interested in constructing a point estimate of the mean μ in the normal distribution $N(\mu, \sigma^2)$ with known variance; this is to be done on the basis of a random sample of size n, $X_1, \ldots, X_n$, drawn from the underlying distribution. This amounts to constructing a suitable statistic of the X_i's, call it $V = V(X_1, \ldots, X_n)$, which for the observed values x_i of X_i, $i = 1, \ldots, n$ is a numerical entity, and declare it to be the (unknown) value of μ. This looks somewhat presumptuous, since from the set of possible values for μ, $-\infty < \mu < \infty$, just one is selected as its value. Thinking along these lines, it might be more reasonable to aim instead at a random interval that will contain the (unknown) value of μ with high (prescribed) probability. This is exactly what a confidence interval does.

To be more precise, casting the problem in a general setting, let $X_1, \ldots, X_n$ be a random sample from the p.d.f. $f(\cdot;\theta)$, $\theta \in \Omega \subseteq \Re$, and let $L = L(X_1, \ldots, X_n)$ and $U = U(X_1, \ldots, X_n)$ be two statistics of the X_i's such that $L < U$. Then the interval with endpoints L and U, $[L, U]$, is called a *random interval*. Let α be a small number in (0, 1), such as 0.005, 0.01, 0.05, and suppose that the random interval $[L, U]$ contains θ with probability equal to $1 - \alpha$ (such as 0.995, 0.99, 0.95) no matter what the true value of θ in Ω is. In other words, suppose that:

$$P_\theta(L \leq \theta \leq U) = 1 - \alpha \quad \text{for all } \theta \in \Omega. \tag{13.1}$$

If relation (13.1) holds, then we say that the random interval $[L, U]$ is a *confidence interval* for θ with *confidence coefficient* $1 - \alpha$.

The significance of a confidence interval is based on the relative frequency interpretation of the concept of probability, and it goes like this: Suppose n independent r.v.'s are drawn from the p.d.f. $f(\cdot;\theta)$, and let $x_1, \ldots, x_n$ be their observed values. Also, let $[L_1, U_1]$ be the interval resulting from the observed values of $L = L(X_1, \ldots, X_n)$ and $U = U(X_1, \ldots, X_n)$; that is, $L_1 = L(x_1, \ldots, x_n)$ and $U_1 = U(x_1, \ldots, x_n)$. Proceed to draw independently a second set of n r.v.'s as above, and let $[L_2, U_2]$ be the resulting interval. Repeat this process independently a large number of times, N, say, with the corresponding interval being $[L_N, U_N]$. Then the interpretation of (13.1) is that on the average, about $100(1 - \alpha)\%$ of the above N intervals will actually contain the true value of θ. For example, for $\alpha = 0.05$ and $N = 1{,}000$, the proportion of such intervals will be 95%; that is, one would expect 950 out of the 1,000 intervals constructed as above to contain the true value of θ. Empirical evidence shows that such an expectation is valid.

We may also define an *upper confidence limit* for θ, $U = U(X_1, \ldots, X_n)$, and a *lower confidence limit* for θ, $L = L(X_1, \ldots, X_n)$, both with *confidence* coefficient $1-\alpha$, if, respectively, the intervals $(-\infty, U]$ and $[L, \infty)$ are confidence intervals for θ with confidence coefficient $1-\alpha$. That is to say:

$$P_\theta(-\infty < \theta \le U) = 1-\alpha, \quad P_\theta(L \le \theta < \infty) = 1-\alpha \quad \text{for all } \theta \in \Omega. \tag{13.2}$$

Confidence intervals and upper and/or lower confidence limits can be sought, for instance, in Examples 5, 6, 8, and 9 (binomial distribution), 11 (Poisson distribution), and 14 (normal distribution) in Chapter 1.

There are some variations of (13.1) and (13.2). For example, when the underlying p.d.f. is discrete, then equalities in (13.1) and (13.2) rarely obtain for a given α and have to be replaced by inequalities $\ge$. Also, except for special cases, equalities in (13.1) and (13.2) are valid only approximately for large values of the sample size n (even in cases where the underlying r.v.'s are continuous). In such cases, we say that the respective confidence intervals (confidence limits) have *confidence coefficient approximately* $1-\alpha$.

Finally, the parameters of interest may be two (or more) rather than one, as we have assumed so far. In such cases, the concept of a confidence interval is replaced by that of a *confidence region* (in the multidimensional parameter space Ω).

13.3 The Basics of Testing Hypotheses

Often, we are not interested in a point estimate of a parameter θ or even a confidence interval for it, but rather whether said parameter lies or does not lie in a specified subset ω of the parameter space Ω. To clarify this point, we refer to some of the examples described in Chapter 1. Thus, in Example 5, all we might be interested in is whether Jones has ESP at all and not to what degree he does. In statistical terms, this amounts to taking n independent observations from a $B(1, \theta)$ distribution and, on the basis of these observations, deciding whether $\theta \in \omega = (0, 0.5]$ (as opposed to $\theta \in \omega^c = (0.5, 1)$); here θ is the probability that Jones correctly identifies the picture. The situation in Example 6 is similar, and the objective might be to decide whether or not $\theta \in \omega = (\theta_0, 1)$; here θ is the true proportion of unemployed workers and θ_0 is a certain desirable or guessed value of θ. Examples 8 and 9 in Chapter 1 fall into the same category.

In Example 11, the stipulated model is a Poisson distribution $P(\theta)$ and, on the basis of n independent observations, we might wish to decide whether or not $\theta \in (\theta_0, \infty)$, where θ_0 is a known value of θ.

In Example 14, the stipulated underlying models may be normal distributions $N(\mu_1, \sigma^2)$ and $N(\mu_2, \sigma^2)$ for the survival times X and Y, respectively, and then the question of interest may be to decide whether or not $\mu_2 \leq \mu_1$; σ^2 may be assumed to be either known or unknown. Of course, we are going to arrive at the desirable decision on the basis of two independent random samples drawn from the underlying distributions.

In Example 16 in Chapter 1, a testing hypothesis problem may be that of testing that the underlying parameters have specified values. Similarly, for Example 3 in Chapter 9.

On the basis of the preceding discussion and examples, we may now proceed with the formulation of the general problem. To this effect, let $X_1, \ldots, X_n$ be i.i.d. r.v.'s with p.d.f. $f(\cdot; \boldsymbol{\theta})$, $\boldsymbol{\theta} \in \Omega \subseteq \Re^r$, $r \geq 1$, and by means of this random sample, suppose we are interested in checking whether $\boldsymbol{\theta} \in \omega$, a proper subset of Ω, or $\boldsymbol{\theta} \in \omega^c$, the complement of ω with respect to Ω. The statements that $\boldsymbol{\theta} \in \omega$ and $\boldsymbol{\theta} \in \omega^c$ are called (*statistical*) *hypotheses* (about $\boldsymbol{\theta}$), and are denoted thus: $H_0 : \boldsymbol{\theta} \in \omega$, $H_A : \boldsymbol{\theta} \in \omega^c$. The hypothesis H_0 is called a *null* hypothesis and the hypothesis H_A is called *alternative* (to H_0) hypothesis. The hypotheses H_0 and H_A are called *simple* if they contain a single point, and *composite* otherwise. More generally, the hypotheses H_0 and H_A are called *simple*, if they completely determine the underlying distribution of the r.v.'s $X_1, \ldots, X_n$, and *composite* otherwise. The procedure of checking whether H_0 is true or not, on the basis of the observed values $x_1, \ldots, x_n$ of $X_1, \ldots, X_n$, is called *testing* the hypothesis H_0 against the alternative H_A.

In the special case that $\Omega \subseteq \Re$, some null hypotheses and the respective alternatives are as follows:

$$H_0 : \theta = \theta_0 \text{ against } H_A : \theta > \theta_0; \quad H_0 : \theta = \theta_0 \text{ against } H_A : \theta < \theta_0;$$
$$H_0 : \theta \leq \theta_0 \text{ against } H_A : \theta > \theta_0; \quad H_0 : \theta \geq \theta_0 \text{ against } H_A : \theta < \theta_0;$$
$$H_0 : \theta = \theta_0 \text{ against } H_A : \theta \neq \theta_0.$$

The testing is carried out by means of a function $\varphi : \Re^n \to [0, 1]$, which is called a *test function* or just a *test*. The number $\varphi(x_1, \ldots, x_n)$ represents the probability of rejecting H_0, given that $X_i = x_i$, $i = 1, \ldots, n$. In its simplest form, φ is the indicator of a set B in $\Re^n$, which is called the *critical* or *rejection region;* its complement B^c is called the *acceptance region*. Thus, $\varphi(x_1, \ldots, x_n) = 1$ if $(x_1, \ldots, x_n)$ is in B, and $\varphi(x_1, \ldots, x_n) = 0$, otherwise. Actually, such a test is called a *nonrandomized* test, as opposed to tests that also take values strictly between 0 and 1, called *randomized* tests. In the case of continuous distributions, nonrandomized tests suffice, but in discrete distributions, a test will typically be required to take on one or two values strictly between 0 and 1.

By using a test φ, suppose that our data $x_1, \ldots, x_n$ lead us to the rejection of H_0. This will happen, for instance, if the test φ is nonrandomized with rejection region B, and $(x_1, \ldots, x_n)$ lies in B. By rejecting the hypothesis H_0, we may be doing the correct thing, because H_0 is false (that is, $\boldsymbol{\theta} \notin \omega$). On the other hand, we may be taking the wrong action, because it may happen that H_0 is indeed true (i.e., $\theta \in \omega$), only the test and the data do not reveal it. Clearly, in so doing, we commit an error, which is referred to as *type I error*. Of course, we would like to find ways of minimizing the frequency of committing this error. To put it more mathematically, this means searching for a rejection region B, which will minimize the above frequency. In our framework, frequencies are measured by probabilities, and this leads to a determination of B so that

$$\begin{aligned} P(\text{of type I error}) &= P(\text{of rejecting } H_0 \text{ whereas } H_0 \text{ is true}) \\ &= P_\theta[(X_1, \ldots, X_n) \text{lies in } B \text{ whereas } \boldsymbol{\theta} \in \omega] \\ &= P_\theta[(X_1, \ldots, X_n) \text{lies in } B |\, \boldsymbol{\theta} \in \omega] \\ &\overset{\text{def}}{=} \alpha(\boldsymbol{\theta}) \text{ is minimum.} \end{aligned} \tag{13.3}$$

Clearly, the probabilities $\alpha(\boldsymbol{\theta})$ in (13.3) must be minimized for each $\boldsymbol{\theta} \in \omega$, since we don't know which value in ω is the true $\boldsymbol{\theta}$. This will happen if we minimize the $\max_{\boldsymbol{\theta}\in\omega} \alpha(\boldsymbol{\theta}) \overset{\text{def}}{=} \alpha$. This maximum probability of type I error is called the *level of significance* of the test employed. Thus, we are led to selecting the rejection region B so that its level of significance α will be minimum. Since $\alpha \geq 0$, its minimum could be 0, and this would happen if (essentially) $B = \varnothing$. But then (essentially) $(x_1, \ldots, x_n)$ would always be in $B^c = \Re^n$, and this would happen with probability

$$P_{\boldsymbol{\theta}}[(X_1, \ldots, X_n) \text{in } \Re^n] = 1 \quad \text{for all } \boldsymbol{\theta}. \tag{13.4}$$

This, however, creates a problem for the following reason. If the rejection region B is $\varnothing$, then the acceptance region is $\Re^n$; that is, we always accept H_0. As long as H_0 is true (i.e., $\boldsymbol{\theta} \in \omega$), this is exactly what we wish to do, but what about the case that H_0 is false (i.e., $\boldsymbol{\theta} \in \omega^c$)? When we accept a false hypothesis H_0, we commit an error, which is called the *type II error*. As in (13.3), this error is also measured in terms of probabilities; namely,

$$\begin{aligned} P(\text{of type II error}) &= P(\text{of accepting } H_0 \text{ whereas } H_0 \text{ is false}) \\ &= P_\theta[(X_1, \ldots, X_n) \text{lies in } B^c \text{ whereas } \boldsymbol{\theta} \in \omega^c] \\ &= P_\theta[(X_1, \ldots, X_n) \text{lies in } B^c |\, \boldsymbol{\theta} \in \omega^c] \\ &\overset{\text{def}}{=} \beta(\boldsymbol{\theta}). \end{aligned} \tag{13.5}$$

According to (13.5), these probabilities would be 1 for all $\boldsymbol{\theta} \in \omega^c$ (actually, for all $\boldsymbol{\theta} \in \Omega$), if $B = \varnothing$. Clearly, this is highly undesirable. The preceding discussion then leads to the conclusion that the rejection region B must be different from $\varnothing$ and then α will be > 0. The objective then becomes that of choosing B so that α will have a preassigned acceptable value (such as 0.005, 0.01, 0.05) and, subject to this restriction, the probabilities of type II error are minimized. That is,

$$\beta(\boldsymbol{\theta}) = P_{\boldsymbol{\theta}}[(X_1, \ldots, X_n) \text{lies in } B^c] \text{ is minimum for each } \boldsymbol{\theta} \in \omega^c. \tag{13.6}$$

Since $P_{\boldsymbol{\theta}}[(X_1, \ldots, X_n) \text{lies in } B^c] = 1 - P_{\boldsymbol{\theta}}[(X_1, \ldots, X_n) \text{lies in } B]$, the minimization in (13.6) is equivalent to the maximization of

$$P_{\boldsymbol{\theta}}[(X_1, \ldots, X_n) \text{lies in } B] = 1 - P_{\boldsymbol{\theta}}[(X_1, \ldots, X_n) \text{lies in } B^c] \quad \text{for all } \boldsymbol{\theta} \in \omega^c.$$

The function $\pi(\boldsymbol{\theta})$, $\boldsymbol{\theta} \in \omega^c$, defined by:

$$\pi(\boldsymbol{\theta}) = P_{\boldsymbol{\theta}}[(X_1, \ldots, X_n) \text{lies in } B], \qquad \boldsymbol{\theta} \in \omega^c, \tag{13.7}$$

is called the *power* of the test employed. So, power of a test = 1− probability of a type II error, and we may summarize our objective as follows: Choose a test with a preassigned level of significance α, which has maximum power among all tests with level of significance $\leq \alpha$. In other words, if φ is the desirable test, then it should satisfy the requirements: The level of significance of φ is α, and its power, to be denoted by $\pi_\varphi(\boldsymbol{\theta})$, $\boldsymbol{\theta} \in \omega^c$, satisfies the inequality $\pi_\varphi(\boldsymbol{\theta}) \geq \pi_{\varphi^*}(\boldsymbol{\theta})$ for all $\boldsymbol{\theta} \in \omega^c$ and any test φ^* with level of significance $\leq \alpha$.

Such a test φ, should it exist, is called *Uniformly Most Powerful* (UMP) for obvious reasons. (The term "most powerful" is explained by the inequality $\pi_\varphi(\boldsymbol{\theta}) \geq \pi_{\varphi^*}(\boldsymbol{\theta})$, and the term "uniformly" is due to the fact that this inequality must hold for *all* $\boldsymbol{\theta} \in \omega^c$.) If ω^c consists of a single point, then the concept of uniformity is void, and we talk simply of a *Most Powerful* (MP) test.

The concepts introduced so far hold for a parameter of any (finite) dimensionality. However, UMP tests can be constructed only when θ is a real-valued parameter, and then only for certain forms of H_0 and H_A and specific p.d.f.'s $f(\cdot;\boldsymbol{\theta})$. If the parameter is multidimensional, desirable tests can still be constructed; they are not going to be, in general, UMP tests, but they are derived, nevertheless, on the basis of principles that are intuitively satisfactory. Preeminent among such tests are the so-called *Likelihood Ratio* (LR) tests. Another class of tests are the so-called *goodness-of-fit* tests, and still others are constructed on the basis of *decision-theoretic* concepts.

On the basis of the random sample $X_1, \ldots, X_n$ with p.d.f. $f(\cdot;\boldsymbol{\theta}), \boldsymbol{\theta} \in \Omega \subseteq \Re^r$, $r \geq 1$, suppose we wish to test the hypothesis $H_0 : \boldsymbol{\theta} \in \omega$ (a proper) subset of Ω. It is understood that the alternative is $H_A : \boldsymbol{\theta} \in \omega^c$, but in the present framework it is not explicitly stated. Let $x_1, \ldots, x_n$ be the observed values of $X_1, \ldots, X_n$ and form the likelihood function $L(\boldsymbol{\theta}) = L(\boldsymbol{\theta} \mid x_1, \ldots, x_n) = \prod_{i=1}^n f(x_i;\boldsymbol{\theta})$. Maximize $L(\boldsymbol{\theta})$ and denote the resulting maximum by $L(\hat{\Omega})$. This maximization happens when $\boldsymbol{\theta}$ is equal to the MLE $\hat{\boldsymbol{\theta}} = \hat{\boldsymbol{\theta}}(x_1, \ldots, x_n)$, of $\boldsymbol{\theta}$, so that $L(\hat{\Omega}) = L(\hat{\boldsymbol{\theta}})$. Next, maximize the likelihood $L(\boldsymbol{\theta})$ under the restriction that $\boldsymbol{\theta} \in \omega$, and denote the resulting maximum by $L(\hat{\omega})$. Denote by $\hat{\boldsymbol{\theta}}_\omega$ the MLE of $\boldsymbol{\theta}$ subject to the restriction that $\boldsymbol{\theta} \in \omega$. Then $L(\hat{\omega}) = L(\hat{\boldsymbol{\theta}}_\omega)$. Assume now that $L(\boldsymbol{\theta})$ is continuous (in $\boldsymbol{\theta}$), and suppose that the true value of $\boldsymbol{\theta}$, call it $\boldsymbol{\theta}_0$, is in ω. It is a property of an MLE that it gets closer and closer to the true parameter as the sample size n increases. Under the assumption that $\boldsymbol{\theta}_0 \in \omega$, it follows that both $\hat{\boldsymbol{\theta}}$ and $\hat{\boldsymbol{\theta}}_\omega$ will be close to $\boldsymbol{\theta}_0$ and therefore close to each other. Then, by the assumed continuity of $L(\boldsymbol{\theta})$, the quantities $L(\hat{\boldsymbol{\theta}})$ and $L(\hat{\boldsymbol{\theta}}_\omega)$ are close together, so that the ratio

$$\lambda(x_1, \ldots, x_n) = \lambda = L(\hat{\boldsymbol{\theta}}_\omega)/L(\hat{\boldsymbol{\theta}}) \tag{13.8}$$

(which is always $\leq$1) is close to 1. On the other hand, if $\boldsymbol{\theta}_0 \in \omega^c$, then $\hat{\boldsymbol{\theta}}$ and $\hat{\boldsymbol{\theta}}_\omega$ are not close together, and therefore $L(\hat{\boldsymbol{\theta}})$ and $L(\hat{\boldsymbol{\theta}}_\omega)$ need not be close to each other either. Thus, the ratio $L(\hat{\boldsymbol{\theta}}_\omega)/L(\hat{\boldsymbol{\theta}})$ need not be close to 1. These considerations lead to the following test:

$$\text{Reject } H_0 \text{ when } \lambda < \lambda_0, \text{ where } \lambda_0 \text{ is a constant to be determined.} \tag{13.9}$$

By the monotonicity of the function $y = \log x$, the inequality $\lambda < \lambda_0$ is equivalent to $-2\log\lambda(x_1, \ldots, x_n) > C(= -2\log\lambda_0)$ (where, as always, log is the natural logarithm). It is seen that an approximate determination of C is made by the fact that, under certain conditions, the distribution of $-2\log\lambda(X_1,\ldots,X_n)$ is χ^2_f, where $f =$ dimension of Ω − dimension of ω. Namely:

$$\text{Reject } H_0 \text{ when } -2\log\lambda > C, \text{ where } C \simeq \chi^2_{f;\alpha}. \tag{13.10}$$

In closing this section, it is to be mentioned that the concept of P-value is another way of looking at a test in an effort to assess how strong (or weak) the rejection of a hypothesis is. The *P-value* (probability value) of a test is defined to be the smallest probability at which the hypothesis tested would be rejected for the data at hand. Roughly put, the P-value of a test is the probability, calculated under the null hypothesis, when the observed value of the test statistic is used as if it were the cut-off point of the test. The P-value of a test (*or significance probability*) often accompanies a null hypothesis that is rejected, as an indication of the strength or weakness of rejection. The smaller the P-value, the stronger the rejection of the null hypothesis and vice versa. Strong rejection of the null hypothesis is also referred to as the result being *highly statistically significant.*

13.4 The Basics of Regression Analysis

In the last three sections, we discussed the general principles of point estimation, interval estimation, and testing hypotheses in a general setup. These principles apply, in particular, in specific models. Two such models are *regression models* and *analysis of variance models.*

A regression model in its simplest form is as follows: At fixed points $x_1,\ldots,x_n$, respective measurements $y_1,\ldots,y_n$ are taken, which may be subject to an assortment of random errors $e_1,\ldots,e_n$. Thus, the y_i's are values of r.v.'s Y_i's, which may often be assumed to have the structure: $Y_i = \beta_1 + \beta_2 x_i + e_i$, $i = 1,\ldots,n$; here β_1 and β_2 are parameters (unknown constants) of the model. For the random errors e_i, it is not unreasonable to assume that $Ee_i = 0$; we also assume that they have the same variance, $Var(e_i) = \sigma^2 \in (0,\infty)$. Furthermore, it is reasonable to assume that the e_i's are i.i.d. r.v.'s, which implies independence of the r.v.'s $Y_1,\ldots,Y_n$. It should be noted, however, that the Y_i's are *not* identically distributed, since, for instance, they have different expectations: $EY_i = \beta_1 + \beta_2 x_i$, $i = 1,\ldots,n$. Putting these assumptions together, we arrive at the following simple *linear regression model*:

$$Y_i = \beta_1 + \beta_2 x_i + e_i, \quad \text{the } e_i\text{'s are i.i.d. r.v.'s with } Ee_i = 0 \text{ and} \\ Var(e_i) = \sigma^2, \quad i = 1,\ldots,n. \tag{13.11}$$

The quantities β_1, β_2, and σ^2 are the *parameters* of the model; the Y_i's are independent but not identically distributed; also, $EY_i = \beta_1 + \beta_2 x_i$ and $Var(Y_i) = \sigma^2$, $i = 1,\ldots,n$.

The term "regression" derives from the way the Y_i's are produced from the x_i's, and the term "linear" indicates that the parameters β_1 and β_2 enter into the model raised to the first power.

The main problems in connection with model (13.11) are to estimate the parameters β_1, β_2, and σ^2; construct confidence intervals for β_1 and β_2; test hypotheses about β_1 and β_2; and predict the expected value EY_{i_0} (or the value itself Y_{i_0}) corresponding to an x_{i_0}, distinct, in general, from $x_1, \ldots, x_n$. Estimates of β_1 and β_2, the LSE's, can be constructed without any further assumptions; the same for an estimate of σ^2. For the remaining parts, however, there is a need to stipulate a distribution for the e_i's. Since the e_i's are random errors, it is reasonable to assume that they are normally distributed; this then implies normal distribution for the Y_i's. Thus, model (13.11) now becomes:

$$Y_i = \beta_1 + \beta_2 x_i + e_i, \quad \text{the } e_i\text{'s are independently distributed as } N(0, \sigma^2), \quad i = 1, \ldots, n. \tag{13.12}$$

Under model (13.12), the MLE's of β_1, β_2, and σ^2 are derived, and their distributions are determined. This allows us to pursue the resolution of the parts of constructing confidence intervals, testing hypotheses, and prediction.

13.5 The Basics of Analysis of Variance

Analysis of Variance (ANOVA) is a powerful technique, which provides the means of assessing and/or comparing several entities. ANOVA can be used effectively in many situations; in particular, it can be used in assessing and/or comparing crop yields corresponding to different soil treatments, or crop yields corresponding to different soils and fertilizers; the comparison of a certain brand of gasoline with or without an additive by using it in several cars; the comparison of different brands of gasoline by using them in several cars; the comparison of the wearing of different materials; the comparison of the effect of different types of oil on the wear of several piston rings, etc.; the comparison of the yields of a chemical substance by using different catalytic methods; the comparison of the strengths of certain objects made of different batches of some material; the comparison of test scores from different schools and different teachers, etc.; and identification of the melting point of a metal by using different thermometers.

Assessment and comparisons are done by way of point estimation, interval estimation, and testing hypotheses, as these techniques apply to the specific ANOVA models to be considered. The more factors involved in producing an outcome, the more complicated the model becomes. However, the basic ideas remain the same throughout.

For the sake of illustrating the issues involved, consider the so-called *one-way layout* or *one-way classification* model. Consider, for example, unleaded regular gasoline, and suppose we supply ourselves with amounts of it purchased from I different companies. The objective is to compare these I brands of gasoline from yield viewpoint. To this end, a car (or several but similar cars) operates under each one of the I brands of gasoline for J runs in each case. Let Y_{ij} be the number of miles per hour for the jth run when the ith brand of gasoline is used. Then the Y_{ij}'s are r.v.'s for which the following structure is assumed: For a given i, the actual number of miles per hour for the jth run varies around a *mean* value μ_i, and these variations are due to an assortment of random errors e_{ij}. In other words, it makes sense to assume that

$Y_{ij} = \mu_i + e_{ij}$. It is also reasonable to assume that the random errors e_{ij} are independent r.v.'s distributed as $N(0, \sigma^2)$, some unknown variance σ^2. Thus, we have stipulated the following model:

$$Y_{ij} = \mu_i + e_{ij}, \quad \text{where the } e_{ij}\text{'s are independently} \\ \sim N(0, \sigma^2), \quad i = 1, \ldots, I(\geq 2), \quad j = 1, \ldots, J(\geq 2). \tag{13.13}$$

The quantities $\mu_i, i = 1, \ldots, I$, and σ^2 are the *parameters* of the model.

It follows that the r.v.'s Y_{ij} are independent and $Y_{ij} \sim N(\mu_i, \sigma^2)$, $j = 1, \ldots, J$, $i = 1, \ldots, I$.

The issues of interest here are those of estimating the μ_i's (mean number of miles per hour for the ith brand of gasoline) and σ^2. Also, we wish to test the hypothesis that there is really no difference between these I different brands of gasoline; in other words, test $H_0 : \mu_1 = \cdots = \mu_I (= \mu$, say, unknown). Should this hypothesis be rejected, we would wish to identify the brands of gasoline that cause the rejection. This can be done by constructing a confidence interval for certain linear combinations of the μ_i's called *contrasts*. That is, $\sum_{i=1}^{I} c_i \mu_i$, where $c_1, \ldots, c_I$ are constants with $\sum_{i=1}^{I} c_i = 0$.

Instead of having one factor (gasoline brand) affecting the outcome (number of miles per hour), there may be two (or more) such factors. For example, there might be some chemical additives meant to enhance the mileage. In this framework, suppose there are J such chemical additives, and let us combine each one of the I brands of gasoline with each one of the J chemical additives. For simplicity, suppose we take just one observation, Y_{ij}, on each one of the IJ pairs. Then it makes sense to assume that the r.v. Y_{ij} is the result of the following additive components: A basic quantity (*grand mean*) μ, the same for all i and j; an *effect* α_i due to the ith brand of gasoline (the jth *row effect*); an *effect* β_j due to the jth chemical additive (the jth *column effect*); and, of course, the random error e_{ij} due to a host of causes. So, the assumed model is then: $Y_{ij} = \mu + \alpha_i + \beta_j + e_{ij}$. As usually, we assume that the e_{ij}'s are independent $\sim N(0, \sigma^2)$ with some (unknown) variance σ^2, which implies that the Y_{ij}'s are independent r.v.'s and $Y_{ij} \sim N(\mu + \alpha_i + \beta_j, \sigma^2)$. We further assume that some of α_i effects are ≥ 0, some are < 0, and on the whole $\sum_{i=1}^{I} \alpha_i = 0$; and likewise for the β_j effects: $\sum_{j=1}^{J} \beta_j = 0$. Summarizing these assumptions, we have then:

$$Y_{ij} = \mu + \alpha_i + \beta_j + e_{ij}, \quad \text{where the } e_{ij}\text{'s are independently} \\ \sim N(0, \sigma^2), \quad i = 1, \ldots, I(\geq 2), \quad j = 1, \ldots, J(\geq 2), \\ \sum_{i=1}^{I} \alpha_i = 0, \quad \sum_{j=1}^{J} \beta_j = 0. \tag{13.14}$$

The quantities $\mu, \alpha_i, i = 1, \ldots, I, \beta_j, j = 1, \ldots, J$ and σ^2 are the *parameters* of the model.

As already mentioned, the implication is that the r.v.'s Y_{ij} are independent and $Y_{ij} \sim N(\mu + \alpha_i + \beta_j, \sigma^2), i = 1, \ldots, I, j = 1, \ldots, J$.

The model described by (13.14) is called *two-way layout* or *two-way classification*, as the observations are affected by two factors.

The main statistical issues are those of estimating the parameters involved and testing irrelevance of either one of the factors involved—that is, testing $H_{0A}: \alpha_1 = \cdots = \alpha_I = 0$, $H_{0B}: \beta_1 = \cdots = \beta_J = 0$.

13.6 The Basics of Nonparametric Inference

All of the problems discussed in the previous sections may be summarized as follows: On the basis of a random sample of size n, $X_1, \ldots, X_n$, drawn from the p.d.f. $f(\cdot; \theta), \theta \in \Omega \subseteq \Re$, construct a point estimate and a confidence interval for θ, and test hypotheses about θ. In other words, the problems discussed were those of making (*statistical*) *inference* about θ. These problems are suitably modified for a multidimensional parameter. The fundamental assumption in this framework is that the functional form of the p.d.f. $f(\cdot; \theta)$ is known; the only thing that does not render $f(\cdot; \theta)$ completely known is the presence of the (unknown constant) parameter θ.

In many situations, stipulating a functional form for $f(\cdot; \theta)$ either is dictated by circumstances or is the product of accumulated experience. In the absence of these, we must still proceed with the problems of estimating important quantities, either by points or by intervals, and testing hypotheses about them. However, the framework now is *nonparametric*, and the relevant inference is referred to as *nonparametric inference*.

Actually, there have been at least three cases so far where *nonparametric estimation* was made without referring to it as such. Indeed, if $X_1, \ldots, X_n$ are i.i.d. r.v.'s with unknown mean μ, then the sample mean $\bar{X}_n$ may be taken as an estimate of μ, regardless of what the underlying distribution of the X_i's is. This estimate is recommended on the basis of at least three considerations. First, it is unbiased, $E\bar{X}_n = \mu$ no matter what the underlying distribution is; second, $\bar{X}_n$ is the moment estimate of μ; and third, by the WLLN, $\bar{X}_n \xrightarrow[n\to\infty]{P} \mu$, so that $\bar{X}_n$ is close to μ, in the sense of probability, for all sufficiently large n. Now suppose that the X_i's also have (an unknown) variance $\sigma^2 \in (0, \infty)$. Define the sample variance S_n^2 by $S_n^2 = \sum_{i=1}^n (X_i - \mu)^2/n$ when μ is known, and by $S_n^2 = \sum_{i=1}^n (X_i - \bar{X}_n)^2/(n-1)$ otherwise. Then S_n^2 can be used as an estimate of σ^2, because it is unbiased (Exercise 3.1(iii) in Chapter 10), and $S_n^2 \xrightarrow[n\to\infty]{P} \sigma^2$ (Theorem 7(i) and Exercise 3.2, both in Chapter 12). Furthermore, by combining $\bar{X}_n$ and S_n^2 and using Theorem 7(ii) in Chapter 12 we have that $\sqrt{n}(\bar{X}_n - \mu)/S_n \simeq N(0, 1)$ for large n. Then, for such n, $[\bar{X}_n - z_{\alpha/2}\frac{S_n}{\sqrt{n}}, \bar{X}_n + z_{\alpha/2}\frac{S_n}{\sqrt{n}}]$ is a confidence interval for μ with confidence coefficient approximately $1 - \alpha$.

Also, the (unknown) d.f. F of the X_i's has been estimated at every point $x \in \Re$ by the empirical d.f. F_n (see, Application 12.2.1(5) in Chapter 12). The estimate F_n has at least two desirable properties. For all $x \in \Re$ and regardless of the form of the d.f. F: $EF_n(x) = F(x)$ and $F_n(x) \xrightarrow[n\to\infty]{P} F(x)$.

What has not been done so far is to estimate the p.d.f. $f(x)$ at each $x \in \Re$, under certain regularity conditions, which do not include postulation of a functional form for f. There are several ways of doing this; the so-called *kernel method* of estimating f being perhaps the most popular. The resulting estimate enjoys several desirable properties.

Regarding testing hypotheses, a hypothesis testing problem could be that of testing the hypothesis that the (unknown) d.f. F is actually equal to a known one, F_0; that is $H_0 : F = F_0$, the alternative H_A being that $F(x) \neq F_0(x)$ for at least one $x \in \Re$. Actually, from a practical viewpoint, it is more important to compare two (unknown) d.f.'s F and G, by stipulating $H_0 : F = G$. The alternative can be any one of the following: $H_A : F \neq G$, $H'_A : F > G$, $H''_A : F < G$, in the sense that $F(x) \geq G(x)$ or $F(x) \leq G(x)$, respectively, for all $x \in \Re$, and strict inequality for at least one x. In carrying out the appropriate tests, one has to use some pretty sophisticated asymptotic results regarding empirical d.f.'s. An alternative approach to using empirical d.f.'s is to employ the concept of a *rank* test or the concept of a *sign* test. In such a context, $F \neq G$ means that either $F > G$ or $F < G$ as defined previously; thus, it cannot be $F(x) > G(x)$ for some x's, and $F(x) < G(x)$ for some other x's. This section is concluded with the concept of regression estimation, but in a nonparametric framework. In such a situation, what is estimated is an entire function rather than a few parameters.

All topics touched upon in this chapter are discussed in considerable detail in Chapters 9 through 15 of the book *An Introduction to Probability Theory and Statistical Inference*, 2nd printing (2005), Academic Press, by G. G. Roussas.

Appendix

Tables A.1–A.6

Table A.1 Cumulative Binomial Distribution

The tabulated quantity is

$$\sum_{j=0}^{k} \binom{n}{j} p^j (1-p)^{n-j}.$$

		p							
n	**k**	**1/16**	**2/16**	**3/16**	**4/16**	**5/16**	**6/16**	**7/16**	**8/16**
2	0	0.8789	0.7656	0.6602	0.5625	0.4727	0.3906	0.3164	0.2500
	1	0.9961	0.9844	0.9648	0.9375	0.9023	0.8594	0.8086	0.7500
	2	1.0000	1.0000	1.0000	1.0000	1.0000	1.0000	1.0000	1.0000
3	0	0.8240	0.6699	0.5364	0.4219	0.3250	0.2441	0.1780	0.1250
	1	0.9888	0.9570	0.9077	0.8437	0.7681	0.6836	0.5933	0.5000
	2	0.9998	0.9980	0.9934	0.9844	0.9695	0.9473	0.9163	0.8750
	3	1.0000	1.0000	1.0000	1.0000	1.0000	1.0000	1.0000	1.0000
4	0	0.7725	0.5862	0.4358	0.3164	0.2234	0.1526	0.1001	0.0625
	1	0.9785	0.9211	0.8381	0.7383	0.6296	0.5188	0.4116	0.3125
	2	0.9991	0.9929	0.9773	0.9492	0.9065	0.8484	0.7749	0.6875
	3	1.0000	0.9998	0.9988	0.9961	0.9905	0.9802	0.9634	0.9375
	4	1.0000	1.0000	1.0000	1.0000	1.0000	1.0000	1.0000	1.0000
5	0	0.7242	0.5129	0.3541	0.2373	0.1536	0.0954	0.0563	0.0312
	1	0.9656	0.8793	0.7627	0.6328	0.5027	0.3815	0.2753	0.1875
	2	0.9978	0.9839	0.9512	0.8965	0.8200	0.7248	0.6160	0.5000
	3	0.9999	0.9989	0.9947	0.9844	0.9642	0.9308	0.8809	0.8125
	4	1.0000	1.0000	0.9998	0.9990	0.9970	0.9926	0.9840	0.9687
	5	1.0000	1.0000	1.0000	1.0000	1.0000	1.0000	1.0000	1.0000
6	0	0.6789	0.4488	0.2877	0.1780	0.1056	0.0596	0.0317	0.0156
	1	0.9505	0.8335	0.6861	0.5339	0.3936	0.2742	0.1795	0.1094
	2	0.9958	0.9709	0.9159	0.8306	0.7208	0.5960	0.4669	0.3437
	3	0.9998	0.9970	0.9866	0.9624	0.9192	0.8535	0.7650	0.6562
	4	1.0000	0.9998	0.9988	0.9954	0.9868	0.9694	0.9389	0.8906
	5	1.0000	1.0000	1.0000	0.9998	0.9991	0.9972	0.9930	0.9844
	6	1.0000	1.0000	1.0000	1.0000	1.0000	1.0000	1.0000	1.0000

(continued)

Introduction to Probability, Second Edition. http://dx.doi.org/10.1016/B978-0-12-800041-0.00020-1

Table A.1 (*Continued*)

		p							
n	***k***	**1/16**	**2/16**	**3/16**	**4/16**	**5/16**	**6/16**	**7/16**	**8/16**
7	0	0.6365	0.3927	0.2338	0.1335	0.0726	0.0373	0.0178	0.0078
	1	0.9335	0.7854	0.6114	0.4449	0.3036	0.1937	0.1148	0.0625
	2	0.9929	0.9537	0.8728	0.7564	0.6186	0.4753	0.3412	0.2266
	3	0.9995	0.9938	0.9733	0.9294	0.8572	0.7570	0.6346	0.5000
	4	1.0000	0.9995	0.9965	0.9871	0.9656	0.9260	0.8628	0.7734
	5	1.0000	1.0000	0.9997	0.9987	0.9952	0.9868	0.9693	0.9375
	6	1.0000	1.0000	1.0000	0.9999	0.9997	0.9990	0.9969	0.9922
	7	1.0000	1.0000	1.0000	1.0000	1.0000	1.0000	1.0000	1.0000
8	0	0.5967	0.3436	0.1899	0.1001	0.0499	0.0233	0.0100	0.0039
	1	0.9150	0.7363	0.5406	0.3671	0.2314	0.1350	0.0724	0.0352
	2	0.9892	0.9327	0.8238	0.6785	0.5201	0.3697	0.2422	0.1445
	3	0.9991	0.9888	0.9545	0.8862	0.7826	0.6514	0.5062	0.3633
	4	1.0000	0.9988	0.9922	0.9727	0.9318	0.8626	0.7630	0.6367
	5	1.0000	0.9999	0.9991	0.9958	0.9860	0.9640	0.9227	0.8555
	6	1.0000	1.0000	0.9999	0.9996	0.9983	0.9944	0.9849	0.9648
	7	1.0000	1.0000	1.0000	1.0000	0.9999	0.9996	0.9987	0.9961
	8	1.0000	1.0000	1.0000	1.0000	1.0000	1.0000	1.0000	1.0000
9	0	0.5594	0.3007	0.1543	0.0751	0.0343	0.0146	0.0056	0.0020
	1	0.8951	0.6872	0.4748	0.3003	0.1747	0.0931	0.0451	0.0195
	2	0.9846	0.9081	0.7707	0.6007	0.4299	0.2817	0.1679	0.0898
	3	0.9985	0.9817	0.9300	0.8343	0.7006	0.5458	0.3907	0.2539
	4	0.9999	0.9975	0.9851	0.9511	0.8851	0.7834	0.6506	0.5000
	5	1.0000	0.9998	0.9978	0.9900	0.9690	0.9260	0.8528	0.7461
	6	1.0000	1.0000	0.9998	0.9987	0.9945	0.9830	0.9577	0.9102
	7	1.0000	1.0000	1.0000	0.9999	0.9994	0.9977	0.9926	0.9805
	8	1.0000	1.0000	1.0000	1.0000	1.0000	0.9999	0.9994	0.9980
	9	1.0000	1.0000	1.0000	1.0000	1.0000	1.0000	1.0000	1.0000
10	0	0.5245	0.2631	0.1254	0.0563	0.0236	0.0091	0.0032	0.0010
	1	0.8741	0.6389	0.4147	0.2440	0.1308	0.0637	0.0278	0.0107
	2	0.9790	0.8805	0.7152	0.5256	0.3501	0.2110	0.1142	0.0547
	3	0.9976	0.9725	0.9001	0.7759	0.6160	0.4467	0.2932	0.1719
	4	0.9998	0.9955	0.9748	0.9219	0.8275	0.6943	0.5369	0.3770
	5	1.0000	0.9995	0.9955	0.9803	0.9428	0.8725	0.7644	0.6230
	6	1.0000	1.0000	0.9994	0.9965	0.9865	0.9616	0.9118	0.8281
	7	1.0000	1.0000	1.0000	0.9996	0.9979	0.9922	0.9773	0.9453
	8	1.0000	1.0000	1.0000	1.0000	0.9998	0.9990	0.9964	0.9893
	9	1.0000	1.0000	1.0000	1.0000	1.0000	0.9999	0.9997	0.9990
	10	1.0000	1.0000	1.0000	1.0000	1.0000	1.0000	1.0000	1.0000

(*continued*)

Table A.1 (*Continued*)

n	k	p = 1/16	2/16	3/16	4/16	5/16	6/16	7/16	8/16
11	0	0.4917	0.2302	0.1019	0.0422	0.0162	0.0057	0.0018	0.0005
	1	0.8522	0.5919	0.3605	0.1971	0.0973	0.0432	0.0170	0.0059
	2	0.9724	0.8503	0.6589	0.4552	0.2816	0.1558	0.0764	0.0327
	3	0.9965	0.9610	0.8654	0.7133	0.5329	0.3583	0.2149	0.1133
	4	0.9997	0.9927	0.9608	0.8854	0.7614	0.6014	0.4303	0.2744
	5	1.0000	0.9990	0.9916	0.9657	0.9068	0.8057	0.6649	0.5000
	6	1.0000	0.9999	0.9987	0.9924	0.9729	0.9282	0.8473	0.7256
	7	1.0000	1.0000	0.9999	0.9988	0.9943	0.9807	0.9487	0.8867
	8	1.0000	1.0000	1.0000	0.9999	0.9992	0.9965	0.9881	0.9673
	9	1.0000	1.0000	1.0000	1.0000	0.9999	0.9996	0.9983	0.9941
	10	1.0000	1.0000	1.0000	1.0000	1.0000	1.0000	0.9999	0.9995
	11	1.0000	1.0000	1.0000	1.0000	1.0000	1.0000	1.0000	1.0000
12	0	0.4610	0.2014	0.0828	0.0317	0.0111	0.0036	0.0010	0.0002
	1	0.8297	0.5467	0.3120	0.1584	0.0720	0.0291	0.0104	0.0032
	2	0.9649	0.8180	0.6029	0.3907	0.2240	0.1135	0.0504	0.0193
	3	0.9950	0.9472	0.8267	0.6488	0.4544	0.2824	0.1543	0.0730
	4	0.9995	0.9887	0.9429	0.8424	0.6900	0.5103	0.3361	0.1938
	5	1.0000	0.9982	0.9858	0.9456	0.8613	0.7291	0.5622	0.3872
	6	1.0000	0.9998	0.9973	0.9857	0.9522	0.8822	0.7675	0.6128
	7	1.0000	1.0000	0.9996	0.9972	0.9876	0.9610	0.9043	0.8062
	8	1.0000	1.0000	1.0000	0.9996	0.9977	0.9905	0.9708	0.9270
	9	1.0000	1.0000	1.0000	1.0000	0.9997	0.9984	0.9938	0.9807
	10	1.0000	1.0000	1.0000	1.0000	1.0000	0.9998	0.9992	0.9968
	11	1.0000	1.0000	1.0000	1.0000	1.0000	1.0000	1.0000	0.9998
	12	1.0000	1.0000	1.0000	1.0000	1.0000	1.0000	1.0000	1.0000
13	0	0.4321	0.1762	0.0673	0.0238	0.0077	0.0022	0.0006	0.0001
	1	0.8067	0.5035	0.2690	0.1267	0.0530	0.0195	0.0063	0.0017
	2	0.9565	0.7841	0.5484	0.3326	0.1765	0.0819	0.0329	0.0112
	3	0.9931	0.9310	0.7847	0.5843	0.3824	0.2191	0.1089	0.0461
	4	0.9992	0.9835	0.9211	0.7940	0.6164	0.4248	0.2565	0.1334
	5	0.9999	0.9970	0.9778	0.9198	0.8078	0.6470	0.4633	0.2905
	6	1.0000	0.9996	0.9952	0.9757	0.9238	0.8248	0.6777	0.5000
	7	1.0000	1.0000	0.9992	0.9944	0.9765	0.9315	0.8445	0.7095
	8	1.0000	1.0000	0.9999	0.9990	0.9945	0.9795	0.9417	0.8666
	9	1.0000	1.0000	1.0000	0.9999	0.9991	0.9955	0.9838	0.9539
	10	1.0000	1.0000	1.0000	1.0000	0.9999	0.9993	0.9968	0.9888
	11	1.0000	1.0000	1.0000	1.0000	1.0000	0.9999	0.9996	0.9983
	12	1.0000	1.0000	1.0000	1.0000	1.0000	1.0000	1.0000	0.9999
	13	1.0000	1.0000	1.0000	1.0000	1.0000	1.0000	1.0000	1.0000

(*continued*)

Table A.1 (*Continued*)

n	k	p = 1/16	2/16	3/16	4/16	5/16	6/16	7/16	8/16
14	0	0.4051	0.1542	0.0546	0.0178	0.0053	0.0014	0.0003	0.0001
	1	0.7833	0.4626	0.2312	0.1010	0.0388	0.0130	0.0038	0.0009
	2	0.9471	0.7490	0.4960	0.2811	0.1379	0.0585	0.0213	0.0065
	3	0.9908	0.9127	0.7404	0.5213	0.3181	0.1676	0.0756	0.0287
	4	0.9988	0.9970	0.8955	0.7415	0.5432	0.3477	0.1919	0.0898
	5	0.9999	0.9953	0.9671	0.8883	0.7480	0.5637	0.3728	0.2120
	6	1.0000	0.9993	0.9919	0.9167	0.8876	0.7581	0.5839	0.3953
	7	1.0000	0.9999	0.9985	0.9897	0.9601	0.8915	0.7715	0.6047
	8	1.0000	1.0000	0.9998	0.9978	0.9889	0.9615	0.8992	0.7880
	9	1.0000	1.0000	1.0000	0.9997	0.9976	0.9895	0.9654	0.9102
	10	1.0000	1.0000	1.0000	1.0000	0.9996	0.9979	0.9911	0.9713
	11	1.0000	1.0000	1.0000	1.0000	1.0000	0.9997	0.9984	0.9935
	12	1.0000	1.0000	1.0000	1.0000	1.0000	1.0000	0.9998	0.9991
	13	1.0000	1.0000	1.0000	1.0000	1.0000	1.0000	1.0000	0.9999
	14	1.0000	1.0000	1.0000	1.0000	1.0000	1.0000	1.0000	1.0000
15	0	0.3798	0.1349	0.0444	0.0134	0.0036	0.0009	0.0002	0.0000
	1	0.7596	0.4241	0.1981	0.0802	0.0283	0.0087	0.0023	0.0005
	2	0.9369	0.7132	0.4463	0.2361	0.1069	0.0415	0.0136	0.0037
	3	0.9881	0.8922	0.6946	0.4613	0.2618	0.1267	0.0518	0.0176
	4	0.9983	0.9689	0.8665	0.6865	0.4729	0.2801	0.1410	0.0592
	5	0.9998	0.9930	0.9537	0.8516	0.6840	0.4827	0.2937	0.1509
	6	1.0000	0.9988	0.9873	0.9434	0.8435	0.6852	0.4916	0.3036
	7	1.0000	0.9998	0.9972	0.9827	0.9374	0.8415	0.6894	0.5000
	8	1.0000	1.0000	0.9995	0.9958	0.9799	0.9352	0.8433	0.6964
	9	1.0000	1.0000	0.9999	0.9992	0.9949	0.9790	0.9364	0.8491
	10	1.0000	1.0000	1.0000	0.9999	0.9990	0.9947	0.9799	0.9408
	11	1.0000	1.0000	1.0000	1.0000	0.9999	0.9990	0.9952	0.9824
	12	1.0000	1.0000	1.0000	1.0000	1.0000	0.9999	0.9992	0.9963
	13	1.0000	1.0000	1.0000	1.0000	1.0000	1.0000	0.9999	0.9995
	14	1.0000	1.0000	1.0000	1.0000	1.0000	1.0000	1.0000	1.0000
	15	1.0000	1.0000	1.0000	1.0000	1.0000	1.0000	1.0000	1.0000
16	0	0.3561	0.1181	0.0361	0.0100	0.0025	0.0005	0.0001	0.0000
	1	0.7359	0.3879	0.1693	0.0635	0.0206	0.0057	0.0014	0.0003
	2	0.9258	0.6771	0.3998	0.1971	0.0824	0.0292	0.0086	0.0021
	3	0.9849	0.8698	0.6480	0.4050	0.2134	0.0947	0.0351	0.0106
	4	0.9977	0.9593	0.8342	0.6302	0.4069	0.2226	0.1020	0.0384
	5	0.9997	0.9900	0.9373	0.8103	0.6180	0.4067	0.2269	0.1051
	6	1.0000	0.9981	0.9810	0.9204	0.7940	0.6093	0.4050	0.2272
	7	1.0000	0.9997	0.9954	0.9729	0.9082	0.7829	0.6029	0.4018

(*continued*)

Table A.1 (*Continued*)

n	k	1/16	2/16	3/16	4/16	5/16	6/16	7/16	8/16
	8	1.0000	1.0000	0.9991	0.9925	0.9666	0.9001	0.7760	0.5982
	9	1.0000	1.0000	0.9999	0.9984	0.9902	0.9626	0.8957	0.7728
	10	1.0000	1.0000	1.0000	0.9997	0.9977	0.9888	0.9609	0.8949
	11	1.0000	1.0000	1.0000	1.0000	0.9996	0.9974	0.9885	0.9616
	12	1.0000	1.0000	1.0000	1.0000	0.9999	0.9995	0.9975	0.9894
	13	1.0000	1.0000	1.0000	1.0000	1.0000	0.9999	0.9996	0.9979
	14	1.0000	1.0000	1.0000	1.0000	1.0000	1.0000	1.0000	0.9997
	15	1.0000	1.0000	1.0000	1.0000	1.0000	1.0000	1.0000	1.0000
	16	1.0000	1.0000	1.0000	1.0000	1.0000	1.0000	1.0000	1.0000
17	0	0.3338	0.1033	0.0293	0.0075	0.0017	0.0003	0.0001	0.0000
	1	0.7121	0.3542	0.1443	0.0501	0.0149	0.0038	0.0008	0.0001
	2	0.9139	0.6409	0.3566	0.1637	0.0631	0.0204	0.0055	0.0012
	3	0.9812	0.8457	0.6015	0.3530	0.1724	0.0701	0.0235	0.0064
	4	0.9969	0.9482	0.7993	0.5739	0.3464	0.1747	0.0727	0.0245
	5	0.9996	0.9862	0.9180	0.7653	0.5520	0.3377	0.1723	0.0717
	6	1.0000	0.9971	0.9728	0.8929	0.7390	0.5333	0.3271	0.1662
	7	1.0000	0.9995	0.9927	0.9598	0.8725	0.7178	0.5163	0.3145
	8	1.0000	0.9999	0.9984	0.9876	0.9484	0.8561	0.7002	0.5000
	9	1.0000	1.0000	0.9997	0.9969	0.9828	0.9391	0.8433	0.6855
	10	1.0000	1.0000	1.0000	0.9994	0.9954	0.9790	0.9323	0.8338
	11	1.0000	1.0000	1.0000	0.9999	0.9990	0.9942	0.9764	0.9283
	12	1.0000	1.0000	1.0000	1.0000	0.9998	0.9987	0.9935	0.9755
	13	1.0000	1.0000	1.0000	1.0000	1.0000	0.9998	0.9987	0.9936
	14	1.0000	1.0000	1.0000	1.0000	1.0000	1.0000	0.9998	0.9988
	15	1.0000	1.0000	1.0000	1.0000	1.0000	1.0000	1.0000	0.9999
	16	1.0000	1.0000	1.0000	1.0000	1.0000	1.0000	1.0000	1.0000
18	0	0.3130	0.0904	0.0238	0.0056	0.0012	0.0002	0.0000	0.0000
	1	0.6885	0.3228	0.1227	0.0395	0.0108	0.0025	0.0005	0.0001
	2	0.9013	0.6051	0.3168	0.1353	0.0480	0.0142	0.0034	0.0007
	3	0.9770	0.8201	0.5556	0.3057	0.1383	0.0515	0.0156	0.0038
	4	0.9959	0.9354	0.7622	0.5187	0.2920	0.1355	0.0512	0.0154
	5	0.9994	0.9814	0.8958	0.7175	0.4878	0.2765	0.1287	0.0481
	6	0.9999	0.9957	0.9625	0.8610	0.6806	0.4600	0.2593	0.1189
	7	1.0000	0.9992	0.9889	0.9431	0.8308	0.6486	0.4335	0.2403
	8	1.0000	0.9999	0.9973	0.9807	0.9247	0.8042	0.6198	0.4073
	9	1.0000	1.0000	0.9995	0.9946	0.9721	0.9080	0.7807	0.5927
	10	1.0000	1.0000	0.9999	0.9988	0.9915	0.9640	0.8934	0.7597
	11	1.0000	1.0000	1.0000	0.9998	0.9979	0.9885	0.9571	0.8811
	12	1.0000	1.0000	1.0000	1.0000	0.9996	0.9970	0.9860	0.9519

(*continued*)

Table A.1 (*Continued*)

		p							
n	*k*	1/16	2/16	3/16	4/16	5/16	6/16	7/16	8/16
	13	1.0000	1.0000	1.0000	1.0000	0.9999	0.9994	0.9964	0.9846
	14	1.0000	1.0000	1.0000	1.0000	1.0000	0.9999	0.9993	0.9962
	15	1.0000	1.0000	1.0000	1.0000	1.0000	1.0000	0.9999	0.9993
	16	1.0000	1.0000	1.0000	1.0000	1.0000	1.0000	1.0000	0.9999
	17	1.0000	1.0000	1.0000	1.0000	1.0000	1.0000	1.0000	1.0000
19	0	0.2934	0.0791	0.0193	0.0042	0.0008	0.0001	0.0000	0.0000
	1	0.6650	0.2938	0.1042	0.0310	0.0078	0.0016	0.0003	0.0000
	2	0.8880	0.5698	0.2804	0.1113	0.0364	0.0098	0.0021	0.0004
	3	0.9722	0.7933	0.5108	0.2631	0.1101	0.0375	0.0103	0.0022
	4	0.9947	0.9209	0.7235	0.4654	0.2440	0.1040	0.0356	0.0096
	5	0.9992	0.9757	0.8707	0.6678	0.4266	0.2236	0.0948	0.0318
	6	0.9999	0.9939	0.9500	0.8251	0.6203	0.3912	0.2022	0.0835
	7	1.0000	0.9988	0.9840	0.9225	0.7838	0.5779	0.3573	0.1796
	8	1.0000	0.9998	0.9957	0.9713	0.8953	0.7459	0.5383	0.3238
	9	1.0000	1.0000	0.9991	0.9911	0.9573	0.8691	0.7103	0.5000
	10	1.0000	1.0000	0.9998	0.9977	0.9854	0.9430	0.8441	0.0672
	11	1.0000	1.0000	1.0000	0.9995	0.9959	0.9793	0.9292	0.8204
	12	1.0000	1.0000	1.0000	0.9999	0.9990	0.9938	0.9734	0.9165
	13	1.0000	1.0000	1.0000	1.0000	0.9998	0.9985	0.9919	0.9682
	14	1.0000	1.0000	1.0000	1.0000	1.0000	0.9997	0.9980	0.9904
	15	1.0000	1.0000	1.0000	1.0000	1.0000	1.0000	0.9996	0.9978
	16	1.0000	1.0000	1.0000	1.0000	1.0000	1.0000	1.0000	0.9996
	17	1.0000	1.0000	1.0000	1.0000	1.0000	1.0000	1.0000	1.0000
	18	1.0000	1.0000	1.0000	1.0000	1.0000	1.0000	1.0000	1.0000
20	0	0.2751	0.0692	0.0157	0.0032	0.0006	0.0001	0.0000	0.0000
	1	0.6148	0.2669	0.0883	0.0243	0.0056	0.0011	0.0002	0.0000
	2	0.8741	0.5353	0.2473	0.0913	0.0275	0.0067	0.0013	0.0002
	3	0.9670	0.7653	0.4676	0.2252	0.0870	0.0271	0.0067	0.0013
	4	0.9933	0.9050	0.6836	0.4148	0.2021	0.0790	0.0245	0.0059
	5	0.9989	0.9688	0.8431	0.6172	0.3695	0.1788	0.0689	0.0207
	6	0.9999	0.9916	0.9351	0.7858	0.5598	0.3284	0.1552	0.0577
	7	1.0000	0.9981	0.9776	0.8982	0.7327	0.5079	0.2894	0.1316
	8	1.0000	0.9997	0.9935	0.9591	0.8605	0.6829	0.4591	0.2517
	9	1.0000	0.9999	0.9984	0.9861	0.9379	0.8229	0.6350	0.4119
	10	1.0000	1.0000	0.9997	0.9961	0.9766	0.9153	0.7856	0.5881
	11	1.0000	1.0000	0.9999	0.9991	0.9926	0.9657	0.8920	0.7483
	12	1.0000	1.0000	1.0000	0.9998	0.9981	0.9884	0.9541	0.8684
	13	1.0000	1.0000	1.0000	1.0000	0.9996	0.9968	0.9838	0.9423
	14	1.0000	1.0000	1.0000	1.0000	0.9999	0.9993	0.9953	0.9793

(*continued*)

Table A.1 (*Continued*)

		p							
n	*k*	1/16	2/16	3/16	4/16	5/16	6/16	7/16	8/16
	15	1.0000	1.0000	1.0000	1.0000	1.0000	0.9999	0.9989	0.9941
	16	1.0000	1.0000	1.0000	1.0000	1.0000	1.0000	0.9998	0.9987
	17	1.0000	1.0000	1.0000	1.0000	1.0000	1.0000	1.0000	0.9998
	18	1.0000	1.0000	1.0000	1.0000	1.0000	1.0000	1.0000	1.0000
	19	1.0000	1.0000	1.0000	1.0000	1.0000	1.0000	1.0000	1.0000
21	0	0.2579	0.0606	0.0128	0.0024	0.0004	0.0001	0.0000	0.0000
	1	0.6189	0.2422	0.0747	0.0190	0.0040	0.0007	0.0001	0.0000
	2	0.8596	0.5018	0.2175	0.0745	0.0206	0.0046	0.0008	0.0001
	3	0.9612	0.7366	0.4263	0.1917	0.0684	0.0195	0.0044	0.0007
	4	0.9917	0.8875	0.6431	0.3674	0.1662	0.0596	0.0167	0.0036
	5	0.9986	0.9609	0.8132	0.5666	0.3172	0.1414	0.0495	0.0133
	6	0.9998	0.9888	0.9179	0.7436	0.5003	0.2723	0.1175	0.0392
	7	1.0000	0.9973	0.9696	0.8701	0.6787	0.4405	0.2307	0.0946
	8	1.0000	0.9995	0.9906	0.9439	0.8206	0.6172	0.3849	0.1917
	9	1.0000	0.9999	0.9975	0.9794	0.9137	0.7704	0.5581	0.3318
	10	1.0000	1.0000	0.9995	0.9936	0.9645	0.8806	0.7197	0.5000
	11	1.0000	1.0000	0.9999	0.9983	0.9876	0.9468	0.8454	0.6682
	12	1.0000	1.0000	1.0000	0.9996	0.9964	0.9799	0.9269	0.8083
	13	1.0000	1.0000	1.0000	0.9999	0.9991	0.9936	0.9708	0.9054
	14	1.0000	1.0000	1.0000	1.0000	0.9998	0.9983	0.9903	0.9605
	15	1.0000	1.0000	1.0000	1.0000	1.0000	0.9996	0.9974	0.9867
	16	1.0000	1.0000	1.0000	1.0000	1.0000	0.9999	0.9994	0.9964
	17	1.0000	1.0000	1.0000	1.0000	1.0000	1.0000	0.9999	0.9993
	18	1.0000	1.0000	1.0000	1.0000	1.0000	1.0000	1.0000	0.9999
	19	1.0000	1.0000	1.0000	1.0000	1.0000	1.0000	1.0000	1.0000
	20	1.0000	1.0000	1.0000	1.0000	1.0000	1.0000	1.0000	1.0000
22	0	0.2418	0.0530	0.0104	0.0018	0.0003	0.0000	0.0000	0.0000
	1	0.5963	0.2195	0.0631	0.0149	0.0029	0.0005	0.0001	0.0000
	2	0.8445	0.4693	0.1907	0.0606	0.0154	0.0031	0.0005	0.0001
	3	0.9548	0.7072	0.3871	0.1624	0.0535	0.0139	0.0028	0.0004
	4	0.9898	0.8687	0.6024	0.3235	0.1356	0.0445	0.0133	0.0022
	5	0.9981	0.9517	0.7813	0.5168	0.2700	0.1107	0.0352	0.0085
	6	0.9997	0.9853	0.8983	0.6994	0.4431	0.2232	0.0877	0.0267
	7	1.0000	0.9963	0.9599	0.8385	0.6230	0.3774	0.1812	0.0669
	8	1.0000	0.9992	0.9866	0.9254	0.7762	0.5510	0.3174	0.1431
	9	1.0000	0.9999	0.9962	0.9705	0.8846	0.7130	0.4823	0.2617
	10	1.0000	1.0000	0.9991	0.9900	0.9486	0.8393	0.6490	0.4159
	11	1.0000	1.0000	0.9998	0.9971	0.9804	0.9220	0.7904	0.5841
	12	1.0000	1.0000	1.0000	0.9993	0.9936	0.9675	0.8913	0.7383

(*continued*)

Table A.1 (*Continued*)

n	k	p 1/16	2/16	3/16	4/16	5/16	6/16	7/16	8/16
	13	1.0000	1.0000	1.0000	0.9999	0.9982	0.9885	0.9516	0.8569
	14	1.0000	1.0000	1.0000	1.0000	0.9996	0.9966	0.9818	0.9331
	15	1.0000	1.0000	1.0000	1.0000	0.9999	0.9991	0.9943	0.9739
	16	1.0000	1.0000	1.0000	1.0000	1.0000	0.9998	0.9985	0.9915
	17	1.0000	1.0000	1.0000	1.0000	1.0000	1.0000	0.9997	0.9978
	18	1.0000	1.0000	1.0000	1.0000	1.0000	1.0000	1.0000	0.9995
	19	1.0000	1.0000	1.0000	1.0000	1.0000	1.0000	1.0000	0.9999
	20	1.0000	1.0000	1.0000	1.0000	1.0000	1.0000	1.0000	1.0000
	21	1.0000	1.0000	1.0000	1.0000	1.0000	1.0000	1.0000	1.0000
23	0	0.2266	0.0464	0.0084	0.0013	0.0002	0.0000	0.0000	0.0000
	1	0.5742	0.1987	0.0532	0.0116	0.0021	0.0003	0.0000	0.0000
	2	0.8290	0.4381	0.1668	0.0492	0.0115	0.0021	0.0003	0.0000
	3	0.9479	0.6775	0.3503	0.1370	0.0416	0.0099	0.0018	0.0002
	4	0.9876	0.8485	0.5621	0.2832	0.1100	0.0330	0.0076	0.0013
	5	0.9976	0.9413	0.7478	0.4685	0.2280	0.0859	0.0247	0.0053
	6	0.9996	0.9811	0.8763	0.6537	0.3890	0.1810	0.0647	0.0173
	7	1.0000	0.9949	0.9484	0.8037	0.5668	0.3196	0.1403	0.0466
	8	1.0000	0.9988	0.9816	0.9037	0.7283	0.4859	0.2578	0.1050
	9	1.0000	0.9998	0.9944	0.9592	0.8507	0.6522	0.4102	0.2024
	10	1.0000	1.0000	0.9986	0.9851	0.9286	0.7919	0.5761	0.3388
	11	1.0000	1.0000	0.9997	0.9954	0.9705	0.8910	0.7285	0.5000
	12	1.0000	1.0000	0.9999	0.9988	0.9895	0.9504	0.8471	0.6612
	13	1.0000	1.0000	1.0000	0.9997	0.9968	0.9806	0.9252	0.7976
	14	1.0000	1.0000	1.0000	0.9999	0.9992	0.9935	0.9686	0.8950
	15	1.0000	1.0000	1.0000	1.0000	0.9998	0.9982	0.9888	0.9534
	16	1.0000	1.0000	1.0000	1.0000	1.0000	0.9996	0.9967	0.9827
	17	1.0000	1.0000	1.0000	1.0000	1.0000	0.9999	0.9992	0.9947
	18	1.0000	1.0000	1.0000	1.0000	1.0000	1.0000	0.9998	0.9987
	19	1.0000	1.0000	1.0000	1.0000	1.0000	1.0000	1.0000	0.9998
	20	1.0000	1.0000	1.0000	1.0000	1.0000	1.0000	1.0000	1.0000
	21	1.0000	1.0000	1.0000	1.0000	1.0000	1.0000	1.0000	1.0000
	22	1.0000	1.0000	1.0000	1.0000	1.0000	1.0000	1.0000	1.0000
24	0	0.2125	0.0406	0.0069	0.0010	0.0001	0.0000	0.0000	0.0000
	1	0.5524	0.1797	0.0448	0.0090	0.0015	0.0002	0.0000	0.0000
	2	0.8131	0.4082	0.1455	0.0398	0.0086	0.0014	0.0002	0.0000
	3	0.9405	0.6476	0.3159	0.1150	0.0322	0.0070	0.0011	0.0001
	4	0.9851	0.8271	0.5224	0.2466	0.0886	0.0243	0.0051	0.0008
	5	0.9970	0.9297	0.7130	0.4222	0.1911	0.0661	0.0172	0.0033
	6	0.9995	0.9761	0.8522	0.6074	0.3387	0.1453	0.0472	0.0113

(*continued*)

Table A.1 (*Continued*)

n	k	p = 1/16	2/16	3/16	4/16	5/16	6/16	7/16	8/16
	7	0.9999	0.9932	0.9349	0.7662	0.5112	0.2676	0.1072	0.0320
	8	1.0000	0.9983	0.9754	0.8787	0.6778	0.4235	0.2064	0.0758
	9	1.0000	0.9997	0.9920	0.9453	0.8125	0.5898	0.3435	0.1537
	10	1.0000	0.9999	0.9978	0.9787	0.9043	0.7395	0.5035	0.2706
	11	1.0000	1.0000	0.9995	0.9928	0.9574	0.8538	0.6618	0.4194
	12	1.0000	1.0000	0.9999	0.9979	0.9835	0.9281	0.7953	0.5806
	13	1.0000	1.0000	1.0000	0.9995	0.9945	0.9693	0.8911	0.7294
	14	1.0000	1.0000	1.0000	0.9999	0.9984	0.9887	0.9496	0.8463
	15	1.0000	1.0000	1.0000	1.0000	0.9996	0.9964	0.9799	0.9242
	16	1.0000	1.0000	1.0000	1.0000	0.9999	0.9990	0.9932	0.9680
	17	1.0000	1.0000	1.0000	1.0000	1.0000	0.9998	0.9981	0.9887
	18	1.0000	1.0000	1.0000	1.0000	1.0000	1.0000	0.9996	0.9967
	19	1.0000	1.0000	1.0000	1.0000	1.0000	1.0000	0.9999	0.9992
	20	1.0000	1.0000	1.0000	1.0000	1.0000	1.0000	1.0000	0.9999
	21	1.0000	1.0000	1.0000	1.0000	1.0000	1.0000	1.0000	1.0000
	22	1.0000	1.0000	1.0000	1.0000	1.0000	1.0000	1.0000	1.0000
	23	1.0000	1.0000	1.0000	1.0000	1.0000	1.0000	1.0000	1.0000
25	0	0.1992	0.0355	0.0056	0.0008	0.0001	0.0000	0.0000	0.0000
	1	0.5132	0.1623	0.0377	0.0070	0.0011	0.0001	0.0000	0.0000
	2	0.7968	0.3796	0.1266	0.0321	0.0064	0.0010	0.0001	0.0000
	3	0.9325	0.6176	0.2840	0.0962	0.0248	0.0049	0.0007	0.0001
	4	0.9823	0.8047	0.4837	0.2137	0.0710	0.0178	0.0033	0.0005
	5	0.9962	0.9169	0.6772	0.3783	0.1591	0.0504	0.0119	0.0028
	6	0.9993	0.9703	0.8261	0.5611	0.2926	0.1156	0.0341	0.0073
	7	0.9999	0.9910	0.9194	0.7265	0.4573	0.2218	0.0810	0.0216
	8	1.0000	0.9977	0.9678	0.8506	0.6258	0.3651	0.1630	0.0539
	9	1.0000	0.9995	0.9889	0.9287	0.7704	0.5275	0.2835	0.1148
	10	1.0000	0.9999	0.9967	0.9703	0.8756	0.6834	0.4335	0.2122
	11	1.0000	1.0000	0.9992	0.9893	0.9408	0.8110	0.5926	0.3450
	12	1.0000	1.0000	0.9998	0.9966	0.9754	0.9003	0.7369	0.5000
	13	1.0000	1.0000	1.0000	0.9991	0.9911	0.9538	0.8491	0.6550
	14	1.0000	1.0000	1.0000	0.9998	0.9972	0.9814	0.9240	0.7878
	15	1.0000	1.0000	1.0000	1.0000	0.9992	0.9935	0.9667	0.8852
	16	1.0000	1.0000	1.0000	1.0000	0.9998	0.9981	0.9874	0.9462
	17	1.0000	1.0000	1.0000	1.0000	1.0000	0.9995	0.9960	0.9784
	18	1.0000	1.0000	1.0000	1.0000	1.0000	0.9999	0.9989	0.9927
	19	1.0000	1.0000	1.0000	1.0000	1.0000	1.0000	0.9998	0.9980
	20	1.0000	1.0000	1.0000	1.0000	1.0000	1.0000	1.0000	0.9995
	21	1.0000	1.0000	1.0000	1.0000	1.0000	1.0000	1.0000	0.9999
	22	1.0000	1.0000	1.0000	1.0000	1.0000	1.0000	1.0000	1.0000
	23	1.0000	1.0000	1.0000	1.0000	1.0000	1.0000	1.0000	1.0000
	24	1.0000	1.0000	1.0000	1.0000	1.0000	1.0000	1.0000	1.0000

Table A.2 Cumulative Poisson Distribution
The tabulated quantity is

$$\sum_{j=0}^{k} e^{-\lambda} \frac{\lambda^j}{j!}.$$

	λ					
k	0.001	0.005	0.010	0.015	0.020	0.025
0	0.9990 0050	0.9950 1248	0.9900 4983	0.9851 1194	0.9801 9867	0.9753 099
1	0.9999 9950	0.9999 8754	0.9999 5033	0.9998 8862	0.9998 0264	0.9996 927
2	1.0000 0000	0.9999 9998	0.9999 9983	0.9999 9945	0.9999 9868	0.9999 974
3		1.0000 0000	1.0000 0000	1.0000 0000	0.9999 9999	1.0000 000
4					1.0000 0000	1.0000 000

	λ					
k	0.030	0.035	0.040	0.045	0.050	0.055
0	0.970 446	0.965 605	0.960 789	0.955 997	0.951 229	0.946 485
1	0.999 559	0.999 402	0.999 221	0.999 017	0.998 791	0.998 542
2	0.999 996	0.999 993	0.999 990	0.999 985	0.999 980	0.999 973
3	1.000 000	1.000 000	1.000 000	1.000 000	1.000 000	1.000 000

	λ					
k	0.060	0.065	0.070	0.075	0.080	0.085
0	0.941 765	0.937 067	0.932 394	0.927 743	0.923 116	0.918 512
1	0.998 270	0.997 977	0.997 661	0.997 324	0.996 966	0.996 586
2	0.999 966	0.999 956	0.999 946	0.999 934	0.999 920	0.999 904
3	0.999 999	0.999 999	0.999 999	0.999 999	0.999 998	0.999 998
4	1.000 000	1.000 000	1.000 000	1.000 000	1.000 000	1.000 000

	λ					
k	0.090	0.095	0.100	0.200	0.300	0.400
0	0.913 931	0.909 373	0.904 837	0.818 731	0.740 818	0.670 320
1	0.996 185	0.995 763	0.995 321	0.982 477	0.963 064	0.938 448
2	0.999 886	0.999 867	0.999 845	0.998 852	0.996 401	0.992 074
3	0.999 997	0.999 997	0.999 996	0.999 943	0.999 734	0.999 224
4	1.000 000	1.000 000	1.000 000	0.999 998	0.999 984	0.999 939
5				1.000 000	0.999 999	0.999 996
6					1.000 000	1.000 000

	λ					
k	0.500	0.600	0.700	0.800	0.900	1.000
0	0.606 531	0.548 812	0.496 585	0.449 329	0.406 329	0.367 879
1	0.909 796	0.878 099	0.844 195	0.808 792	0.772 482	0.735 759
2	0.985 612	0.976 885	0.965 858	0.952 577	0.937 143	0.919 699
3	0.998 248	0.996 642	0.994 247	0.990 920	0.986 541	0.981 012
4	0.999 828	0.999 606	0.999 214	0.998 589	0.997 656	0.996 340
5	0.999 986	0.999 961	0.999 910	0.999 816	0.999 657	0.999 406
6	0.999 999	0.999 997	0.999 991	0.999 979	0.999 957	0.999 917
7	1.000 000	1.000 000	0.999 999	0.999 998	0.999 995	0.999 990
8			1.000 000	1.000 000	1.000 000	0.999 999
9						1.000 000

(*continued*)

Table A.2 (*Continued*)

	λ							
k	**1.20**	**1.40**	**1.60**	**1.80**	**2.00**	**2.50**	**3.00**	**3.50**
0	0.3012	0.2466	0.2019	0.1653	0.1353	0.0821	0.0498	0.0302
1	0.6626	0.5918	0.5249	0.4628	0.4060	0.2873	0.1991	0.1359
2	0.8795	0.8335	0.7834	0.7306	0.6767	0.5438	0.4232	0.3208
3	0.9662	0.9463	0.9212	0.8913	0.8571	0.7576	0.6472	0.5366
4	0.9923	0.9857	0.9763	0.9636	0.9473	0.8912	0.8153	0.7254
5	0.9985	0.9968	0.9940	0.9896	0.9834	0.9580	0.9161	0.8576
6	0.9997	0.9994	0.9987	0.9974	0.9955	0.9858	0.9665	0.9347
7	1.0000	0.9999	0.9997	0.9994	0.9989	0.9958	0.9881	0.9733
8		1.0000	1.0000	0.9999	0.9998	0.9989	0.9962	0.9901
9				1.0000	1.0000	0.9997	0.9989	0.9967
10						0.9999	0.9997	0.9990
11						1.0000	0.9999	0.9997
12							1.0000	0.9999
13								1.0000

	λ							
k	**4.00**	**4.50**	**5.00**	**6.00**	**7.00**	**8.00**	**9.00**	**10.00**
0	0.0183	0.0111	0.0067	0.0025	0.0009	0.0003	0.0001	0.0000
1	0.0916	0.0611	0.0404	0.0174	0.0073	0.0030	0.0012	0.0005
2	0.2381	0.1736	0.1247	0.0620	0.0296	0.0138	0.0062	0.0028
3	0.4335	0.3423	0.2650	0.1512	0.0818	0.0424	0.0212	0.0103
4	0.6288	0.5321	0.4405	0.2851	0.1730	0.0996	0.0550	0.0293
5	0.7851	0.7029	0.6160	0.4457	0.3007	0.1912	0.1157	0.0671
6	0.8893	0.8311	0.7622	0.6063	0.4497	0.3134	0.2068	0.1301
7	0.9489	0.9134	0.8666	0.7440	0.5987	0.4530	0.3239	0.2202
8	0.9786	0.9597	0.9319	0.8472	0.7291	0.5925	0.4577	0.3328
9	0.9919	0.9829	0.9682	0.9161	0.8305	0.7166	0.5874	0.4579
10	0.9972	0.9933	0.9863	0.9574	0.9015	0.8159	0.7060	0.5830
11	0.9991	0.9976	0.9945	0.9799	0.9467	0.8881	0.8030	0.6968
12	0.9997	0.9992	0.9980	0.9912	0.9730	0.9362	0.8758	0.7916
13	0.9999	0.9997	0.9993	0.9964	0.9872	0.9658	0.9261	0.8645
14	1.0000	0.9999	0.9998	0.9986	0.9943	0.9827	0.9585	0.9165
15		1.0000	0.9999	0.9995	0.9976	0.9918	0.9780	0.9513
16			1.0000	0.9998	0.9990	0.9963	0.9889	0.9730
17				0.9999	0.9996	0.9984	0.9947	0.9857
18				1.0000	0.9999	0.9993	0.9976	0.9928
19						0.9997	0.9989	0.9965
20					1.0000	0.9999	0.9996	0.9984
21						1.0000	0.9998	0.9993
22							0.9999	0.9997
23							1.0000	0.9999
24								1.0000

Table A.3 Cumulative Normal Distribution

The tabulated quantity is

$$\Phi(x) = \frac{1}{\sqrt{2\pi}} \int_{-\infty}^{x} e^{-t^2/2} dt.$$
$$[\Phi(-x) = 1 - \Phi(x)].$$

x	$\Phi(x)$	x	$\Phi(x)$	x	$\Phi(x)$	x	$\Phi(x)$
0.00	0.500000	0.45	0.673645	0.90	0.815940	1.35	0.911492
0.01	0.503989	0.46	0.677242	0.91	0.818589	1.36	0.913085
0.02	0.507978	0.47	0.680822	0.92	0.821214	1.37	0.914657
0.03	0.511966	0.48	0.684386	0.93	0.823814	1.38	0.916207
0.04	0.515953	0.49	0.687933	0.94	0.826391	1.39	0.917736
0.05	0.519939	0.50	0.691462	0.95	0.828944	1.40	0.919243
0.06	0.523922	0.51	0.694974	0.96	0.831472	1.41	0.920730
0.07	0.527903	0.52	0.698468	0.97	0.833977	1.42	0.922196
0.08	0.531881	0.53	0.701944	0.98	0.836457	1.43	0.923641
0.09	0.535856	0.54	0.705401	0.99	0.838913	1.44	0.925066
0.10	0.539828	0.55	0.708840	1.00	0.841345	1.45	0.926471
0.11	0.543795	0.56	0.712260	1.01	0.843752	1.46	0.927855
0.12	0.547758	0.57	0.715661	1.02	0.846136	1.47	0.929219
0.13	0.551717	0.58	0.719043	1.03	0.848495	1.48	0.930563
0.14	0.555670	0.59	0.722405	1.04	0.850830	1.49	0.931888
0.15	0.559618	0.60	0.725747	1.05	0.853141	1.50	0.933193
0.16	0.563559	0.61	0.729069	1.06	0.855428	1.51	0.934478
0.17	0.567495	0.62	0.732371	1.07	0.857690	1.52	0.935745
0.18	0.571424	0.63	0.735653	1.08	0.859929	1.53	0.936992
0.19	0.575345	0.64	0.738914	1.09	0.862143	1.54	0.938220
0.20	0.579260	0.65	0.742154	1.10	0.864334	1.55	0.939429
0.21	0.583166	0.66	0.745373	1.11	0.866500	1.56	0.940620
0.22	0.587064	0.67	0.748571	1.12	0.868643	1.57	0.941792
0.23	0.590954	0.68	0.751748	1.13	0.870762	1.58	0.942947
0.24	0.594835	0.69	0.754903	1.14	0.872857	1.59	0.944083
0.25	0.598706	0.70	0.758036	1.15	0.874928	1.60	0.945201
0.26	0.602568	0.71	0.761148	1.16	0.876976	1.61	0.946301
0.27	0.606420	0.72	0.764238	1.17	0.879000	1.62	0.947384
0.28	0.610261	0.73	0.767305	1.18	0.881000	1.63	0.948449
0.29	0.614092	0.74	0.770350	1.19	0.882977	1.64	0.949497
0.30	0.617911	0.75	0.773373	1.20	0.884930	1.65	0.950529
0.31	0.621720	0.76	0.776373	1.21	0.886861	1.66	0.951543
0.32	0.625516	0.77	0.779350	1.22	0.888768	1.67	0.952540
0.33	0.629300	0.78	0.782305	1.23	0.890651	1.68	0.953521
0.34	0.633072	0.79	0.785236	1.24	0.892512	1.69	0.954486
0.35	0.636831	0.80	0.788145	1.25	0.894350	1.70	0.955435
0.36	0.640576	0.81	0.791030	1.26	0.896165	1.71	0.956367
0.37	0.644309	0.82	0.793892	1.27	0.897958	1.72	0.957284
0.38	0.648027	0.83	0.796731	1.28	0.899727	1.73	0.958185
0.39	0.651732	0.84	0.799546	1.29	0.901475	1.74	0.959070
0.40	0.655422	0.85	0.802337	1.30	0.903200	1.75	0.959941

(*continued*)

Table A.3 (*Continued*)

x	$\Phi(x)$	x	$\Phi(x)$	x	$\Phi(x)$	x	$\Phi(x)$
0.41	0.659097	0.86	0.805105	1.31	0.904902	1.76	0.960796
0.42	0.662757	0.87	0.807850	1.32	0.906582	1.77	0.961636
0.43	0.666402	0.88	0.810570	1.33	0.908241	1.78	0.962462
0.44	0.670031	0.89	0.813267	1.34	0.909877	1.79	0.963273
1.80	0.964070	2.30	0.989276	2.80	0.997445	3.30	0.999517
1.81	0.964852	2.31	0.989556	2.81	0.997523	3.31	0.999534
1.82	0.965620	2.32	0.989830	2.82	0.997599	3.32	0.999550
1.83	0.966375	2.33	0.990097	2.83	0.997673	3.33	0.999566
1.84	0.967116	2.34	0.990358	2.84	0.997744	3.34	0.999581
1.85	0.967843	2.35	0.990613	2.85	0.997814	3.35	0.999596
1.86	0.968557	2.36	0.990863	2.86	0.997882	3.36	0.999610
1.87	0.969258	2.37	0.991106	2.87	0.997948	3.37	0.999624
1.88	0.969946	2.38	0.991344	2.88	0.998012	3.38	0.999638
1.89	0.970621	2.39	0.991576	2.89	0.998074	3.39	0.999651
1.90	0.971283	2.40	0.991802	2.90	0.998134	3.40	0.999663
1.91	0.971933	2.41	0.992024	2.91	0.998193	3.41	0.999675
1.92	0.972571	2.42	0.992240	2.92	0.998250	3.42	0.999687
1.93	0.973197	2.43	0.992451	2.93	0.998305	3.43	0.999698
1.94	0.973810	2.44	0.992656	2.94	0.998359	3.44	0.999709
1.95	0.974412	2.45	0.992857	2.95	0.998411	3.45	0.999720
1.96	0.975002	2.46	0.993053	2.96	0.998462	3.46	0.999730
1.97	0.975581	2.47	0.993244	2.97	0.998511	3.47	0.999740
1.98	0.976148	2.48	0.993431	2.98	0.998559	3.48	0.999749
1.99	0.976705	2.49	0.993613	2.99	0.998605	3.49	0.999758
2.00	0.977250	2.50	0.993790	3.00	0.998650	3.50	0.999767
2.01	0.977784	2.51	0.993963	3.01	0.998694	3.51	0.999776
2.02	0.978308	2.52	0.994132	3.02	0.998736	3.52	0.999784
2.03	0.978822	2.53	0.994297	3.03	0.998777	3.53	0.999792
2.04	0.979325	2.54	0.994457	3.04	0.998817	3.54	0.999800
2.05	0.979818	2.55	0.994614	3.05	0.998856	3.55	0.999807
2.06	0.980301	2.56	0.994766	3.06	0.998893	3.56	0.999815
2.07	0.980774	2.57	0.994915	3.07	0.998930	3.57	0.999822
2.08	0.981237	2.58	0.995060	3.08	0.998965	3.58	0.999828
2.09	0.981691	2.59	0.995201	3.09	0.998999	3.59	0.999835
2.10	0.982136	2.60	0.995339	3.10	0.999032	3.60	0.999841
2.11	0.982571	2.61	0.995473	3.11	0.999065	3.61	0.999847
2.12	0.982997	2.62	0.995604	3.12	0.999096	3.62	0.999853
2.13	0.983414	2.63	0.995731	3.13	0.999126	3.63	0.999858
2.14	0.983823	2.64	0.995855	3.14	0.999155	3.64	0.999864
2.15	0.984222	2.65	0.995975	3.15	0.999184	3.65	0.999869
2.16	0.984614	2.66	0.996093	3.16	0.999211	3.66	0.999874
2.17	0.984997	2.67	0.996207	3.17	0.999238	3.67	0.999879
2.18	0.985371	2.68	0.996319	3.18	0.999264	3.68	0.999883
2.19	0.985738	2.69	0.996427	3.19	0.999289	3.69	0.999888
2.20	0.986097	2.70	0.996533	3.20	0.999313	3.70	0.999892

(*continued*)

Table A.3 ***(Continued)***

x	$\Phi(x)$	x	$\Phi(x)$	x	$\Phi(x)$	x	$\Phi(x)$
2.21	0.986447	2.71	0.996636	3.21	0.999336	3.71	0.999896
2.22	0.986791	2.72	0.996736	3.22	0.999359	3.72	0.999900
2.23	0.987126	2.73	0.996833	3.23	0.999381	3.73	0.999904
2.24	0.987455	2.74	0.996928	3.24	0.999402	3.74	0.999908
2.25	0.987776	2.75	0.997020	3.25	0.999423	3.75	0.999912
2.26	0.988089	2.76	0.997110	3.26	0.999443	3.76	0.999915
2.27	0.988396	2.77	0.997197	3.27	0.999462	3.77	0.999918
2.28	0.988696	2.78	0.997282	3.28	0.999481	3.78	0.999922
2.29	0.988989	2.79	0.997365	3.29	0.999499	3.79	0.999925
3.80	0.999928	3.85	0.999941	3.90	0.999952	3.95	0.999961
3.81	0.999931	3.86	0.999943	3.91	0.999954	3.96	0.999963
3.82	0.999933	3.87	0.999946	3.92	0.999956	3.97	0.999964
3.83	0.999936	3.88	0.999948	3.93	0.999958	3.98	0.999966
3.84	0.999938	3.89	0.999950	3.94	0.999959	3.99	0.999967

Table A.4 Critical Values for Chi-Square Distribution

Let χ_r^2 be a random variable having the chi-square distribution with r degrees of freedom. Then the tabulated quantities are the numbers x for which:

$$P(\chi_r^2 \leq x) = \gamma.$$

	γ					
r	**0.005**	**0.01**	**0.025**	**0.05**	**0.10**	**0.25**
1	–	–	0.001	0.004	0.016	0.102
2	0.010	0.020	0.051	0.103	0.211	0.575
3	0.072	0.115	0.216	0.352	0.584	1.213
4	0.207	0.297	0.484	0.711	1.064	1.923
5	0.412	0.554	0.831	1.145	1.610	2.675
6	0.676	0.872	1.237	1.635	2.204	3.455
7	0.989	1.239	1.690	2.167	2.833	4.255
8	1.344	1.646	2.180	2.733	3.490	5.071
9	1.735	2.088	2.700	2.325	4.168	5.899
10	2.156	2.558	3.247	3.940	4.865	6.737
11	2.603	3.053	3.816	4.575	5.578	7.584
12	3.074	3.571	4.404	5.226	6.304	9.438
13	3.565	4.107	5.009	5.892	7.042	9.299
14	4.075	4.660	5.629	6.571	7.790	10.165
15	4.601	5.229	6.262	7.261	8.547	11.037
16	5.142	5.812	6.908	7.962	9.312	11.912
17	5.697	6.408	7.564	8.672	10.085	12.792
18	6.265	7.015	8.231	8.390	10.865	13.675
19	6.844	7.633	8.907	10.117	11.651	14.562
20	7.434	8.260	9.591	10.851	12.443	15.452

(continued)

Table A.4 (*Continued*)

	γ					
r	0.005	0.01	0.025	0.05	0.10	0.25
21	8.034	8.897	10.283	11.591	13.240	16.344
22	8.643	9.542	10.982	12.338	14.042	17.240
23	9.260	10.196	11.689	13.091	14.848	18.137
24	9.886	10.856	12.401	13.848	15.659	19.037
25	10.520	11.524	13.120	14.611	16.473	19.939
26	11.160	12.198	13.844	13.379	17.292	20.843
27	11.808	12.879	14.573	16.151	18.114	21.749
28	12.461	13.565	15.308	16.928	18.939	22.657
29	13.121	14.257	16.047	17.708	19.768	23.567
30	13.787	14.954	16.791	18.493	20.599	24.478
31	14.458	15.655	17.539	19.281	21.434	25.390
32	15.134	16.362	18.291	20.072	22.271	26.304
33	15.815	17.074	19.047	20.867	23.110	27.219
34	16.501	17.789	19.806	21.664	23.952	28.136
35	17.192	18.509	20.569	22.465	24.797	29.054
36	17.887	19.233	21.336	23.269	25.643	29.973
37	18.586	19.960	22.106	24.075	26.492	30.893
38	19.289	20.691	22.878	24.884	27.343	31.815
39	19.996	21.426	23.654	25.695	28.196	32.737
40	20.707	22.164	24.433	26.509	29.051	33.660
41	21.421	22.906	25.215	27.326	29.907	34.585
42	22.138	23.650	25.999	28.144	30.765	35.510
43	22.859	24.398	26.785	28.965	31.625	36.436
44	23.584	25.148	27.575	29.787	32.487	37.363
45	24.311	25.901	28.366	30.612	33.350	38.291
	γ					
r	0.75	0.90	0.95	0.975	0.99	0.995
1	1.323	2.706	3.841	5.024	6.635	7.879
2	2.773	4.605	5.991	7.378	9.210	10.597
3	4.108	6.251	7.815	9.348	11.345	12.838
4	5.385	7.779	9.488	11.143	13.277	14.860
5	6.626	9.236	11.071	12.833	15.086	16.750
6	7.841	10.645	12.592	14.449	16.812	18.548
7	9.037	12.017	14.067	16.013	18.475	20.278
8	10.219	13.362	15.507	17.535	20.090	21.955
9	11.389	14.684	16.919	19.023	21.666	23.589
10	12.549	15.987	18.307	20.483	23.209	25.188
11	13.701	17.275	19.675	21.920	24.725	26.757
12	14.845	18.549	21.026	23.337	26.217	28.299
13	15.984	19.812	23.362	24.736	27.688	29.819
14	17.117	21.064	23.685	26.119	29.141	31.319
15	18.245	22.307	24.996	27.488	30.578	32.801

(*continued*)

Table A.4 (*Continued*)

	γ					
r	0.75	0.90	0.95	0.975	0.99	0.995
16	19.369	23.542	26.296	28.845	32.000	34.267
17	20.489	24.769	27.587	30.191	33.409	35.718
18	21.605	25.989	28.869	31.526	34.805	37.156
19	22.718	27.204	30.144	32.852	36.191	38.582
20	23.828	28.412	31.410	34.170	37.566	39.997
21	24.935	29.615	32.671	35.479	38.932	41.401
22	26.039	30.813	33.924	36.781	40.289	42.796
23	27.141	32.007	35.172	38.076	41.638	44.181
24	28.241	33.196	36.415	39.364	42.980	45.559
25	29.339	34.382	37.652	40.646	44.314	46.928
26	30.435	35.563	38.885	41.923	45.642	48.290
27	31.528	36.741	40.113	43.194	46.963	49.645
28	32.620	37.916	41.337	44.641	48.278	50.993
29	33.711	39.087	42.557	45.722	49.588	52.336
30	34.800	40.256	43.773	46.979	50.892	53.672
31	35.887	41.422	44.985	48.232	51.191	55.003
32	36.973	42.585	46.194	49.480	53.486	56.328
33	38.058	43.745	47.400	50.725	54.776	57.648
34	39.141	44.903	48.602	51.966	56.061	58.964
35	40.223	46.059	49.802	53.203	57.342	60.275
36	41.304	47.212	50.998	54.437	58.619	61.581
37	42.383	48.363	52.192	55.668	59.892	62.883
38	43.462	49.513	53.384	56.896	61.162	64.181
39	44.539	50.660	54.572	58.120	62.428	65.476
40	45.616	51.805	55.758	59.342	63.691	66.766
41	46.692	52.949	56.942	60.561	64.950	68.053
42	47.766	54.090	58.124	61.777	66.206	69.336
43	48.840	55.230	59.304	62.990	67.459	70.616
44	49.913	56.369	60.481	64.201	68.710	71.893
45	50.985	57.505	61.656	65.410	69.957	73.166

Table A.5

Selected Discrete and Continuous Distributions and Some of Their Characteristics

	Probability density functions in one variable		
Distribution	**Probability Density Function**	**Mean**	**Variance**
Binomial, $B(n,p)$	$f(x) = \binom{n}{x} p^x q^{n-x}, x = 0, 1, \ldots, n;$ $0 < p < 1, q = 1 - p$	np	npq
Bernoulli, $B(1, p)$	$f(x) = p^x q^{1-x}, x = 0, 1$	p	pq
Geometric	$f(x) = pq^{x-1}, x = 1, 2, \ldots;$ $0 < p < 1, q = 1 - p$	$\frac{1}{p}$	$\frac{q}{p^2}$
Poisson, $P(\lambda)$	$f(x) = e^{-\lambda} \frac{\lambda^x}{x!}, x = 0, 1, \ldots; \lambda > 0$	λ	λ
Hypergeometric	$f(x) = \frac{\binom{m}{x}\binom{n}{r-x}}{\binom{m+n}{r}}$, where $x = 0, 1, \ldots, r \left(\binom{m}{r} = 0, r > m\right)$	$\frac{mr}{m+n}$	$\frac{mnr(m+n-r)}{(m+n)^2(m+n-1)}$
Gamma	$f(x) = \frac{1}{\Gamma(\alpha)\beta^\alpha} x^{\alpha-1} \exp\left(-\frac{x}{\beta}\right), x > 0;$ $\alpha, \beta > 0$	$\alpha\beta$	$\alpha\beta^2$
Negative exponential	$f(x) = \lambda \exp(-\lambda x), x > 0; \lambda > 0$; or $f(x) = \frac{1}{\mu} e^{-x/\mu}, x > 0; \mu > 0$	$\frac{1}{\lambda}$ μ	$\frac{1}{\lambda^2}$ μ^2
Chi-square, χ_r^2	$f(x) = \frac{1}{\Gamma(\frac{r}{2})2^{r/2}} x^{\frac{r}{2}-1} \exp\left(-\frac{x}{2}\right), x > 0;$ $r > 0$ integer	r	$2r$
Normal, $N(\mu, \sigma^2)$	$f(x) = \frac{1}{\sqrt{2\pi}\sigma} \exp\left[-\frac{(x-\mu)^2}{2\sigma^2}\right],$ $x \in \Re; \mu \in \Re, \sigma > 0$	μ	σ^2
Standard normal, $N(0, 1)$	$f(x) = \frac{1}{\sqrt{2\pi}} \exp\left(-\frac{x^2}{2}\right), x \in \Re$	0	1

Table A.5
(*Continued*)

	Probability density functions in one variable		
Distribution	**Probability Density Function**	**Mean**	**Variance**
Uniform, $U(\alpha, \beta)$	$f(x) = \frac{1}{\beta-\alpha}, \alpha \le x \le \beta;$ $-\infty < \alpha < \beta < \infty$	$\frac{\alpha+\beta}{2}$	$\frac{(\alpha-\beta)^2}{12}$
Multinomial $M(n; p_1, \ldots, p_k)$	$f(x_1, \ldots, x_k) = \frac{n!}{x_1! x_2! \cdots x_k!} \times$ $p_1^{x_1} p_2^{x_2} \cdots p_k^{x_k}, x_i \ge 0$ integers, $x_1 + x_2 + \cdots + x_k = n; p_j > 0, j = 1, 2, \ldots, k, p_1 + p_2 + \cdots + p_k = 1$	$np_1, \ldots, np_k$	$np_1 q_1, \ldots, np_k q_k$ $q_i = 1 - p_i, j = 1, \ldots, k$
Bivariate normal $N(\mu_1, \mu_2, \sigma_1^2, \sigma_2^2, \rho)$	$f(x_1, x_2) = \frac{1}{2\pi\sigma_1\sigma_2\sqrt{1-\rho^2}} \exp\left(-\frac{q}{2}\right),$ $q = \frac{1}{1-\rho^2}\left[\left(\frac{x_1-\mu_1}{\sigma_1}\right)^2 - 2\rho\left(\frac{x_1-\mu_1}{\sigma_1}\right) \times \left(\frac{x_2-\mu_2}{\sigma_2}\right) + \left(\frac{x_2-\mu_2}{\sigma_2}\right)^2\right],$ $x_1, x_2, \in \Re; \mu_1, \mu_2 \in \Re, \sigma_1, \sigma_2 > 0, -1 \le \rho \le 1, \rho =$ correlation coefficient	μ_1, μ_2	σ_1^2, σ_2^2
***k*-Variate normal $N(\mu, \Sigma)$**	$f(\mathbf{x}) = (2\pi)^{-k/2}\lvert\mathbf{\Sigma}\rvert^{-1/2} \times$ $\exp\left[-\frac{1}{2}(\mathbf{x}-\boldsymbol{\mu})'\mathbf{\Sigma}^{-1}(\mathbf{x}-\boldsymbol{\mu})\right],$ $\mathbf{x} \in \Re^k; \boldsymbol{\mu} \in \Re^k, \mathbf{\Sigma} : \mathrm{k} \times \mathrm{k}$ nonsingular symmetric matrix	$\mu_1, \ldots, \mu_k$	Covariance matrix: $\mathbf{\Sigma}$

Table A.5 (*Continued*)

Distribution	Moment Generating Function
Binomial, $B(n, p)$	$M(t) = (pe^t + q)^n,\ t \in \Re$
Bernoulli, $B(1,p)$	$M(t) = pe^t + q,\ t \in \Re$
Geometric	$M(t) = \frac{pe^t}{1 - qe^t},\ t < -\log q$
Poisson, $P(\lambda)$	$M(t) = \exp(\lambda e^t - \lambda),\ t \in \Re$
Hypergeometric	—
Gamma	$M(t) = \frac{1}{(1 - \beta t)^\alpha},\ t < \frac{1}{\beta}$
Negative exponential	$M(t) = \frac{1}{1 - t/\lambda},\ t < \lambda;$ or $M(t) = \frac{1}{1 - \mu t},\ t < \frac{1}{\mu}$
Chi-square, χ_r^2	$M(t) = \frac{1}{(1 - 2t)^{r/2}},\ t < \frac{1}{2}$
Normal, $N(\mu, \sigma^2)$	$M(t) = \exp\left(\mu t + \frac{\sigma^2 t^2}{2}\right),\ t \in \Re$
Standard normal, $N(0, 1)$	$M(t) = \exp\left(\frac{t^2}{2}\right),\ t \in \Re$
Uniform, $U(\alpha, \beta)$	$M(t) = \frac{e^{t\beta} - e^{t\alpha}}{t(\beta - \alpha)},\ t \in \Re$
Multinomial, $M(n; p_1, \ldots, p_k)$	$M(t_1, \ldots, t_k) = (p_1 e^{t_1} + \cdots + p_k e^{t_k})^n,$ $t_1, \ldots, t_k \in \Re$
Bivariate normal, $N(\mu_1, \mu_2, \sigma_1^2, \sigma_2^2, \rho)$	$M(t_1, t_2) = \exp\left[\mu_1 t_1 + \mu_2 t_2 + \frac{1}{2}(\sigma_1^2 t_1^2 + 2\rho\sigma_1\sigma_2 t_1 t_2 + \sigma_2^2 t_2^2)\right],$ $t_1, t_2 \in \Re$
k-Variate normal, $N(\boldsymbol{\mu}, \boldsymbol{\Sigma})$	$M(t) = \exp\left(\mathbf{t}'\boldsymbol{\mu} + \frac{1}{2}\mathbf{t}'\boldsymbol{\Sigma}\mathbf{t}\right),$ $\mathbf{t} \in \Re^k$

Table A.6 Handy Reference to Some Formulas Used in the Text

1. $\sum_{k=1}^{n} k = \frac{n(n+1)}{2}$, $\sum_{k=1}^{n} k^2 = \frac{n(n+1)(2n+1)}{6}$, $\sum_{k=1}^{n} k^3 = \left(\frac{n(n+1)}{2}\right)^2$.
2. $(a+b)^n = \sum_{x=0}^{n} \binom{n}{x} a^x b^{n-x}$.
3. $(a_1 + \cdots + a_k)^n = \sum \frac{n!}{x_1! \ldots x_k!} a_1^{x_1} \ldots a_k^{x_k}$ where the summation is over all ≥ 0 integers $x_1, \ldots, x_k$ with $x_1 + \cdots + x_k = n$.
4. $\sum_{n=k}^{\infty} r^n = \frac{r^k}{1-r}$, $k = 0, 1, \ldots, |r| < 1$.
5. $\sum_{n=1}^{\infty} nr^n = \frac{r}{(1-r)^2}$, $\sum_{n=2}^{\infty} n(n-1)r^n = \frac{2r^2}{(1-r)^3}$, $|r| < 1$.
6. $e^x = lim_{n\to\infty}(1 + \frac{x}{n})^n$, $e^x = lim_{n\to\infty}(1 + \frac{x_n}{n})^n$, $x_n \underset{n\to\infty}{\to} x$, $e^x = \sum_{n=0}^{\infty} \frac{x^n}{n!}$, $x \in \Re$.
7. $(a\,u(x) + b\,v(x))' = a\,u'(x) + b\,v'(x)$, $(u(x)v(x))' = u'(x)v(x) + u(x)v'(x)$, $\left(\frac{u(x)}{v(x)}\right)' = \frac{u'(x)v(x) - u(x)v'(x)}{v^2(x)}$, $\frac{d}{dx}u(v(x)) = \left(\frac{d}{v(x)}u((x))\right)\left(\frac{d}{dx}v(x)\right)$.
8. $\frac{\partial^2}{\partial x \partial y} w(x,y) = \frac{\partial^2}{\partial y \partial x} w(x,y)$ (under certain conditions).
9. $\frac{d}{dt} \sum_{n=1}^{\infty} w(n,t) = \sum_{n=1}^{\infty} \frac{\partial}{\partial t} w(n,t)$ (under certain conditions).
10. $\frac{d}{dt} \int_a^b w(x,t)dx = \int_a^b (\frac{\partial}{\partial t} w(x,t))dx$ $(-\infty \leq a < b \leq \infty)$ (under certain conditions).
11. $\sum_{n=1}^{\infty}(c\,x_n + d\,y_n) = c\sum_{n=1}^{\infty} x_n + d\sum_{n=1}^{\infty} y_n$.
12. $\int_a^b (c\,u(x) + d\,v(x))dx = c\int_a^b u(x)dx + d\int_a^b v(x)dx$ $(-\infty \leq a < b \leq \infty)$.
13. $\int_a^b u(x)dv(x) = u(x)v(x)\,|_a^b - \int_a^b v(x)du(x)(-\infty \leq a < b \leq \infty)$.
    In particular: $\int_a^b x^n dx = \frac{x^{n+1}}{n+1}\,|_a^b = \frac{b^{n+1} - a^{n+1}}{n+1} (n \neq -1)$,
    $\int_a^b \frac{dx}{x} = log\,x\,|_a^b = log\,b - log\,a$ $(0 < a < b$, $log\,x$ is the natural logarithm of $x)$,
    $\int_a^b e^x dx = e^x\,|_a^b = e^b - e^a$.
14. If $u'(x_0) = 0$, then x_0 maximizes $u(x)$ if $u''(x_0) < 0$, and x_0 minimizes $u(x)$ if $u''(x_0) > 0$.
15. Let $\frac{\partial}{\partial x} w(x,y)\,|_{\substack{x=x_0 \\ y=y_0}} = 0$, $\frac{\partial}{\partial y} w(x,y)\,|_{\substack{x=x_0 \\ y=y_0}} = 0$, and set
    $c_{11} = \frac{\partial^2}{\partial x^2} w(x,y)\,|_{\substack{x=x_0 \\ y=y_0}}$, $c_{12} = \frac{\partial^2}{\partial x \partial y} w(x,y)\,|_{\substack{x=x_0 \\ y=y_0}} = c_{21} = \frac{\partial^2}{\partial y \partial x} w(x,y)\,|_{\substack{x=x_0 \\ y=y_0}}$,
    $c_{22} = \frac{\partial^2}{\partial y^2} w(x,y)\,|_{\substack{x=x_0 \\ y=y_0}}$, $\mathbf{C} = \begin{pmatrix} c_{11} & c_{12} \\ c_{21} & c_{22} \end{pmatrix}$.
    Then (x_0, y_0) maximizes $w(x,y)$ if the matrix $\mathbf{C}$ is negative definite; i.e., for all real λ_1, λ_2 not both 0, it holds:
    $(\lambda_1, \lambda_2)\begin{pmatrix} c_{11} & c_{12} \\ c_{21} & c_{22} \end{pmatrix}\begin{pmatrix} \lambda_1 \\ \lambda_2 \end{pmatrix} = \lambda_1^2 c_{11} + 2\lambda_1\lambda_2 c_{12} + \lambda_2^2 c_{22} < 0$;
    (x_0, y_0) minimizes $w(x,y)$ if the matrix $\mathbf{C}$ is positive definite; i.e., $\lambda_1^2 c_{11} + 2\lambda_1\lambda_2 c_{12} + \lambda_2^2 c_{22} > 0$ with λ_1, λ_2 as above.
16. Criteria analogous to those in #15 hold for a function in k variables $w(x_1, \ldots, x_k)$.

Some Notation and Abbreviations

$\Re$	real line
$\Re^k, k \geq 1$	k-dimensional Euclidean space
$\uparrow, \downarrow$	increasing (nondecreasing) and decreasing (nonincreasing), respectively
$\mathcal{S}$	sample space; also, sure (or certain) event
$\varnothing$	empty set; also, impossible event
$A \subseteq B$	event A is contained in event B (event A implies event B)
A^c	complement of event A
$A \cup B$	union of events A and B
$A \cap B$	intersection of events A and B
$A - B$	difference of events A and B (in this order)
r.v.	random variable
I_A	indicator of the set A: $I_A(x) = 1$ if $x \in A$, $I_A(x) = 0$ if $x \notin A$
$(X \in B) = X^{-1}(B)$	inverse image of the set B under $X : X^{-1}(B) = \{s \in \mathcal{S}; X(s) \in B\}$
$X(\mathcal{S})$	range of X
P	probability function (measure)
$P(A)$	probability of the event A
P_X	probability distribution of X (or just distribution of X)
F_X	distribution function (d.f.) of X
f_X	probability density function (p.d.f.) of X
$P(A\|B)$	conditional probability of A, given B
$\binom{n}{k}$	combinations of n objects taken k at a time
$P_{n,k}$	permutations of n objects taken k at a time
$n!$	n factorial
EX or $\mu(X)$ or μ_X or just μ	expectation (mean value, mean) of X
$Var(X)$ or $\sigma^2(X)$ or σ_X^2 or just σ^2	variance of X
$\sqrt{Var(X)}$ or $\sigma(X)$ or σ_X or just σ	standard deviation (s.d.) of X
M_X or just M	moment generating function (m.g.f.) of X
$B(n, p)$	binomial distribution with parameters n and p
$P(\lambda)$	Poisson distribution with parameter λ
χ_r^2	chi-square distribution with r degrees of freedom (d.f.)
$N(\mu, \sigma^2)$	Normal distribution with parameters μ and σ^2
Φ	distribution function (d.f.) of the standard $N(0, 1)$ distribution

Introduction to Probability, Second Edition. http://dx.doi.org/10.1016/B978-0-12-800041-0.00021-3

$U(\alpha, \beta)$ or $R(\alpha, \beta)$	Uniform (or rectangular) distribution with parameters α and β
$X \sim B(n, p)$ etc.	the r.v. X has the distribution indicated
$\chi^2_{r;\alpha}$	the point for which $P(X > \chi^2_{r;\alpha}) = \alpha$, $X \sim \chi^2_r$
z_α	the point for which $P(Z > z_\alpha) = \alpha$, where $Z \sim N(0, 1)$
$P_{X_1,\ldots,X_n}$ or $P_{\mathbf{X}}$	joint probability distribution of the r.v.'s $X_1, \ldots, X_n$ or probability distribution of the random vector $\mathbf{X}$
$F_{X_1,\ldots,X_n}$ or $F_{\mathbf{X}}$	joint d.f. of the r.v.'s $X_1, \ldots, X_n$ or d.f. of the random vector $\mathbf{X}$
$f_{X_1,\ldots,X_n}$ or $f_{\mathbf{X}}$	joint p.d.f. of the r.v.'s $X_1, \ldots, X_n$ or p.d.f. of the random vector $\mathbf{X}$
$M_{X_1,\ldots,X_n}$ or $M_{\mathbf{X}}$	joint m.g.f. of the r.v.'s $X_1, \ldots, X_n$ or m.g.f. of the random vector $\mathbf{X}$
i.i.d. (r.v.'s)	independent identically distributed (r.v.'s)
$f_{X\mid Y}(\cdot\mid Y = y)$ or $f_{X\mid Y}(\cdot\mid y)$	conditional p.d.f. of X, given $Y = y$
$E(X\mid Y = y)$	conditional expectation of X, given $Y = y$
$Var(X\mid Y = y)$ or $\sigma^2(X\mid Y = y)$	conditional variance of X, given $Y = y$
$Cov(X, Y)$	covariance of X and Y
$\rho(X, Y)$ or $\rho_{X,Y}$	correlation coefficient of X and Y
t_r	(Student's) t distribution with r degrees of freedom (d.f.)
$t_{r;\alpha}$	the point for which $P(X > t_{r;\alpha}) = \alpha$, $X \sim t_r$
F_{r_1,r_2}	F distribution with r_1 and r_2 degrees of freedom (d.f.)
$F_{r_1,r_2;\alpha}$	the point for which $P(X > F_{r_1,r_2;\alpha}) = \alpha$, $X \sim F_{r_1,r_2}$
$X_{(j)}$ or Y_j	jth order statistic of $X_1, \ldots, X_n$
$\xrightarrow{P}$, $\xrightarrow{d}$, $\xrightarrow{\text{q.m.}}$	convergence in probability, distribution, quadratic mean, respectively
WLLN	Weak Law of Large Numbers
CLT	Central Limit Theorem
θ	letter used for a one-dimensional parameter
$\boldsymbol{\theta}$	symbol used for a multidimensional parameter
Ω	letter used for a parameter space
ML	maximum likelihood
MLE	maximum likelihood estimate
UMV	uniformly minimum variance
UMVU	uniformly minimum variance unbiased
LS	least squares
LSE	least squares estimate
H_0	null hypothesis

H_A	alternative hypothesis
φ	letter used for a test function
α	letter used for level of significance
$\beta(\theta)$ or $\beta(\boldsymbol{\theta})$	probability of type II error at $\theta(\boldsymbol{\theta})$
$\pi(\theta)$ or $\pi(\boldsymbol{\theta})$	power of a test at $\theta(\boldsymbol{\theta})$
MP	most powerful (test)
UMP	uniformly most powerful (test)
LR	likelihood ratio
$\lambda = \lambda(x_1, \ldots, x_n)$	likelihood ratio test function
$\log\ x$	the logarithm of $x(> 0)$ with base always e whether it is so explicitly stated or not

Answers to Even-Numbered Exercises

Chapter 2

Section 2.2

2.2

(i) $\mathcal{S} = \{(r,r,r), (r,r,b), (r,r,g), (r,b,r), (r,b,b), (r,b,g), (r,g,r),$
$(r,g,b), (r,g,g), (b,r,r), (b,r,b), (b,r,g), (b,b,r), (b,b,b),$
$(b,b,g), (b,g,r), (b,g,b), (b,g,g), (g,r,r), (g,r,b), (g,r,g),$
$(g,b,r), (g,b,b), (g,b,g), (g,g,r), (g,g,b), (g,g,g)\}.$

(ii) $A = \{(r,b,g), (r,g,b), (b,r,g), (b,g,r), (g,r,b), (g,b,r)\},$
$B = \{(r,r,b), (r,r,g), (r,b,r), (r,b,b), (r,g,r), (r,g,g), (b,r,r),$
$(b,r,b), (b,b,r), (b,b,g), (b,g,b), (b,g,g), (g,r,r),$
$(g,r,g), (g,b,b), (g,b,g), (g,g,r), (g,g,b)\},$
$C = A \cup B = \mathcal{S} - \{(r,r,r), (b,b,b), (g,g,g)\}.$

2.4

(i) Denoting by (x_1, x_2) the cars sold in the first and the second sale, we have:

$$\mathcal{S} = \{(a_1,a_1), (a_1,a_2), (a_1,a_3), (a_2,a_1), (a_2,a_2), (a_2,a_3), (a_3,a_1),$$
$$(a_3,a_2), (a_3,a_3), (a_1,b_1), (a_1,b_2), (a_2,b_1), (a_2,b_2), (a_3,b_1),$$
$$(a_3,b_2), (a_1,c), (a_2,c), (a_3,c), (b_1,a_1), (b_1,a_2), (b_1,a_3),$$
$$(b_2,a_1), (b_2,a_2), (b_2,a_3), (b_1,b_1), (b_1,b_2), (b_2,b_1), (b_2,b_2),$$
$$(b_1,c), (b_2,c), (c,a_1), (c,a_2), (c,a_3), (c,b_1), (c,b_2), (c,c)\}.$$

(ii) $A = \{(a_1,a_1), (a_1,a_2), (a_1,a_3), (a_2,a_1), (a_2,a_2), (a_2,a_3), (a_3,a_1),$
$(a_3,a_2), (a_3,a_3)\},$
$B = \{(a_1,b_1), (a_1,b_2), (a_2,b_1), (a_2,b_2), (a_3,b_1), (a_3,b_2)\},$
$C = B \cup \{(b_1,a_1), (b_1,a_2), (b_1,a_3), (b_2,a_1), (b_2,a_2), (b_2,a_3)\},$
$D = \{(c,b_1), (c,b_2), (b_1,c), (b_2,c)\}.$

Introduction to Probability, Second Edition. http://dx.doi.org/10.1016/B978-0-12-800041-0.00022-5

2.6

$$E = A^c,\ F = C - D = C \cap D^c,\ G = B - C = B \cap C^c,$$
$$H = A^c - B = A^c \cap B^c = (A \cup B)^c,\ I = B^c.$$

2.8

(i) $B_0 = A_1^c \cap A_2^c \cap A_3^c$.
(ii) $B_1 = (A_1 \cap A_2^c \cap A_3^c) \cup (A_1^c \cap A_2 \cap A_3^c) \cup (A_1^c \cap A_2^c \cap A_3)$.
(iii) $B_2 = (A_1 \cap A_2 \cap A_3^c) \cup (A_1 \cap A_2^c \cap A_3) \cup (A_1^c \cap A_2 \cap A_3)$.
(iv) $B_3 = A_1 \cap A_2 \cap A_3$.
(v) $C = B_0 \cup B_1 \cup B_2 = (A_1 \cap A_2 \cap A_3)^c$.
(vi) $D = B_1 \cup B_2 \cup B_3 = A_1 \cup A_2 \cup A_3$.

2.10 If $A = \varnothing$, then $A \cap B^c = \varnothing$, $A^c \cap B = \mathcal{S} \cap B = B$, so that $(A \cap B^c) \cup (A^c \cap B) = B$ for every B. Next, let $(A \cap B^c) \cup (A^c \cap B) = B$ and take $B = \varnothing$ to obtain $A \cap B^c = A$, $A^c \cap B = \varnothing$, so that $A = \varnothing$.

2.12 $A \subseteq B$ implies that, for every $s \in A$, we have $s \in B$, whereas $B \subseteq C$ implies that, for every $s \in B$, we have $s \in C$. Thus, for every $s \in A$, we have $s \in C$, so that $A \subseteq C$.

2.14 For $s \in \cup_j A_j$, let $j_0 \geq 1$ be the first j for which $s \in A_{j_0}$. Then, if $j_0 = 1$, it follows that $s \in A_1$ and therefore s belongs in the right-hand side of the relation. If $j_0 > 1$, then $s \notin A_j$, $j = 1, \ldots, j_0 - 1$, but $s \in A_{j_0}$, so that $s \in A_1^c \cap \cdots \cap A_{j_0-1}^c \cap A_{j_0}$ and hence s belongs to the right-hand side of the relation. Next, let s belong to the right-hand side event. Then, if $s \in A_1$, it follows that $s \in \cup_j A_j$. If $s \notin A_j$ for $j = 1, \ldots, j_0 - 1$ but $s \in A_{j_0}$, it follows that $s \in \cup_j A_j$. The identity is established.

2.16

(i) Since $-5 + \frac{1}{n+1} < -5 + \frac{1}{n}$ and $20 - \frac{1}{n} < 20 - \frac{1}{n+1}$, it follows that $(-5 + \frac{1}{n}, 20 - \frac{1}{n}) \subset (-5 + \frac{1}{n+1}, 20 - \frac{1}{n+1})$, or $A_n \subset A_{n+1}$, so that $\{A_n\}$ is increasing. Likewise, $7 + \frac{3}{n+1} < 7 + \frac{3}{n}$, so that $(0, 7 + \frac{3}{n+1}) \subset (0, 7 + \frac{3}{n})$, or $B_{n+1} \subset B_n$; thus, $\{B_n\}$ is decreasing.
(ii) $\cup_{n=1}^{\infty} A_n = \cup_{n=1}^{\infty}(-5 + \frac{1}{n}, 20 - \frac{1}{n}) = (-5, 20)$, and $\cap_{n=1}^{\infty} B_n = \cap_{n=1}^{\infty}(0, 7 + \frac{3}{n}) = (0, 7]$.

Section 2.3

3.2 Each one of the r.v.'s X and Y takes on the values: 0, 1, 2, 3 and $X + Y = 3$.

3.4 X takes on the values: $-3, -2, -1, 0, 1, 2, 3, 4, 5, 6, 7$,

$$\begin{aligned}(X \leq 2) = \{&(-3, 0), (-3, 1), (-3, 2), (-3, 3), (-3, 4), (-2, 0), (-2, 1),\\ &(-2, 2), (-2, 3), (-2, 4), (-1, 0), (-1, 1), (-1, 2), (-1, 3),\\ &(0, 0), (0, 1), (0, 2), (1, 0), (1, 1), (2, 0)\},\end{aligned}$$

$$\begin{aligned}(3 < X \leq 5) &= (4 \leq X \leq 5) = (X = 4 \text{ or } X = 5)\\ &= \{(0,4),(1,3),(1,4),(2,2),(2,3),(3,1),(3,2)\},\\ (X > 6) &= (X \geq 7) = \{(3,4)\}.\end{aligned}$$

3.6

(i) $\mathcal{S} = \{(1,1),(1,2),(1,3),(1,4),(2,1),(2,2),(2,3),(2,4),(3,1),(3,2),$ $(3,3),(3,4),(4,1),(4,2),(4,3),(4,4)\}$.

(ii) The values of X are: 2, 3, 4, 5, 6, 7, 8.

(iii) $(X \leq 3) = (X = 2 \text{ or } X = 3) = \{(1,1),(1,2),(2,1)\}$,
$(2 \leq X < 5) = (2 \leq X \leq 4) = (X = 2 \text{ or } X = 3 \text{ or } X = 4) =$
$\{(1,1),(1,2),(2,1),(1,3),(2,2),(3,1)\}, \quad (X > 8) = \varnothing$.

3.8

(i) $\mathcal{S} = [8{:}00, 8{:}15]$.
(ii) The values of X consist of the interval [8:00, 8:15].
(iii) The event described is the interval [8:10, 8:15].

Section 2.4

4.2 (i) $3 \times 4 \times 5 = 60$; (ii) $1 \times 2 \times 5 = 10$; (iii) $3 \times 4 \times 1 = 12$.
4.4 (i) $3 \times 2 \times 3 \times 2 \times 3 = 108$; (ii) $3 \times 2 \times 2 \times 1 \times 1 = 12$.
4.6 2^n; $2^5 = 32$, $2^{10} = 1,024$, $2^{15} = 32,768$, $2^{20} = 1,048,576$, $2^{25} = 33,554,432$.
4.8 Start with $\binom{n+1}{m+1} / \binom{n}{m}$, expand in terms of factorial, do the cancellations, and you end up with $(n+1)/(m+1)$.
4.10 Selecting r out of $m+n$ in $\binom{m+n}{r}$ ways is equivalent to selecting x out of m in $\binom{m}{x}$ ways and $r-x$ out of n in $\binom{n}{r-x}$ ways where $x = 0, 1, \ldots, r$. Then $\binom{m+n}{r} = \sum_{x=0}^{r} \binom{m}{x}\binom{n}{r-x}$.
4.12 The required number is $\binom{n}{3}$, which for $n = 10$ becomes $\binom{10}{3} = 120$.
4.14 The required number is found by pairing out the n countries in $\binom{n}{2}$ ways and then multiplying this by 2 to account for the 2 ambassadors involved. Thus, $\binom{n}{2} \times 2 = \frac{n(n-1)}{2!} \times 2 = \frac{n(n-1)}{2} \times 2 = n(n-1)$. For the given values of n, we have, respectively, $9 \times 10 = 90$, $49 \times 50 = 2,450$, $99 \times 100 = 9,900$.

Chapter 3

Section 3.2

2.2 Since $A \cup B \supseteq A$, we have $P(A \cup B) \geq P(A) = \frac{3}{4}$. Also, $A \cap B \subseteq B$ implies $P(A \cap B) \leq P(B) = \frac{3}{8}$. Finally, $P(A \cap B) = P(A) + P(B) - P(A \cup B) = \frac{3}{4} + \frac{3}{8} - P(A \cup B) = \frac{9}{8} - P(A \cup B) \geq \frac{9}{8} - 1 = \frac{1}{8}$.

2.4 We have: $A^c \cap B = B \cap A^c = B - A$ and $A \subset B$. Therefore $P(A^c \cap B) = P(B - A) = P(B) - P(A) = \frac{5}{12} - \frac{1}{4} = \frac{1}{6} \simeq 0.167$. Likewise, $A^c \cap C = C - A$ with $A \subset C$, so that $P(A^c \cap C) = P(C - A) = P(C) - P(A) = \frac{7}{12} - \frac{1}{4} = \frac{1}{3} \simeq 0.333$, $B^c \cap C = C - B$ with $B \subset C$, so that $P(B^c \cap C) = P(C - B) = P(C) - P(B) = \frac{7}{12} - \frac{5}{12} = \frac{1}{6} \simeq 0.167$. Next, $A \cap B^c \cap C^c = A \cap (B^c \cap C^c) = A \cap (B \cup C)^c = A \cap C^c = A - C = \varnothing$, so that $P(A \cap B^c \cap C^c) = 0$, and $A^c \cap B^c \cap C^c = (A \cup B \cup C)^c = C^c$, so that $P(A^c \cap B^c \cap C^c) = P(C^c) = 1 - P(C) = 1 - \frac{7}{12} = \frac{5}{12} \simeq 0.417$.

2.6 The event A is defined as follows: $A =$ "$x = 7n, n = 1, \ldots, 28$," so that $P(A) = \frac{28}{200} = \frac{7}{50} = 0.14$. Likewise, $B =$ "$x = 3n + 10, n = 1, \ldots, 63$," so that $P(B) = \frac{63}{200} = 0.315$, and $C =$ "$x^2 + 1 \leq 375$" = "$x^2 \leq 374$" = "$x \leq \sqrt{374}$" = "$x \leq 19$," and then $P(C) = \frac{19}{200} = 0.095$.

2.8 Denote by A, B, and C the events that a student reads news magazines A, B, and C, respectively. Then the required probability is $P(A^c \cap B^c \cap C^c)$. However,

$$\begin{aligned} P(A^c \cap B^c \cap C^c) &= P((A \cup B \cup C)^c) = 1 - P(A \cup B \cup C) \\ &= 1 - [P(A) + P(B) + P(C) - P(A \cap B) - P(A \cap C) \\ &\quad - P(B \cap C) + P(A \cap B \cap C)] \\ &= 1 - (0.20 + 0.15 + 0.10 - 0.05 - 0.04 - 0.03 + 0.02) \\ &= 1 - 0.35 = 0.65. \end{aligned}$$

2.10 From the definition of A, B, and C, we have:

$$\begin{aligned} A &= \{(0,4), (0,6), (1,3), (1,5), (1,9), (2,2), (2,4), (2,8), (3,1), (3,3), \\ &\quad (3,7), (4,0), (4,2), (4,6), (5,1), (5,5), (6,0), (6,4)\}, \\ B &= \{(0,0), (1,2), (2,4), (3,6), (4,8)\}, \\ C &= \{(0,1), (0,2), (0,3), (0,4), (0,5), (0,6), (0,7), (0,8), (0,9), (1,0), \\ &\quad (1,2), (1,3), (1,4), (1,5), (1,6), (1,7), (1,8), (1,9), (2,0), (2,1), \\ &\quad (2,3), (2,4), (2,5), (2,6), (2,7), (2,8), (2,9), (3,0), (3,1), (3,2), \\ &\quad (3,4), (3,5), (3,6), (3,7), (3,8), (3,9), (4,0), (4,1), (4,2), (4,3), \\ &\quad (4,5), (4,6), (4,7), (4,8), (4,9), (5,0), (5,1), (5,2), (5,3), (5,4), \\ &\quad (5,6), (5,7), (5,8), (5,9), (6,0), (6,1), (6,2), (6,3), (6,4), (6,5), \\ &\quad (6,7), (6,8), (6,9)\}, \end{aligned}$$

or

$$C^c = \{(0,0), (1,1), (2,2), (3,3), (4,4), (5,5), (6,6)\}.$$

Therefore, since the number of points in $\mathcal{S}$ is $7 \times 10 = 70$, we have:

$$P(A) = \frac{18}{70} = \frac{9}{35} \simeq 0.257, \qquad P(B) = \frac{5}{70} = \frac{1}{14} \simeq 0.071,$$
$$P(C) = \frac{63}{70} = \frac{9}{10} = 0.9, \quad \text{or} \quad P(C) = 1 - P(C^c) = 1 - \frac{7}{70} = \frac{63}{70} = 0.9.$$

2.12 The required probability is: $\frac{1}{360} \simeq 0.003$.

2.14 The required probability is: $\dfrac{\binom{n-1}{m}}{\binom{n}{m}} = 1 - \frac{m}{n}$.

2.16 The 500 bulbs can be chosen in $\binom{2000}{500}$ ways, and x defective can be chosen in $\binom{200}{x}$ ways, whereas the $500 - x$ good bulbs can be chosen in $\binom{1800}{500-x}$ ways. Since the probability of having exactly x defective bulbs among the 500 chosen is: $\binom{200}{x}\binom{1800}{500-x}/\binom{2000}{500}$, the required probability is given by: $\dfrac{1}{\binom{2000}{500}} \sum_{x=0}^{25} \binom{200}{x}\binom{1800}{500-x}$.

2.18

GENDER			
Class	Male	Female	Totals
Freshmen	40	60	100
Sophomore	30	50	80
Junior	24	46	70
Senior	12	28	40
Totals	106	184	290

The number of ways to select 12 students is: $\binom{290}{12}$. Next, the number of ways to select the committee members under each one of the requirements in parts (i)–(iv), and the respective probabilities are:

(i) $\dfrac{\binom{184}{7}\binom{106}{5}}{\binom{290}{12}} \simeq 0.218.$

(ii) $\dfrac{\binom{100}{3}\binom{80}{3}\binom{70}{3}\binom{40}{3}}{\binom{290}{12}} \simeq 0.012.$

(iii) $$\frac{\binom{60}{2}\binom{40}{1}\binom{50}{2}\binom{30}{1}\binom{46}{2}\binom{24}{1}\binom{28}{2}\binom{12}{1}}{\binom{290}{12}} \simeq 0.0005.$$

(iv) $$1 - P(\text{no seniors}) = 1 - \frac{\binom{250}{12}}{\binom{290}{12}} \simeq 1 - 0.162 = 0.838.$$

2.20 We have:

$$\begin{aligned}
\textbf{(i) } P(S \cap L) &= 1 - P[(S \cap L)^c] = 1 - P(S^c \cup L^c) \\
&= 1 - [P(S^c) + P(L^c) - P(S^c \cap L^c)] \\
&= 1 - [1 - P(S) + 1 - P(L) - P(S^c \cap L^c)] \\
&= P(S) + P(L) + P(S^c \cap L^c) - 1 \\
&= 0.25 + 0.35 + 0.45 - 1 = 0.05. \\
\textbf{(ii) } P(S \cup L) &= P(S) + P(L) - P(S \cap L) \\
&= 0.25 + 0.35 - 0.05 = 0.55.
\end{aligned}$$

Section 3.3

3.2

(i) For $0 < x \le 2$, $f(x) = \frac{d}{dx}(2c(x^2 - \frac{1}{3}x^3)) = 2c(2x - x^2)$. Thus, $f(x) = 2c(2x - x^2)$, $0 < x \le 2$ (and 0 elsewhere).

(ii) From $\int_0^2 2c(2x - x^2)dx = 1$, we get $\frac{8c}{3} = 1$, so that $c = 3/8$.

3.4

(i)

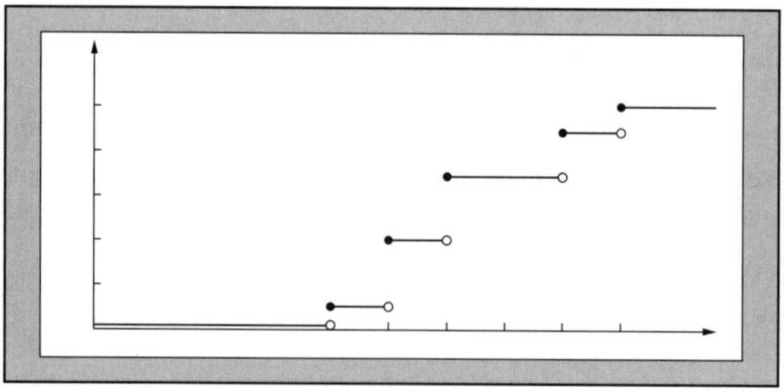

(ii) $P(X \le 6.5) = 0.7$, $P(X > 8.1) = 1 - P(X \le 8.1) = 1 - 0.9 = 0.1$, $P(5 < X < 8) = P(X < 8) - P(X \le 5) = 0.7 - 0.4 = 0.3$.

3.6

(i) We need two relations which are provided by: $\int_0^1 (cx + d)dx = 1$ and $\int_{1/2}^1 (cx + d)dx = 1/3$, or: $c + 2d = 2$ and $9c + 12d = 8$, and hence $c = -\frac{4}{3}, d = \frac{5}{3}$.

(ii) For $0 \le x \le 1$, $F(x) = \int_0^x (-\frac{4}{3}t + \frac{5}{3})dt = -\frac{2x^2}{3} + \frac{5x}{3}$. Thus,

$$F(x) = \begin{cases} 0, & x < 0 \\ -\frac{2x^2}{3} + \frac{5x}{3}, & 0 \le x \le 1 \\ 1, & x > 1. \end{cases}$$

3.8 From $\sum_{x=0}^{\infty} c\alpha^x = c\sum_{x=0}^{\infty} \alpha^x = c \times \frac{1}{1-\alpha} = 1$, we get $c = 1 - \alpha$.

3.10

(i) $\sum_{x=0}^{\infty} c(\frac{1}{3})^x = c[1 + \frac{1}{3} + (\frac{1}{3})^2 + \cdots] = \frac{c}{1-\frac{1}{3}} = \frac{3c}{2} = 1$ and $c = \frac{2}{3}$.

(ii) $P(X \ge 3) = \frac{2}{3}\sum_{x=3}^{\infty}(\frac{1}{3})^x = \frac{2}{3} \times \frac{1/3^3}{2/3} = \frac{1}{27} \simeq 0.037$.

3.12

(i) $\int_0^{\infty} ce^{-cx}\,dx = -\int_0^{\infty} de^{-cx} = -e^{-cx}|_0^{\infty} = -(0-1) = 1$ for every $c > 0$.

(ii) $P(X \ge 10) = \int_{10}^{\infty} ce^{-cx}\,dx = -e^{-cx}|_{10}^{\infty} = -(0 - e^{-10c}) = e^{-10c}$.

(iii) $P(X \ge 10) = 0.5$ implies $e^{-10c} = \frac{1}{2}$, so that $-10c = -\log 2$ and $c = \frac{1}{10}\log 2 \simeq \frac{0.693}{10} \simeq 0.069$.

3.14

(i) From $\sum_{j=0}^{\infty} \frac{c}{3^j} = c\sum_{j=0}^{\infty} \frac{1}{3^j} = c \times \frac{1}{1-\frac{1}{3}} = \frac{3c}{2} = 1$, we get $c = \frac{2}{3}$.

(ii) $P(X \ge 3) = c\sum_{j\ge 3} \frac{1}{3^j} = c \times \frac{1/3^3}{1-\frac{1}{3}} = c \times \frac{1}{2\times 3^2} = \frac{2}{3} \times \frac{1}{2\times 3^2} = \frac{1}{3^3} = \frac{1}{27} \simeq 0.037$.

(iii) $P(X = 2k+1, k = 0, 1, \ldots) = c\sum_{k=0}^{\infty} \frac{1}{3^{2k+1}} = c(\frac{1}{3} + \frac{1}{3^3} + \frac{1}{3^5} + \cdots) = c \times \frac{1/3}{1-\frac{1}{9}} = c \times \frac{3}{8} = \frac{2}{3} \times \frac{3}{8} = 0.25$.

(iv) $P(X = 3k+1, k = 0, 1, \ldots) = c\sum_{k=0}^{\infty} \frac{1}{3^{3k+1}} = c(\frac{1}{3} + \frac{1}{3^4} + \frac{1}{3^7} + \cdots) = c \times \frac{1/3}{1-\frac{1}{27}} = c \times \frac{9}{26} = \frac{2}{3} \times \frac{9}{26} = \frac{3}{13} \simeq 0.231$.

3.16

(i) P(no items are sold) $= f(0) = \frac{1}{2} = 0.5$.

(ii) P(more than 3 items are sold) $= \sum_{x=4}^{\infty}(\frac{1}{2})^{x+1} = (\frac{1}{2})^5 \times \frac{1}{1-\frac{1}{2}} = \frac{1}{16} = 0.0625$.

(iii) P(an odd number of items are sold) $= (\frac{1}{2})^2 + (\frac{1}{2})^4 + (\frac{1}{2})^6 + \cdots = (\frac{1}{2})^2 \times \frac{1}{1-\frac{1}{4}} = \frac{1}{3} \simeq 0.333$.

3.18

(i) Since $\int_0^\infty c^2 x e^{-cx} dx = -cxe^{-cx}|_0^\infty - e^{-cx}|_0^\infty = 1$ for all $c > 0$, the given function is a p.d.f. for all $c > 0$.

(ii) From part (i),

$$P(X \geq t) = -c\,x\,e^{-cx}|_t^\infty - e^{-cx}|_t^\infty = cte^{-ct} + e^{-ct} = \frac{ct+1}{e^{ct}}.$$

(iii) Here $ct + 1 = 2 + 1 = 3$, so that $\frac{ct+1}{e^{ct}} = \frac{3}{e^2} \simeq 0.405$.

3.20 We have:

$$P(X > x_0) = \int_{x_0}^{1} n(1-x)^{n-1} dx = -\int_{x_0}^{1} d(1-x)^n$$

$$= -(1-x)^n|_{x_0}^1 = (1-x_0)^n,$$ and it is given that this probability is $1/10^{2n}$. Thus,

$$(1-x_0)^n = \frac{1}{10^{2n}}, \text{ or } 1 - x_0 = \frac{1}{100} \text{ and } x_0 = 0.99.$$

3.22 If $s \in \cup_i (X \in B_i)$, then $s \in (X \in B_i)$ for at least one i, so that $X(s) \in B_i$. Then $X(s) \in (\cup_i B_i)$ and $s \in (X \in (\cup_i B_i))$. Thus, the left-hand side is contained in the right-hand side. Next, let $s \in (X \in (\cup_i B_i))$. Then $X(s) \in \cup_i B_i$, so that $X(s) \in B_i$ for at least one i. Hence $s \in (X \in B_i)$, and then $s \in \cup_i (X \in B_i)$. Thus, the right-hand side is contained in the left-hand side. The proof is completed.

3.24 For $x \in \Re$, $F_X(x) = P_X((-\infty, x])$ and hence $0 \leq F_X(x) \leq 1$; for $x_1 < x_2$, we have $(-\infty, x_1] \subset (-\infty, x_2]$, so that $P_X((-\infty, x_1]) \leq P_X((-\infty, x_2])$, or $F_X(x_1) \leq F_X(x_2)$; as $n \to \infty$, $x_n \downarrow x$ implies $(-\infty, x_n] \downarrow (-\infty, x]$, so that $P_X((-\infty, x_n]) \downarrow P_X((-\infty, x])$ (by Theorem 2), or $F_X(x_n) \downarrow F_X(x)$, which shows that F_X is right-continuous; next, as $n \to \infty$ and $x_n \uparrow \infty$, it follows that $(-\infty, x_n] \uparrow \Re$, so that $P_X((-\infty, x_n]) \uparrow P_X(\Re) = 1$ (by Theorem 2, again), or $F_X(x_n) \uparrow 1$, or $F_X(\infty) = 1$; finally, as $x_n \downarrow -\infty$, we have $(-\infty, x_n] \downarrow \varnothing$, hence $P_X((-\infty, x_n]) \downarrow P_X(\varnothing) = 0$, or $F_X(x_n) \downarrow 0$, or $F_X(-\infty) = 0$.

3.26 There is no contradiction between $f_X(x) = P(X = x) = 0$ for all $x \in \Re$, and $\int_\Re f_X(x)dx = 1$, as $\Re = \cup_{x\in\Re}\{x\}$ and this union consists of *uncountably* many terms. Recall that property (P3) of the Axiomatic Definition of Probability stipulates *additivity* of probabilities for *countably* many terms only.

Chapter 4

Section 4.1

1.2 We have: $P(A|A \cup B) = \frac{P(A\cap(A\cup B))}{P(A\cup B)} = \frac{P(A)}{P(A\cup B)} = \frac{P(A)}{P(A)+P(B)}$ (since $A \cap B = \varnothing$), and likewise, $P(B|A \cup B) = \frac{P(B\cap(A\cup B))}{P(A\cup B)} = \frac{P(B)}{P(A)+P(B)}$.

1.4

(i) $P(b_2|b_1) = 15/26 \simeq 0.577$; (ii) $P(g_2|g_1) = 13/24 \simeq 0.542$;
(iii) $P(b_2) = 0.52$; (iv) $P(b_1 \cap g_2) = 0.22$.

1.6 Parts (i) and (ii) follow without any calculations by using the fact that $P(\cdot|B)$ and $P(\cdot|C)$ are probability functions, or directly as follows:

(i) $P(A^c|B) = \frac{P(A^c \cap B)}{P(B)} = \frac{P(B - A\cap B)}{P(B)} = \frac{P(B)-P(A\cap B)}{P(B)} = 1 - \frac{P(A\cap B)}{P(B)} = 1 - P(A|B)$.

(ii) $P(A \cup B|C) = \frac{P((A\cup B)\cap C)}{P(C)} = \frac{P((A\cap C)\cup(B\cap C))}{P(C)} = \frac{P(A\cap C)+P(B\cap C)-P(A\cap B\cap C)}{P(C)} = \frac{P(A\cap C)}{P(C)} + \frac{P(B\cap C)}{P(C)} - \frac{P((A\cap B)\cap C)}{P(C)} = P(A|C) + P(B|C) - P(A\cap B|C)$.

(iii) In the sample space $\mathcal{S}$ = {*HHH, HHT, HTH, THH, HTT, THT, TTH, TTT*} with all outcomes being equally likely, define the events:

$$A = \text{“the \# of } H\text{'s is} \leq 2\text{”} = \{TTT, TTH, THT, HTT, THH, HTH, HHT\},$$
$$B = \text{“the \# of } H\text{'s is} > 1\text{”} = HHT, HTH, THH, HHH\}.$$

Then $B^c = \{HTT, THT, TTH, TTT\}$, $A\cap B^c = B^c$, $A\cap B = \{HHT, HTH, THH\}$, so that:

$$P(A|B^c) = \frac{P(A\cap B^c)}{P(B^c)} = \frac{P(B^c)}{P(B^c)} = 1 \quad \text{and}$$
$$1 - P(A|B) = 1 - \frac{P(A\cap B)}{P(B)} = 1 - \frac{3/8}{4/8} = 1 - \frac{3}{4} = \frac{1}{4}.$$

Thus, $P(A|B^c) \neq 1 - P(A|B)$.

(iv) In the sample space $\mathcal{S} = \{1, 2, 3, 4, 5\}$ with all outcomes being equally likely, consider the events $A = \{1, 2\}$, $B = \{3, 4\}$, and $C = \{2, 3\}$, so that $A\cap B = \varnothing$ and $A\cup B = \{1, 2, 3, 4\}$, Then:

$$P(C|A\cup B) = \frac{P(C\cap(A\cup B))}{P(A\cup B)} = \frac{2/5}{4/5} = \frac{2}{4} = \frac{1}{2}, \quad \text{whereas}$$
$$P(C|A) = \frac{P(A\cap C)}{P(A)} = \frac{1/5}{2/5} = \frac{1}{2},$$
$$P(C|B) = \frac{P(B\cap C)}{P(B)} = \frac{1/5}{2/5} = \frac{1}{2}, \quad \text{so that}$$
$$P(C|A\cup B) \neq P(C|A) + P(C|B).$$

1.8 For $n = 2$, the theorem is true since $P(A_2|A_1) = \frac{P(A_1\cap A_2)}{P(A_1)}$ yields $P(A_1\cap A_2) = P(A_2|A_1)P(A_1)$. Next, assume

$$P(A_1\cap\cdots\cap A_k) = P(A_k|A_1\cap\cdots\cap A_{k\ 1})\cdots P(A_2|A_1)P(A_1)$$

and show that

$$P(A_1 \cap \cdots \cap A_{k+1}) = P(A_{k+1}|A_1 \cap \cdots \cap A_k)$$
$$P(A_k|A_1 \cap \cdots \cap A_{k-1}) \cdots P(A_2|A_1)P(A_1).$$

Indeed,

$$\begin{aligned} &P(A_1 \cap \cdots \cap A_{k+1}) \\ &= P((A_1 \cap \cdots \cap A_k) \cap A_{k+1}) \\ &= P(A_{k+1}|A_1 \cap \cdots \cap A_k)P(A_1 \cap \cdots \cap A_k) \\ &\quad \text{(by applying the theorem for two events } A_1 \cap \cdots \cap A_k \text{ and } A_{k+1}) \\ &= P(A_{k+1}|A_1 \cap \cdots \cap A_k)P(A_k|A_1 \cap \cdots \cap A_{k-1}) \cdots P(A_2|A_1)P(A_1) \end{aligned}$$

(by the induction hypothesis).

1.10 With obvious notation, we have:

$$\begin{aligned} &P(\text{1st red and 4th red}) \\ &= P(R_1 \cap R_2 \cap R_3 \cap R_4) + P(R_1 \cap R_2 \cap B_3 \cap R_4) \\ &+ P(R_1 \cap B_2 \cap R_3 \cap R_4) + P(R_1 \cap B_1 \cap B_2 \cap R_4) \\ &= P(R_4|R_1 \cap R_2 \cap R_3)P(R_3|R_1 \cap R_2)P(R_2|R_1)P(R_1) \\ &+ P(R_4|R_1 \cap R_2 \cap B_3)P(B_3|R_1 \cap R_2)P(R_2|R_1)P(R_1) \\ &+ P(R_4|R_1 \cap B_2 \cap R_3)P(R_3|R_1 \cap B_2)P(B_2|R_1)P(R_1) \\ &+ P(R_4|R_1 \cap B_1 \cap B_2)P(B_2|R_1 \cap B_1)P(B_1|R_1)P(R_1) \\ &= \frac{7}{12} \times \frac{8}{13} \times \frac{9}{14} \times \frac{10}{15} + \frac{8}{12} \times \frac{5}{13} \times \frac{9}{14} \times \frac{10}{15} + \frac{8}{12} \times \frac{9}{13} \times \frac{5}{14} \times \frac{10}{15} \\ &+ \frac{9}{12} \times \frac{4}{13} \times \frac{5}{14} \times \frac{10}{15} \\ &= \frac{1}{12 \times 13 \times 14 \times 15} \\ &\quad \times (7 \times 8 \times 9 \times 10 + 5 \times 8 \times 9 \times 10 \times 2 + 4 \times 5 \times 9 \times 10) \\ &= \frac{9 \times 10 \times 156}{12 \times 13 \times 14 \times 15} = \frac{3}{7} \simeq 0.429. \end{aligned}$$

1.12

(i) $P(+) = 0.01188$; **(ii)** $P(D|+) = \frac{190}{1188} \simeq 0.16$.

1.14 Let I = "switch I is open," II = "switch II is open," S = "signal goes through." Then: (i) $P(S) = 0.48$; (ii) $P(\text{I}^c|S^c) = \frac{5}{13} \simeq 0.385$; (iii) $P(\text{II}^c|S^c) = \frac{10}{13} \simeq 0.769$.

1.16 With F = "an individual is female," M = "an individual is male," C = "an individual is color-blind," we have:

$P(\text{F}) = 0.52$, $P(\text{M}) = 0.48$, $P(\text{C}|\text{F}) = 0.25$, $P(\text{C}|\text{M}) = 0.05$, and therefore

$P(\text{C}) = 0.154$, $P(\text{M}|\text{C}) = \dfrac{12}{77} \simeq 0.156$.

1.18 With obvious notation, we have:

$$\text{(i) } P(\text{D}) = 0.029; \text{ (ii) } P(\text{I}|\text{D}) = \frac{12}{29} \simeq 0.414;$$

$$\text{(iii) } P(\text{II}|\text{D}) = \frac{9}{29} \simeq 0.310, \quad \text{and} P(\text{III}|\text{D}) = \frac{8}{29} \simeq 0.276.$$

1.20

(i) $P(X > t) = \int_t^\infty \lambda e^{-\lambda x}\, dx = -\int_t^\infty de^{-\lambda x} = -e^{-\lambda x}|_t^\infty = e^{-\lambda t}$.

(ii)

$$P(X > s+t|X > s) = \frac{P(X > s+t, X > s)}{P(X > s)} = \frac{P(X > s+t)}{P(X > s)}$$

$$= \frac{e^{-\lambda(s+t)}}{e^{-st}} \quad \text{(by part (i))}$$

$$= e^{-\lambda t}.$$

(iii) The conditional probability that X is greater than t units beyond s, given that it has been greater than s, does not depend on s and is the same as the (unconditional) probability that X is greater than t. That is, this distribution has some sort of "memoryless" property.

1.22 For $i = 1, 2, 3$, let S_i = "the card drawn the ith time is a spade." Then the required probability is: $P(S_1 \mid S_2 \cap S_3) = \frac{P(S_1 \cap S_2 \cap S_3)}{P(S_2 \cap S_3)} = \frac{11}{50} = 0.22$.

1.24 It is given that: $P(D) = 0.01$, $P(+|D) = 0.95$, $P(+|D^c) = 0.005$. Then:

$$\text{(i) } P(+) = 0.01445; \text{ (ii) } P(D|+) = \frac{190}{289} \simeq 0.657.$$

Section 4.2

2.2 Here $P(A) = P(A \cap A) = P(A)P(A) = [P(A)]^2$, and this happens if $P(A) = 0$, whereas, if $P(A) \neq 0$, it happens only if $P(A) = 1$.

2.4 Since $P(A_1 \cap A_2) = P(A_1)P(A_2)$, we have to show that:

$$\begin{aligned} P(A_1 \cap (B_1 \cup B_2)) &= P(A_1)P(B_1 \cup B_2),\ P(A_2 \cap (B_1 \cup B_2)) \\ &= P(A_2)P(B_1 \cup B_2),\ P(A_1 \cap A_2 \cap (B_1 \cup B_2)) \\ &= P(A_1)P(A_2)P(B_1 \cup B_2). \end{aligned}$$

Indeed,

$$\begin{aligned} &P(A_1 \cap (B_1 \cup B_2)) \\ &= P((A_1 \cap B_1) \cup (A_1 \cap B_2)) \\ &= P(A_1 \cap B_1) + P(A_1 \cap B_2) = P(A_1)P(B_1) + P(A_1)P(B_2) \\ &= P(A_1)P(B_1 \cup B_2), \text{ and similarly for } P(A_2 \cap (B_1 \cup B_2)). \end{aligned}$$

Finally,

$$
\begin{aligned}
P(A_1 \cap A_2 \cap (B_1 \cup B_2)) &= P((A_1 \cap A_2 \cap B_1) \cup (A_1 \cap A_2 \cap B_2)) \\
&= P(A_1 \cap A_2 \cap B_1) + P(A_1 \cap A_2 \cap B_2) \\
&= P(A_1)P(A_2)P(B_1) + P(A_1)P(A_2)P(B_2) \\
&= P(A_1)P(A_2)P(B_1 \cup B_2).
\end{aligned}
$$

2.6

(i) Clearly, $A = (A \cap B \cap C) \cup (A \cap B^c \cap C) \cup (A \cap B \cap C^c) \cup (A \cap B^c \cap C^c)$ and hence $P(A) = 0.6875$. Likewise, $P(B) = 0.4375$, $P(C) = 0.5625$.
(ii) A, B, and C are not independent.
(iii) $P(A \cap B) = \frac{4}{16}$, and then $P(A|B) = \frac{4}{7} \simeq 0.571$.
(iv) A and B are not independent.

2.8

(i) $S = \{HHH, HHT, HTH, THH, HTT, THT, TTH, TTT\}$, $A = \{HHH, TTT\}$ with $P(A) = p^3 + q^3 \quad (q = 1 - p)$.
(ii) $P(A) = 0.28$.

2.10

(i) $c = 1/25$.
(ii) See figure.

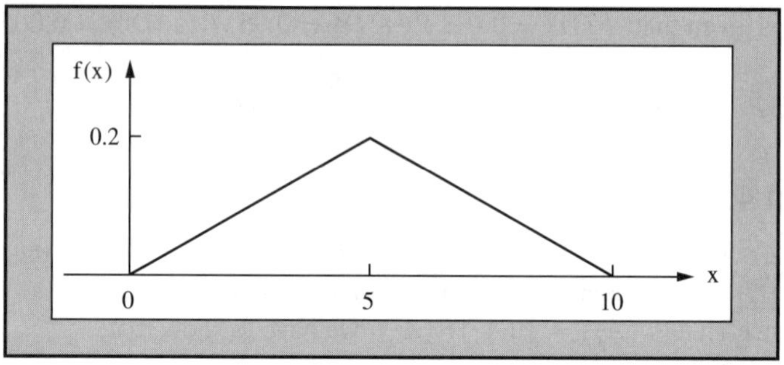

(iii) $P(A) = P(X > 5) = 0.50$, $P(B) = P(5 < X < 7.5) = 0.375$.
(iv) $P(B|A) = 0.75$.
(v) A and B are not independent.

2.12

(a) (i) $(\frac{n_R}{n})^3$; (ii) $1 - (\frac{n_B + n_W}{n})^3$; (iii) $\frac{6 n_R n_B n_W}{n^3}$.

(b) (i) $\dfrac{\binom{n_R}{3}}{\binom{n}{3}}$; (ii) $1 - \dfrac{\binom{n_B + n_W}{3}}{\binom{n}{3}}$; (iii) $\dfrac{\binom{n_R}{1}\binom{n_B}{1}\binom{n_W}{1}}{\binom{n}{3}}$.

2.14 With obvious notations, we have:

(i) $P(B) = (m_1n_2 + n_1m_2 + 2n_1n_2)/2(m_1 + n_1)(m_2 + n_2)$.
(ii) $P(W) = (m_1n_2 + n_1m_2 + 2m_1m_2)/2(m_1 + n_1)(m_2 + n_2)$.

For the given values of m_1, m_2, n_1, and n_2, we have:

$$P(B) = \frac{61}{120} \simeq 0.508, \quad P(W) = \frac{59}{120} \simeq 0.492.$$

2.16

(i) $P(\text{no circuit is closed}) = (1 - p_1)\cdots(1 - p_n)$.
(ii) $P(\text{at least 1 circuit is closed}) = 1 - (1 - p_1)\cdots(1 - p_n)$.
(iii) $P(\text{exactly 1 circuit is closed}) = p_1(1 - p_2)\cdots(1 - p_n) + (1 - p_1)p_2 \times (1 - p_3)\cdots(1 - p_n) + \cdots + (1 - p_1)\cdots(1 - p_{n-1})p_n$.
(iv) The answers above are: $(1 - p)^n, 1 - (1 - p)^n, np(1 - p)^{n-1}$.
(v) The numerical values are: 0.01024, 0.98976, 0.0768.

2.18 For $i \geq 1$, we have $P(A_{1i}) = \frac{2}{8} = \frac{1}{4}$ and $P(B_i) = \frac{2}{8} = \frac{1}{4}$. Then:

$$P(A_1) = \frac{1}{4} + \left(\frac{1}{4}\right)^2 + \left(\frac{1}{4}\right)^3 + \cdots = \frac{1/4}{1 - \frac{1}{4}} = \frac{1}{3} \simeq 0.333.$$

2.20

(i) It is shown that: $P(A \cap (B \cup C)) = P(A)P(B \cup C)$, so that A and $B \cup C$ are independent.
(ii) Independence of A, B, C implies independence of A^c, B, C^c; then $P(A^c \cap (B \cap C^c)) = P(A^c)P(B \cap C^c)$, so that A^c and $B \cap C^c$ are independent.
(iii) By (i), $B \cup C$ and A are independent. Then $P(B \cup C|A) = P(B \cup C) = P(B) + P(C) - P(B)P(C)$.

2.22 Parts (i) and (ii) are argued easily.

(iii) $P(A) = \lim_{n\to\infty} P(A_n) = \lim_{n\to\infty} p^n = 0$.

Chapter 5

Section 5.1

1.2

(i) $EX = 0,\ EX^2 = c^2$, and $Var(X) = c^2$.
(ii) $P(|X - EX| \leq c) = P(-c \leq X \leq c) = P(X = -c, X = c) = 1 = \frac{c^2}{c^2} = \frac{Var(X)}{c^2}$.

1.4 If Y is the net loss to the company, then $EY = \$600$, and if P is the premium to be charged, then $P = \$700$.

1.6 $Var(X) = EX^2 - (EX)^2$, by expanding and taking expectations. Also, $E[X(X-1)] = Var(X) + (EX)^2 - EX$ by expanding, taking expectations, and using the first result. That $Var(X) = E[X(X-1)] + EX - (EX)^2$ follows from the first two results.

1.8 (i) $EX = 2$, $E[X(X-1)] = 4$; (ii) $Var(X) = 2$.

1.10 $EX = \frac{4}{3}$, $EX^2 = 2$, so that $Var(X) = \frac{2}{9}$ and s.d. of $X = \frac{\sqrt{2}}{3} \simeq 0.471$.

1.12 $c_1 = -1/12$, $c_2 = 5/3$.

1.14

(i) By adding and subtracting μ, we get: $E(X-c)^2 = Var(X) + (\mu - c)^2$.
(ii) Immediate from part (i).

1.16

(i) Setting $x = \tan u = \frac{\sin u}{\cos u}$, $-\frac{\pi}{2} < u < \frac{\pi}{2}$, and observing that $\frac{d\tan u}{du} = \frac{1}{\cos^2 u}$, we get $\int_{-\infty}^{\infty} \frac{dx}{1+x^2} = \pi$, so that $\frac{1}{\pi}\int_{-\infty}^{\infty} \frac{dx}{1+x^2} = 1$.
(ii) $\frac{1}{\pi}\int_{-\infty}^{\infty} x \times \frac{dx}{1+x^2} = \frac{1}{2\pi}\int_{-\infty}^{\infty} \frac{d(1+x^2)}{1+x^2} = \frac{1}{2\pi}\log(1+x^2)\big|_{-\infty}^{\infty} = \frac{1}{2\pi}(\infty - \infty)$.

1.18 For the discrete case, $X \geq c$ means $x_i \geq c$ for all values x_i of X. Then $x_i f_X(x_i) \geq c f_X(x_i)$ and hence $\sum_{x_i} x_i f_X(x_i) \geq \sum_{x_i} c f_X(x_i)$. But $\sum_{x_i} x_i f_X(x_i) = EX$ and $\sum_{x_i} c f_X(x_i) = c \sum_{x_i} f_X(x_i) = c$. Thus, $EX \geq c$. The particular case follows, of course, by taking $c = 0$. In the continuous case, summation signs are replaced by integrals.

1.20

(i) $P(40 \leq X \leq 60) = 0.32$
(ii) $\mu = EX = 10 \times 0.13 + 20 \times 0.12 + 30 \times 0.10 + 40 \times 0.13 + 50 \times 0.10 + 60 \times 0.09 + 70 \times 0.11 + 80 \times 0.08 + 90 \times 0.08 + 100 \times 0.06 = 49.6$, $EX^2 = 3,232$, so that $Var(X) = 3,232 - (49.6)^2 = 3,232 - 2,460.16 = 771.84$ and $\sigma \simeq 27.782$.
(iii) $[\mu - \sigma, \mu + \sigma] = [49.6 - 27.782, 49.6 + 27.782] = [21.818, 77.382]$, $[\mu - 1.5\sigma, \mu + 1.5\sigma] = [49.6 - 41.673, 49.6 + 41.673] = [7.927, 91.273]$, $[\mu - 2\sigma, \mu + 2\sigma] = [49.6 - 55.564, 49.6 + 55.564] = [-5.964, 105.164]$. It follows that the respective proportions of claims are: 53%, 94% and 100%.

Section 5.2

2.2

(i) $c = \sigma/\sqrt{1-\alpha}$; **(ii)** $c = \frac{1}{\sqrt{0.05}} \simeq 4.464$.

2.4

(i) By the Tchebichev inequality, $P(|X-\mu| \geq c) = 0$ for all $c > 0$.

(ii) Consider a sequence $0 < c_n \downarrow 0$ as $n \to \infty$. Then $P(|X-\mu| \geq c_n) = 0$ for all n, or equivalently, $P(|X-\mu| < c_n) = 1$ for all n, whereas, clearly, $\{(|X-\mu| < c_n)\}$ is a nonincreasing sequence of events and its limits is $\cap_{n=1}^{\infty}(|X-\mu| < c_n)$. Then, by Theorem 2 in Chapter 2, $1 = \lim_{n\to\infty} P(|X-\mu|<c_n) = P(\cap_{n=1}^{\infty}(|X-\mu|<c_n))$. However, it is clear that

$$\cap_{n=1}^{\infty}(|X-\mu|<c_n)=(|X-\mu| \leq 0) = (X=\mu).$$

Thus, $P(X=\mu)=1$, as was to be seen.

Section 5.3

3.2 (i) $x_p = [(n+1)p]^{1/(n+1)}$; (ii) For $p = 0.5$ and $n = 3$, we have $x_{0.5} = 2^{1/4} \simeq 1.189$.

3.4 (i) $c_1 = c_2 = 1$; (ii) $x_{1/3} = 0$.

3.6 (i)

x	2	3	4	5	6	7	8	9	10	11	12
$f(x)$	1/36	2/36	3/36	4/36	5/36	6/36	5/36	4/36	3/36	2/36	1/36

(ii) $EX = 7$; (iii) median = mode = mean = 7.

3.8 By the Hint, $P(X \leq c) = \int_{-\infty}^{c} f(x)dx = \int_0^{\infty} f(c-y)dy$, and $P(X \geq c) = \int_c^{\infty} f(x)dx = \int_0^{\infty} f(c+y)dy$. Since $f(c-y) = f(c+y)$, it follows that $P(X \leq c) = P(X \geq c)$, and hence c is the median.

3.10

(i) $p = P(Y \leq y_p) = P[g(X) \leq y_p] = P[X \leq g^{-1}(y_p)]$, so that $g^{-1}(y_p) = x_p$ and $y_p = g(x_p)$.

(ii) $x_p = -\log(1-p)$.

(iii) $y_p = g(-\log(1-p))$.

Chapter 6

Section 6.1

1.2

(i) It follows by using the identity $\binom{n+1}{x} = \binom{n}{x} + \binom{n}{x-1}$.

(ii) $B(26, 0.25; 10) = 0.050725$.

1.4 If X is the number of those favoring the proposal, then $X \sim B(15, 0.4375)$. Therefore: (i) $P(X \geq 5) = 0.859$; (ii) $P(X \geq 8) = 0.3106$.

1.6 If X is the number of times the bull's eye is hit, then $X \sim B(100, p)$. Therefore:

(i) $P(X \geq 40) = \sum_{x=40}^{100} \binom{100}{x} p^x q^{100-x} \quad (q = 1 - p)$.

(ii) $P(X \geq 40) = \sum_{x=40}^{100} \binom{100}{x} (0.25)^x (0.75)^{100-x}$.

(iii) $EX = np = 100p$, $Var(X) = npq = 100pq$, and for $p = 0.25$, $EX = 25$, $Var(X) = 18.75$, s.d. of $X = \sqrt{18.75} \simeq 4.330$.

1.8 From the Tchebichev inequality, $n = 8,000$.

1.10

(i) Writing $\binom{n}{x}$ in terms of factorials, and after cancellations, we get:

$$EX = np \sum_{y=0}^{n-1} \binom{n-1}{y} p^y q^{(n-1)-y} = np \times 1 = np.$$

Likewise,

$$E[X(X-1)] = n(n-1)p^2 \sum_{y=0}^{n-2} \binom{n-2}{y} p^y q^{(n-2)-y} = n(n-1)p^2 \times 1$$
$$= n(n-1)p^2.$$

(ii) From Exercise 1.6, $Var(X) = n(n-1)p^2 + np - (np)^2 = npq$.

1.12 Mode = 25 and $f(25) = \binom{100}{25} (\frac{1}{4})^{25} (\frac{3}{4})^{75}$; one would bet on $X = 25$.

1.14 Here X has the geometric distribution with $p = 0.01$, so that $f(x) = (0.01)(0.99)^{x-1}$, $x = 1, 2, \ldots$ Then:

$$P(X \leq 10) = 1 - P(X \geq 11) = 1 - (0.01)(0.99)^{10}[1 + 0.99 + (0.99)^2 + \cdots]$$
$$= 1 - (0.01)(0.99)^{10} \times \frac{1}{1 - 0.99} = 1 - (0.99)^{10} \simeq 0.096.$$

1.16 The r.v. X has the geometric distribution with parameter p. Then:

(i) $n \geq \log(1 - \alpha)/\log q$.

(ii) For $\alpha = 0.95$ and $p = 0.25$, we have $n = 11$, and for $\alpha = 0.95$ and $p = 0.50$, we have $n = 5$.

1.18

(i) With $q = 1 - p$, $M_X(t) = \frac{pe^t}{1-qe^t}$, $t < -\log q$, and for the given p, $M_X(t) = \frac{0.01e^t}{1-0.99e^t}$, $t < -\log(0.99)$ $(\simeq 0.01)$.

(ii) $\frac{d}{dt}\left(\frac{pe^t}{1-qe^t}\right)|_{t=0} = \frac{1}{p} = EX$, $\frac{d^2}{dt^2}\left(\frac{pe^t}{1-qe^t}\right)|_{t=0} = \frac{1+q}{p^2} = EX^2$, so that $Var(X) = \frac{q}{p^2}$.

1.20 $\lambda \simeq 2.3$, $P(X = 5) \simeq 0.054$.

1.22

(i) $EX = \sum_{x=0}^{\infty} x \times e^{-\lambda}\frac{\lambda^x}{x!} = \lambda$, $E[X(X-1)] = \sum_{x=2}^{\infty} x(x-1)e^{-\lambda}\frac{\lambda^x}{x!} = \lambda^2$, so that:

(ii) $Var(X) = E[X(X-1)] + EX - (EX)^2 = \lambda$.

1.24

(i) An appropriate probability model is the Poisson distribution with parameter 3, $P(3)$.

(ii) Here we wish to find the mode(s) of the distribution. There are two of them, and they are 2 and 3. The respective probability is 0.2241.

1.26

(i) $\frac{\binom{70}{5}\binom{10}{0}}{\binom{80}{5}} = \frac{\binom{70}{5}}{\binom{80}{5}} = \frac{78,591}{156,104} \simeq 0.503$.

(ii) $\frac{1}{\binom{80}{5}}\left[\binom{70}{3}\binom{10}{2} + \binom{70}{4}\binom{10}{1} + \binom{70}{5}\binom{10}{0}\right] \simeq 0.987$.

1.28 Starting out with $f(x+1) = \frac{\binom{m}{x+1}\binom{n}{r-x-1}}{\binom{m+n}{x+1}}$, writing our the right-hand side in terms of factorials, and effecting the obvious cancellations, we arrive at $\frac{(m-x)(r-x)}{(m-r+x+1)(x+1)} f(x)$.

1.30 From $P(X = x) = \binom{n}{x}(0.15)^x(0.85)^{n-x}$, we get:

$$P(X = 0) = (0.85)^n < n(0.15)(0.85)^{n-1}$$

if and only if $n > \frac{85}{15} \simeq 5.667$, so that $n = 6$.

1.32 The number of defective items has the hypergeometric distribution with $m = 3$, $n = 997$, and $r = 100$. Therefore:

(i) $P(X \le 1) = \frac{538,501}{553,890} \simeq 0.972$,

(ii) $EX = 0.3$; (iii) $Var(X) = \frac{2,991}{100 \times 111} \simeq 0.269$, and $\sigma(X) \simeq 0.519$.

1.34 Here $r = 100$, $\frac{m}{m+n} = 0.003$, so that $r \times \frac{m}{m+n} = 0.3$. Therefore the required probability is (from Poisson tables) 0.963064.

1.36

(i) The relation $P(X = 2) = 3P(X = 4)$ yields $\lambda = 2 = EX = Var(X)$.
(ii) $P(2 \leq X \leq 4) = 0.0527$ (from Poisson tables).

Section 6.2

2.2

(i) By integration, using the definition of $\Gamma(\alpha)$ and the recursive relation for $\Gamma(\alpha + 1)$, we get $EX = \frac{\beta}{\Gamma(\alpha)}\Gamma(\alpha + 1) = \alpha\beta$. Likewise, $EX^2 = \frac{\beta^2}{\Gamma(\alpha)}\Gamma(\alpha + 2) = \alpha(\alpha + 1)\beta^2$, so that $Var(X) = \alpha\beta^2$.
(ii) $EX = 1/\lambda$, $Var(X) = 1/\lambda^2$ from part (i).
(iii) $EX = r$, $Var(X) = 2r$ from part (i).

2.4

(i) (a) With $g(X) = cX$, we have $Eg(X) = c/\lambda$.
(b) With $g(X) = c(1 - 0.5e^{-\alpha X})$, we have $Eg(X) = \frac{(\alpha+0.5\lambda)c}{\alpha+\lambda}$.
(ii) (a) 10; (b) 1.5.

2.6 Indeed, if T_1 is the time from 0 to the occurrence of the 1st event, then for $t > 0$,

$$P(T_1 > t) = P(0 \text{ events occurred in the time interval } (0, t]) = e^{-\lambda t},$$

so that $P(T_1 \leq t) = 1 - e^{-\lambda t}$ and $f_{T_1}(t) = \lambda e^{-\lambda t}$. For $k \geq 2$, and t_{k-1} denoting the time of occurrence of the $(k - 1)$-st event, we have
$P(T_k > t) = P(0 \text{ events occurred in the time interval} (t_{k-1}, t_{k-1} + t] = 0) = e^{-\lambda t}$, and then $f_{T_k}(t) = e^{-\lambda t}$ as above.

2.8

(i) $\int_0^\infty \alpha\beta x^{\beta-1} e^{-\alpha x^\beta} dx = -\int_0^\infty de^{-\alpha x^\beta} = -e^{-\alpha x^\beta}|_0^\infty = 1$.
(ii) $\beta = 1$ and any $\alpha > 0$.

2.10

(i) $\int_0^{x_p} e^{-x} dx = p$ yields $x_p = -\log(1 - p)$.
(ii) $y_p = e^{x_p}$, since the function $y = e^x$ $(x > 0)$ is strictly increasing, and $y_p = 1/(1 - p)$.
(iii) For $p = 0.5$, we get $x_{0.5} = \log 2$ $(\simeq 0.693)$, and $y_{0.5} = 2$.

2.12 By part (iv) of Exercise 2.11:

(i) $P(-1 < Z < 1) = 2\Phi(1) - 1 = 2 \times 0.841345 - 1 = 0.68269.$
(ii) $P(-2 < Z < 2) = 2\Phi(2) - 1 = 2 \times 0.977250 - 1 = 0.9545.$
(iii) $P(-3 < Z < 3) = 2\Phi(3) - 1 = 2 \times 0.998650 - 1 = 0.9973.$

2.14 We have:

$$P(|X - \mu| < k\sigma) \geq 1 - \frac{\sigma^2}{(k\sigma)^2} = 1 - \frac{1}{k^2}.$$

Therefore, for $k = 1, 2, 3$, the respective bounds are: 0 (meaningless); $\frac{3}{4} = 0.75$, which is about 78.6% of the exact probability (0.9545) in the normal case; $\frac{8}{9} \simeq 0.8889$, which is about 89.1% of the exact probability (0.9973) in the normal case.

2.16 Assuming independence in testing, the number of defective items is $Y \sim B(25, p)$ with $p = P(X < 1{,}800) = 0.158655$. Then:

$$P(Y \leq 15) = \sum_{y=0}^{15} 25y(0.158655)^y(0.841345)^{25-y}.$$

2.18

(i) From $f(x) = \frac{1}{\sqrt{2\pi}\sigma}\exp\left[-\frac{(x-\mu)^2}{2\sigma^2}\right]$, we get

$$f(\mu + y) = \frac{1}{\sqrt{2\pi}\sigma}\exp\left(-\frac{y^2}{2}\right) = f(\mu - y).$$

(ii) $\frac{d}{dx}f(x) = 0$ yields $x = \mu$, and $\frac{d^2}{dx^2}f(x)\mid_{x=\mu} = -\frac{1}{\sqrt{2\pi}\sigma^3} < 0$, so that $x = \mu$ maximizes $f(x)$.

2.20 From $f(x) = \frac{1}{\sqrt{2\pi}\sigma}\exp[-\frac{(x-\mu)^2}{2\sigma^2}]$, we get

$$f''(x) = -\frac{1}{\sqrt{2\pi}\sigma^3}\exp\left[-\frac{(x-\mu)^2}{2\sigma^2}\right]\left[1 - \left(\frac{x-\mu}{\sigma}\right)^2\right].$$

Then, for $x = \mu \pm \sigma$, $f''(\mu \pm \sigma) = 0$.

2.22 $\Gamma(\frac{1}{2}) = \int_0^\infty y^{-\frac{1}{2}}e^{-y}dy$, and by setting $y^{1/2} = t/\sqrt{2}$, we get $\Gamma(\frac{1}{2}) = \frac{\sqrt{2}}{2}\int_{-\infty}^{\infty} e^{-t^2/2}dt = \sqrt{\pi} \times \frac{1}{\sqrt{2\pi}}\int_{-\infty}^{\infty} e^{-t^2/2}dt = \sqrt{\pi}.$

2.24 Observe that $M_X(t) = e^{\mu t + \frac{\sigma^2 t^2}{2}}$ with $\mu = \alpha$ and $\sigma^2 = 2\beta$, which is the m.g.f. of the $N(\mu, \sigma^2)$ distribution. Thus, $X \sim N(\alpha, 2\beta)$.

2.26

(i) By Exercise 1.17 in Chapter 5, and with $t \in \Re$, $M_X(t) = \sum_{n=0}^{\infty}(EX^n)\frac{t^n}{n!}$, which here becomes by replacing EX^n, $\sum_{n=0}^{\infty}\frac{(t^2/2)^k}{k!} = e^{t^2/2}$, which is the m.g.f. of $N(0, 1)$.

(ii) Thus, $X \sim N(0, 1)$.

2.28 We have:

$$EX = \frac{\alpha+\beta}{2}, \quad EX^2 = \frac{\beta^2+\alpha\beta+\alpha^2}{3}, \text{ so that}$$

$$Var(X) = EX^2 - (EX)^2 = \frac{(\alpha-\beta)^2}{12}.$$

2.30

(i) $P(X \le 1) = 0.982477$ (from Poisson tables).

(ii) $P(Y > 5) = e^{-0.2} \simeq 0.818$.

2.32 Let Y be defined by:

$$Y = \begin{cases} C, & 0 < X \le \mu/2 \\ C/2, & \mu/2 < X \le \mu \\ 0, & X > \mu. \end{cases}$$

Then: (i) $EY = \frac{C}{2}(2 - e^{-1/2} - e^{-1})$. (ii) \$1,230.

2.34

(i) From $\int_0^m \lambda e^{-\lambda x}\,dx = 0.5$, we get $m = \frac{0.693}{\lambda}$, and for $\lambda = 0.005$, $m = 138.6$.

(ii) The actual claim size Y is: $Y = X$ if $X < M$, and $Y = M$ if $X \ge M$. Then, for $0 < y < M$, $F_Y(y) = 1 - e^{-\lambda y}$, so that $f_Y(y) = \lambda e^{-\lambda y}$, and $f_Y(M) = P(Y = M) = P(X \ge M) = e^{-\lambda M}$. Thus, $f_Y(y) = \lambda e^{-\lambda y}, 0 < y < M$, and $f_Y(y) = e^{-\lambda M}$ for $y = M$.

(iii) $\lambda M = 1$, so that $f_Y(y) = 0.005e^{-0.005y}, 0 < y < 200$, and $f_Y(y) = e^{-1}, y = 200$.

Chapter 7

Section 7.1

1.2 $P(X = 0, Y = 1) = P(X = 0, Y = 2) = P(X = 1, Y = 2) = 0$,
$P(X = 0, Y = 0) = 0.3$, $P(X = 1, Y = 0) = 0.2$, $P(X = 1, Y = 1) = 0.2$,
$P(X = 2, Y = 0) = 0.075$, $P(X = 2, Y = 1) = 0.15$,
$P(X = 2, Y = 2) = 0.075$.

1.4 (i) $\int_0^2 \int_0^1 (x^2 + \frac{xy}{2}) dx\, dy = \frac{6}{7} \times \frac{7}{6} = 1$; (ii) $P(X > Y) = \frac{15}{56} \simeq 0.268$.

1.6 (i) $P(X \le x) = 1 - e^{-x},\ x > 0$; (ii) $P(Y \le y) = 1 - e^{-y},\ y > 0$;
(iii) $P(X < Y) = 0.5$; (iv) $P(X + Y < 3) = 1 - 4e^{-3} \simeq 0.801$.

1.8 $c = 1/\sqrt{2\pi}$.

1.10 $c = 6/7$.

1.12 Here $f(x, y) = \frac{2}{C^2}$, for (x, y) in the triangle OAC; i.e., $0 \le x \le y \le C$. Then:
(i) $EU = \frac{2C^2}{3}$; (ii) $EU = \frac{2}{3}$.

Section 7.2

2.2

$$f_X(0) = 0.3,\ f_X(1) = 0.4,\ f_X(2) = 0.3;$$
$$f_Y(0) = 0.575,\ f_Y(1) = 0.35,\ f_Y(2) = 0.075.$$

2.4

(i)

$$f_X(1) = 7/36,\ \ f_X(2) = 17/36,\ \ f_X(3) = 12/36;$$
$$f_Y(1) = 7/36,\ \ f_Y(2) = 14/36,\ \ f_Y(3) = 15/36.$$

(ii)

$$f_{X|Y}(1|1) = 2/7,\ \ f_{X|Y}(2|1) = 2/7,\ \ f_{X|Y}(3|1) = 3/7;$$
$$f_{X|Y}(1|2) = 1/14,\ \ f_{X|Y}(2|2) = 10/14,\ \ f_{X|Y}(3|2) = 3/14;$$
$$f_{X|Y}(1|3) = 4/15,\ \ f_{X|Y}(2|3) = 5/15,\ \ f_{X|Y}(3|3) = 6/15;$$
$$f_{Y|X}(1|1) = 2/7,\ \ f_{Y|X}(2|1) = 2/7,\ \ f_{Y|X}(3|1) = 3/7;$$
$$f_{Y|X}(1|2) = 2/17,\ \ f_{Y|X}(2|2) = 10/17,\ \ f_{Y|X}(3|2) = 5/17;$$
$$f_{Y|X}(1|3) = 3/12,\ \ f_{Y|X}(2|3) = 3/12,\ \ f_{Y|X}(3|3) = 6/12.$$

2.6

(i) $f_X(x) = \frac{2x}{n(n+1)},\ x = 1, \ldots, n$; $f_Y(y) = \frac{2(n-y+1)}{n(n+1)},\ y = 1, \ldots, n$.

(ii) $f_{X|Y}(x|y) = \frac{1}{n-y+1},\ y = 1, \ldots, x,\ x = 1, \ldots, n$; $f_{Y|X}(y|x) = \frac{1}{x}$,
$y = 1, \ldots, x;\quad x = 1, \ldots, n.$

(iii)

$$E(X|Y = y) = \frac{n(n+1) - (y-1)y}{2(n-y+1)},\ y = 1, \ldots, n;$$
$$E(Y|X = x) = \frac{x+1}{2},\ x = 1, \ldots, n.$$

2.8 $f_X(x) = \frac{2}{5}(3x + 1),\ 0 \le x \le 1;\quad f_Y(y) = \frac{3}{5}(2y^2 + 1),\ 0 \le y \le 1.$

2.10

(i) $f_X(x) = xe^{-x},\ x > 0$; $f_Y(y) = e^{-y},\ y > 0$.

(ii) $f_{Y|X}(y|x) = e^{-y},\ x > 0,\ y > 0$.

(iii) $P(X > \log 4) = \frac{1+\log 4}{4} \simeq 0.597$.

2.12

(i) $f_X(x) = \frac{6x}{7}(2x+1),\ 0 < x \le 1;\ f_Y(y) = \frac{3y}{14} + \frac{2}{7},\ 0 \le y \le 2;$
$f_{Y|X}(y|x) = \frac{2x+y}{4x+2},\ 0 < x \le 1,\ 0 \le y \le 2.$

(ii) $EY = \frac{8}{7};\ E(Y|X = x) = \frac{2}{3} \times \frac{3x+2}{2x+1},\ 0 < x \le 1.$

(iii) It follows by a direct integration.

(iv) $P(Y > \frac{1}{2}|X < \frac{1}{2}) = \frac{483}{560} = 0.8625.$

2.14 From Exercise 1.8, $f_{X,Y}(x, y) = \frac{1}{\sqrt{2\pi}} ye^{-\frac{xy}{2}},\ 0 < y < x$, and it has been found in the discussion of Exercise 1.8 that $f_Y(y) = 2ce^{-y^2/2} = \frac{2}{\sqrt{2\pi}} e^{-y^2/2},\ y > 0.$ Hence

$$f_{X|Y}(x|y) = \frac{1}{2} ye^{\frac{y^2}{2}} e^{-\frac{y}{2}x},\quad 0 < y < x.$$

2.16

(i)

$$f_X(x) = \begin{cases} 6x/7, & 0 < x \le 1 \\ 6x(2-x)/7, & 1 < x < 2 \\ 0, & \text{elsewhere.} \end{cases}$$

(ii) $f_Y(y|x)$ is 1 for $0 < x \le 1$, and is $1/(2-x)$ for $1 < x < 2$ (and 0 otherwise), whereas $1 \le x + y < 2$.

2.18

(i) $f_{X|Y}(\cdot|y)$ is the Poisson p.d.f. with parameter y.

(ii) $f_{X,Y}(x, y) = e^{-2y} \frac{y^x}{x!},\ x = 0, 1, \ldots.$

(iii) $f_X(x) = \frac{1}{2^{x+1}},\ x = 0, 1, \ldots.$

2.20 (i), (ii) follow by applying the definitions.

2.22

(i) $f_{X|Y}(x|y) = \frac{2x}{y^2}, 0 < x \le y < 1;\ f_{Y|X}(y|x) = \frac{2y}{1-x^2}, 0 < x \le y < 1.$

(ii) $E(X|Y = y) = \frac{2y}{3}, 0 < y < 1;\ E(Y|X = x) = \frac{2(1-x^3)}{3(1-x^2)}, 0 < x < 1.$

2.24 $Var(X|Y = y) = E(X^2|Y = y) - [E(X|Y = y)]^2 = \frac{y^2}{18},\ 0 < y < 1.$

2.26 Here:

$$f_X(0) = 0.14,\ f_X(1) = 0.36,\ f_X(2) = 0.30,\ f_X(3) = 0.20;$$
$$f_Y(0) = 0.40,\ f_Y(1) = 0.35,\ f_Y(2) = 0.21,\ f_Y(3) = 0.04.$$

Then:

(i) $P(X \ge 2) = 0.50,\ P(Y \le 2) = 0.96,\ P(X \ge 2, Y \le 2) = 0.49.$

(ii) $EX = 1.56,\ EY = 0.89,\ EX^2 = 3.36,\ EY^2 = 1.55$, so that: $Var(X) = 0.9264,\ Var(Y) = 0.7579$, and $\sigma(X) \simeq 0.962,\ \sigma(Y) \simeq 0.871.$

2.28

(i) From $\int_0^{1/\sqrt{c}} cx^2\,dx = 1$, we get $c = 1/9$, so that $f_X(x) = x^2/9, 0 < x < 3$.

(ii) $f_{X,T}(x,t) = \frac{x}{18}$, $0 < x < 3$, $0.5x < t < 2.5x$, and $f_T(t) = \frac{3.84}{36}t^2$, $0 < t < 1.5$, and $f_T(t) = 0.25 - \frac{0.16}{36}t^2$, $1.5 \le t < 7.5$.

(iii) From (ii), $P(2 < T < 4) = \frac{11.26}{27} \simeq 0.417$.

Chapter 8

Section 8.1

1.2

(i) Consider the continuous case, as in the discrete case integrals are simply replaced by summation signs. So, for relation (8.3), we have:

$$\begin{aligned} E[c\,g(X,Y)] &= \int_{-\infty}^{\infty}\int_{-\infty}^{\infty} cg(x,y)f_{X,Y}(x,y)dx\,dy \\ &= c\int_{-\infty}^{\infty}\int_{-\infty}^{\infty} g(x,y)f_{X,Y}(x,y)dx\,dy = cEg(X,Y); \\ E[c\,g(X,Y)+d] &= \int_{-\infty}^{\infty}\int_{-\infty}^{\infty} [c\,g(X,Y)+d]f_{X,Y}(x,y\,dx\,dy \\ &= c\int_{-\infty}^{\infty}\int_{-\infty}^{\infty} g(X,Y)f_{X,Y}(x,y)dx\,dy + \\ & \quad d\int_{-\infty}^{\infty}\int_{-\infty}^{\infty} f_{X,Y}(x,y)dx\,dy \\ &= cEg(X,Y)+d. \end{aligned}$$

For relation (8.4) we have:

$$\begin{aligned} E[g_1(X,Y)+g_2(X,Y)] &= \int_{-\infty}^{\infty}\int_{-\infty}^{\infty} [g_1(x,y) \\ &\quad +g_2(x,y)]f_{X,Y}(x,y)dx\,dy \\ &= \int_{-\infty}^{\infty}\int_{-\infty}^{\infty} g_1(x,y)f_{X,Y}(x,y)dx\,dy \\ &\quad + \int_{-\infty}^{\infty}\int_{-\infty}^{\infty} g_2(x,y)f_{X,Y}(x,y)dx\,dy \\ &= E\,g_1(X,Y) + E\,g_2(X,Y). \end{aligned}$$

As for relation (8.6), we have:

For the discrete case, let x_i and y_j be the values of the r.v.'s X, and Y, respectively, and let $f_{X,Y}$ be their joint p.d.f. Then, for all x_i and y_j, we have $g(x_i, y_j) \leq h(x_i, y_j)$, hence $g(x_i, y_j) f_{X,Y}(x_i, y_j) \leq h(x_i, y_j) f_{X,Y}(x_i, y_j)$, and by summing over x_i and y_j, we get: $Eg(X, Y) \leq Eh(X, Y)$. In particular, $g(X) \leq h(Y)$ means that, if $X(s) = x_i$ and $Y(s) = y_j$, then $g(x_i) \leq h(y_j)$, hence $g(x_i) f_{X,Y}(x_i, y_j) \leq h(y_j) f_{X,Y}(x_i, y_j)$, and by summing over x_i and y_j, we get:

$$\begin{aligned}
\sum_{x_i}\sum_{y_j} g(x_i) f_{X,Y}(x_i, y_j) &= \sum_{x_i} g(x_i) \sum_{y_j} f_{X,Y}(x_i y_j) \\
&= \sum_{x_i} g(x_i) f_X(x_i) = EX, \\
\sum_{x_i}\sum_{y_j} h(y_j) f_{X,Y}(x_i, y_j) &= \sum_{y_j}\sum_{x_i} h(y_j) \times f_{X,Y}(x_i, y_i) \\
&= \sum_{y_j} h(y_j) \sum_{x_i} f_{X,Y}(x_i, y_j) \\
&= \sum_{y_j} h(y_j) f_Y(y_j) = EY.
\end{aligned}$$

The result follows. For the continuous case, summations are replaced by integrations.

(ii) For relation (8.8), it sufficies to justify the second part only. Indeed,

$$\begin{aligned}
Var[c\, g(X, Y) + d] &= E\{[c\, g(X, Y) + d] - E[c\, g(X, Y) + d]\}^2 \\
&= E[c\, g(X, Y) + d - c\, E\, g(X, Y) - d]^2 \\
&= E\{c[g(X, Y) - E\, g(X, Y)]\}^2 \\
&= E\{c^2[g(X, Y) - E\, g(X, Y)]^2\} \\
&= c^2 E[g(X, Y) - E\, g(X, Y)]^2 = c^2 Var[g(X, Y)].
\end{aligned}$$

1.4 For relation (8.16), we have:

$$\begin{aligned}
\frac{\partial}{\partial t_1} M_{X,Y}(t_1, t_2) \mid_{t_1=t_2=0} &= \frac{\partial}{\partial t_1} E e^{t_1 X + t_2 Y} \mid_{t_1=t_2=0} \\
&= E\left(\frac{\partial}{\partial t_1} e^{t_1 X + t_2 Y}\right) \mid_{t_1=t_2=0} \\
&= E(X^{t_1 X + t_2 Y}) \mid_{t_1=t_2=0} = EX,
\end{aligned}$$

and similarly for the second relation. As for relation (8.17), we have:

$$\begin{aligned}\frac{\partial^2}{\partial t_1\,\partial t_2} M_{X,Y}(t_1,t_2)\mid_{t_1=t_2=0} &= \frac{\partial^2}{\partial t_1\,\partial t_2} Ee^{t_1X+t_2Y}\mid_{t_1=t_2=0}\\ &= E\left(\frac{\partial^2}{\partial t_1\partial t_2}e^{t_1X+t_2Y}\right)\mid_{t_1=t_2=0}\\ &= E(XYe^{t_1X+t_2Y})\mid_{t_1=t_2=0}= E(XY).\end{aligned}$$

Section 8.2

2.2 Apply the exercise cited in the Hint with $Z = X - Y$ and $Z = X + Y$.

2.4

(i) $EX = 1, EY = 0.5, EX^2 = 1.6, EY^2 = 0.65$, so that $Var(X) = 0.6$ and $Var(Y) = 0.4$.

(ii) $E(XY) = 0.8$, so that $Cov(X, Y) = 0.3$ and $\rho(X, Y) = 1.25\sqrt{0.24} \simeq 0.613$.

(iii) The r.v.'s X and Y are positively correlated.

2.6

(i) $EX = \frac{77}{36}, EY = \frac{20}{9}, EX^2 = \frac{183}{36}, EY^2 = \frac{99}{18}$, so that $Var(X) = 659/36^2$ and $Var(Y) = 728/36^2$.

(ii) $E(XY) = \frac{171}{36}$, so that $Cov(X, Y) = -\frac{4}{36^2}$, and $\rho(X, Y) = -\frac{2}{\sqrt{182\times 659}} \simeq -0.006$.

2.8 $EX = 0, Var(X) = 10/4, EY = 5/2, EY^2{=}34/4, Var(Y){=}9/4, E(XY){=}0$, so that $Cov(X, Y) = 0$ and $\rho(X, Y) = 0$. The r.v.'s X and Y are not anywhere close to being *linearly* related.

2.10 By employing the marginal p.d.f.'s found in Exercise 2.8 of Chapter 7, we obtain:

(i) $EX = \frac{3}{5}, \quad EY = \frac{3}{5}$.

(ii)

$$EX^2 = \frac{13}{30}, \quad EY^2 = \frac{11}{25}, \text{ so that :}$$
$$Var(X) = \frac{11}{150} \quad Var(Y) = \frac{2}{25}.$$

(iii)

$$E(XY) = \frac{7}{20}, \text{ so that } Cov(X, Y) = -\frac{1}{100},$$
$$\rho(X, Y) = -\frac{\sqrt{3,300}}{440} \simeq -0.13.$$

(iv) The r.v.'s X and Y are negatively correlated.

2.12

(i)

$$\begin{aligned} E[E(Y^2|X)] &= \int_{-\infty}^{\infty} \left[\int_{-\infty}^{\infty} y^2 f_{Y|X}(y|x)dy \right] f_X(x)dx \\ &= \int_{-\infty}^{\infty} \int_{-\infty}^{\infty} y^2 f_{X,Y}(x,y)dy\,dx \\ &= \int_{-\infty}^{\infty} y^2 f_Y(y)dy = EY^2. \end{aligned}$$

(ii) Likewise

$$\begin{aligned} E[XE(Y|X)] &= E\left[X \int_{-\infty}^{\infty} y f_{Y|X}(y|X)dy \right] \\ &= \int_{-\infty}^{\infty} \left[x \int_{-\infty}^{\infty} y f_{Y|X}(y|x)dy \right] f_X(x)dx \\ &= \int_{-\infty}^{\infty} \int_{-\infty}^{\infty} xy f_{Y|X}(y|x) f_X(x)dy\,dx \\ &= \int_{-\infty}^{\infty} \int_{-\infty}^{\infty} xy f_{X,Y}(x,y)dx\,dy = E(XY), \\ E[E(XY|X)] &= E\left[\int_{-\infty}^{\infty} Xy f_{Y|X}(y|X)dy \right] \\ &= \int_{-\infty}^{\infty} \left[\int_{-\infty}^{\infty} xy f_{Y|X}(y|x)dy \right] f_X(x)dx \\ &= \int_{-\infty}^{\infty} \int_{-\infty}^{\infty} xy f_{Y|X}(y|x) f_X(x)dy\,dx \\ &= \int_{-\infty}^{\infty} \int_{-\infty}^{\infty} xy f_{X,Y}(x,y)dx\,dy = E(XY). \end{aligned}$$

So, $E(XY) = E[E(XY \mid X)] = E[XE(Y|X)]$.

Section 8.3

3.2 With $Var(X) = \sigma^2$, we get $Cov(X,Y) = a\sigma^2$ and $\rho(X,Y) = \frac{a}{|a|}$. Thus, $|\rho(X,Y)| = 1$, and $\rho(X,Y) = 1$ if and only if $a > 0$, and $\rho(X,Y) = -1$ if and only if $a < 0$.

3.4 By differentiation, with respect to α and β, of the function $g(\alpha,\beta) = E[Y-(\alpha X+\beta)]^2$, and by equating the derivatives to 0, we find: $\hat{\alpha} = \frac{\sigma_Y}{\sigma_X}\rho(X,Y)$, $\hat{\beta} = EY - \hat{\alpha}EX$. The 2×2 matrix M of the second-order derivatives is given by:

$M = 4\begin{pmatrix} EX^2 & EX \\ EX & 1 \end{pmatrix}$, which is positive definite. Then $\hat{\alpha}$ and $\hat{\beta}$ are minimizing values.

Chapter 9

Section 9.1

1.2 $M_{X_1,X_2,X_3}(t_1, t_2, t_3) = c^3/(c - t_1)(c - t_2)(c - t_2)$, provided t_1, t_2, t_3 are $<c$.
1.4 Follows by applying properties of expectations.

Section 9.2

2.2 If X_1, X_2, and X_3 are the numbers of customers buying brand A, brand B, or just browsing, then X_1, X_2, X_3 have the Multinomial distribution with parameters $n = 10$, $p_1 = 0.25$, $p_2 = 0.40$, and $p_3 = 0.35$. Therefore:

(i) $P(X_1 = 2, X_2 = 3, X_3 = 5) = \frac{10!}{2!3!5!}(0.25)^2 \times (0.40)^3 \times (0.35)^5 \simeq 0.053$.
(ii) $P(X_1 = 1, X_2 = 3 | X_3 = 6) = \frac{4!}{1!3!}(\frac{5}{13})^1(\frac{8}{13})^3 \simeq 0.358$.

2.4 They follow by taking the appropriate derivatives of the m.g.f. in (9.12) and evaluating them at 0.

2.6

(i) $$P(X_1 = 4, X_2 = 1, X_3 = 1) = \frac{\binom{20}{4}\binom{10}{1}\binom{2}{1}}{\binom{32}{6}} = \frac{8{,}075}{75{,}516} \simeq 0.10693.$$

(ii) $P(X_1 = 4, X_2 = 1, X_3 = 1) = \frac{6!}{4!1!1!}(\frac{10}{16})^4(\frac{5}{16})(\frac{1}{6}) \simeq 0.08941$.
(iii) The probability 0.08941 is about 83.6% of the probability 0.10693.

2.8

(i) The r.v.'s X_1, X_2, X_3 have the multinomial distribution with parameters n and $p_1 = p^2$, $p_2 = 2p(1 - p)$, $p_3 = (1 - p)^2$.
(ii) $P(X_1 = 8, X_2 = 6, X_3 = 1) = \frac{15!}{8!6!1!}[(0.75)^2]^8(2 \times 0.75 \times 0.25)^6 \times [(0.25)^2]^1 = 45{,}045 \times (0.5625)^8 \times (0.375)^6 \times (0.0625) \simeq 0.07847$.
(iii) Since $X_2 \sim B(15, 0.375) = B(15, 6/16)$, we have:
$P(X_2 = 6) = \binom{15}{6}(\frac{6}{16})^6(\frac{10}{16})^9 = 0.2025$ (from the binomial tables).
(iv) $P(X_1 = 8, X_3 = 1 \mid X_2 = 6) = \frac{P(X_1=8, X_2=6, X_3=1)}{P(X_2=6)} \simeq \frac{0.07847}{0.2025} \simeq 0.38751$.

2.10 Let X_1, X_2, and X_3 be the r.v.'s denoting the numbers, among the 15 selected, watching news, a documentary program, and other programs, respectively.

Then these r.v.'s have the multinomial distribution with $n = 15, k = 3, p_1 = 0.3125 (= 5/16), p_2 = 0.25\ (= 4/16), p_3 = 0.4375\ (= 7/16)$. Therefore:

(i) $P(X_1 = 5, X_2 = 4, X_3 = 6) = \frac{15!}{5!4!6!}(0.3125)^5(0.25)^4(0.4375)^6 \simeq 0.05148$.

(ii) Here $X_3 \sim B(15, 0.25) = B(15, 4/16)$, so that $P(X_3 \geq 3) = 1 - P(X_3 \leq 2) = 1 - 0.2361 = 0.7639$ (from the binomial tables).

(iii) Let X be the r.v. denoting the number of those among the 15 selected, who do not watch the news. Then $X = 15 - X_1$, so that

$$\begin{aligned} P(X \leq 5) &= P(15 - X_1 \leq 5) = P(X_1 \geq 10) \\ &= 1 - P(X_1 \leq 9) = 1 - 0.9949 = 0.0051 \end{aligned}$$

(from the binomial tables), since $X_1 \sim B(15, 0.3125) = B(15, 5/16)$.

Section 9.3

3.2 Indeed,

$$\begin{aligned} E(XY) &= \int_{-\infty}^{\infty}\int_{-\infty}^{\infty} xyf_{X,Y}(x,y)dx\,dy \\ &= \int_{-\infty}^{\infty}\int_{-\infty}^{\infty} xyf_{Y|X}(y|x)f_X(x)dx\,dy \\ &= \int_{-\infty}^{\infty}\int_{-\infty}^{\infty} xyf_{Y|X}(y|x)f_X(x)dy\,dx \\ &= \int_{-\infty}^{\infty} xf_X(x)\left[\int_{-\infty}^{\infty} yf_{Y|X}(y|x)dy\right]dx \\ &= \int_{-\infty}^{\infty} xf_X(x)\left[\mu_2 + \frac{\rho\sigma_2}{\sigma_1}(x - \mu_1)\right]dx \\ &\quad \text{(because, by(54), } Y|X = x \sim N(b_x, (\sigma_2\sqrt{1-\rho^2})^2), \\ &\quad \text{so that} E(Y|X = x) = b_x = \mu_2 + \frac{\rho\sigma_2}{\sigma_1}(x - \mu_1)) \\ &= \mu_2\int_{-\infty}^{\infty} xf_X(x)dx + \frac{\rho\sigma_2}{\sigma_1}\left[\int_{-\infty}^{\infty} x^2f_X(x)dx - \mu_1\int_{-\infty}^{\infty} xf_X(x)dx\right] \\ &= \mu_2\mu_1 + \frac{\rho\sigma_2}{\sigma_1}\left(EX^2 - \mu_1^2\right) \\ &= \mu_1\mu_2 + \frac{\rho\sigma_2}{\sigma_1} \times \sigma_1^2 = \mu_1\mu_2 + \rho\sigma_1\sigma_2. \end{aligned}$$

3.4

(i) Follows by applying the definition.

(ii) Straightforward by the properties cited.

(iii) In (ii) here, the quantities $\mu_1, \mu_2, \sigma_1, \sigma_2$, and ρ will have to be computed by using the p.d.f. $f_{X,Y}$ (and its marginals), whereas in Example 4(i), they are read out of the formula for $f_{X,Y}$ given there.

3.6

(i) For $t \in \Re$, we have:

$$
\begin{aligned}
M_{c_1X+c_2Y}(t) &= Ee^{t(c_1X+c_2Y)} = Ee^{(c_1t)X+(c_2t)Y} \\
&= \exp\left\{\mu_1(c_1t) + \mu_2(c_2t) + \right.\\
&\left.\frac{1}{2}\left[\sigma_1^2(c_1t)^2 + 2\rho\sigma_1\sigma_2(c_1t)(c_2t) + \sigma_2^2(c_2t)^2\right]\right\} \\
&= \exp\left[(c_1\mu_1 + c_2\mu_2)t\right.\\
&\left.+\frac{1}{2}(c_1^2\sigma_1^2 + 2c_1c_2\rho\sigma_1\sigma_2 + c_2^2\sigma_2^2)t^2\right],
\end{aligned}
$$

which is the m.g.f. of a r.v. distributed as normal with parameters $c_1\mu_1 + c_2\mu_2$ and $c_1^2\sigma_1^2+2c_1c_2\rho\sigma_1\sigma_2+c_2^2\sigma_2^2$. Thus, c_1X+c_2Y has this distribution.

(ii) For $t_1, t_2 \in \Re$,we have:

$$M_{X,Y}(t_1, t_2) = Ee^{t_1X+t_2Y} = Ee^{(t_1X+t_2Y)\times 1} = M_{t_1X+t_2Y}(1).$$

On the other hand, if $Z \sim N(\mu, \sigma^2)$, then $M_Z(t) = e^{\mu t+\frac{\sigma^2t^2}{2}}$ and $M_Z(1) = e^{\mu+\frac{\sigma^2}{2}}$. Since it is assumed that $t_1X + t_2Y$ is normally distributed, we have then:

$$
\begin{aligned}
M_{X,Y}(t_1, t_2) &= M_{t_1X+t_2Y}(1) \\
&= \exp\left\{E(t_1X + t_2Y) + \frac{[Var(t_1X + t_2Y)]^2}{2}\right\}.
\end{aligned}
$$

But

$$
\begin{aligned}
E(t_1X + t_2Y) &= \mu_1t_1 + \mu_2t_2, \\
Var(t_1X + t_2Y) &= \sigma_1^2t_1^2 + \sigma_2^2t_2^2 + 2\rho\sigma_1\sigma_2t_1t_2,
\end{aligned}
$$

where $\mu_1 = EX, \mu_2 = EY, \sigma_1^2 = Var(X), \sigma_2^2 = Var(Y)$, and $\rho = \rho(X, Y)$. Therefore

$$M_{X,Y}(t_1, t_2) = \exp\left(\mu_1t_1 + \mu_2t_2 + \frac{\sigma_1^2t_1^2 + 2\rho\sigma_1\sigma_2t_1t_2 + \sigma_2^2t_2^2}{2}\right),$$

which is the m.g.f. of the bivariate normal distribution with parameters $\mu_1, \mu_2, \sigma_1^2, \sigma_2^2$ and ρ.

Therefore the joint distribution of X and Y is the bivariate normal with parameters just described.

3.8

(i) The joint m.g.f. of X and Y is given by:

$$M_{X,Y}(t_1,t_2)=\exp\left[\mu_1 t_1+\mu_2 t_2+\frac{1}{2}\left(\sigma_1^2 t_1^2+2\rho\sigma_1\sigma_2 t_1 t_2+\sigma_2^2 t_2^2\right)\right],\quad t_1,t_2\in\Re.$$

Therefore,

$$\begin{aligned}M_{U,V}(t_1,t_2)&=Ee^{t_1U+t_2V}=Ee^{t_1(X+Y)+t_2(X-Y)}\\&=Ee^{(t_1+t_2)X+(t_1-t_2)Y}=M_{X,Y}(t_1+t_2,t_1-t_2)\\&=\exp\Big\{\mu_1(t_1+t_2)+\mu_2(t_1-t_2)+\frac{1}{2}\big[\sigma_1^2(t_1+t_2)^2\\&\quad+2\rho\sigma_1\sigma_2(t_1+t_2)(t_1-t_2)]+\sigma_2^2(t_1-t_2)^2\big]\Big\}\\&=\exp\Big[(\mu_1+\mu_2)t_1+(\mu_1-\mu_2)t_2\\&\quad+\frac{1}{2}\left(\tau_1^2t_1^2+2\rho_0\tau_1\tau_2t_1t_2+\tau_2^2t_2^2\right)\Big],\end{aligned}$$

since

$$\begin{aligned}&\sigma_1^2(t_1+t_2)^2+2\rho\sigma_1\sigma_2(t_1+t_2)(t_1-t_2)+\sigma_2^2(t_1-t_2)^2\\&=\sigma_1^2t_1^2+2\sigma_1^2t_1t_2+\sigma_1^2t_2^2+2\rho\sigma_1\sigma_2t_1^2-2\rho\sigma_1\sigma_2t_2^2\\&\quad+\sigma_2^2t_1^2-2\sigma_2^2t_1t_2+\sigma_2^2t_2^2\\&=(\sigma_1^2+2\rho\sigma_1\sigma_2+\sigma_2^2)t_1^2+(\sigma_1^2-2\rho\sigma_1\sigma_2+\sigma_2^2)t_2^2+2(\sigma_1^2-\sigma_2^2)t_1t_2\\&=\tau_1^2t_1^2+2\frac{\sigma_1^2-\sigma_2^2}{\tau_1\tau_2}\tau_1\tau_2t_1t_2+\tau_2^2t_2^2\\&=\tau_1^2t_1^2+2\rho_0\tau_1\tau_2t_1t_2+\tau_2^2t_2^2.\end{aligned}$$

However, the last expression above; namely,

$$\exp\left[(\mu_1+\mu_2)t_1+(\mu_1-\mu_2)t_2+\frac{1}{2}\left(\tau_1^2t_1^2+2\rho_0\tau_1\tau_2t_1t_2+\tau_2^2t_2^2\right)\right]$$

is the m.g.f. of the bivariate normal distribution with parameters $\mu_1+\mu_2$, $\mu_1-\mu_2$, τ_1^2, τ_2^2, and ρ_0. Therefore this is the joint distribution of the r.v.'s U and V.

(ii) That $U\sim N(\mu_1+\mu_2,\tau_1^2)$ and $V\sim N(\mu_1-\mu_2,\tau_2^2)$ follows by the fact that the distributions of U and V are the marginal distributions of the bivariate normal distribution as given above.

(iii) The r.v.'s U and V are uncorrelated if and only if $\rho(U, V) = 0$. However, $\rho(U, V) = \rho_0 = (\sigma_1^2 - \sigma_2^2)/\tau_1\tau_2$. Therefore $\rho(U, V) = 0$ if and only if $\sigma_1^2 = \sigma_2^2$.

3.10

(i) $EX = 3.2, EY = 12, \sigma^2(X) = 1.44, \sigma^2(Y) = 16, \rho(X, Y) = 0.7$, and $Cov(X, Y) = 3.36$.

(ii)

$$E(X|Y = 10) = 2.78, \quad E(Y|X = 3.8) = 13.4,$$
$$\sigma^2(X|Y = 10) = 0.7344, \quad \sigma^2(Y|X = 3.8) = 8.16.$$

(iii) $X \sim N(3.2, 1.44), \quad Y \sim N(12, 16)$.

(iv) $P(0.8 < X < 4.2) = 0.773981, \quad P(Y > 14) = 0.308538$.

(v) By part (ii), $X|Y{=}10{\sim}N(2.78, 0.7344)$, $Y|X{=}3.8{\sim}N(13.4, 8.16)$.

(vi) $P(X > 3.2|Y = 10) = 0.312067, \quad P(Y < 12|X = 3.8) = 0.312067$.

3.12

(i) $\sigma^2(U) = \sigma_1^2 + c^2\sigma_2^2 + 2c\rho\sigma_1\sigma_2$.

(ii) $c = -\rho\sigma_1/\sigma_2$.

(iii) $\min_c \sigma^2(U) = (1 - \rho^2)\sigma_1^2$.

(iv) X and Y are independent if and only if $\rho = 0$. Then $c = 0$, so that $\sigma^2(U) = \sigma_1^2 = \min_c \sigma^2(U)$.

3.14 (Proof by differentiation)

(i) Set $h(a, b) = E(\hat{Y} - Y)^2 = a^2 - 2\mu_2 a + \sigma_1^2 b^2 - 2\rho\sigma_1\sigma_2 b + (\mu_2^2 + \sigma_2^2)$. Then $\frac{\partial h}{\partial a} = 0, \ \frac{\partial h}{\partial b} = 0$ yield $a = \mu_2, \ b = \rho\sigma_2/\sigma_1$. Furthermore, $\frac{\partial^2 h}{\partial a^2} = 2, \ \frac{\partial^2 h}{\partial a \partial b} = \frac{\partial^2 h}{\partial b \partial a} = 0, \ \frac{\partial^2 h}{\partial b^2} = 2\sigma_1^2$, and

$$(\lambda_1 \ \lambda_2)\begin{pmatrix} 1 & 0 \\ 0 & \sigma_1^2 \end{pmatrix}\begin{pmatrix} \lambda_1 \\ \lambda_2 \end{pmatrix} = \lambda_1^2 + \sigma_1^2\lambda_2^2 > 0(\lambda_1^2 + \lambda_2^2 \neq 0),$$

so that $a = \mu_2, b = \rho\sigma_2/\sigma_1$ minimize $h(a, b)$.

(ii) $\hat{Y} = \mu_2 + \frac{\rho\sigma_2}{\sigma_1}(X - \mu_1)$, and this is equal to $E(Y|X = x)$.

Chapter 10

Section 10.1

1.2 The relation $f_{X,Y}(x, y) = f_X(x) f_Y(y)$ holds true for all values of x and y, and therefore X and Y are independent.

1.4 The r.v.'s X and Y are not independent, since, e.g., $f_{X,Y}(0.1, 0.1) = 0.132 \neq 0.31824 = 0.52 \times 0.612 = f_X(0.1) f_Y(0.1)$.

1.6

(i) $f_X(x) = \frac{6}{5}(x^2 + \frac{1}{2}), 0 \leq x \leq 1$; $f_Y(y) = \frac{6}{5}(y + \frac{1}{3}), 0 \leq y \leq 1$.

(ii) The r.v.'s are not independent, since, e.g., $f_{X,Y}(\frac{1}{2}, \frac{1}{4}) = \frac{3}{5} \neq \frac{9}{10} \times \frac{7}{10} = f_X(\frac{1}{2}) f_Y(\frac{1}{4})$.

1.8

(i) $f_X(x) = 2x, 0 < x < 1$; $f_Y(y) = 2y, 0 < y < 1$; $f_Z(z) = 2z$, $0 < z < 1$.

(ii) The r.v.'s are independent because, clearly,
$f_{X,Y,Z}(x, y, z) = f_X(x) f_Y(y) f_Z(z)/$ for all x, y, and z.

(iii) $P(X < Y < Z) = 1/6$.

1.10

(i) c can be any positive constant.

(ii) $f_{X,Y}(x, y) = c^2 e^{-cx-cy}$, $x > 0$, $y > 0$, and likewise for $f_{X,Z}$ and $f_{Y,Z}$.

(iii) $f_X(x) = ce^{-cx}$, $x > 0$, and likewise for f_Y and f_Z.

(iv) The r.v.'s X and Y are independent, and likewise for the r.v.'s X, Z and Y, Z. Finally, from part (iii), it follows that the r.v.'s X, Y, and Z are also independent.

1.12

(i) $EX = 200$ days; (ii) $M_{X+Y}(t) = 1/(1 - 200t)^2, t < 0.005$, and $f_{X+Y}(t) = (0.005)^2 te^{-0.005t}$, $t > 0$.

(iii) $P(X + Y > 500) = 2.5e^{-2.5} + e^{-2.5} \simeq 0.287$.

1.14

(i) $M_U(t) = \exp[(a\mu_1 + b)t + \frac{(a\sigma_1)^2 t^2}{2}]$ which is the m.g.f. of the $N(a\mu_1 + b, (a\sigma_1)^2)$ distribution. Likewise for V.

(ii) $M_{U,V}(t_1, t_2) = \exp[(a\mu_1 + b)t_1 + \frac{(a\sigma_1)^2 t_1^2}{2} + (c\mu_2 + d)t_2 + \frac{(c\sigma_2)^2 t_2}{2}]$.

(iii) Follows from parts (i) and (ii), since $M_U(t_1) M_V(t_2) = M_{U,V}(t_1, t_2)$ for all t_1, t_2.

1.16 $M_{\bar{X}}(t) = [M(\frac{t}{n})]^n$.

1.18

(i) $E\bar{X} = p$ and $Var(\bar{X}) = pq/n$; (ii) $n = 10{,}000$.

1.20

(i)

$$f_X(-1) = 2\alpha + \beta,\ f_X(0) = 2\beta,\ f_X(1) = 2\alpha + \beta;$$
$$f_Y(-1) = 2\alpha + \beta,\ f_Y(0) = 2\beta,\ f_Y(1) = 2\alpha + \beta.$$

(ii) $EX = EY = 0$, and $E(XY) = 0$; (iii) $Cov(X, Y) = 0$.
(iv) The r.v.'s are not independent, since, e.g., $f(0, 0) = 0 \neq (2\beta) \times (2\beta) = f_X(0) f_Y(0)$.

1.22

(i) $E\bar{X} = \mu$ and $Var(\bar{X}) = \sigma^2/n$.
(ii) For $k = 1, n = 100$; for $k = 2, n = 25$; and for $k = 3, n = 12$.

1.24

(i) $E\bar{X} = \mu$ and $Var(\bar{X}) = \sigma^2/n$.
(ii) The smallest integer n which is $\geq 1/(1-\alpha)c^2$.
(iii) For $c = 0.1$, the required n is its smallest value $\geq 100/(1-\alpha)$. For $\alpha = 0.90, n = 1{,}000$; for $\alpha = 0.95, n = 2{,}000$; for $\alpha = 0.99, n = 10{,}000$.

1.26 By Theorem 1(iii), it suffices to show that: $M_{Y,Z}(t_1, t_2) = M_Y(t_1)M_Z(t_2)$ for all t_1, t_2 belonging in a non-degenerate interval containing 0. Indeed,

$$
\begin{aligned}
M_{Y,Z}(t_1, t_2) &= Ee^{t_1 Y + t_2 Z} \\
&= Ee^{t_1 g(X_{i_1},\ldots,X_{i_m}) + t_2 h(X_{j_1},\ldots,X_{j_n})} \\
&= E[e^{t_1 g(X_{i_1},\ldots,X_{i_m})} \times e^{t_2 h(X_{j_1},\ldots,X_{j_n})}] \\
&= \int_{-\infty}^{\infty} \cdots \int_{-\infty}^{\infty} e^{t_1 g(x_{i_1},\ldots,x_{i_m})} \times e^{t_2 h(x_{j_1},\ldots,x_{j_n})} f_{X_1,\ldots,X_k}(x_1,\ldots,x_k)dx_1\ldots dx_k \\
&\quad \text{(for the continuous case)} \\
&= \int_{-\infty}^{\infty} \cdots \int_{-\infty}^{\infty} e^{t_1 g(x_{i_1},\ldots,x_{i_m})} \times e^{t_2 h(x_{j_1},\ldots,x_{j_n})} f_{X_1}(x_1)\cdots f_{X_k}(x_k)dx_1\ldots dx_k \\
&\quad \text{(by independence of the} X_i\text{'s)} \\
&= \int_{-\infty}^{\infty} \cdots \int_{-\infty}^{\infty} [e^{t_1 g(x_{i_1},\ldots,x_{i_m})} f_{X_{i_1}}(x_{i_1})\ldots f_{X_{i_m}}(x_{i_m}) \times e^{t_2 h(x_{j_1},\ldots,x_{j_n})} \times \\
&\quad f_{X_{j_1}}(x_{j_1})\ldots f_{X_{j_n}}(x_{j_n})dx_{i_1}\ldots dx_{i_m}dx_{j_1}\ldots dx_{j_n}] \\
&= \Bigg[\underbrace{\int_{-\infty}^{\infty} \cdots \int_{-\infty}^{\infty}}_{m \text{ integrals}} e^{t_1 gh(x_{i_1},\ldots,x_{i_m})} f_{X_{i_1}}(x_{i_1})\ldots f_{X_{i_m}}(x_{i_m})dx_{i_1}\ldots dx_{i_m}\Bigg] \times \\
&\quad \Bigg[\underbrace{\int_{-\infty}^{\infty} \cdots \int_{-\infty}^{\infty}}_{n \text{ integrals}} e^{t_2 h(x_{j_1},\ldots,x_{j_n})} f_{X_{j_1}}(x_{j_1})\ldots f_{X_{j_n}}(x_{j_n})dx_{j_1}\ldots dx_{j_n}\Bigg] \\
&= Ee^{t_1 g(X_{i_1},\ldots,X_{i_m})} \times Ee^{t_2 h(X_{j_1},\ldots,X_{j_n})} \\
&= Ee^{t_1 Y} \times Ee^{t_2 Z} = M_Y(t_1)M_Z(t_2),
\end{aligned}
$$

as was to be seen. The integrals are replaced by summation signs in the discrete case.

1.28

(i) $U \sim N(0, 2)$.
(ii) $Var(U) = 2, \quad Cov(X, U) = 1, \quad \rho(X, Y) = \sqrt{2}/2$.
(iii) $M_{X,U}(t_1, t_2) = \exp(\frac{t_1^2+2t_2+2t_1t_2}{2}), t_1, t_2 \in \Re$.
(iv) $\frac{\partial^2}{\partial t_1 \partial t_2} M_{X,U}(t_1, t_2) \mid_{t_1=t_2=0} = 1 = E(XU)$, so that $Cov(X, U) = 1$.

1.30

(i) $c = \frac{1}{4}$.
(ii) $f_X(x) = \frac{1}{2}, -1 \leq x \leq 1, f_Y(y) = \frac{1}{2}, -1 \leq y \leq 1$.
(iii) They are independent, because $f_{X,Y}(x, y) = f_X(x) f_Y(y)$ for all x, and y.
(iv) $P(X^2 + Y^2 \leq 1) = \frac{\pi}{4}$.

Section 10.2

2.2

(i) By independence, $X + Y \sim B(m + n, p)$.
(ii) Here $X + Y \sim B$ (20, 0.25), and hence:

$$P(5 \leq X + Y \leq 15) = \sum_{t=5}^{15} \binom{20}{t} (0.25)^t (0.75)^{20-t}$$
$$= 1.0000 - 0.4148 = 0.5852.$$

2.4

(i) Clearly, $p = P(X_1 \in I) = \int_I f(x)dx$, for the continuous case, and $p = \sum_{x \in I} f(x)$, for the discrete case.
(ii) Define the r.v.'s $Y_1, \ldots, Y_n$ as follows: $Y_i = 1$ if $X_i \in I$, and $Y_i = 0$ otherwise. Then the r.v.'s $Y_1, \ldots, Y_n$ are also independent and their common distribution is $B(1, p)$. Thus, if $Y = Y_1 + \cdots + Y_n$, then $Y \sim B(n, p)$, and the question is rephrazed as follows: P(at least k of $X_1, \ldots, X_k$ take values in I) $= P(Y \geq k) = 1 - P(Y \leq k-1) = 1 - \sum_{y=0}^{k-1} \binom{n}{y} p^y q^{n-y}$ $(q = 1 - p)$.
(iii) Here $p = \int_{1/\lambda}^{\infty} \lambda e^{-\lambda x} dx = -e^{-\lambda x} |_{1/\lambda}^{\infty} = e^{-1} = 1/e$, and hence the probability in part (ii) is equal to: $1 - \sum_{y=0}^{k-1} \binom{n}{y} \left(\frac{1}{e}\right)^y \left(1 - \frac{1}{e}\right)^{n-y}$.
(iv) The required probability is:

$$1 - \left(1 - \frac{1}{e}\right)^4 - 4\left(\frac{1}{e}\right)\left(1 - \frac{1}{e}\right)^3 \simeq 0.469.$$

2.6

(i) Since $X_i \sim B(1, p)$, $i = 1, \ldots, n$, and these r.v.'s are independent, it follows that $X = X_1 + \cdots + X_n \sim B(n, p)$.

(ii) From $f(x) = \binom{n}{x} p^x q^{n-x} \simeq e^{-np} \frac{(np)^x}{x!} = e^{-2} \frac{2^x}{x!}$, $x = 0, 1, \ldots$, we get from the Poisson tables:

$$\begin{aligned} f(0) &\simeq 0.1353 = 0.1353 \\ f(1) &\simeq 0.4060 - 0.1353 = 0.2707 \\ f(2) &\simeq 0.6767 - 0.4060 = 0.2707 \\ f(3) &\simeq 0.8571 - 0.6767 = 0.1804 \\ f(4) &\simeq 0.9473 - 0.8571 = 0.0902. \end{aligned}$$

2.8 Let X_i be the r.v. denoting the number of no-shows in the i th flight, and assume (reasonably enough) that the r.v.'s $X_1, \ldots, X_5$ are independent. Then the r.v. $X = X_1 + \cdots + X_5 \sim P(5\lambda)$, and since $\lambda = 2$, we have that $X \sim P(10)$. Therefore, we obtain from the Poisson tables:

(i) $P(X = 0) = e^{-10} \simeq 0$.
(ii) $P(X \leq 5) = 0.0671$.
(iii) $P(X = 5) = 0.0671 - 0.0293 = 0.0378$.
(iv) $P(X \geq 5) = 1 - P(X \leq 4) = 1 - 0.0293 = 0.9707$.
(v) $P(X \leq 10) = 0.538$.
(vi) $P(X = 10) = 0.5838 - 0.4579 = 0.1259$.
(vii) $P(X \geq 10) = 1 - P(X \leq 9) = 1 - 0.4579 = 0.5421$.
(viii) $P(X \leq 15) = 0.9513$.
(ix) $P(X = 15) = 0.9513 - 0.9165 = 0.0348$.
(x) $P(X \geq 15) = 1 - P(X \leq 14) = 1 - 0.9165 = 0.0835$.

2.10 By Exercise 1.14(i) here, $-Y \sim N(-\mu_2, \sigma_2^2)$. Then:

(i) By independence of X and $-Y$, we have that $X - Y \sim N(\mu_1 - \mu_2, \sigma_1^2 + \sigma_2^2)$.
Therefore:

(ii)

$$\begin{aligned} P(X > Y) &= P(X - Y > 0) \\ &= P\left[\frac{(X - Y) - (\mu_1 - \mu_2)}{\sqrt{\sigma_1^2 + \sigma_2^2}} > -\frac{\mu_1 - \mu_2}{\sqrt{\sigma_1^2 + \sigma_2^2}}\right] \\ &= P\left(Z > -\frac{\mu_1 - \mu_2}{\sqrt{\sigma_1^2 + \sigma_2^2}}\right) = 1 - \Phi\left(-\frac{\mu_1 - \mu_2}{\sqrt{\sigma_1^2 + \sigma_2^2}}\right). \end{aligned}$$

(iii) When $\mu_1 = \mu_2$, then $\Phi(-\frac{\mu_1-\mu_2}{\sqrt{\sigma_1^2+\sigma_2^2}}) = \Phi(0) = 0.5$, and hence $P(X > Y) = 0.5$ (regardless of σ_1 and σ_2).

2.12

(i) $X \sim N(\mu \sum_{i=1}^n \alpha_i,\ \sigma^2 \sum_{i=1}^n \alpha_i^2)$, $Y \sim N(\mu \sum_{j=1}^n \beta_j,\ \sigma^2 \sum_{j=1}^n \beta_j^2)$.

(ii) Starting with $M_{X,Y}(t_1, t_2)$, using independence of the X_i's, inserting the m.g.f. of the X_i's, and using the notation indicated, we obtain the desired result.

(iii) It is immediate by the fact that $M_{X,Y}(t_1, t_2)$, given in (ii), is the m.g.f. of the bivariate normal distribution with parameters $\mu_1, \mu_2, \sigma_1, \sigma_2$, and $\rho = \rho(X, Y)$.

(iv) Immediate from part (iii).

2.14

(i) n is the smallest integer such that $n \geq 1/c^2(1-p)$.

(ii) Let $p = 0.95$. Then, for $c = 1,\ c = 0.5$, and $c = 0.25$, we get, respectively: $n = 20,\ n = 80$, and $n = 320$.

2.16 Here $\frac{X+Y}{2} \sim N(\mu, \frac{3\sigma^2}{4})$, so that $P(|\frac{X+Y}{2} - \mu| \leq 1.5\sigma) = 2\Phi(\sqrt{3}) - 1 = 0.91637$.

Chapter 11

Section 11.1

1.2

(i) $X \sim N(\frac{\mu-160}{9}, \frac{25\sigma^2}{81})$; (ii) $a \simeq 32.222$, $b = 35$.

(iii) $a_k = \frac{5\mu-160}{9} - k\frac{5\sigma}{9}$, $b_k = \frac{5\mu-160}{9} + k\frac{5\sigma}{9}$.

1.4 $f_Y(y) = \lambda y^{-(\lambda+1)},\ y > 1$; $f_Z(z) = \lambda e^{z-\lambda e^z},\ z \in \Re$.

1.6

(i) $f_Y(y) = \frac{1}{2}e^{-y/2},\ y > 0$, which is the p.d.f. of a χ_2^2.

(ii) $\sum_{i=1}^n Y_i \sim \chi_{2n}^2$, since $Y_i \sim \chi_2^2,\ i = 1, \ldots, n$, independent.

1.8 $f_Y(y) = \frac{1}{\Gamma(\frac{3}{2})m^{3/2}} y^{\frac{3}{2}-1} e^{-y/m},\ y > 0$.

1.10 Restrict ourselves to $y>0$. Then $F_Y(y)=F_X\left(\sqrt{\frac{2y}{m}}\right)$, so that $f_Y(y) = \frac{d}{dy}F_X\left(\sqrt{\frac{2y}{m}}\right) = \frac{1}{\Gamma(\frac{3}{2})m^{3/2}} y^{\frac{3}{2}-1} e^{-\frac{y}{m}}$, since $\Gamma(\frac{3}{2}) = \frac{1}{2}\Gamma(\frac{1}{2}) = \frac{1}{2}\sqrt{\pi}$. The last expression is the p.d.f. of Gamma with $\alpha = \frac{3}{2},\ \beta = m$.

Section 11.2

2.2

(i) $f_{U,V}(u,v) = \frac{u}{(1+v)^2} e^{-u},\ u > 0,\ v > 0.$

(ii) $f_U(u) = u\,e^{-u},\ u > 0;\ f_V(v) = 1/(1+v)^2,\ v > 0.$

(iii) U and V are independent.

2.4

(i) $f_{U,V}(u,v) = \frac{1}{|ac|} f_X(\frac{u-b}{a}) f_Y(\frac{v-d}{c})$, $(u,v) \in T$.

(ii) $f_{U,V}(u,v) = \frac{1}{\sqrt{2\pi}|a|\sigma_1} \exp\{-\frac{[u-(a\mu_1+b)]^2}{2(a\sigma_1)^2}\} \times \frac{1}{\sqrt{2\pi}|c|\sigma_2} \times \exp\{-\frac{[v-(c\mu_2+d)]^2}{2(c\sigma_2)^2}\}$, and therefore U and V are independently distributed as $N(a\mu_1+b, (a\sigma_1)^2)$ and $N(c\mu_2+d, (c\sigma_2)^2)$, respectively.

2.6

(i) $f_{U,V}(u,v) = \frac{1}{\sqrt{2\pi}} e^{-u^2/2} \times \frac{1}{\sqrt{2\pi}} e^{-v^2/2}$, $u, v \in \Re$.

(ii) $U \sim N(0,1)$, $V \sim N(0,1)$.

(iii) U and V are independent.

(iv) By parts (ii) and (iii), $X+Y \sim N(0,2)$ and $X-Y \sim N(0,2)$.

2.8 $f_U(u) = 1, 0 \le u \le 1.$

2.10

(i) For $r = 1$, $f_{X_1}(t) = \frac{1}{\pi} \times \frac{1}{1+t^2}$, since $\Gamma(\frac{1}{2}) = \sqrt{\pi}$. Then, by Exercise 1.16 in Chapter 5, the $\int_{-\infty}^{\infty} f_{X_1}(t)dt = \infty - \infty$, so that EX_1 does not exist.

(ii) For $r \ge 2$, $EX_r = 0$ by a simple integration.

(iii) Next, $\int_{-\infty}^{\infty} t^2 \times (1+\frac{t^2}{r})^{-\frac{r+1}{2}} dt = r\sqrt{r}\int_0^1 y^{\frac{r-2}{2}-1}(1-y)^{\frac{3}{2}-1}dy$, by setting first $\frac{t^2}{2} = x$, and $\frac{1}{1+x} = y$ next. In the last expression above, implement the relation given in the Hint to get that this last expression on the right-hand side above, is equal to:

$$r\sqrt{r}\frac{\Gamma(\frac{r-2}{2})\Gamma(\frac{3}{2})}{\Gamma(\frac{r+1}{2})} \quad (\text{for } r \ge 3).$$

By means of this, the recursive relation of the Gamma function, and the fact that $\Gamma(\frac{1}{2}) = \sqrt{\pi}$, we get: $EX_r^2 = \frac{r}{r-2}$, so that $Var(X_r) = \frac{r}{r-2}$ $(r \ge 3)$.

Section 11.3

3.2 It follows by forming the inner products of the row vectors.

3.4 It follows from the joint p.d.f. $f_{X,Y}$, the transformations $u = \frac{x-\mu_1}{\sigma_1}$, $v = \frac{y-\mu_2}{\sigma_2}$, and the fact that the Jacobian $J = \sigma_1\sigma_2$.

3.6

(i) It follows from the joint p.d.f. $f_{X,Y}$, the transformations $u = x + y$, $v = x - y$, and the fact that the Jacobian $J = -1/2$.
(ii) U and V are independent by the fact that they have the bivariate normal distribution and their correlation coefficient is 0.
(iii) It follows from part (i) as marginals of the bivariate normal.

3.8

(i) $P(a\mu < \bar{X} < b\mu, 0 < S^2 < c\sigma^2) = [\Phi(k(b-1)\sqrt{n}) - \Phi(k(a-1)\sqrt{n})] \times P(\chi^2_{n-1} < c(n-1))$; (ii) The probability is 0.89757.

Section 11.5

5.2 $EY_1 = \frac{1}{n+1}$, $EY_n = \frac{n}{n+1}$, and $EY_1 \to 0$, $EY_n \to 1$, as $n \to \infty$.
5.4 $E(Y_1Y_n) = \frac{1}{n+2}$. Therefore, by Exercise 5.2, $Cov(Y_1, Y_n) = \frac{1}{(n+1)^2(n+2)}$.
5.6 $f_Z(z) = \lambda e^{-\lambda z}$, $z > 0$.
5.8

(i) $g_n(y_n) = n\lambda e^{-\lambda y_n}(1 - e^{-\lambda y_n})^{n-1}$, $y_n > 0$.
(ii) For $n = 2$, $EY_2 = 3/2\lambda$, and for $n = 3$, $EY_3 = 11/6\lambda$.

5.10 $g_{1n}(y_1, y_n) = n(n-1)[F(y_n) - F(y_1)]^{n-2} f(y_1)f(y_n)$, $a < y_1 < y_n < b$.

Chapter 12

Section 12.1

1.2 For every $\varepsilon > 0$, $P(|X_n| > \varepsilon) = P(X_n = 1) = p_n$, and therefore $X_n \xrightarrow{P} 0$ if and only if $p_n \to 0$ as $n \to \infty$.
1.4

(i) $P(|Y_{1,n}| > \varepsilon) = (1-\varepsilon)^n \to 0$, as $n \to \infty$.
(ii) $P(|Y_{n,n} - 1| > \varepsilon) = 1 - P(|Y_{n,n} - 1| \le \varepsilon)$ and $P(|Y_{n,n} - 1| \le \varepsilon) = 1 - (1-\varepsilon)^n \to 1$, so that $P(|Y_{n,n} - 1| > \varepsilon) \to 0$, as $n \to \infty$.

1.6 $E\bar{X}_n = \mu$ and $E(\bar{X}_n - \mu)^2 = Var(\bar{X}_n) = \frac{\sigma^2}{n} \to 0$, as $n \to \infty$.
1.8 $E(Y_n - X)^2 = E(Y_n - X_n)^2 + E(X_n - X)^2 + 2E[(Y_n - X_n)(X_n - X)] \to 0$, as $n \to \infty$, by the assumptions made, and the fact that $|E[(Y_n - X_n)(X_n - X)]| \le E^{1/2}|X_n - Y_n|^2 \times E^{1/2}|X_n - X|^2$.

Section 12.2

2.2

(i) $M_X(t) = (1-\alpha)/(1-\alpha e^t)$, $t < -\log\alpha$.

(ii) $EX = \alpha/(1-\alpha)$.

(iii) $M_{\bar{X}_n}(t) = \left(\frac{1-\alpha}{1-\alpha e^{t/n}}\right)^n = \{1 - \frac{\alpha t/(1-\alpha)+[\alpha/(1-\alpha)]nR(\frac{t}{n})}{n}\}^{-n} \underset{n\to\infty}{\longrightarrow} e^{\alpha t/(1-\alpha)}$, since $\frac{n}{t}R(\frac{t}{n}) \underset{n\to\infty}{\longrightarrow} 0$ for fixed t, and $e^{\alpha t/(1-\alpha)}$ is the m.g.f. of $\frac{\alpha}{1-\alpha}$.

2.4 Since $X \sim B(1{,}000, p)$, we have:

(i) $P(1,000p - 50 \le X \le 1,000p+50) = \sum_{x=1000p-50}^{1000p+50} \binom{1,000}{x} p^x q^{1,000-x}$, $q = 1-p$. For $p = \frac{1}{2}$ and $p = \frac{1}{4}$:

$$P(450 \le X \le 550) = \sum_{x=450}^{550} \binom{1,000}{x} (0.5)^{1000},$$

$$P(200 \le X \le 300) = \sum_{x=200}^{300} \binom{1,000}{x} (0.25)^x \times (0.75)^{1000-x}.$$

(ii) For $p = \frac{1}{2}$ and $p = \frac{1}{4}$, the approximate probabilities are, respectively: $\Phi(3.16)+\Phi(3.22)-1 = 0.99857$, $\Phi(3.65)+\Phi(3.72)-1 = 0.999769$.

2.6 $EX_i = \frac{7}{2}$, $EX_i^2 = \frac{91}{6}$, so that $Var(X_i) = \frac{35}{12}$. Therefore $P(150 \le X \le 200) = P(149 < X \le 200) \simeq \Phi(2.07) - \Phi(2.15) = 0.964996$ without continuity correction. With continuity correction, $P(149 < X \le 200) = P(149.5 < X \le 200.5) \simeq 2\Phi(2.11) - 1 = 0.965142$.

2.8 Since $X \sim B(1,000, 0.03)$, the required approximate probability is: $P(X \le 50) = P(-0.5 < X \le 50) \simeq \Phi(3.71) - \Phi(5.65) - 1 = \Phi(3.71) = 0.999896$ without continuity correction. With continuity correction, $P(-0.5 < X \le 50) = P(0 < X \le 50.5) \simeq \Phi(3.80) = 0.999928$.

2.10 $P(|\frac{X}{n} - 0.53| \le 0.02) \simeq 2\Phi(\frac{0.02\sqrt{n}}{\sqrt{0.2491}}) - 1 = 0.99$, so that $n = 4, 146$.

2.12 Since $EX_i = Var(X_i) = \lambda$, setting $S_n = \sum_{i=1}^n X_i$, we have:

(i) $P(S_n \le n\lambda) = P(-0.5 < S_n \le n\lambda) \simeq \Phi(\frac{0.5+n\lambda}{\sqrt{n\lambda}}) - 0.5$.

(ii) For the cases that $n\lambda$ is not an integer, and it is an integer, we have, respectively:

$$P(S_n \ge n\lambda) \simeq 1 + \Phi\left(\frac{n\lambda - [n\lambda]}{\sqrt{n\lambda}}\right) - \Phi\left(\frac{0.5 - n\lambda}{\sqrt{n\lambda}}\right)$$

(where $[n\lambda]$ is the integer part of $n\lambda$),

$$P(S_n \ge n\lambda) \simeq 1 + \Phi\left(\frac{1}{\sqrt{n\lambda}}\right) - \Phi\left(\frac{0.5 + n\lambda}{\sqrt{n\lambda}}\right).$$

(iii) Also, for the cases that $\frac{n\lambda}{2}$ is not an integer, and it is an integer, we have, respectively:

$$P(\frac{n\lambda}{2} \le S_n \le \frac{3n\lambda}{4}) \simeq \Phi\left(\frac{n\lambda - [\frac{n\lambda}{2}]}{\sqrt{n\lambda}}\right) - \Phi\left(\frac{\sqrt{n\lambda}}{4}\right),$$

$$P(\frac{n\lambda}{2} \le S_n \le \frac{3n\lambda}{4}) \simeq \Phi\left(\frac{n\lambda + 2}{2\sqrt{n\lambda}}\right) - \Phi\left(\frac{\sqrt{n\lambda}}{4}\right).$$

(iv) For $n\lambda = 100$, parts (i)–(iii) become: $P(S_n \le 100) \simeq 0.5$; $P(S_n \ge 100) \simeq 0.539828$; $P(50 \le S_n \le 75) \simeq 0.00621$.

2.14 The total life time is $X = \sum_{i=1}^{50} X_i$, where X_i's are independently distributed as negative exponential with $\lambda = 1/1500$. Then $P(X \ge 80{,}000) \simeq 1 - \Phi(0.47) = 0.319178$.

2.16

(i) $P(a \le \bar{X} \le b) \simeq \Phi((2b-1)\sqrt{3n}) - \Phi((2a-1)\sqrt{3n})$.

(ii) Here $(2b-1)\sqrt{3n} = 0.75$, $(2a-1)\sqrt{3n} = -0.75$, and the above probability is: $2\Phi(0.75) - 1 = 0.546746$.

2.18 $P(|\bar{X} - \mu| \le 0.0001) \simeq 2\Phi(0.2\sqrt{n}) - 1 = 0.99$, and then $n = 167$.

2.20

(i) $P(|\bar{X}_n - \mu| < k\sigma) \simeq 2\Phi(k\sqrt{n}) - 1 = p$, so that n is the smallest integer $\ge [\frac{1}{k}\Phi^{-1}(\frac{1+p}{2})]^2$.

(ii) Here n is the smallest integer $\ge 1/(1-p)k^2$.

(iii) For $p = 0.90$, $p = 0.95$, and $p = 0.99$, and the respective values of k, we determine the values of n by means of the CLT and the Tchebichev inequality.

Then, for the various values of k, the respective values of n are given in the following table for part (i).

k\p	**0.90**	**0.95**	**0.99**
0.50	11	16	27
0.25	44	62	107
0.10	271	385	664

For the Tchebichev inequality, the values of n are given by the entries of the table below.

k\p	**0.90**	**0.95**	**0.99**
0.50	40	80	400
0.25	160	320	1,600
0.10	1,000	2,000	10,000

2.22

(i) $P(|\bar{X}-\bar{Y}| \leq 0.25\sigma) = P(|\bar{Z}| \leq 0.25\sigma) \simeq 2\Phi(\frac{0.25\sqrt{n}}{\sqrt{2}}) - 1 = 0.95$, and then $n = 123$.

(ii) From $1 - \frac{2}{0.0625n} \geq 0.95$, we find $n = 640$.

2.24 We have $EX_i = \frac{1}{\lambda}$, $Var(X_i) = \frac{1}{\lambda^2}$, so that $ES_n = \frac{n}{\lambda}$, $Var(S_n) = \frac{n}{\lambda^2}$, $\sigma(S_n) = \frac{\sqrt{n}}{\lambda}$. Then:

(i) $P(S_n \leq nP) = P(0 \leq S_n \leq nP) \simeq \Phi[\sqrt{n}(\lambda P - 1)]$ (since n is expected to be large), and $P = \frac{1}{\lambda}[1 + \frac{1}{\sqrt{n}}\Phi^{-1}(p)]$.

(ii) For the given values, $P = 1{,}000(1 + \frac{1}{100} \times 2.33) = 1{,}023.3$.

2.26 Since $EX = 0$ and $Var(X) = 1/12$, we have:
$P(|\bar{X}_{100}| \leq 0.1) \simeq 2\Phi(\sqrt{12}) - 1 = 0.99946$.

Section 12.3

3.2 From $\sum_{i=1}^{n}(X_i - \bar{X}_n)^2 = \sum_{i=1}^{n} X_i^2 - n\bar{X}_n^2$, we get
$\frac{1}{n-1}\sum_{i=1}^{n}(X_i - \bar{X}_n)^2 = \frac{n}{n-1} \times \frac{1}{n}\sum_{i=1}^{n} X_i^2 - \frac{n}{n-1}\bar{X}_n^2 \xrightarrow{P} 1\times(\sigma^2+\mu^2) - 1\times\mu^2 = \sigma^2$ (by Theorems 1, 5(i), (ii), and 6(ii)).

Index

D

J

K

L

Q

R

S

T

U

V